ANALYTICAL METHODS
IN
CONDUCTION
HEAT TRANSFER

Glen E. Myers

Professor
Department of Mechanical Engineering
University of Wisconsin–Madison

ANALYTICAL METHODS IN CONDUCTION HEAT TRANSFER
Second Edition

AMCHT Publications
4733 Lafayette Drive
Madison, WI 53705–4827

Computer: Power Macintosh 7100/80
Fonts: Times, Helvetica, Courier
Word processing: Word 5.1a
Equations: MathType 3.0
Drawings: Canvas 3.5.4

This book was printed on acid-free paper with soy-based ink.
Printed and bound by Thomson-Shore, Inc.

ANALYTICAL METHODS IN CONDUCTION HEAT TRANSFER
Second Edition

Printed in the United States of America.

AMCHT Publications (e-mail: myers@engr.wisc.edu).

ISBN 978-0-9666065-0-8

to

SUSAN

in memory of

GREGORY

PREFACE
TO THE
FIRST EDITION

This book has evolved from a set of notes that were originally prepared as a replacement of lectures for off-campus students who could not attend class. Consequently the material is written for the student who is studying without the aid of an instructor. In the early stages of development the notes were used in conjunction with lectures on campus. During the later stages the campus lectures were discontinued when it became apparent that the notes had progressed to where they could be used successfully without assistance from the instructor. The notes were also used at Stanford University and at the University of Arizona to enhance the clarity of the final version of the book.

The book has been designed to meet the needs of students taking a course in conduction heat transfer at the first-year graduate level or as a senior elective. (The student is expected to have already taken a first course in heat transfer.) One purpose in taking a course in conduction is to gain the ability to solve more advanced heat-transfer problems. Gaining this ability will require an increased facility in engineering problem formulation and a higher level of mathematics than is usually achieved at the undergraduate level.

The engineering involved in solving conduction problems is contained in the mathematical modeling and in the interpretation of the final mathematical results. This type of material is most effectively taught by the instructor in the classroom and most effectively learned by working comprehensive problems. The instructor can select specific applications which he feels will be of particular interest to his students and then build mathematical models in class where he must defend his assumptions in the face of student questions. Challenging problems can be assigned to be worked on outside of class. Discussion of the assumptions and results is an excellent way to develop an engineering appreciation of conduction.

The optimum situation in a heat-transfer course is for the entire class to have the necessary mathematical background to solve the advanced problems that are being discussed. The instructor can then devote his efforts to the engineering aspects of problem solving. It has been the author's experience, however, that the instructor is often faced with a class that does not yet have enough mathematical tools. Either the student has not yet had the necessary mathematics or, at best, is currently enrolled in a course that may provide some of the information he needs (although several weeks too late perhaps). This usually means that the conduction course must teach mathematics as well as engineering.

One of the intents of this book is to remove the burden of having to discuss mathematical techniques in the heat-transfer classroom, allowing the instructor to concentrate on engineering aspects while leaving the mathematical details to be learned by the student from the textbook outside of class. The book is written for the student who has absolutely no background in the mathematical topics required to solve the conduction heat transfer examples contained in Chaps. 2 through 9. It is designed to give the engineer a palatable exposure to enough additional mathematics to enable him to solve problems that were beyond his undergraduate heat-transfer course. The problems selected for discussion have been chosen because they elucidate mathematical procedures and because they also have significant engineering applications. An engineering approach to mathematics is taken throughout the book. Existence and uniqueness theorems are not presented because, if the model is physically correct, there must be a unique solution. Indeed, the engineer's physical insight into the problem gives him a valuable tool that is not often used by the pure mathematician. The book attempts to bridge the gap between mathematician and engineer.

The book is not intended to be a compilation of existing classical solutions to conduction problems. An extensive supply of such solutions may be found in *Conduction of Heat in Solids,* 2d ed., by Carslaw and Jaeger.[1] An understanding of the material in the first seven chapters of this book is essential for proper and efficient use of Carslaw and Jaeger's treatise.

It is important that the engineer have a balanced view of methods for solving problems. An overemphasis of exact, analytical methods at the expense of approximate, computer oriented techniques would not reflect the trends in technology. Consequently this book does not attempt to cover all possible analytical methods that might be used, nor does it cover the topics in as great a depth as if it were a mathematics textbook. The variety and depth of presentation of the exact methods of solution have been chosen to give the engineer an appreciation of the kinds of problems that can be solved exactly and the amount of work required to do so. A good portion of the text is devoted to approximate methods that are suitable for the digital computer.

It is expected that the student will find the first chapter to be almost entirely a review of topics he was exposed to in his first course in heat transfer. Familiarity with this material is essentially the only prerequisite

---

[1] Oxford University Press, Fair Lawn, NJ, 1959.

required for the remainder of the book. The amount of time devoted to this chapter will depend upon the student's background.

Bessel functions arise in the mathematical solution of circular fin problems and from the solution of partial differential equations in cylindrical coordinates. Chapter 2 is intended to serve as an introduction to these functions. In so doing, it should also be of value when considering other types of special functions which often appear in heat-transfer analyses.

Separation of variables is the most common technique for solving partial differential equations and is discussed in Chap. 3. Emphasis is given to the types of boundary conditions found in engineering practice. These are often omitted in a mathematics course.

The extension of basic solutions to more complicated problems using the notion of superposition is considered in Chap. 4. This idea is discussed in detail due to its usefulness in convection as well as in conduction and in other engineering systems.

Problems with periodic boundary conditions often occur in engineering. The method of complex combination discussed in Chap. 5 is very useful in obtaining the sustained solution to such problems.

The Laplace transformation is introduced in Chap. 6 primarily because of the ease with which it can handle the semi-infinite solid. This technique can also be used as an alternative to separation of variables.

Although the use of nondimensional quantities is widespread in heat transfer and other engineering fields, it is rarely considered as an analytical tool in its own right. Chapter 7 discusses normalization as a mathematical tool for simplifying problems and for obtaining deeper insights without going through the complete mathematical solution of a problem. This chapter should be read in parallel with the earlier chapters of the book.

The first seven chapters should provide the student with a good idea of the types of problems that can be solved exactly and some of the possible methods of obtaining solutions. There are many problems which arise in engineering practice, however, that can not be handled exactly. Therefore the remainder of the book is devoted to the discussion of numerical methods that can be carried out on the digital computer. The finite-difference method is the most well known of these techniques and is given a fairly comprehensive treatment in Chap. 8.

During the 1960s there was widespread development of the finite-element method for obtaining numerical solutions to problems. These developments have been primarily in the field of solid-body mechanics but have spilled over into heat transfer because of the thermal-stress problem. The finite-element method is presented in Chap. 9. The examples that are discussed are identical to those in Chap. 8 so that comparisons with the finite difference method can be made.

Several appendixes have been included to enable a student to take a conduction heat-transfer course on his own without benefit of an instructor. A selection of comprehensive problems is given in Appendix F to help provide experience in building mathematical models, using the mathematical techniques discussed in the text, and then drawing some engineering conclusions. Appendix G shows how these comprehensive problems could be used in a one-semester course. The comments in Appendix H should help the student select the most important points in each section of the text. The exercises at the end of each chapter can be worked as the student desires to provide a check on his understanding of the text material. Answers to many of these exercises are given in Appendix I.

ACKNOWLEDGMENTS FOR THE FIRST EDITION　　The people most responsible for this book are the students for whom it is written. Their questions, suggestions, and criticisms have served as an invaluable stimulus in the development of the book.

A special debt of gratitude is due Prof. E. F. Obert of the University of Wisconsin for inspiring the original set of notes for his AIM program for off-campus students and for his continuing advice during the development of the notes into a textbook.

Professors W. M. Kays of Stanford University and H. C. Perkins of the University of Arizona have each used the notes in their conduction courses and have given many helpful suggestions. Lectures by Prof. W. Weaver of Stanford University on the finite-element method in structural analysis were useful in preparing Chap. 9.

The comments and criticisms of the finite-difference and the finite-element chapters given by C. B. Moyer and F. C. Weiler of the Aerotherm Corporation, P. S. Andersen of the Danish Atomic Energy Commission Research Establishment Risö (currently a Ph.D. candidate at Stanford), and B. F. Blackwell of the Sandia Corporation (currently a Ph.D. candidate at Stanford), based on their practical experience with computer techniques, have greatly improved this portion of the book.

This book could never have been produced without the assistance of the ladies who did the typing. Mary Huber has borne the brunt of all the typing, cutting, pasting, and retyping that occurred during the years of developing the final manuscript. Her patience, attention to detail, and typing skill have been a valued contribution. Jan Elliott, Bobbe Fowlie, Ditter Peschcke-Koedt, and Betsy Emory all contributed expertly in the final stages of preparation of the manuscript.

Finally, this book could never have been written if it were not for the support of my wife. Her encouragement and sacrifice have been an invaluable contribution.

GLEN E. MYERS

PREFACE
TO THE
SECOND EDITION

It has been 27 years since the first edition of this book was printed. A major change during this period has been the increase in computing power available to students and to engineers. The importance of numerical methods relative to analytical methods has increased. As a result, the one-semester course in heat conduction that I teach has evolved until it is about one-half analytical (primarily Bessel equations/functions, separation of variables and superposition) and one-half numerical (finite differences and finite elements). The class schedule in Appendix G shows the time I spend and the sections I cover. Complex combination and Laplace transforms are barely mentioned in my course to free up more time for numerical methods. Matrix notation and linear algebra are additional skills that are developed in studying numerical methods in this edition.

Although the major modifications of the second edition are in the finite-difference and finite-element chapters, there is a significant change in the emphasis in the analytical chapters. In the first edition, most of the mathematical solutions used normalized variables and parameters. Students were encouraged to first derive the governing equation (and its boundary conditions) in dimensional form and then normalize the problem before obtaining the mathematical solution. Unfortunately, students often thought that they *must* first normalize a problem before solving it (and sometimes normalization was a struggle). During the past 27 years I have come to believe that it is better to first solve the problem[1] and then normalize the solution for presentation. Dimensions, units and familiar parameter groups (or lack thereof) can often be used to help uncover math errors. It is easier to make wise normalization choices when you already have the solution. Therefore, in the second edition, most of the solutions are carried out in dimensional form. The benefits of normalization without solving the differential equation are still discussed in Chapter 7.

The treatment of separation of variables in Chapter 3 has been revised. Since most real-life heat-conduction problems are not homogeneous, Chapter 3 now approaches *all* problems as if they were nonhomogeneous rather than first discussing the special case of homogeneous problems. Section 3•1 first discusses one-dimensional, transient problems to illustrate the solution method[2] when the nonhomogeneous terms are independent of time. When the nonhomogeneous terms are time dependent another method (variation of parameters) is discussed. The discussion of two-dimensional problems in Section 3•2 is more than was covered in the first edition to set the stage for the discussion of numerical methods. Solutions for a steady-state problem and for a transient problem are found. These two analytical solutions are the test cases for the finite-difference and finite-element solutions in chapters 8 and 9. Section 3•3 treats four cylindrical-coordinate problems, one problem in spherical coordinates and then discusses the reduction of two- and three-dimensional transients to one-dimensional transients.

Most students now taking a graduate course in conduction have already had a reasonable exposure to one-dimensional, finite-difference solutions in their previous heat-transfer course(s). Thus the material on numerical methods in the second edition of this book primarily concentrates on two-dimensional problems. As in the first edition, the finite-difference method is discussed first (Chapter 8) because most students already have some background that makes it easier to introduce matrix notation, derive the governing finite-difference equations, review solution techniques (Gauss elimination, Cholesky decomposition, Euler and Crank-Nicolson) and understand numerically induced oscillations. It is also helpful to show some of the difficulties that arise in using finite differences for two-dimensional problems to motivate the need to study finite elements.

Chapter 8 now contains a major new section on the analysis of transient solutions. In this section the discrete finite-difference system of ordinary differential equations for transient problems is solved exactly (*i.e.*, without discretizing the time variable). The solution technique and notation used in Chapter 8 to obtain the exact solution of the transient finite-difference equations is similar to the method used in Chapter 3. Although the exact solution of the finite-difference system of differential equations is not yet practical for large conduction problems, it is instructive to compare it to solutions obtained by separation of variables in Chapter 3. The "exact" finite-difference solution is also compared to the Euler and Crank-Nicolson solutions to gain an understanding of the critical-time-step problem.

Chapter 9, on finite elements, has been completely rewritten. In the first edition, finite-element theory was based on variational calculus (which was new to most students). In this second edition, Galerkin's method is used to develop finite elements because it is easier for students to understand. The governing finite-element system of ordinary differential equations uses notation introduced in Chapter 8 for finite differences. Chapter 9 is written

---

[1] This assumes that a solution is obtainable.

[2] The solution method was called "partial solutions" in the first edition.

for easy comparison to the more familiar finite-difference material in Chapter 8.

Many of the exercises at the end of Chapter 9 now ask the student to use the computer program *FEHT*[1] to construct finite-element models and make calculations. *FEHT* has both Macintosh and Windows versions.

This second edition has been developed primarily over the past eight summers on a Macintosh computer. The large page size (to reduce the number of pages and to provide room for the large matrix equations in chapters 8 and 9) and the page format (wide margins for figures and student notes) were chosen early so the revised class notes produced every other year would closely resemble the final version of the book. Students have been able to study from the latest version of the developing text and could offer comments to improve the book. An index was included with the last set of revised class notes to obtain student feedback. Printer's plates were produced directly from the author's PDF files. Hopefully this process of development and production will enhance the accuracy of the final version of the book.

COMPUTER COMMUNICATION WITH THE AUTHOR

If you have questions, concerns or suggestions about this book, you can contact the author via e-mail at: *myers@engr.wisc.edu*.

ACKNOWLEDGMENTS FOR THE SECOND EDITION

I am indebted to Professors S. A. Klein and W. A. Beckman of the University of Wisconsin-Madison for writing program *FEHT* to solve heat-conduction problems using the finite-element techniques discussed in Chapter 9. It is now convenient to assign problems for students to work that involve irregular geometries requiring hundreds of nodes to model. Many interesting questions may be studied with *FEHT*.

The computations and original plots for most of the graphs in the book and the tables in the appendixes were done using *EES*[2], another program written by Professor S. A. Klein.

Professors G. H. Golub of Stanford University and B. Noble of the University of Wisconsin-Madison provided many helpful ideas about matrix theory, linear algebra and numerical methods.

L. L. Litzkow's creativity, computer knowledge, graphics capabilities and her willingness to provide assistance saved me many times when I had trouble getting the computer to do what I wanted.

GLEN E. MYERS

[1] Klein, S. A., W. A. Beckman and G. E. Myers: *FEHT – Finite Element Analysis*, F-Chart Software <http://www.fchart.com/>.

[2] Klein, S. A. and F. L. Alvarado: *EES – Engineering Equation Solver*, F-Chart Software <http://www.fchart.com/>.

CONTENTS

REVIEW OF ELEMENTARY PROBLEMS

1•0 INTRODUCTION

This chapter is intended to assist you by indicating the subject matter with which you should be familiar before continuing with the remainder of the textbook. Section 1•1 reviews the basic principles of heat transfer and how they can be used to formulate differential equations. Section 1•2 is essentially a review of the mathematics background expected of you. Several elementary problems of heat-transfer interest are discussed. Some ideas presented here may be new to you and should be considered thoroughly since they are rather important to good engineering analysis.

A thorough understanding of Chapter 1 is the prerequisite for the remainder of the material in this book. If you are unfamiliar with or rusty in applying the material in this chapter, you should brush up by studying a more elementary heat-transfer textbook [1, 2, 3][1] and by working as many of the exercises at the end of this chapter as possible.

---

[1] Numbers in brackets denote references listed at the end of each chapter.

1•1 PROBLEM FORMULATION

This section reviews the use of energy balances and rate equations in arriving at differential equations to describe heat-transfer problems. It is expected that you are acquainted with this procedure from an undergraduate exposure to heat transfer. This material should provide a good review for those who are a little uncertain in this area.

1•1•1 The energy balance

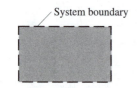

Figure 1•1 *Clearly marked system boundary.*

The first step in making an energy balance is to define a system. The careful definition of a system is crucial to a correct analysis. To be sure that this is carefully done, use a red pencil (shown in Figure 1•1 as a dashed line) to indicate exactly the boundaries of the system you wish to consider.

The next step is to indicate, by means of arrows, the energy flows you are going to consider crossing the system boundary. Figure 1•2 shows one arrow labeled q_{in} to represent heat transfer into the system and a second arrow q_{out} for heat transfer out of the system. The rate of thermal-energy generation within the system is g and rate of energy storage within the system is $dE/d\theta$. Be sure to include all energy terms you want to consider. Do not include any terms that you are not going to consider. Figure 1•2 should serve as a visual reminder of the energy terms to be considered in the next step.

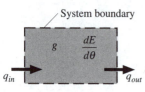

Figure 1•2 *System boundary with important energy terms indicated.*

The final step in making an energy balance is to look at the system with its arrows and symbols and make a mathematical statement that the energies represented by these terms must be conserved. This is simply the first law of thermodynamics. In this case the mathematical statement is that

$$q_{in} + g = q_{out} + \frac{dE}{d\theta} \qquad (1\cdot1\cdot1)$$

The notation in the energy-balance equation should agree with the notation in your system sketch.

1•1•2 Rate equations

In addition to being sure that energy is being conserved, the heat-transfer analyst is also interested in the rates at which the energy transfer occurs. It is then necessary to state rate equations to describe rates of energy flow. The more common ones in heat transfer are listed below.

Conduction Conduction is the transfer of heat through materials without net mass motion of the material. The rate equation which describes this mechanism is (see Figure 1•3)

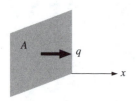

Figure 1•3 *Conduction in the x direction, normal to area A.*

$$q = -kA\frac{\partial t}{\partial x} \qquad (1\cdot1\cdot2)$$

where q = rate of heat flow in x direction by conduction; E/Θ
$ k$ = thermal conductivity; $E/\Theta\text{-}L\text{-}T$
$ A$ = area normal to x direction through which heat flows; L^2
$ t$ = temperature; T
$ x$ = length variable; L

Equation (1·1·2) may be considered to be the defining equation for the thermal conductivity. The conductivity is a thermodynamic property that may be looked up in Appendix K or in other references.

Convection Convection is the name given to the process by which thermal energy is transferred between a solid and a fluid flowing past it. The rate equation used to describe this mechanism is (see Figure 1•4)

$$q_c = hA(t_w - t_\infty) \qquad (1\cdot1\cdot3)$$

where q_c = rate of heat flow by convection; E/Θ
    h = heat-transfer coefficient; $E/\Theta\text{-}L^2\text{-}T$
    $t_w - t_\infty$ = temperature potential for heat flow away from surface; T
    A = surface area through which heat flows; L^2

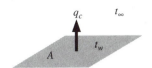

Figure 1•4 *Convection from surface area A to ambient surroundings.*

It should be recalled that (1·1·3) is really the definition of h. The heat-transfer coefficient is found from recipes which contain such information as fluid properties, dimensions of the solid and fluid velocity [1, 2, 3]. You should already be familiar with how this was done in elementary courses.

Radiation Radiation heat transfer is the net exchange of thermal energy between two surfaces obeying the laws of electromagnetics. In conduction problems, radiation from a solid to its infinite surroundings must often be considered. In this case the rate equation used to describe the process is (see Figure 1•5)

$$q_r = \sigma\varepsilon A(T_w^4 - T_\infty^4) \qquad (1\cdot1\cdot4)$$

where q_r = rate of heat flow by radiation; E/Θ
    σ = Stefan-Boltzmann constant; $E/\Theta\text{-}L^2\text{-}T^4$
    ε = emissivity of the surface, dimensionless
    A = surface area through which heat flows; L^2
    T_w = absolute surface temperature; T
    T_∞ = absolute ambient temperature; T

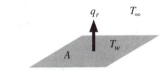

Figure 1•5 *Radiation from surface area A to ambient surroundings.*

Storage in a solid Thermal-energy storage in a solid occurs when its temperature increases with time. The equation describing this process is (see Figure 1•6)

$$\frac{dE}{d\theta} = \rho V c \frac{\partial t}{\partial \theta} \qquad (1\cdot1\cdot5)$$

where $\dfrac{dE}{d\theta}$ = rate of energy storage; E/Θ
    ρ = density; M/L^3
    V = volume; L^3
    c = specific heat; $E/M\text{-}T$
    θ = time; Θ

Figure 1•6 *Energy storage within volume V.*

Observe that ρ, V and c have been assumed to be constant and removed from the derivative in (1·1·5). The $\rho V c$ product is the thermal *capacitance* of the solid.

Generation in a solid Energy generation in solids occurs when other forms of energy (*e.g.,* chemical, electrical, nuclear) are converted into thermal energy. This process can be described by (see Figure 1•7)

$$g = g'''V \qquad (1\cdot1\cdot6)$$

where g = rate of energy generation; E/Θ
    g''' = generation rate per unit volume; $E/\Theta\text{-}L^3$
    V = volume; L^3

Figure 1•7 *Energy generation within volume V.*

All these rate equations should be familiar to you from your previous exposure to heat transfer. They are the ones which will be the most important in this book. Equations (1·1·2), (1·1·5) and (1·1·6) will be especially important since we will be speaking mainly of conduction in solids. Often it may be more convenient to alter the symbols in a given problem, but the basic ideas are stated by the above.

1•1•3 Differential equations

The previous two sections each discussed a different idea—the energy balance and the rate equation. In this section these two ideas will be combined to arrive at differential equations describing several systems commonly found in engineering practice. This will be done by discussing the thin-rod equation, the lumped-system equation and the heat equation.

Thin-rod equation In the design of heat exchangers, finned surfaces are often employed to improve performance. An example of such a geometry is the pin fin which is a thin circular pin, or rod, that extends between two prime surfaces in a heat exchanger. One typical pin is considered in Figure 1•8. If each prime surface is at temperature t_0 and the fluid passing the rod is at temperature t_∞, it is of interest to calculate the steady-state heat transfer between the thin rod and the fluid. To do this, it is necessary to determine the temperature distribution along the rod.

From the observed symmetry of the problem it is only necessary to consider the left-hand half of the rod as shown in Figure 1•9. The maximum, or minimum, temperature in the rod must occur at the midpoint $x = L$. Therefore the x derivative is taken to be zero there. The rod is assumed to be slim enough so that the temperature variation over the cross section of the rod may be neglected in comparison with the axial variation. The rod has a cross-sectional area A and a perimeter p.

As always, the first step in the analysis is to define a system and indicate the important energy flows. In this case, only axial conduction along the rod and convection from the perimeter will be considered as shown in Figure 1•10. The symbol $q|_x$ denotes the rate of energy flow by conduction at position x, while $q|_{x+\Delta x}$ denotes the conduction at position $x + \Delta x$. The symbol q_c refers to the convective energy transfer from the system to the fluid flowing past the rod.

The energy balance can now be written directly from Figure 1•10 as

$$q|_x = q|_{x+\Delta x} + q_c$$

Next, the appropriate rate equations should be written. Thus

$$q|_x = -\left[kA\frac{dt}{dx}\right]_x$$

$$q|_{x+\Delta x} = -\left[kA\frac{dt}{dx}\right]_{x+\Delta x}$$

$$q_c = h(p\Delta x)(t - t_\infty)$$

The rate equations are now substituted into the energy balance to give

$$-\left[kA\frac{dt}{dx}\right]_x = -\left[kA\frac{dt}{dx}\right]_{x+\Delta x} + h(p\Delta x)(t - t_\infty)$$

The reduction to a differential equation begins by combining the terms evaluated at x and at $x + \Delta x$ and then dividing by Δx to give

$$\frac{\left[kA\dfrac{dt}{dx}\right]_{x+\Delta x} - \left[kA\dfrac{dt}{dx}\right]_x}{\Delta x} = hp(t - t_\infty)$$

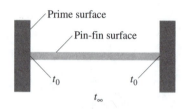

Figure 1•8 *Typical pin-fin surface in a heat exchanger.*

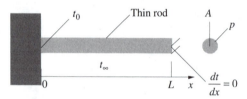

Figure 1•9 *The thin-rod problem.*

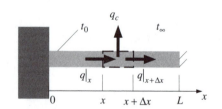

Figure 1•10 *Thin-rod system with important energy terms indicated.*

From calculus the definition of a derivative is

$$\frac{df}{dx} = \lim_{\Delta x \to 0} \frac{f(x + \Delta x) - f(x)}{\Delta x}$$

This is exactly the form of the term on the left of the above equation where the function f is taken to be $kA(dt/dx)$. Then, in the limit as $\Delta x \to 0$, the thin-rod equation becomes

$$\frac{d}{dx}\left[kA\frac{dt}{dx}\right] - hp(t - t_\infty) = 0$$

If k and A are taken to be constants, they can be brought outside the differentiation. Upon dividing by kA, the equation then becomes

$$\frac{d^2t}{dx^2} - \frac{hp}{kA}(t - t_\infty) = 0 \tag{1·1·7}$$

Lumped-system equation The analysis of many engineering transients does not require complete description of the temperature as a function of position as well as time. Rather, the variation with time is much more important and the variation with position may be overlooked.[1] In this case, the problem is said to be "lumpable."

As an example (Figure 1•11), consider a small metallic object of volume V and surface area A which is suddenly immersed in a hot bath at temperature t_∞. Once again, the first step in the analysis is to define a system and indicate the important energy terms. Here the entire object will be taken to be the system rather than only a portion of it as in the previous example. The only energy transfers that will be considered are (1) the convection between ambient and object and (2) the energy storage in the object. The system and these energy terms are shown in Figure 1•12. The energy balance can now be written as

$$q_c = \frac{dE}{d\theta}$$

Next, the appropriate rate equations are written.

$$q_c = hA(t_\infty - t)$$

$$\frac{dE}{d\theta} = \rho Vc \frac{\partial t}{\partial \theta}$$

Observe that the temperature difference in the convection rate equation has been chosen so that when it is positive heat flows in the direction shown on the diagram.[2] Substituting the rate equations into the energy balance gives

$$hA(t_\infty - t) = \rho Vc \frac{dt}{d\theta}$$

or

$$\frac{dt}{d\theta} + \frac{hA}{\rho Vc}(t - t_\infty) = 0 \tag{1·1·8}$$

Figure 1•11 *Small metal object immersed in a bath.*

Figure 1•12 *Lumped system with important energy terms indicated.*

---

[1] The circumstances in which this is permissible are discussed in detail in Section 7•2•1. This section can be read most advantageously after Section 3•1•4 has been studied. The result is that a lumped analysis is usually acceptable for $hL/k < 0.1$, where L is related to the size of the system. The group hL/k is called the *Biot number*.

[2] The convection arrow can be drawn in either direction on the diagram as long as the energy balance and the rate equation are written consistently with the arrow direction. The same differential equation will be obtained.

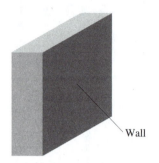

Figure 1•13 *Plane-wall fuel element in a nuclear reactor.*

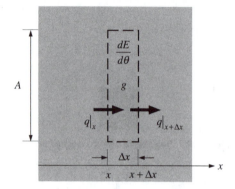

Figure 1•14 *Plane-wall system with important energy terms indicated.*

Heat equation This equation, in one-dimensional form, considers the flow of thermal energy in a plane wall. An example in engineering practice might be the start up of a nuclear reactor in which the fuel elements are in the form of large slabs. As the control rods are withdrawn and the nuclear reaction begins, thermal energy is generated and the slab (or plane wall) will begin to warm up. The engineer is often interested in how fast the wall will approach its steady operating temperature (Figure 1•13).

The first step in the analysis of this problem is to define a system and indicate the energy terms to be considered.[1] In this case consider a system of length Δx in the x direction with cross-sectional area A normal to the x direction. The important energy terms are the conduction into and out of the system, generation and storage. The system and the energy terms are shown in Figure 1•14.

The next step is to look at the diagram and write the energy balance in terms of the arrows and symbols shown on the diagram. Thus

$$q|_x + g = q|_{x+\Delta x} + \frac{dE}{d\theta}$$

The rate equations describing these energy transfers are given by

$$q|_x = -\left[kA\frac{\partial t}{\partial x}\right]_x$$

$$g = g'''(A\Delta x)$$

$$q|_{x+\Delta x} = -\left[kA\frac{\partial t}{\partial x}\right]_{x+\Delta x}$$

$$\frac{dE}{d\theta} = \rho(A\Delta x)c\frac{\partial t}{\partial \theta}$$

These rate equations are then substituted into the energy balance to get

$$-\left[kA\frac{\partial t}{\partial x}\right]_x + g'''(A\Delta x) = -\left[kA\frac{\partial t}{\partial x}\right]_{x+\Delta x} + \rho(A\Delta x)c\frac{\partial t}{\partial \theta}$$

To arrive at a differential equation, the terms evaluated at x and $x + \Delta x$ in the above expression must be combined and division by Δx performed as shown below.

$$\frac{\left[kA\frac{\partial t}{\partial x}\right]_{x+\Delta x} - \left[kA\frac{\partial t}{\partial x}\right]_x}{\Delta x} + g'''A = \rho Ac\frac{\partial t}{\partial \theta}$$

The first term in the above equation is of the proper form to become a derivative as $\Delta x \to 0$. In this case, however, since both x and θ are independent variables, the derivative is a partial derivative rather than an ordinary derivative. Thus, as $\Delta x \to 0$, the equation becomes

$$\frac{\partial}{\partial x}\left[kA\frac{\partial t}{\partial x}\right] + g'''A = \rho Ac\frac{\partial t}{\partial \theta}$$

---

[1] In engineering practice, some idealizations would already have been made to obtain the above plane-wall model of the actual engineering system. These idealizations are really the first step in the analysis and not the derivation of the governing differential equation.

For the problem considered here, A is a constant and thus can be divided out. Thus,

$$\frac{\partial}{\partial x}\left[k\frac{\partial t}{\partial x}\right] + g''' = \rho c \frac{\partial t}{\partial \theta}$$

If, in addition, it is also assumed that k is a constant, then the above can be rewritten as

$$k\frac{\partial^2 t}{\partial x^2} + g''' = \rho c \frac{\partial t}{\partial \theta} \qquad (1\cdot1\cdot9)$$

The assumption that k is a constant is an idealization that may not always be valid. In some problems the solid material may not be homogeneous and consequently the conductivity, as well as the density and the specific heat, might be functions of position. In other cases the temperature changes can be so great as to necessitate a consideration of the dependence of k (and perhaps ρ and c) upon temperature. Problems of this nature will not be discussed until chapters 8 and 9.

If we had been more general and allowed conduction in the y and z directions as well, a small volume element, Δx by Δy by Δz, would have been considered in the above manner. The resulting equation would then be given by

$$\frac{\partial}{\partial x}\left[k_x\frac{\partial t}{\partial x}\right] + \frac{\partial}{\partial y}\left[k_y\frac{\partial t}{\partial y}\right] + \frac{\partial}{\partial z}\left[k_z\frac{\partial t}{\partial z}\right] + g''' = \rho c \frac{\partial t}{\partial \theta} \qquad (1\cdot1\cdot10)$$

where k_x, k_y and k_z are the conductivities in the x-, y- and z-directions, respectively. For a material in which $k_x = k_y = k_z = k$ (a constant) $(1\cdot1\cdot10)$ reduces to

$$k\left[\frac{\partial^2 t}{\partial x^2} + \frac{\partial^2 t}{\partial y^2} + \frac{\partial^2 t}{\partial z^2}\right] + g''' = \rho c \frac{\partial t}{\partial \theta}$$

This further reduces to $(1\cdot1\cdot9)$ when conduction is only in the x-direction.

In cylindrical coordinates, r, ϕ and z as shown in Figure 1•15, an energy balance on the small volume shown in Figure 1•16 and rate equations in the r-, ϕ- and z-directions with directional conductivities k_r, k_ϕ and k_z, the resulting equation, instead of $(1\cdot1\cdot10)$, is given by

$$\frac{1}{r}\frac{\partial}{\partial r}\left[k_r r\frac{\partial t}{\partial r}\right] + \frac{1}{r^2}\frac{\partial}{\partial \phi}\left[k_\phi\frac{\partial t}{\partial \phi}\right] + \frac{\partial}{\partial z}\left[k_z\frac{\partial t}{\partial z}\right] + g''' = \rho c \frac{\partial t}{\partial \theta} \qquad (1\cdot1\cdot11)$$

The transformation between Cartesian and cylindrical coordinates is given by:

$$x = r\cos(\phi) \qquad\qquad y = r\sin(\phi) \qquad\qquad z = z$$

Using spherical coordinates, r, ψ and ϕ shown in Figure 1•17, an energy balance on the small volume shown in Figure 1•18 and rate equations in the r-, ψ-, and ϕ-directions with directional conductivities k_r, k_ψ and k_ϕ, the resulting equation, instead of $(1\cdot1\cdot10)$, is given by

$$\frac{1}{r^2}\frac{\partial}{\partial r}\left[k_r r^2\frac{\partial t}{\partial r}\right] + \frac{1}{r^2\sin(\psi)}\frac{\partial}{\partial \psi}\left[k_\psi \sin(\psi)\frac{\partial t}{\partial \psi}\right]$$
$$+ \frac{1}{r^2\sin^2(\psi)}\frac{\partial}{\partial \phi}\left[k_\phi\frac{\partial t}{\partial \phi}\right] + g''' = \rho c \frac{\partial t}{\partial \theta} \qquad (1\cdot1\cdot12)$$

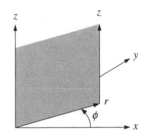

Figure 1•15 *Cylindrical coordinate system.*

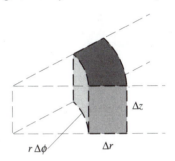

Figure 1•16 *Small volume in cylindrical coordinate system.*

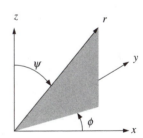

Figure 1•17 *Spherical coordinate system.*

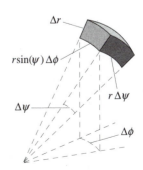

Figure 1•18 *Small volume in spherical coordinate system.*

The transformation between Cartesian and spherical coordinates is given by:

$$x = r\sin(\psi)\cos(\phi) \qquad y = r\sin(\psi)\sin(\phi) \qquad z = r\cos(\psi)$$

For one-dimensional heat transfer with constant thermal conductivity, the three heat equations, (1·1·10), (1·1·11) and (1·1·12), simplify to give

Plane wall:

$$k\frac{\partial^2 t}{\partial x^2} + g''' = \rho c \frac{\partial t}{\partial \theta} \tag{1·1·13}$$

Infinitely long, circular cylinder:

$$k\left[\frac{\partial^2 t}{\partial r^2} + \frac{1}{r}\frac{\partial t}{\partial r}\right] + g''' = \rho c \frac{\partial t}{\partial \theta} \tag{1·1·14}$$

Sphere:

$$k\left[\frac{\partial^2 t}{\partial r^2} + \frac{2}{r}\frac{\partial t}{\partial r}\right] + g''' = \rho c \frac{\partial t}{\partial \theta} \tag{1·1·15}$$

Often, one of these one-dimensional models is all that is needed for engineering purposes. We will consider the solution of these equations in Chapter 3.

1·1·4 Boundary and initial conditions

As one knows from a course in differential equations, the differential equation alone is not enough to completely specify the solution. Additional restrictions at the boundaries or at a given time are required. A condition that is specified at a boundary is called a *boundary condition*. A condition that is given at a particular time, usually at the start, is called an *initial condition*.

The temperature distribution along the thin rod discussed earlier is not determined by (1·1·7) alone, for example. Some information about the temperature or its derivative must be given at the ends of the rod. One common case is where the temperature at one end is given (*e.g.*, $t = t_0$ at $x = 0$) and the other end is adiabatic (*i.e.*, $dt/dx = 0$ at $x = L$). This, as we shall see, is enough information to solve the problem. Observe that the differential equation is *second* order in x (since the second derivative with respect to x is the highest that appears) and requires *two* boundary conditions.

In the lumped system discussed in Section 1•1•3, the temperature-time curve is desired. Again, the differential equation (1·1·8) alone is not enough. One must know the temperature of the object at the start of the transient. Observe that the equation is *first* order in time and requires *one* initial condition.

In general, the number of boundary and initial conditions required to completely formulate any problem can be determined from the differential equation. If the differential equation is second order in x, it needs two x boundary conditions. If the differential equation is first order in θ, it needs one θ condition. Therefore (1·1·13) requires two x-boundary conditions and one initial condition.

1•2 SOLUTIONS TO ELEMENTARY PROBLEMS

There are many elementary problems whose solutions should already be familiar to you before reading this book. Several representative examples are discussed in this section as a review of the mathematics with which you should be acquainted. Additional comments pertaining to the solution of ordinary differential equations are given in Appendix A.

1•2•1 Simple plane wall

The term *plane wall* is applied to a system which is finite in one direction but large enough in the other two directions that edge effects can be neglected for engineering purposes. Here, the important temperature changes are assumed to occur only in the finite direction.

An example in engineering practice might be to calculate the steady-state heat loss through the brick wall in a home as a function of the outside surface temperature when the inside is maintained at temperature t_0. This is shown in Figure 1•19.

If it is assumed that the wall is large enough so that edge effects may be neglected and that the wall is homogeneous, the governing differential equation may be obtained from (1·1·10) as

$$\frac{d^2t}{dx^2} = 0 \qquad (1\cdot2\cdot1)$$

Upon integrating once,

$$\frac{dt}{dx} = a_1 \qquad (1\cdot2\cdot2)$$

The required heat flux through the wall has been given as q''. From the conduction rate equation,

$$q'' = -k\frac{dt}{dx} = -ka_1$$

or

$$a_1 = -\frac{q''}{k}$$

Therefore (1·2·2) becomes

$$\frac{dt}{dx} = -\frac{q''}{k}$$

Integrating again gives

$$t = -\frac{q''}{k}x + a_2$$

Next, the boundary condition at $x = 0$ is applied to give $t_0 = a_2$. The solution thus becomes

$$t = t_0 - \frac{q''}{k}x \qquad (1\cdot2\cdot3)$$

The temperature at $x = L$ is given simply as

$$t_L = t_0 - \frac{q''}{k}L$$

This solution is shown in Figure 1•20.

The notion of a *thermal circuit* is extremely useful in one-dimensional, steady-state problems without energy generation, such as the one we are

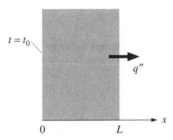

Figure 1•19 *Steady-state heat loss through a brick wall.*

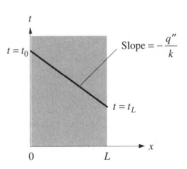

Figure 1•20 *Steady-state temperature distribution in a plane wall.*

discussing. A thermal-circuit representation of this problem can be determined by rearranging the expression for t_L as

$$q'' = \frac{t_0 - t_L}{L/k}$$

or, for a cross-sectional area A,

$$q = \frac{t_0 - t_L}{L/kA}$$

Here q *is* the thermal current, $t_0 - t_L$ is the thermal potential and L/kA is called the *thermal resistance*. By analogy to electrical circuits, this may be pictured as shown in Figure 1•21.

At this point it is well to observe that the convection rate equation (1·1·3) can also be cast into a thermal-circuit form by writing it as

$$q_c = \frac{t_w - t_\infty}{1/hA}$$

Here the thermal resistance is given by $1/hA$.

In the problem considered in this section, it was assumed that the surface temperatures of the wall were the same as the adjacent fluid temperatures. This is only the case when the heat-transfer coefficient is large. Observe how simply the problem can be handled for finite heat-transfer coefficients by using the thermal-circuit idea. The thermal circuit is shown in Figure 1•22. Then

$$q = \frac{t_{\infty 0} - t_{\infty L}}{\dfrac{1}{h_0 A} + \dfrac{L}{kA} + \dfrac{1}{h_L A}}$$

will give the desired relation between the temperatures and the heat flow.

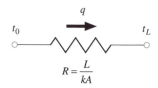

$$R = \frac{L}{kA}$$

Figure 1•21 *Thermal circuit for a plane wall.*

$$R_0 = \frac{1}{h_0 A} \qquad R_w = \frac{L}{kA} \qquad R_L = \frac{1}{h_L A}$$

Figure 1•22 *Thermal circuit for a plane wall with convection on both surfaces.*

1•2•2 Radial heat flow in a cylinder

Consider the radial flow of heat from the inside to the outside of a thick-walled cylinder (Figure 1•23). The inside surface temperature is given as t_i (at r_i) and the outside surface temperature is t_o (at r_o). This is a common geometry in any pipe-flow situation.

For the steady-state case in which only radial conduction is important and for no energy generation, (1·1·11) reduces to the ordinary differential equation

$$\frac{d^2 t}{dr^2} + \frac{1}{r}\frac{dt}{dr} = 0 \tag{1·2·4}$$

Since only derivatives of t appear and not t alone, a simplification is obtained by setting

$$p = \frac{dt}{dr}$$

Thus (1·2·4) becomes

$$\frac{dp}{dr} + \frac{1}{r}p = 0 \tag{1·2·5}$$

The variables can be separated to give

$$\frac{dp}{p} = -\frac{dr}{r}$$

Upon integrating,

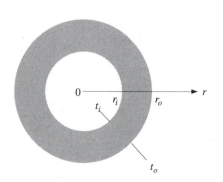

Figure 1•23 *Thick-walled cylinder.*

$$\ln(p) = -\ln(r) + \ln(a_1)$$

or after taking antilogarithms of both sides,

$$p = \frac{dt}{dr} = \frac{a_1}{r}$$

Separating variables once again,

$$dt = a_1 \frac{dr}{r}$$

and integrating,

$$t = a_1 \ln(r) + a_2 \qquad (1 \cdot 2 \cdot 6)$$

The next step is to apply the boundary conditions. Thus

$$t_i = a_1 \ln(r_i) + a_2 \qquad \text{at } r = r_i$$

$$t_o = a_1 \ln(r_o) + a_2 \qquad \text{at } r = r_o$$

These two equations can be solved for a_1 and a_2 and then substituted into (1·2·6) to give

$$t = t_i - \frac{t_i - t_o}{\ln(r_o / r_i)} \ln(r / r_i) \qquad (1 \cdot 2 \cdot 7)$$

A pictorial representation of (1·2·7) is shown in Figure 1•24. The temperature distribution is no longer linear as in the plane-wall problem previously considered. This is due to the increasing area for conduction as r increases. Since the heat-transfer rate is a constant, this means that dt / dr must decrease as r increases.

The heat-transfer rate through the cylinder can be found by substituting (1·2·7) into the rate equation. Thus

$$q = -\left[kA\frac{dt}{dr}\right]_{r=r_o} = -\left[k 2\pi r L \frac{-(t_i - t_o)}{r\ln(r_o / r_i)}\right]_{r=r_o} = \frac{2\pi k L}{\ln(r_o / r_i)}(t_i - t_o)$$

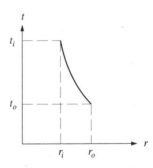

Figure 1•24 *Steady-state temperature distribution in a thick-walled cylinder.*

It should again be pointed out that a thermal circuit may be introduced by rewriting the solution for the heat loss as

$$q = \frac{t_i - t_o}{\dfrac{1}{2\pi k L}\ln(r_o / r_i)}$$

In this case the thermal resistance is given by

$$R = \frac{1}{2\pi k L}\ln(r_o / r_i)$$

An insulated pipe with both internal and external convection resistance would be represented by the series circuit shown in Figure 1•25. Here

$$R_i = \frac{1}{h_i A_i} \qquad\qquad R_p = \frac{1}{2\pi k_p L}\ln(r_o / r_i)_{pipe}$$

$$R_o = \frac{1}{h_o A_o} \qquad\qquad R_{ins} = \frac{1}{2\pi k_{ins} L}\ln(r_o / r_i)_{ins}$$

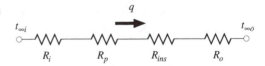

Figure 1•25 *Thermal circuit for an insulated pipe with internal and external convection.*

The heat loss is then computed as

$$q = \frac{t_{\infty i} - t_{\infty o}}{R_i + R_p + R_{ins} + R_o}$$

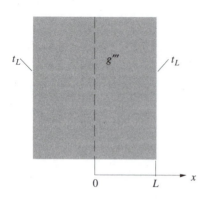

Figure 1•26 *Plane wall with uniform energy generation.*

1•2•3 Energy generation in a plane wall

Consider a plane wall with uniform energy generation due to a nuclear reaction or an electrical current (Figure 1•26). The outside surface temperatures are uniform and equal to t_L. If the generation rate per unit volume is g''' and the thermal conductivity is k, the steady-state-governing ordinary differential equation resulting from (1·1·10) for this case is given by

$$\frac{d^2 t}{dx^2} + \frac{g'''}{k} = 0 \tag{1·2·8}$$

Once again, only derivatives of t appear so that the substitution $p = dt/dx$ will provide a simplification. (Another method is simply to integrate the equation as it stands.) Thus (1·2·8) becomes

$$\frac{dp}{dx} = -\frac{g'''}{k}$$

Separating variables and integrating,

$$p = \frac{dt}{dx} = -\frac{g'''}{k}x + a_1 \tag{1·2·9}$$

Although no explicit information regarding dt/dx is given, the symmetry involved indicates that $dt/dx = 0$ at $x = 0$. This can, and will, be used as a boundary condition in place of $t = t_L$ at $x = -L$. Thus (1·2·9) says

$$0 = a_1$$

Equation (1·2·9) then becomes

$$\frac{dt}{dx} = -\frac{g'''}{k}x$$

Separating variables and integrating,

$$t = -\frac{g'''}{k}\frac{x^2}{2} + a_2 \tag{1·2·10}$$

Now $t = t_L$ at $x = +L$. Thus

$$t_L = -\frac{g'''L^2}{2k} + a_2 \tag{1·2·11}$$

Equations (1·2·10) and (1·2·11) can be subtracted to give

$$t - t_L = \frac{g'''}{2k}(L^2 - x^2) \tag{1·2·12}$$

Thus the temperature distribution is parabolic as shown in Figure 1•27. Equation (1·2·12) can be differentiated to show that the maximum temperature occurs at $x = 0$. Thus, from (1·2·12), the maximum temperature difference is given by $t_o - t_L = g'''L^2/2k$.

The total energy-generation rate g can be written directly as $g'''2AL$ for a portion of the plane wall with cross-sectional area A. It can also be obtained from evaluating the conduction at $x = L$ by substituting (1·2·12) into the conduction rate equation (1·1·2) and multiplying by 2 due to symmetry. Thus, for example, at $x = L$,

$$g = 2q\big|_{x=L} = -2kA\frac{dt}{dx}\bigg|_{x=L} = -2kA\left[\frac{g'''}{2k}(-2x)\right]_{x=L} = g'''2AL$$

as expected.

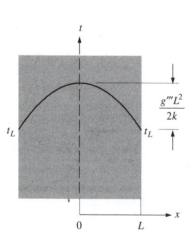

Figure 1•27 *Steady-state temperature distribution in a plane wall with uniform energy generation.*

1·2·4 The thin rod

Consider a thin rod extending from a wall at temperature t_0 into an ambient at temperature t_∞ (Figure 1•28). The rod is long enough so that it is reasonable to assume its free end is adiabatic. For a rod of cross section A, perimeter p and thermal conductivity k with a heat-transfer coefficient h, the governing differential equation (1·1·7) may be restated here as

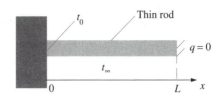

Figure 1·28 *The thin-rod problem.*

$$\frac{d^2t}{dx^2} - m^2 t = -m^2 t_\infty \tag{1·2·13}$$

where $m^2 = hp/kA$ has been used for simplicity.

Observe that (1·2·13) is nonhomogeneous because Ct (where C is an arbitrary constant) will not satisfy it. Usually it is easier to obtain the general solution of homogeneous differential equations. To solve (1·2·13) we will assume the solution can be written as the sum of a particular solution $a(x)$ and a homogeneous solution $u(x)$ as given by

$$t(x) = a(x) + u(x) \tag{1·2·14}$$

Upon substituting (1·2·14) into (1·2·13),

$$\frac{d^2a}{dx^2} + \frac{d^2u}{dx^2} - m^2[a + u] = -m^2 t_\infty$$

This can be satisfied if $u(x)$ satisfies the homogeneous equation

$$\frac{d^2u}{dx^2} - m^2 u = 0 \tag{1·2·15}$$

and the function $a(x)$ satisfies the nonhomogeneous equation

$$\frac{d^2a}{dx^2} - m^2 a = -m^2 t_\infty \tag{1·2·16}$$

We don't need the general solution for $a(x)$; we will be happy with any particular solution that satisfies (1·2·16).

When the right-hand side of (1·2·16) is a constant, $a(x)$ will also be a constant.[1] For constant t_∞, (1·2·16) is satisfied by

$$a(x) = t_\infty \tag{1·2·17}$$

One method of solving (1·2·15) is to assume that it can be satisfied by a solution of the form $u = c\exp(\lambda x)$. The constants c and λ can be evaluated by substituting this form of the solution into (1·2·15) to get

$$c\lambda^2 \exp(\lambda x) - m^2 c \exp(\lambda x) = 0$$

Note that c divides out and thus remains arbitrary. Upon also dividing by $\exp(\lambda x)$, we see that $\lambda^2 = m^2$ or $\lambda = \pm m$. There are two acceptable values of λ ($\lambda = m$ and $\lambda = -m$). Therefore both $c_1 \exp(mx)$ and $c_2 \exp(-mx)$ will satisfy (1·2·15). Since (1·2·15) is *linear* in u and also *homogeneous*, it can be shown that when these solutions are added together their sum will also be a solution. Thus a more general solution of (1·2·15) is

$$u(x) = c_1 \exp(mx) + c_2 \exp(-mx) \tag{1·2·18}$$

This can be verified by direct substitution of (1·2·18) into (1·2·15).

---

[1] When the right-hand side is not a constant, the determination of a particular solution is more complicated. If the right-hand side involves only terms of the form x^m (where m is an integer) or $\sin(qx)$, $\cos(qx)$, $\exp(qx)$, $\sinh(qx)$, $\cosh(qx)$ and/or products of two or more such functions, the method of *undetermined coefficients* may be used [4]. A more general method called *variation of parameters* [4] may also be used.

The general solution to (1·2·13) may be found by substituting (1·2·17) and (1·2·18) into (1·2·14) to give

$$t(x) = t_\infty + c_1 \exp(mx) + c_2 \exp(-mx) \tag{1·2·19}$$

Since (1·2·13) is second order, two boundary conditions are expected. Notice also that two arbitrary constants, c_1 and c_2, have appeared in the solution, (1·2·19). These two constants will be determined next by using two boundary conditions.

Since $t = t_0$ at $x = 0$, (1·2·19) gives (at $x = 0$),

$$t_0 = t_\infty + c_1 + c_2 \tag{1·2·20}$$

At $x = L$, the heat flow is taken to be zero. Thus $q = -kA(dt/dx) = 0$. Therefore $dt/dx = 0$ at $x = L$. When this is substituted into (1·2·19), one obtains

$$0 = c_1 m \exp(mL) - c_2 m \exp(-mL) \tag{1·2·21}$$

Equations (1·2·20) and (1·2·21) can be solved simultaneously for c_1 and c_2 and then substituted back into (1·2·19) to give

$$t(x) = t_\infty + (t_0 - t_\infty) \frac{\exp(-mL)\exp(mx) + \exp(mL)\exp(-mx)}{\exp(-mL) + \exp(mL)} \tag{1·2·22}$$

Calculations are rather cumbersome using (1·2·22) since numerous exponentials must be computed and then multiplied together, added and divided to obtain the temperature. The computations can be simplified by introducing the hyperbolic functions. Recall that

$$\sinh(x) = \frac{\exp(x) - \exp(-x)}{2} \qquad \frac{d}{dx}\sinh(x) = \cosh(x)$$

$$\cosh(x) = \frac{\exp(x) + \exp(-x)}{2} \qquad \frac{d}{dx}\cosh(x) = \sinh(x)$$

$$\tanh(x) = \frac{\sinh(x)}{\cosh(x)}$$

These three functions are shown in Figure 1·29 and are tabulated in most mathematical tables (*e.g.*, [5, 6]). Observe that $\sinh(x)$ is zero at $x = 0$ just as $\sin(x)$ is. Also, $\cosh(x)$ is 1 at $x = 0$ as is $\cos(x)$. Both functions increase rapidly as x increases. The hyperbolic tangent becomes 1 as x increases. In addition to the relations shown above, there are numerous other relations corresponding to the trigonometric identities with which you are familiar. Hyperbolic functions are standard on many calculators.

Using these definitions, (1·2·22) can be rewritten as

$$t(x) = t_\infty + (t_0 - t_\infty) \frac{\cosh(mL - mx)}{\cosh(mL)} \tag{1·2·23}$$

Computations using (1·2·23) are greatly simplified compared with (1·2·22). For convenience in plotting, (1·2·23) may be written in normalized form as

$$\frac{t - t_\infty}{t_0 - t_\infty} = \frac{\cosh\left[mL\left(1 - \dfrac{x}{L}\right)\right]}{\cosh(mL)}$$

This solution is presented in Figure 1·30. As mL increases the temperature decays more rapidly with distance from the wall. Note that

$$mL = \sqrt{\frac{hp}{kA}}\,L = \sqrt{\frac{hpL^2}{kA}} = \sqrt{\frac{hpL}{kA/L}} = \sqrt{\frac{\text{Convection conductance}}{\text{Conduction conductance}}}$$

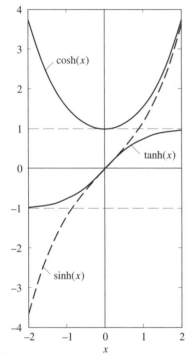

Figure 1·29 *Hyperbolic functions.*

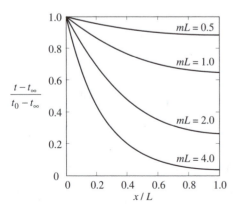

Figure 1·30 *Temperature distribution along a thin rod for several values of mL.*

Often an engineer is interested in the total heat-transfer rate from the rod. Since there is no generation or storage of energy within the rod, the conduction into the rod at $x = 0$ must be equal to the convection from the surface of the rod as shown in Figure 1•31. The conduction into the rod at $x = 0$ can be evaluated by using the conduction rate equation. Thus

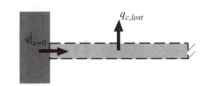

Figure 1•31 *Total heat transfer from a thin rod.*

$$q|_{x=0} = -\left[kA \frac{dt}{dx} \right]_{x=0}$$

The temperature derivative may be found by differentiating (1·2·23) with respect to x. Upon substituting this result, one finds that

$$q|_{x=0} = \frac{-kA(t_0 - t_\infty)}{\cosh(mL)} \left[\sinh\{m(L-x)\}(-m) \right]_{x=0} = kAm(t_0 - t_\infty)\tanh(mL)$$

Then using the definition of m ($m^2 = hp/kA$),

$$q|_{x=0} = \sqrt{hpkA}\,(t_0 - t_\infty)\tanh(mL)$$

The heat transfer from the rod can also be obtained by evaluating the convection. This is found by writing the convection lost from a differential length of rod and then integrating over the entire length of the rod.

$$q_{c,lost} = \int_{x=0}^{L} hp(t - t_\infty)\,dx$$

Using (1·2·23) for $t - t_\infty$,

$$q_{c,lost} = \int_{x=0}^{L} \frac{hp(t_0 - t_\infty)\cosh[m(L-x)]}{\cosh(mL)}\,dx$$

Integrating,

$$q_{c,lost} = \frac{hp(t_0 - t_\infty)}{\cosh(mL)} \left[\frac{\sinh\{m(L-x)\}}{-m} \right]_{x=0}^{L}$$

Substituting the limits of integration,

$$q_{c,lost} = \frac{hp(t_0 - t_\infty)}{m}\tanh(mL)$$

Then, using the definition of m,

$$q_{c,lost} = \sqrt{hpkA}\,(t_0 - t_\infty)\tanh(mL)$$

Observe that, as expected, $q|_{x=0} = q_{c,lost}$.

The heat-transfer rate from a fin is often compared to an ideal fin of the same geometry that is uniformly at the temperature of the prime surface. The heat transfer from the ideal thin rod would be given by

$$q_{ideal} = hpL(t_0 - t_\infty)$$

The ratio of the actual heat-transfer rate to the ideal heat-transfer rate is called the *fin efficiency* and is given by

$$\eta = \frac{q_{actual}}{q_{ideal}} = \frac{\sqrt{hpkA}\,(t_0 - t_\infty)\tanh(mL)}{hpL(t_0 - t_\infty)}$$

Using the definition of m ($m^2 = hp/kA$), this simplifies to give

$$\eta = \frac{\tanh(mL)}{mL}$$

The thin-rod fin efficiency is shown in Figure 1•32. Fin efficiency increases as mL decreases.

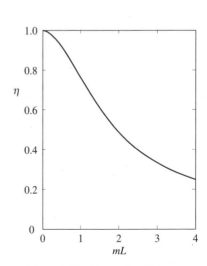

Figure 1•32 *Thin-rod fin efficiency.*

Figure 1•33 *The lumped system.*

1•2•5 Lumped-system transients

A small object of volume V, surface area A, density ρ, specific heat c, initially at temperature $t^{(0)}$ is suddenly placed in an ambient at temperature t_∞ (Figure 1•33).

One engineering application of this model might be the transient response of a thermocouple that is suddenly inserted into a flow stream. The governing differential equation is given by (1·1·8) which is restated here as[1]

$$\frac{dt}{d\theta} + \frac{hA}{\rho Vc}t = \frac{hA}{\rho Vc}t_\infty \tag{1·2·24}$$

To solve this equation we will assume the solution can be written as

$$t(\theta) = a + u(\theta) \tag{1·2·25}$$

where a is a constant. Upon substituting (1·2·25) into (1·2·24) we obtain

$$\frac{du}{d\theta} + \frac{hA}{\rho Vc}[a + u] = \frac{hA}{\rho Vc}t_\infty \tag{1·2·26}$$

If t_∞ is constant, (1·2·26) can be made homogeneous by taking

$$a = t_\infty \tag{1·2·27}$$

Equation (1·2·26) then reduces to

$$\frac{du}{d\theta} + \frac{hA}{\rho Vc}u = 0 \tag{1·2·28}$$

One way[2] to solve (1·2·28) is to separate variables

$$\frac{du}{u} = -\frac{hA}{\rho Vc}d\theta$$

and then integrate to obtain

$$\ln(u) = -\frac{hA}{\rho Vc}\theta + \ln(c_1)$$

Taking antilogarithms,

$$u = c_1 \exp\left(-\frac{hA}{\rho Vc}\theta\right) = c_1 \exp[-(hA/\rho Vc)\theta] \tag{1·2·29}$$

Upon substituting (1·2·27) and (1·2·29) into (1·2·25) we obtain

$$t(\theta) = t_\infty + c_1 \exp[-(hA/\rho Vc)\theta] \tag{1·2·30}$$

The initial condition is that $t = t^{(0)}$ at $\theta = 0$. Thus

$$t^{(0)} = t_\infty + c_1 \quad \text{or} \quad c_1 = t^{(0)} - t_\infty$$

and the solution (1·2·30) becomes

$$t(\theta) = t_\infty + (t^{(0)} - t_\infty)\exp[-(hA/\rho Vc)\theta] \tag{1·2·31}$$

---

[1] Recall that this equation was derived for systems having a small Biot number. For systems in which the Biot number is not small, the spatial variation of temperature must also be considered. This leads to a partial differential equation for temperature as a function of position and time. This book is written for the student who has no familiarity with the solution of partial differential equations. Indeed, the desire of the student to learn how to solve the partial differential equations which arise in more advanced heat-transfer problems is probably the main reason for studying this textbook.

[2] Another way would be to use the integrating factor technique.

For convenience in plotting, (1·2·31) may be written in normalized form as

$$\frac{t(\theta) - t_\infty}{t^{(0)} - t_\infty} = \exp(-hA\theta / \rho Vc)$$

This result is shown in Figure 1•34. A much more simplified presentation of this result can be made if (1·2·31) is stated in normalized form. This is discussed in detail in Chapter 7 because of the importance of normalization as an analytical tool. At this point Section 7•1•1 should be thoroughly read.

The notion of a thermal circuit can also be used in this lumped-system transient. We may rewrite (1·2·24) as

$$\rho Vc \frac{dt}{d\theta} + \frac{t - t_\infty}{1 / hA} = 0$$

The term $1/hA$ is the thermal resistance of the problem which has been assumed to be entirely due to convection. The term ρVc is the thermal capacitance of the problem. The analogy to electrical circuits is shown in Figure 1•35. At time zero the switch is closed and the capacitor discharges through the resistor to the ground state. The RC product is the *time constant* θ_c of the system. That is, $\theta_c = RC = \rho Vc / hA$.

We may rewrite (1·2·24) in terms of the time constant θ_c as

$$\frac{dt}{d\theta} + \frac{1}{\theta_c} t = \frac{1}{\theta_c} t_\infty \qquad (1\cdot2\cdot32)$$

The solution (1·2·31) may be rearranged and rewritten in terms of θ_c as

$$\frac{t(\theta) - t(0)}{t(\infty) - t(0)} = 1 - \exp(-\theta / \theta_c)$$

The left-hand side is called the *response* of the system. Substitution of numerical values shows that when $\theta = \theta_c$ the response is 0.6321 and when $\theta = 3\theta_c$ the response is 0.9502. Sometimes an engineer may only need the time constant of a transient. This can be obtained from the differential equation (1·2·32) or from $\theta_c = RC$ without having to obtain the solution.

Occasionally a thermocouple might be used to measure a fluctuating ambient temperature. An example might be measuring the gas temperature in an engine cylinder. In this case, t_∞ would no longer be constant but a function of time. Equation (1·2·24) could not be made homogeneous as was done to obtain (1·2·28). For example, suppose

$$t_\infty = t_\infty^{(0)} + T \sin(\omega\theta) \qquad (1\cdot2\cdot33)$$

where $t_\infty^{(0)}$ is the value of t_∞ at $\theta = 0$ and T is a constant amplitude of the disturbance. Equation (1·2·24) then becomes

$$\frac{dt}{d\theta} + \frac{hA}{\rho Vc} t = \frac{hA}{\rho Vc} \left[t_\infty^{(0)} + T \sin(\omega\theta) \right] \qquad (1\cdot2\cdot34)$$

To solve (1·2·34) for $t(\theta)$ we will first assume that

$$t(\theta) = a + u(\theta) \qquad (1\cdot2\cdot35)$$

where a is a constant. Substituting (1·2·35) into (1·2·34),

$$\frac{du}{d\theta} + \frac{hA}{\rho Vc} [a + u] = \frac{hA}{\rho Vc} \left[t_\infty^{(0)} + T \sin(\omega\theta) \right]$$

Two of the terms can be eliminated by choosing

$$a = t_\infty^{(0)} \qquad (1\cdot2\cdot36)$$

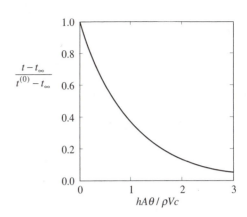

Figure 1•34 *Normalized temperature-time response of a lumped system.*

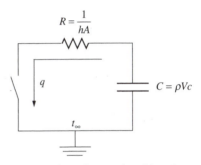

Figure 1•35 *Thermal circuit for a lumped-system transient.*

The differential equation then reduces to

$$\frac{du}{d\theta} + \frac{hA}{\rho Vc} u = \frac{hA}{\rho Vc} T \sin(\omega\theta) \tag{1·2·37}$$

One way to solve (1·2·37) is to recall the integrating factor which is useful in solving equations of the form

$$\frac{dy}{dx} + P(x)y = Q(x)$$

If one multiplies by the integrating factor $\exp[\int P(x)\,dx]$, the left-hand side of the equation becomes a total derivative and the right-hand side is still a function of x only. The integrating factor for (1·2·37) is $\exp(\frac{hA}{\rho Vc}\theta)$. Thus (1·2·37) can be written as

$$\frac{d}{d\theta}\left[u\exp\left(\frac{hA}{\rho Vc}\theta\right)\right] = \frac{hA}{\rho Vc}\exp\left(\frac{hA}{\rho Vc}\theta\right)T\sin(\omega\theta)$$

Separating variables and integrating,

$$u\exp\left(\frac{hA}{\rho Vc}\theta\right) = \frac{hAT}{\rho Vc}\frac{\exp\left(\frac{hA}{\rho Vc}\theta\right)\left[\frac{hA}{\rho Vc}\sin(\omega\theta) - \omega\cos(\omega\theta)\right]}{\left(\frac{hA}{\rho Vc}\right)^2 + \omega^2} + c_1$$

(This may be verified by differentiating both sides with respect to θ.) Multiplying through by $\exp(-\frac{hA}{\rho Vc}\theta)$,

$$u = \frac{hAT}{\rho Vc}\frac{\left[\frac{hA}{\rho Vc}\sin(\omega\theta) - \omega\cos(\omega\theta)\right]}{\left(\frac{hA}{\rho Vc}\right)^2 + \omega^2} + c_1\exp\left(-\frac{hA}{\rho Vc}\theta\right) \tag{1·2·38}$$

Upon substituting (1·2·36) and (1·2·38) into (1·2·35) we obtain

$$t(\theta) = t_\infty^{(0)} + \frac{hAT}{\rho Vc}\frac{\left[\frac{hA}{\rho Vc}\sin(\omega\theta) - \omega\cos(\omega\theta)\right]}{\left(\frac{hA}{\rho Vc}\right)^2 + \omega^2} + c_1\exp\left(-\frac{hA}{\rho Vc}\theta\right) \tag{1·2·39}$$

If the initial condition is $t = t^{(0)}$ at $\theta = 0$, (1·2·39) gives

$$t^{(0)} = t_\infty^{(0)} - \frac{\frac{hAT}{\rho Vc}\omega}{\left(\frac{hA}{\rho Vc}\right)^2 + \omega^2} + c_1 \tag{1·2·40}$$

Solving (1·2·40) for c_1 and then substituting into (1·2·39) gives

$$t = t_\infty^{(0)} + \frac{hAT}{\rho Vc}\frac{\left[\frac{hA}{\rho Vc}\sin(\omega\theta) - \omega\cos(\omega\theta)\right]}{\left(\frac{hA}{\rho Vc}\right)^2 + \omega^2} + \left[t^{(0)} - t_\infty^{(0)} + \frac{\frac{hAT}{\rho Vc}\omega}{\left(\frac{hA}{\rho Vc}\right)^2 + \omega^2}\right]\exp\left(-\frac{hA}{\rho Vc}\theta\right)$$

$$\underbrace{\hphantom{t = t_\infty^{(0)} + \frac{hAT}{\rho Vc}\frac{\left[\frac{hA}{\rho Vc}\sin(\omega\theta) - \omega\cos(\omega\theta)\right]}{\left(\frac{hA}{\rho Vc}\right)^2 + \omega^2}}}_{\text{Sustained}} \qquad \underbrace{\hphantom{\left[t^{(0)} - t_\infty^{(0)} + \frac{\frac{hAT}{\rho Vc}\omega}{\left(\frac{hA}{\rho Vc}\right)^2 + \omega^2}\right]\exp\left(-\frac{hA}{\rho Vc}\theta\right)}}_{\text{Transient}}$$

$$\tag{1·2·41}$$

Observe that the solution (1·2·41) is the sum of two terms. The first term is called the *sustained solution* and is periodic. The second term contains the initial condition and is seen to decay to zero as time increases. It is called the *transient portion* of the solution. Observe that the effect of the initial condition decays out of the problem as time goes on. Thus at long enough times the solution is independent of the initial condition.

It is of interest to consider only the sustained solution which can be rewritten as

$$t = t_\infty^{(0)} + \frac{\dfrac{hAT}{\rho Vc}}{\sqrt{\left(\dfrac{hA}{\rho Vc}\right)^2 + \omega^2}}\left[\frac{\dfrac{hA}{\rho Vc}}{\sqrt{\left(\dfrac{hA}{\rho Vc}\right)^2 + \omega^2}}\sin(\omega\theta) - \frac{\omega}{\sqrt{\left(\dfrac{hA}{\rho Vc}\right)^2 + \omega^2}}\cos(\omega\theta)\right]$$

Next it is convenient to define

$$\beta = \tan^{-1}\left[\frac{\omega}{hA / \rho Vc}\right] \tag{1·2·42}$$

This can be visualized by the triangle shown in Figure 1•36, where

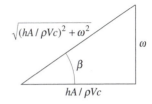

Figure 1•36 *Definition of phase angle.*

$$\sin(\beta) = \frac{\omega}{\sqrt{(hA / \rho Vc)^2 + \omega^2}} \quad \text{and} \quad \cos(\beta) = \frac{hA / \rho Vc}{\sqrt{(hA / \rho Vc)^2 + \omega^2}}$$

The expression for the sustained solution now becomes

$$t = \frac{(hA / \rho Vc)T}{\sqrt{(hA / \rho Vc)^2 + \omega^2}}[\cos(\beta)\sin(\omega\theta) - \sin(\beta)\cos(\omega\theta)]$$

Using the trigonometric relation for the sine of the difference between two angles reduces the above to

$$t = t_\infty^{(0)} + \frac{(hA / \rho Vc)T}{\sqrt{(hA / \rho Vc)^2 + \omega^2}}\sin(\omega\theta - \beta) \tag{1·2·43}$$

Observe from (1·2·42) that as ω approaches zero the phase angle β will also approach zero and the sustained solution (1·2·43) reduces to

$$t = t_\infty^{(0)} + T\sin(\omega\theta) \tag{1·2·44}$$

Thus, for small frequencies of ambient temperature fluctuation, the object (1·2·44) follows right along with the same phase and amplitude as the ambient (1·2·33).

As ω becomes large the phase angle approaches $\pi / 2$ and the sustained solution approaches

$$t = t_\infty^{(0)} + \frac{hAT}{\rho Vc\omega}\sin\left(\omega\theta - \frac{\pi}{2}\right) \tag{1·2·45}$$

Thus, as the frequency of ambient fluctuations increases, the temperature of the object is out of phase by 90° and the amplitude of the oscillations becomes smaller. As ω becomes infinite, the amplitude of the oscillations in (1·2·45) disappears and $t = t_\infty^{(0)}$. The object is unable to follow the ambient temperature.

It is evident from this analysis that thermocouples can be used to measure fluctuating temperatures providing that the frequencies are slow enough to allow the thermocouple to follow along. By comparing ω with $hA / \rho Vc$, the engineer can make an estimate of whether the thermocouple will be suitable for this application. Observe that good design means obtaining as large a value of $hA / \rho Vc$ as possible. This is often accomplished by making the thermocouple small.

1•3 INTRODUCTION TO ADVANCED PROBLEMS

The problems whose solutions were discussed in Section 1•2 were rather elementary. They were solved by relatively simple techniques which should have been familiar to you from an elementary course in heat transfer. You should observe that in every case there was only one independent variable (that is, they involved ordinary differential equations, not partial differential equations) and, except for radial flow in a cylinder, they had constant coefficients. If either of these restrictions is removed, the problem becomes much more involved. The remainder of the book is devoted to the acquisition of a variety of mathematical skills which can be used to solve these "nonelementary" problems.

At this point in your study you should read the rest of Section 7•1 regarding normalization. You should note that although the mathematical analysis may have been carried out in dimensional form the results are often more conveniently presented in nondimensional form. You should also be aware that a differential equation in dimensional form (as derived from basic principles) can be normalized into a nondimensional form. This normalized differential equation is simpler than its dimensional counterpart in that it contains fewer parameters. The boundary conditions are also more simply written.

The engineer who has reached this point and wants to learn about the mathematical skills presented in chapters 2 to 9 should be familiar with normalization and should be able to normalize final results. One should be able to normalize a differential equation and its boundary (and/or initial) conditions. You should be acquiring the ability to think about problems in normalized form as easily as you can think about problems in dimensional form.

Chapters 2 to 6 are designed to teach the mathematical skills necessary to solve more advanced problems in heat transfer. A simplified notation will also often be used to denote differentiation. Thus u' will denote the first derivative and u'' will denote the second derivative of u when u is a function of only one variable. If u is a function of two variables, for example $u(x,\theta)$, then differentiation with respect to x would be written as u_x and u_{xx}. Differentiation with respect to θ would be denoted by u_θ.

SELECTED REFERENCES

1. Holman, J. P.: *Heat Transfer*, 6th ed., McGraw-Hill Book Company, New York, 1986.

2. Incropera, F. P. and D. P. DeWitt: *Introduction to Heat Transfer*, 3rd ed., John Wiley & Sons, New York, 1996.

3. Kreith, F. and M. S. Bohn: *Principles of Heat Transfer*, 4th ed., Harper & Row, New York, 1986.

4. Hildebrand, F. B.: *Advanced Calculus for Applications*, 2nd ed., Prentice-Hall, Inc., Englewood Cliffs, NJ, 1976.

5. Flügge, W.: *Four-Place Tables of Transcendental Functions*, McGraw-Hill Book Company, New York, 1954.

6. Abramowitz, M. and I. A. Stegun (eds.): *Handbook of Mathematical Functions*, Applied Mathematics Series 55, National Bureau of Standards, 1964. [Dover, 1965].

EXERCISES

1•1 Derive (1·1·10) from basic principles.

1•2 Compute the heat flux [Btu/hr-ft^2] through a 4-ft-thick furnace wall whose conductivity is 0.50 Btu/hr-ft-F. The inside wall temperature is 1670 F and the outside wall temperature is 70 F.

1•3 An 0.8-ft-thick wall is constructed from a material which has an average conductivity of 0.75 Btu/hr-ft-F. The wall is to be insulated with a material having an average thermal conductivity of 0.2 Btu/hr-ft-F. If the inner and outer surface temperatures of the composite wall are to be 2400 and 80 F, respectively, determine the thickness of insulation required. Assume the heat flow is 580 Btu/hr-ft^2.

1•4 Derive (1·2·4) by setting up a system and proceeding from basic principles.

1•5 A hollow spherical shell of conductivity k conducts q Btu/hr of heat through it. The sphere is at steady state and there is no generation. Derive the governing differential equation from basic principles. Then, assuming $t = t_o$ at $r = r_o$, solve the differential equation for the temperature distribution.

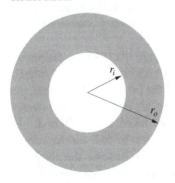

1•6 Derive (1·2·8) by setting up a system and proceeding from basic principles.

1•7 Consider uniform energy generation in a solid. Compare the temperature distributions in a plane wall, a cylinder and a sphere by plotting temperature as a function of position for each case on the coordinates shown. Here g''' is the generation rate and k the thermal conductivity for the material in each case.

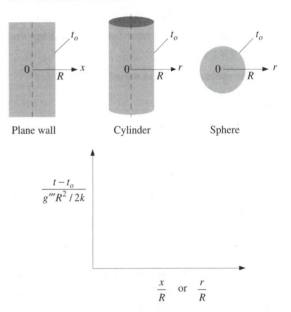

Can the differences in centerline temperature be explained by comparing the volume-to-surface-area ratio for each case? Explain.

Based on the above, which shape would be the best for use as a fuel element in a nuclear reactor? Explain.

1•8 A 2-in-thick steel plate, inserted in an adiabatic wall, generates heat at a rate of 200 Btu/hr-ft^3. *Estimate* the maximum plate temperature.

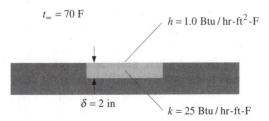

1•9 Calculate the heat transfer along the perfectly insulated rod shown below. The temperature at one end of the rod is 500 F, while at the other end it is 0 F. The cross-sectional area of the bar is 1.0 ft^2. Its thermal conductivity is a function of the temperature of the material and is given by

$$k = k_o(1 + at^2)$$

where $k_o = 40$ Btu/hr-ft-F and $a = 3.0 \times 10^{-6}$ F^{-2}.

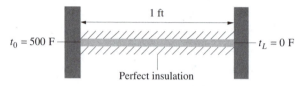

1•10 A thin rod is held so that one end is at 100 F and the other end is at 0 F. The thermal conductivity is constant. The rod is thin enough so that its temperature may be assumed uniform at any cross section (*i.e.*, the rod temperature varies only with distance along the rod). Qualitatively sketch the temperature distribution along the rod for the following cases:

(a) The outside surface of the rod is perfectly insulated
(b) $t_\infty = 0$ F, rod uninsulated
(c) $t_\infty = 300$ F, rod uninsulated
(d) $t_\infty = -100$ F, rod uninsulated

1•11 Obtain the thin-rod solution for the case of an infinitely long rod.

1•12 Obtain the thin-rod solution for a convective free end at $x = L$. The heat-transfer coefficient on the free end is not the same as the heat-transfer coefficient along the rod.

1•13 As an engineer, you will usually be expected to be able to justify assumptions made during the analysis. Since the reason for making the assumption in the first place was to simplify the analysis and save time, the justification should not be long and involved. Certainly, the solution to the simplified problem plus any justification necessary should be more economical in time than the solution to the original problem would have been.

In the thin rod it was assumed that the temperature was uniform at any cross section of the rod. How valid is this assumption?

One possible approach would be to make an estimate of the maximum temperature difference between the center of the rod and the outside surface by assuming that the radial temperature distribution is parabolic. The slope can be found by equating the convective heat-transfer rate at any given axial position to the radial conduction rate at the same position. This could then be compared with the difference in temperature between the rod surface and the ambient. What conclusions can you make?

1•14 The thin rod shown in the sketch is electrically heated by a current I. The electrical resistance of the rod per foot of length is R'. The rod surface is cooled with a uniform and constant heat-transfer coefficient h by an ambient at t_∞. The ends of the rod are maintained at temperature t_0.

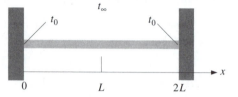

(a) Derive the differential equation for the temperature distribution along the rod from basic principles.
(b) State the boundary conditions for the problem.
(c) Derive an expression for the temperature as a function of position along the rod.
(d) Determine the location and magnitude of the temperature extremes along the rod.
(e) Determine the length of rod for which the maximum temperature difference between one position and another along the rod the greatest. How much is it?

1•15 The differential equation for a thin rod of diameter D with an exponential generation term, $g''' = g_o'''\exp(-x/D)$, is given by

$$\frac{d^2t}{dx^2} - m^2(t - t_\infty) = -\frac{g_o'''}{k}\exp(-x/D)$$

where $m^2 = hp/kA$. The boundary conditions are that

$$t(0) = t_0 \qquad \text{and} \qquad t(\infty) = t_\infty$$

The quantities h, p, k, A, t_∞, g_o''', D and t_0 are constant. Solve for the temperature distribution along the rod for

(a) $m^2D^2 \neq 1$.
(b) $m^2D^2 = 1$.

Check to be sure that your final answer in each case satisfies the differential equation and both boundary conditions.

1•16 A current-carrying 10-gage (0.1019-in diameter) copper wire is insulated with a 0.05-in-thick layer of rubber. The maximum temperature of the rubber is 200 F when the temperature of the surrounding air is 120 F. The external heat-transfer coefficient can be taken as 4 Btu/hr-ft^2-F. For the copper wire, $k_w = 220$ Btu/hr-ft-F and for rubber insulation, $k_i = 0.08$ Btu/hr-ft-F. Assume the electrical resistance of the wire is 1.288 ohms/1000 linear feet.

(a) Determine the temperature [F] of the outside surface of the insulation.
(b) Assuming uniform energy generation inside the wire, determine the temperature [F] at the center of the wire.
(c) Determine the electrical current [A] in the wire.
(d) Assuming the heat-transfer coefficient and the current are constant, determine whether the temperature at the

inside of the insulation will increase or decrease as insulation thickness increases.

(e) Determine the insulation thickness [in] for which the inside insulation temperature is a minimum. Explain, physically, why any other thickness would make the inside insulation temperature higher.

Note: *Derive* all equations from basic principles. Do not just plug into some formula.

1·17 A thin, 0.5-in-diameter copper ($k = 220$ Btu/hr-ft-F) rod spans the distance between two parallel plates 6 inches apart. Air flows in the space between the plates, providing a heat-transfer coefficient of 40 Btu/hr-ft^2-F at the surface of the rod. The surface temperature of the plates exceeds that of the air by 100 F. By how much does the temperature at the midpoint of the rod exceed the temperature of the air passing by the rod? Solve the problem analytically. *Derive* all equations you use.

1·18 A 0.125-in-diameter spherical thermocouple is placed in a bath of ice water. The heat-transfer coefficient between the couple and the water may be taken as 6 Btu/hr-ft^2-F, the specific heat of the metal is 0.1 Btu/lbm-F and the density of the junction is 550 lbm/ft^3. If the original temperature of the thermocouple before immersion is 80 F, how many seconds would it take for the temperature reading of the thermocouple to change from 60 to 40 F?

1·19 The use of a lumped-system analysis assumes that the internal temperature difference is small. An estimate of the maximum temperature difference can be made by assuming a quadratic internal temperature distribution that has zero slope at the center of the lumped system. At the surface of the lumped system, equate the convection heat-transfer rate to the conduction heat-transfer rate. Compare the internal and external temperature differences.

Show that the Biot number appears naturally in the above analysis as a criterion for using the lumped-system approximation. Discuss its relation to the maximum internal temperature difference.

1·20 The exhaust pipe of an automobile engine, originally at the ambient temperature t_∞ is suddenly subjected to a blast of gas through the interior of the pipe. The gas temperature t_g can be assumed essentially constant because of the large amount of gas flow and because of the mixing action in the muffler tending to smoothen out unevenness in temperature. If the heat-transfer coefficient on the inside of the pipe (area = A_i) is assumed constant at some value h_i and the outside (area = A_o) heat-transfer coefficient constant at some value h_o, derive an expression suitable for the determination of the exhaust pipe temperature t at any time θ. Radiation effects as well as the temperature drop across the small thickness of the exhaust pipe may be neglected.

1·21 A bare 12 gage (0.081-in diameter) copper wire is at 120 F in 120 F air when a current of 40 amp suddenly starts flowing through it. How long will it take the wire to reach 63 percent of its steady-state operating temperature rise if the surface heat-transfer coefficient is 4 Btu/hr-ft^2-F? Assume that the internal resistance to heat flow in the wire is negligible and that the wire resistance remains constant at 2 ohms/1000 ft; the product of the mass of 1000 ft of the wire and its specific heat is 1.87 Btu/F.

1·22 The composite plane wall shown below is initially at a uniform temperature. Suddenly the fluid on the left-hand side is stepped up by 100 F. The fluid on the right-hand side is held at its original temperature.

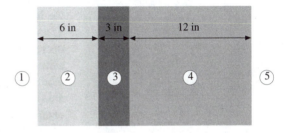

No.	k Btu/hr-ft-F	ρ lbm/ft^3	c Btu/lbm-F	h Btu/hr-ft^2-F
1				2
2	0.2	10	0.2	
3	25	600	0.1	
4	2	10	0.01	
5				200

How long will it take for steady state to be attained (within about 5 percent)?

1·23 Fill in the steps between (1·2·34) and (1·2·39) using the method of undetermined coefficients rather than an integrating factor.

1·24 A sphere of ice, 2-in diameter initially, is suspended in a room and rotated so that the heat-transfer coefficient around the ice remains constant with time. If the room is at a temperature of 80 F and the heat-transfer coefficient is 2 Btu/hr-ft^2-F, how long does it take the sphere to lose half its original volume?

The density of ice may be taken as 58 lbm/ft^3 and the latent heat of fusion as 144 Btu/lbm.

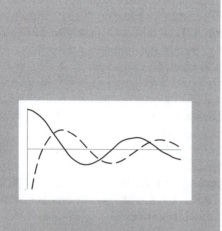

CHAPTER

2

BESSEL'S EQUATION AND RELATED FUNCTIONS

2.0 INTRODUCTION

Thin circular fins are often found on tubes when it is important to improve the heat transfer from the tube. Since the convective thermal resistance $1/hA$ is usually the controlling one, the addition of fins is an attempt to decrease the resistance to heat flow by increasing the surface area. The additional surface area will not all be at the same temperature as the finless tube because of the temperature drop in the fin due to conduction radially along the fin. To evaluate the performance of the fin, it is necessary to obtain an expression for the temperature distribution in the fin.

This problem may be studied by considering a fin of uniform thickness δ with inner radius r_i and outer radius r_o as shown in Figure 2•1. The temperature at the root of the fin (r_i) is taken to be t_i and the ambient is at temperature t_∞. It will also be assumed that there is negligible heat loss from the end at r_o.

A suitable system for analysis is the ring indicated in Figure 2•1. By taking the entire thickness of the fin as the system, temperature variations normal to the fin surface will be overlooked. This corresponds to the thin-rod equation developed in Section 1•1•3. The important energy terms have been shown to be the radial conduction along the fin and the convection out the sides. Energy generation and storage are both zero.

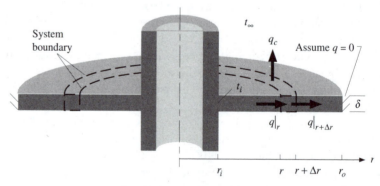

Figure 2•1 *Circular fin showing system boundary and important energy terms.*

Referring to the three energy arrows in Figure 2•1, the energy balance becomes

$$q|_r = q|_{r+\Delta r} + q_c$$

The appropriate rate equations are

$$q|_r = -\left[k2\pi r\delta \frac{dt}{dr} \right]_r$$

$$q|_{r+\Delta r} = -\left[k2\pi r\delta \frac{dt}{dr}\right]_{r+\Delta r}$$

$$q_c = 2h(2\pi r\Delta r)(t - t_\infty)$$

Observe the handling of the variable conduction area in this problem. The area $2\pi r\delta$ is a function of r and must be kept within the brackets until after proceeding to the limit a little later on. Also note the extra 2 in the expression for q_c to account for the fin losing heat from both sides.

The rate equations are now substituted into the energy balance to give

$$-\left[k2\pi r\delta \frac{dt}{dr}\right]_r = -\left[k2\pi r\delta \frac{dt}{dr}\right]_{r+\Delta r} + 2h(2\pi r\Delta r)(t - t_\infty)$$

The usual procedure is now followed to reduce the above to a differential equation. The terms at r and $r + \Delta r$ are combined, and the equation is divided by Δr to give

$$\frac{\left[k2\pi r\delta \frac{dt}{dr}\right]_{r+\Delta r} - \left[k2\pi r\delta \frac{dt}{dr}\right]_r}{\Delta r} = 4\pi h r(t - t_\infty)$$

Taking the limit as $\Delta r \to 0$ gives

$$\frac{d}{dr}\left[2\pi k\delta r \frac{dt}{dr}\right] = 4\pi h r(t - t_\infty)$$

For constant thermal conductivity and fin thickness, $2\pi k\delta$ may be taken outside the differentiation. Upon dividing by $2\pi k\delta$, the equation becomes

$$\frac{d}{dr}\left[r \frac{dt}{dr}\right] = \frac{2h}{k\delta} r(t - t_\infty)$$

Performing the differentiation [treating $r(dt/dr)$ as a product], rearranging, and dividing by r gives

$$\frac{d^2 t}{dr^2} + \frac{1}{r}\frac{dt}{dr} - \frac{2h}{k\delta} t = -\frac{2h}{k\delta} t_\infty \tag{2·0·1}$$

The boundary conditions are

$$t = t_i \text{ at } r = r_i \qquad \text{and} \qquad \frac{dt}{dr} = 0 \text{ at } r = r_o$$

Observe that t appears in (2·0·1). This prohibits the simplification used to solve (1·2·4). Due to the nonconstant coefficient $1/r$, a solution of the form $a\exp(\lambda r)$ does not work either. Therefore "nonelementary" methods are required for solution. If the right-hand side of (2·0·1) were 0 the equation would be a Bessel equation. The solution of (2·0·1) will involve Bessel functions. Its solution will be in terms of what are called modified Bessel functions.

Bessel equations also result from the analysis of straight fins whose cross-sectional area varies in certain prescribed ways. The solutions of partial differential equations in cylindrical coordinates also yield Bessel equations. Thus a general study of Bessel equations and their solutions will be valuable in more advanced analysis.

The mathematical background necessary for solving the circular-fin problem will be slowly developed in this chapter. The solutions to other forms of Bessel's equation are also considered. By the end of the chapter you should be able to handle Bessel equations and the related functions with very little difficulty.

2•1 SERIES SOLUTION TO ORDINARY DIFFERENTIAL EQUATIONS

Often the engineering student is bewildered by the solution of differential equations by means of a series. This section is intended to remove some of this bewilderment so that the student will feel more comfortable with series solutions. It is not intended to provide a thorough background in the mathematics involved in obtaining series solutions to ordinary differential equations. References [6, 7] discuss this subject in considerable detail.

2•1•1 Introduction

Before beginning a study of Bessel functions, it might be helpful to consider a differential equation that should already be familiar to you. For example, the differential equation

$$\frac{d^2u}{dx^2} + m^2u = 0 \qquad\qquad (2 \cdot 1 \cdot 1)$$

with the boundary conditions that $u(0) = 0$ and $u(1) = 1$, is a type that almost every engineer has solved as an undergraduate. The standard procedure is to assume a solution of the form $u = a\exp(\lambda x)$, substitute it into the differential equation, and thereby determine λ as was done in Section 1•2•4. This is satisfactory for equations with constant coefficients as in the above, but it will not work for Bessel's equation. Therefore the above will now be solved by a more general series solution. The series method will also be applicable to Bessel's equation.

The series solution is obtained by assuming a solution of the form

$$u(x) = a_0 + a_1x + a_2x^2 + \cdots = \sum_{n=0}^{\infty} a_nx^n \qquad\qquad (2 \cdot 1 \cdot 2)$$

It is important to realize in working with the summation notation that the index (n in this equation) is actually only a *dummy index*.[1] The problem now is to find suitable values of the a_n so that the assumed series solution (2·1·2) will satisfy (2·1·1). This is much the same idea as for finding λ. The same procedure is attempted—substitute (2·1·2) into (2·1·1) and see what happens. First, (2·1·2) will be differentiated twice:

$$\frac{du}{dx} = a_1 + 2a_2x + 3a_3x^2 + \cdots = \sum_{n=0}^{\infty} na_nx^{n-1}$$

$$\frac{d^2u}{dx^2} = 2a_2 + 3(2)a_3x + \cdots = \sum_{n=0}^{\infty} n(n-1)a_nx^{n-2}$$

Observe that the first two terms in the series expression for the second derivative do not really appear since they are zero; thus the series can equally well be written as

---

[1] This is analogous to a dummy variable of integration. For example, in

$$\int_{x=0}^{1} x^2\,dx \qquad \text{or} \qquad \int_{y=0}^{1} y^2\,dy$$

the variables of integration, x and y, are dummy variables since the value of the integral does not depend upon what symbols are used in the integration. Both the above integrals have the same numerical value, $1/3$. A change of variables can also be made by letting $z = x - 2$, for example. Then

$$\int_{x=0}^{1} x^2\,dx = \int_{z=-2}^{-1} (z+2)^2\,dz$$

The resulting integration over z will still have the same numerical value as the original integral.

$$\frac{d^2u}{dx^2} = \sum_{n=2}^{\infty} n(n-1)a_n x^{n-2}$$

Upon substituting (2·1·2) into (2·1·1), one thus obtains

$$\sum_{n=2}^{\infty} n(n-1)a_n x^{n-2} + \sum_{n=0}^{\infty} m^2 a_n x^n = 0$$

The next step is to combine the two summations into one. To do this, the summations must run over the same values of the summation index. This can be accomplished by a change of the dummy index in the first summation to $\ell = n - 2$. Then

$$\sum_{\ell=0}^{\infty} (\ell+2)(\ell+1)a_{\ell+2} x^{\ell} + \sum_{n=0}^{\infty} m^2 a_n x^n = 0$$

Observe that the limits of the summation were also changed just as the limits of an integral are changed by a substitution of variable. Since ℓ is just a dummy index, it may be called anything we wish—n again, for example. Then

$$\sum_{n=0}^{\infty} (n+2)(n+1)a_{n+2} x^n + \sum_{n=0}^{\infty} m^2 a_n x^n = 0$$

Combining the two summations into one,

$$\sum_{n=0}^{\infty} \left[(n+2)(n+1)a_{n+2} + m^2 a_n \right] x^n = 0$$

Observe that the only way for this to be true is for the bracketed term to be zero for any (and all) n.[1] Thus

$$(n+2)(n+1)a_{n+2} + m^2 a_n = 0 \quad \text{or} \quad a_{n+2} = -\frac{m^2 a_n}{(n+2)(n+1)}$$

This is called a *recursion relation* and provides a restriction on the coefficients a_n. This restriction is necessary in order that the series (2·1·2) satisfy (2·1·1).

This recursion relation may be examined more closely.

$$a_2 = -\frac{m^2 a_0}{2(1)} \qquad\qquad\qquad \text{for } n = 0$$

$$a_3 = -\frac{m^2 a_1}{3(2)} \qquad\qquad\qquad \text{for } n = 1$$

$$a_4 = -\frac{m^2 a_2}{4(3)} = -\frac{m^2}{4(3)}\left[-\frac{m^2 a_0}{2(1)}\right] = \frac{m^4 a_0}{4!} \qquad \text{for } n = 2$$

$$a_5 = -\frac{m^2 a_3}{5(4)} = -\frac{m^2}{5(4)}\left[-\frac{m^2 a_1}{3(2)}\right] = \frac{m^4 a_1}{5!} \qquad \text{for } n = 3$$

$$a_6 = -\frac{m^2 a_4}{6(5)} = -\frac{m^2}{6(5)}\left[\frac{m^4 a_0}{4!}\right] = -\frac{m^6 a_0}{6!} \qquad \text{for } n = 4$$

---

[1] If this were not so, the only way the series could sum to zero would be if the terms canceled one another. While this might be possible for a specific value of x, it would not be possible for every value of x.

By observation of these relations, the generalizations may be made that

1. The coefficients of the even-powered terms can be represented by

$$a_{2n} = \frac{(-1)^n m^{2n} a_0}{(2n)!}$$

2. The coefficients of the odd-powered terms can be represented by

$$a_{2n+1} = \frac{(-1)^n m^{2n} a_1}{(2n+1)!}$$

The assumed series solution (2·1·2) may then be rewritten as a sum of the even powers of x plus a sum of the odd powers of x.

$$u(x) = \sum_{n=0}^{\infty} a_{2n} x^{2n} + \sum_{n=0}^{\infty} a_{2n+1} x^{2n+1}$$

The expressions for a_{2n} and a_{2n+1} may now be substituted to yield

$$u(x) = \sum_{n=0}^{\infty} \frac{(-1)^n m^{2n}}{(2n)!} a_0 x^{2n} + \sum_{n=0}^{\infty} \frac{(-1)^n m^{2n}}{(2n+1)!} a_1 x^{2n+1}$$

$$= a_0 \sum_{n=0}^{\infty} \frac{(-1)^n (mx)^{2n}}{(2n)!} + \frac{a_1}{m} \sum_{n=0}^{\infty} \frac{(-1)^n (mx)^{2n+1}}{(2n+1)!}$$

To avoid writing these summations each time, two new functions $C(mx)$ and $S(mx)$ can be defined as follows:

$$C(mx) \equiv \sum_{n=0}^{\infty} \frac{(-1)^n (mx)^{2n}}{(2n)!} = 1 - \frac{(mx)^2}{2!} + \frac{(mx)^4}{4!} - \cdots \tag{2·1·3}$$

$$S(mx) \equiv \sum_{n=0}^{\infty} \frac{(-1)^n (mx)^{2n+1}}{(2n+1)!} = \frac{mx}{1!} - \frac{(mx)^3}{3!} + \frac{(mx)^5}{5!} - \cdots \tag{2·1·4}$$

Here $\equiv$ means *defined as*. The solution then becomes

$$u(x) = a_0 C(mx) + a_1' S(mx) \tag{2·1·5}$$

where a_0 and $a_1' = a_1 / m$ are the two arbitrary constants necessary to fit the boundary conditions, given below (2·1·1), that $u(0) = 0$ and $u(1) = 1$. Thus

$$u(0) = a_0 C(0) + a_1' S(0) = 0$$

and

$$u(1) = a_0 C(1) + a_1' S(1) = 1$$

Observe from (2·1·3) and (2·1·4) that $C(0) = 1$ and $S(0) = 0$. Thus the boundary condition at $x = 0$ gives

$$u(0) = a_0 1 + a_1' 0 = 0$$

This requires that $a_0 = 0$. The boundary condition at $x = 1$ then gives

$$u(1) = a_1' S(m) = 1$$

The second constant is thus found to be $a_1' = 1/S(m)$. The final solution follows from (2·1·5) with $a_0 = 0$ and $a_1' = 1/S(m)$. Thus

$$u(x) = \frac{S(mx)}{S(m)} \tag{2·1·6}$$

A plot of $S(x)$ and $C(x)$ is shown in Figure 2•2. The functions are observed to oscillate and have an infinite number of zeros. If the series for $S(x)$ is differentiated,

$$\frac{dS}{dx} = \sum_{n=0}^{\infty} \frac{(-1)^n (2n+1) x^{2n}}{(2n+1)!} = \sum_{n=0}^{\infty} \frac{(-1)^n x^{2n}}{(2n)!} = C(x)$$

Thus there is a relation between the functions $S(x)$ and $C(x)$.

These functions could be studied in great detail to learn more about them. There is no need to in this case since they are recognized simply as sine and cosine. Their properties are already quite familiar. A similar procedure will be used in the next section to find a series solution to one form of Bessel's equation.

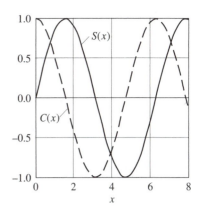

Figure 2•2 *Series solutions $S(x)$ and $C(x)$.*

2•1•2 The method of Frobenius

Needless to say, mathematicians have studied the power-series solutions of ordinary differential equations and have developed some general rules, proofs, and theorems. These are often discussed as the *method of Frobenius*. Rather than go into detail, only that portion of the theory which is of use here will be stated without proof.

To solve an equation of the form

$$\frac{d^2 u}{dx^2} + p(x)\frac{du}{dx} + q(x)u = 0 \tag{2·1·7}$$

where $xp(x)$ and $x^2 q(x)$ are *regular* functions[1] at $x = 0$ (they generally are in engineering problems), assume that the solution is of the form

$$u(x) = x^s \sum_{n=0}^{\infty} a_n x^n \tag{2·1·8}$$

Here s is a parameter which will have to be determined (just as the a_n will have to be determined) so that the differential equation will be satisfied. In the solution to (2·1·1), s was equal to zero.

In general, for a second-order equation such as (2·1·7), there will be two different values (s_1 and s_2) of s that will appear. The problem will be handled a little differently depending upon what these values are. In this section, only a special case in which $s_1 = s_2$ will be considered.

During the course of determining s_1 and s_2 ($= s_1$), a function $u(x;s)$ will be generated. (This will be shown later.) Mathematicians have shown the two independent solutions (u_1 and u_2) to be

$$u_1 = u(x;s)\big|_{s=s_1} \tag{2·1·9}$$

and

$$u_2 = \frac{\partial u(x;s)}{\partial s}\bigg|_{s=s_1} \tag{2·1·10}$$

where $s_1 = s_2$ is a special value of s determined so that the differential equation will be satisfied.

There is a lot more to the method of Frobenius than the above, but this will suffice for the present. Having seen one example of how a series solution can be used to solve a new problem, an engineer should be content to accept further solutions without having to go through all the details.

---

[1] A regular function is one that can be expanded as a series in powers of x. For further discussion of regular functions, see [6].

2•2 ORDINARY BESSEL FUNCTIONS

The details of a series solution of Bessel's equation will be presented only once to give you an idea of what is involved. The engineer should appreciate the work it takes to determine a solution and be thankful that it is not necessary to go through the entire procedure every time; shortcuts have been developed that can be used.

After observing the method of series solution, the properties of the solution and other related functions will be discussed. As the discussion continues, it will be helpful to compare it with the series solution of Section 2•1•1 that turned out to be sines and cosines.

2•2•1 Series solution of a zero-order Bessel equation

During the discussion of partial differential equations in cylindrical coordinates, the following equation will arise:

$$\frac{d^2u}{dx^2} + \frac{1}{x}\frac{du}{dx} + u = 0 \tag{2·2·1}$$

Notice how similar (2·2·1) is to the circular-fin equation (2·0·1) which will be discussed later.

The first step is to multiply (2·2·1) by x to arrive at a more convenient form:

$$x\frac{d^2u}{dx^2} + \frac{du}{dx} + xu = 0 \tag{2·2·2}$$

Following the method of Frobenius outlined in Section 2•1•2, a solution is assumed of the form

$$u(x) = \sum_{n=0}^{\infty} a_n x^{n+s} \tag{2·2·3}$$

Then

$$\frac{du}{dx} = \sum_{n=0}^{\infty}(n+s)a_n x^{n+s-1}$$

$$\frac{d^2u}{dx^2} = \sum_{n=0}^{\infty}(n+s)(n+s-1)a_n x^{n+s-2}$$

Substituting these into (2·2·2) gives

$$\sum_{n=0}^{\infty}(n+s)(n+s-1)a_n x^{n+s-1} + \sum_{n=0}^{\infty}(n+s)a_n x^{n+s-1} + \sum_{n=0}^{\infty}a_n x^{n+s+1} = 0$$

Observe that the first two series begin with a power of x that is 2 lower than the third series. It is convenient to "split off" the first two terms of these series so that all the series summations begin with the same power of x. Thus

$$s(s-1)a_0 x^{s-1} + sa_0 x^{s-1} + (s+1)sa_1 x^s + (s+1)a_1 x^s$$

$$+ \sum_{n=2}^{\infty}(n+s)(n+s-1)a_n x^{n+s-1} + \sum_{n=2}^{\infty}(n+s)a_n x^{n+s-1} + \sum_{n=0}^{\infty}a_n x^{n+s+1} = 0$$

Letting $\ell = n - 2$ in the first two series so that the summations will run over the same indices, and combining the leading terms,

$$s^2 a_0 x^{s-1} + (s+1)^2 a_1 x^s$$

$$+\sum_{\ell=0}^{\infty}[(\ell+s+2)(\ell+s+1)+(\ell+s+2)]a_{\ell+2}x^{\ell+s+1}+\sum_{n=0}^{\infty}a_nx^{n+s+1}=0$$

Thus, upon replacing ℓ by n in the first summation, recognizing that

$$(\ell+s+2)(\ell+s+1)+(\ell+s+2)=(\ell+s+2)^2$$

and combining the two summations, the above reduces to

$$s^2a_0x^{s-1}+(s+1)^2a_1x^s+\sum_{n=0}^{\infty}\left[(n+s+2)^2a_{n+2}+a_n\right]x^{n+s+1}=0$$

Notice that to satisfy the above it is necessary that the following conditions be satisfied:

1. $s^2a_0=0$

2. $(s+1)^2a_1=0$

3. $(n+s+2)^2a_{n+2}+a_n=0$

In the method of Frobenius, the procedure is to set s^2 equal to zero to satisfy condition 1. The equation $s^2=0$ is called the *indicial equation*. The roots of the indicial equation are $s_1=0$ and $s_2=0$.[1] When this result is substituted into condition 2, it follows that a_1 must be taken to be zero.

The value of s will not be set equal to zero in condition 3 until after a relation between a_{n+2} and a_n has been developed. Condition 3 can be satisfied without first setting s equal to zero by requiring that

$$a_{n+2}=\frac{-a_n}{(n+s+2)^2}$$

As will be seen later, it is essential that s not be set equal to zero yet in order to form the function $u(x;s)$ that is needed to find the two solutions u_1 and u_2.

Next, a general term for the coefficients must be found.

$$a_2=\frac{-a_0}{(s+2)^2} \qquad\qquad \text{for n = 0}$$

$$a_3=\frac{-a_1}{(s+3)^2} \qquad\qquad \text{for n = 1}$$

Since a_1 is zero, note that a_3 and all the succeeding odd terms will be zero.

$$a_4=\frac{-a_2}{(s+4)^2}=\frac{a_0}{(s+4)^2(s+2)^2} \qquad\qquad \text{for n = 2}$$

$$a_6=\frac{-a_4}{(s+6)^2}=\frac{-a_0}{(s+6)^2(s+4)^2(s+2)^2} \qquad\qquad \text{for n = 4}$$

The general term can then be written by inspection as

$$a_{2n}=\frac{(-1)^na_0}{(s+2n)^2(s+2n-2)^2\cdots(s+4)^2(s+2)^2}$$

Upon rearranging the denominator, the expression for a_{2n} becomes

---

[1] When the roots of the indicial equation are not equal, the method of Frobenius as discussed in Section 2•1•2 must be modified. These cases are explained in [6, 7]. Since the method of Frobenius is only being introduced here to provide background for the study of Bessel equations and is not of primary interest in heat transfer, these other cases will not be considered.

$$a_{2n} = \frac{(-1)^n a_0}{[(s+2)(s+4)\cdots(s+2n-2)(s+2n)]^2}$$

For convenience, we will define the *product symbol* as

$$\prod_{i=1}^{n} a_i \equiv \begin{cases} a_1 \cdot a_2 \cdot a_3 \cdot \ldots \cdot a_n & \text{for } n \geq 1 \\ 1 & \text{for } n < 1 \end{cases} \tag{2·2·4}$$

Defining the product to be 1 for $n < 1$ (the lower limit) may seem somewhat arbitrary at first but is necessary to be completely correct, as will be seen shortly. This product definition can be used to replace the denominator in a_{2n} to give

$$a_{2n} = \frac{(-1)^n a_0}{\left[\displaystyle\prod_{m=1}^{n} (s+2m)\right]^2} \tag{2·2·5}$$

Observe that for $n = 0$, the product in the denominator is, by definition, equal to unity, and the above reduces to $a_0 = a_0$. Any other definition of the product for $n < 1$ would not give this result.

This choice for the a will identically satisfy condition 3. Observe that the a are functions of s. They are zero for odd subscripts and equal to the above for even subscripts. In the method of Frobenius the function $u(x;s)$ is formed by using the original assumption (2·2·3) for $u(x)$ and substituting the value for $a_n(s)$. Thus

$$u(x;s) \equiv \sum_{n=0}^{\infty} a_n(s) x^{n+s}$$

or, since the odd terms are zero, the series can be rewritten as

$$u(x;s) = \sum_{n=0}^{\infty} a_{2n}(s) x^{2n+s}$$

Upon substituting (2·2·5) into the above,

$$u(x;s) = \sum_{n=0}^{\infty} \frac{(-1)^n a_0 x^{2n+s}}{\left[\displaystyle\prod_{m=1}^{n} (s+2m)\right]^2}$$

For simplification, we can define

$$b_n(s) = \frac{1}{\left[\displaystyle\prod_{m=1}^{n} (s+2m)\right]^2} \tag{2·2·6}$$

Then

$$u(x;s) = x^s \sum_{n=0}^{\infty} (-1)^n a_0 b_n(s) x^{2n} \tag{2·2·7}$$

One solution, u_1, is obtained when $s = s_1 = 0$ is substituted into (2·2·7). Thus

$$u_1(x) = u(x;s)\big|_{s=s_1=0} = \sum_{n=0}^{\infty} (-1)^n a_0 b_n(0) x^{2n}$$

Setting $s = 0$ in (2·2·6) gives

$$b_n(0) = \frac{1}{\left[\prod_{m=1}^{n}(2m)\right]^2} \qquad (2\cdot2\cdot8)$$

The first solution is then given by

$$u_1(x) = a_0 \sum_{n=0}^{\infty} \frac{(-1)^n x^{2n}}{\left[\prod_{m=1}^{n}(2m)\right]^2} \qquad (2\cdot2\cdot9)$$

To get the second solution, u_2, $\partial u/\partial s$ must be found. Differentiating (2·2·7),[1]

$$\frac{\partial u}{\partial s} = x^s \sum_{n=0}^{\infty} (-1)^n a_0 \frac{db_n(s)}{ds} x^{2n} + x^s \ln(x) \sum_{n=0}^{\infty} (-1)^n a_0 b_n(s) x^{2n} \qquad (2\cdot2\cdot10)$$

The simplest way to find $db_n(s)/ds$ is to start from the definition of b_n (2·2·6), and take logs.

$$\ln[b_n(s)] = -2\ln\left[\prod_{m=1}^{n}(s+2m)\right] = -2\sum_{m=1}^{n}\ln(s+2m)$$

where the summation is defined to be 0 whenever $n < 1$ since the product was defined (2·2·4) to be 1 for $n < 1$, and $\log(1) = 0$. Differentiating,[2]

$$\frac{1}{b_n(s)}\frac{db_n(s)}{ds} = \frac{b_n'(s)}{b_n(s)} = -2\sum_{m=1}^{n}\frac{1}{s+2m}$$

Thus

$$b_n'(s) = -2b_n(s)\sum_{m=1}^{n}\frac{1}{s+2m}$$

For $s = 0$ this reduces to

$$b_n'(0) = -2b_n(0)\sum_{m=1}^{n}\frac{1}{2m} \qquad (2\cdot2\cdot11)$$

The second solution, u_2, is obtained by setting $s = 0$ in (2·2·10) and then substituting (2·2·8) and (2·2·11) into it. Thus

$$u_2 = \frac{\partial u}{\partial s}\bigg|_{s=0} = -2a_0 \sum_{n=0}^{\infty} \frac{(-1)^n \sum_{m=1}^{n}\frac{1}{2m}}{\left[\prod_{m=1}^{n}2m\right]^2} x^{2n} + a_0 \ln(x) \sum_{n=0}^{\infty} \frac{(-1)^n x^{2n}}{\left[\prod_{m=1}^{n}2m\right]^2} \qquad (2\cdot2\cdot12)$$

Equations (2·2·9) and (2·2·12) are the two independent solutions necessary for a complete solution. Any constant times them will also work. These solutions can be simplified by noting that

$$\prod_{m=1}^{n}2m = 2\cdot4\cdot6\cdots2n = 2^n[1\cdot2\cdot3\cdots n] = 2^n n!$$

---

[1] Recall that $\dfrac{d}{dx}a^x = a^x \ln(a)$

[2] Recall that $\dfrac{d}{dx}\ln(y) = \dfrac{1}{y}\dfrac{dy}{dx}$

Then

$$u_1(x) = a_0 \sum_{n=0}^{\infty} \frac{(-1)^n x^{2n}}{2^{2n}(n!)^2} \tag{2·2·13}$$

$$u_2(x) = -2a_0 \sum_{n=0}^{\infty} \frac{(-1)^n x^{2n}}{2^{2n}(n!)^2} \left(\sum_{m=1}^{n} \frac{1}{2m} \right) + a_0 \ln(x) \sum_{n=0}^{\infty} \frac{(-1)^n x^{2n}}{2^{2n}(n!)^2} \tag{2·2·14}$$

A new function $J_0(x)$ is defined for convenience by

$$J_0(x) = \sum_{n=0}^{\infty} \frac{(-1)^n x^{2n}}{2^{2n}(n!)^2} = 1 - \frac{x^2}{2^2 \cdot 1} + \frac{x^4}{2^4 (2!)^2} - \cdots \tag{2·2·15}$$

The function $J_0(x)$ is called a *Bessel function of first kind of order zero*. Then

$$u_1(x) = a_0 J_0(x)$$

and

$$u_2 = a_0 \ln(x) J_0(x) - a_0 \sum_{n=0}^{\infty} \frac{(-1)^n x^{2n}}{2^{2n}(n!)^2} \left(\sum_{m=1}^{n} \frac{1}{m} \right)$$

Observe that the lower index on the summation over n can be changed from 0 to 1 since

$$\sum_{m=1}^{n} \frac{1}{m} = 0 \qquad \text{for } n = 0$$

Another function, $N_0(x)$ is conveniently defined as

$$N_0(x) = J_0(x)\ln(x) - \sum_{n=1}^{\infty} \frac{(-1)^n x^{2n}}{2^{2n}(n!)^2} \left(\sum_{m=1}^{n} \frac{1}{m} \right)$$

The function $N_0(x)$ is called a *Neumann function*. A function $Y_0(x)$ is more commonly used as the second solution to Bessel's equation. It can be defined as a linear combination of $N_0(x)$ and $J_0(x)$ in the following way:

$$Y_0(x) = \frac{2}{\pi} N_0(x) + \frac{2}{\pi} [\gamma - \ln(2)] J_0(x) \tag{2·2·16}$$

The constant γ is called the *Euler constant* and is defined by

$$\gamma = \lim_{n \to \infty} \left[1 + \frac{1}{2} + \frac{1}{3} + \cdots + \frac{1}{n} - \ln(n) \right] = 0.5772 \tag{2·2·17}$$

The function $Y_0(x)$ is a *Bessel function of second kind of order zero*.

Thus the two independent solutions may now be written as

$$u_1 = J_0(x)$$

and

$$u_2 = Y_0(x)$$

The general solution to (2·2·2) is any linear combination of these. Thus

$$u = A J_0(x) + B Y_0(x) \tag{2·2·18}$$

where A and B are arbitrary constants to be determined from boundary conditions in the usual manner.

If this solution had been carried out in more general form, the solution to

$$u'' + \frac{1}{x}u' + m^2 u = 0 \tag{2.2.19}$$

would be given by

$$u = AJ_0(mx) + BY_0(mx) \tag{2.2.20}$$

This can be verified by making a change of independent variable from x to $\xi = mx$ in (2·2·19). This reduces the equation to (2·2·1), whose solution is given by (2·2·18).

It should also be remarked that if the right-hand side of the differential equation had been other than zero, the equation could not be classified as a Bessel equation. However, its solution would still involve Bessel functions because the equation could be solved in two parts: a homogeneous solution (leading to Bessel functions) and a nonhomogeneous (or particular) solution which could be obtained by using the homogeneous solution in a method called *variation of parameters*.[1]

The Bessel functions $J_0(x)$ and $Y_0(x)$ are two of a class of Bessel functions called ordinary Bessel functions. As shown in this section, they arise naturally from a series solution of (2·2·2) if one persists through the necessary algebra. Obviously the engineer is more interested in learning the characteristics of these functions and how to use them than in the details of obtaining them. Consequently the remainder of this section will discuss the features of ordinary Bessel functions of interest to the engineer.

2•2•2 Behavior of zero-order Bessel functions

It is difficult to imagine the behavior of these functions from their series representations. This can also be said for the sine and cosine functions. Evaluation of the series for $J_0(x)$ will show that $J_0(x)$ is unity at $x = 0$ and oscillates with decreasing amplitude as x increases. A plot is shown in Figure 2•3, and a tabulation can be found in [10].

As will be shown later, the ordinary Bessel functions are close relatives to the sines and cosines. Observe that both $J_0(x)$ and $\sin(x)$ oscillate and have an infinite number of zeros. This fact will be quite important in the discussion of partial differential equations. The first six zeros of $J_0(x)$ as obtained from Table C•3 in Appendix C are tabulated in Table 2•1. Another similarity is that, as x increases, the interval between zeros of $J_0(x)$ approaches π as shown in Table 2•1.

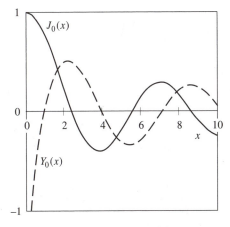

Figure 2•3 *Zero-order Bessel functions.*

Table 2•1 *First five zeros of $J_0(x)$.*

Zeros of $J_0(x)$	Interval between zeros
2.4048	
5.5201	3.1153
8.6537	3.1336
11.7915	3.1378
14.9309	3.1394
18.0711	3.1402
⋮	⋮
	$3.1416 = \pi$

Bessel functions of the second kind, $Y_0(x)$, have a much different behavior at the origin. As a result of the logarithmic term in its definition

---

[1] See [6], for example. See also Appendix A•2•3.

(2·2·16), $Y_0(x)$ approaches minus infinity as x approaches zero. This function is shown in Figure 2•3 and tabulated in [10]. Once again, observe that $Y_0(x)$ oscillates with decreasing amplitude as x increases. It, too, has an infinite number of zeros.

The fact that $Y_0(x)$ is infinite at $x = 0$ often leads to a simplification of the solution (2·2·18) since in many physical situations the solution cannot be allowed to be infinite at $x = 0$. Consequently this fact is used as a boundary condition, and B must therefore be zero. The solution then contains only $J_0(x)$.

The Bessel functions should become a part of your "analytical tool kit," just as are the trigonometric functions, the exponential functions, and the hyperbolic functions. It will be helpful to know the general behavior of the Bessel functions just as you do any of the other functions. Anytime you need a numerical value of a Bessel function, there are tables available just as with the other functions. As you would expect, Bessel functions can be differentiated and integrated. In summary, Bessel functions are just like any other function you have run into—a little mysterious at first, but quite normal as you gain experience with them.

2•2•3 Bessel functions of positive integer order *k*

Before proceeding with the discussion of the properties of Bessel functions, it is convenient to generalize a bit and introduce Bessel functions of integer order. The discussion of $J_0(x)$ and $Y_0(x)$ is then just a special case of the integer-order functions.

The integer-order functions arise from the solution of

$$x^2 \frac{d^2u}{dx^2} + x\frac{du}{dx} + (x^2 - k^2)u = 0 \tag{2·2·21}$$

where k is an integer. Observe the similarity between (2·2·21) and (2·2·2). One extra term has been added.

The method of Frobenius will work, and a series solution to (2·2·21) can be found. More knowledge of the method is required than has been presented in this book, however (see [6, 7]). One of the independent solutions is found to be

$$u(x) = a_0 x^k \sum_{n=0}^{\infty} \frac{(-1)^n x^{2n}}{2^{2n}\left[\prod_{m=1}^{n}(m+k)\right]n!} \tag{2·2·22}$$

For k being a positive integer, (2·2·22) can be simplified by noting that

$$\prod_{m=1}^{n}(m+k) = \frac{(k!)(1+k)(2+k)(3+k)\cdots(n+k)}{k!} = \frac{(n+k)!}{k!} \tag{2·2·23}$$

Then (2·2·22) becomes

$$u(x) = a_0 x^k \sum_{n=0}^{\infty} \frac{(-1)^n x^{2n} k!}{2^{2n}(n+k)!n!}$$

For convenience, replace a_0 by $a_0'/2^k k!$ and define a new function $J_k(x)$ such that $u(x) = a_0' J_k(x)$. The definition of $J_k(x)$ is then given by

$$J_k(x) = \sum_{n=0}^{\infty} \frac{(-1)^n x^{2n+k}}{2^{2n+k} n!(n+k)!} \tag{2·2·24}$$

This is one of the two independent solutions and is called a *Bessel function of first kind of order k*. The corresponding second solution would be called $Y_k(x)$, and the general solution would be

$$u = AJ_k(x) + BY_k(x)$$

The form of $Y_k(x)$ would be

$$Y_k(x) = J_k(x)\ln(x) + \text{(an infinite series)} \tag{2.2.25}$$

Observe the similarity between the Bessel functions of order k and those of zero order. The behavior of $J_1(x)$ and $Y_1(x)$ is shown in Figure 2•4. These functions are tabulated in [10]. $J_1(x)$ also oscillates with decreasing amplitude as x increases, but it starts out at zero for $x = 0$ as do all the higher-order functions. $Y_1(x)$ contains the logarithmic term which becomes infinite as x goes to zero just as $Y_0(x)$ did.

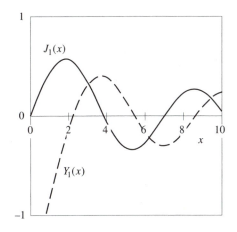

Figure 2•4 *First-order Bessel functions.*

2•2•4 Relations between Bessel functions and their derivatives

There are many relationships between the Bessel functions of various orders and their derivatives. Some of the more common ones are presented in [10] and in Appendix B.

An example of how these might be obtained can be indicated by considering the definition of $J_0(x)$. Thus from (2·2·15)

$$J_0(x) = \sum_{n=0}^{\infty} \frac{(-1)^n x^{2n}}{2^{2n}(n!)^2}$$

This can next be differentiated to give

$$\frac{dJ_0(x)}{dx} = J_0'(x) = \sum_{n=0}^{\infty} \frac{(-1)^n 2n x^{2n-1}}{2^{2n}(n!)^2} = \sum_{n=1}^{\infty} \frac{(-1)^n x^{2n-1}}{2^{2n-1}n!(n-1)!}$$

To return to a summation with a lower index of zero, let $\ell = n - 1$. Then the above becomes

$$J_0'(x) = \sum_{\ell=0}^{\infty} \frac{(-1)^{\ell+1} x^{2\ell+1}}{2^{2\ell+1}(\ell+1)!\ell!}$$

By comparison with (2·2·24), the above summation is seen to be just the negative of $J_1(x)$. Thus

$$J_0'(x) = -J_1(x) \tag{2.2.26}$$

More generally, one can show the following *recurrence relations*:

$$J_{k-1}(x) + J_{k+1}(x) = \frac{2k}{x} J_k(x) \tag{2.2.27}$$

and

$$J_{k-1}(x) - J_{k+1}(x) = 2J_k'(x) \tag{2.2.28}$$

Equation (2·2·26) can be obtained from (2·2·27) and (2·2·28) by setting k equal to zero and subtracting.

Similar relations also hold for Bessel functions of second kind, $Y_k(x)$. Recurrence relations are presented in [10] and in Appendix B.

2•2•5 The gamma function

In the discussion of Bessel functions of noninteger order it will be convenient to employ a function called the *gamma function*, $\Gamma(x)$, that can be defined as follows:

$$\Gamma(x) = \int_{t=0}^{\infty} t^{x-1} \exp(-t)\, dt \qquad (2 \cdot 2 \cdot 29)$$

Values of $\Gamma(x)$ are given in tables [8, 10] for $1 \le x \le 2$. Values of $\Gamma(x)$ for other x may be found from (2·2·29) by integrating (2·2·29) by parts to give

$$\Gamma(x) = [t^{x-1}\{-\exp(-t)\}]_{t=0}^{\infty} - \int_{t=0}^{\infty} \{-\exp(-t)\}(x-1)t^{x-2}\, dt$$

For $x > 1$, the integrated terms vanish and the above reduces to

$$\Gamma(x) = (x-1)\int_{t=0}^{\infty} t^{x-2} \exp(-t)\, dt$$

Comparison with (2·2·29) shows that the integral is $\Gamma(x-1)$. Therefore,

$$\Gamma(x) = (x-1)\Gamma(x-1) \qquad (2 \cdot 2 \cdot 30)$$

This result may be used to evaluate $\Gamma(x)$ for $x > 2$ from tabulated values. To determine $\Gamma(4.2)$ we can repeatedly apply (2·2·30) to give

$$\Gamma(4.2) = 3.2 \times \Gamma(3.2) = 3.2 \times 2.2 \times \Gamma(2.2) = 3.2 \times 2.2 \times 1.2 \times \Gamma(1.2)$$

From [10] we can find $\Gamma(1.2) = 0.91817$. Thus

$$\Gamma(4.2) = 3.2 \times 2.2 \times 1.2 \times 0.91817 = 7.7567$$

The gamma function[1] is related to the factorial function. This can be seen by repeated application of (2·2·30) for $x = n$ (an integer).

$$\Gamma(n) = (n-1)(n-2)\cdots(2)(1)\Gamma(1) = (n-1)!\,\Gamma(1)$$

Next, $\Gamma(1)$ can be found by setting $x = 1$ in (2·2·29) and integrating to give

$$\Gamma(1) = \int_{t=0}^{\infty} \exp(-t)\, dt = -\exp(-t)\big|_{t=0}^{\infty} = 1$$

Therefore,

$$\Gamma(n) = (n-1)! \qquad (2 \cdot 2 \cdot 31)$$

A plot of the gamma function $\Gamma(x)$ is given in Figure 2•5. For points (a), (b) and (c) we see from (2·2·31) that $\Gamma(1) = 0! = 1$, $\Gamma(2) = 1! = 1$ and $\Gamma(3) = 2! = 2$. Thus, for $x > 1$, the gamma function is just filling in the spaces between the factorial function which is defined only for integer arguments. The gamma function is extended to values of $x < 1$ by using (2·2·30) rewritten as $\Gamma(x-1) = \Gamma(x)/(x-1)$. By repeated application,

$$\Gamma(x-1) = \frac{1}{x-1} \times \frac{1}{x} \times \frac{1}{x+1} \times \cdots \times \frac{1}{x+n} \times \Gamma(x+n+1)$$

The number of terms is such that $1 \le x+n+1 \le 2$ so that $\Gamma(x+n+1)$ can be looked up in a table. When $x-1$ is equal to zero or a negative integer the gamma function is infinite because a zero will occur in the denominator during the calculation. This extension is also shown in Figure 2•5.

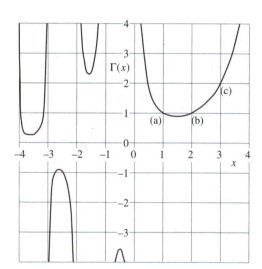

Figure 2•5 *The gamma function.*

---

[1] A relative of the gamma function is the beta function which is defined as follows:

$$\beta(p,q) = \int_{t=0}^{1} t^{p-1}(1-t)^{q-1}\, dt$$

It can be shown that

$$\beta(p,q) = \frac{\Gamma(p)\Gamma(q)}{\Gamma(p+q)}$$

Another function is the incomplete beta function defined by

$$\beta_v(p,q) = \int_{t=0}^{v} t^{p-1}(1-t)^{q-1}\, dt$$

The value for $\Gamma(\frac{1}{2})$ can be evaluated as shown below:

$$\Gamma(\tfrac{1}{2}) = \int_{t=0}^{\infty} t^{-\frac{1}{2}} \exp(-t)\, dt$$

Let $\tau = t^{\frac{1}{2}}$. Then

$$\Gamma(\tfrac{1}{2}) = \int_{\tau=0}^{\infty} \tau^{-1} \exp(-\tau^2) 2\tau \, d\tau = 2\int_{\tau=0}^{\infty} \exp(-\tau^2)\, d\tau$$

This definite integral is found in most integral tables and is equal to $\sqrt{\pi}/2$. Thus

$$\Gamma(\tfrac{1}{2}) = \sqrt{\pi}$$

2•2•6 Bessel functions of arbitrary order *k*

The solution to $x^2 u'' + xu' + (x^2 - k^2)u = 0$ has one solution given by (2·2·22) as

$$u_1(x) = a_0 x^k \sum_{n=0}^{\infty} \frac{(-1)^n x^{2n}}{2^{2n} n! \prod_{m=1}^{n}(m+k)}$$

In this section k will no longer be restricted to integer values. This means that the product function can no longer be simplified in terms of the factorial function as in Section 2•2•3. However, by multiplying and dividing by $\Gamma(1+k)$, the product function in the denominator can be expressed in terms of gamma functions as

$$\prod_{m=1}^{n}(m+k) = \underbrace{\underbrace{\underbrace{\Gamma(1+k)(1+k)}_{\Gamma(2+k)}(2+k)\cdots(n+k)}_{\Gamma(3+k)}}_{\Gamma(n+k+1)} \frac{1}{\Gamma(1+k)} = \frac{\Gamma(n+k+1)}{\Gamma(k+1)}$$

$$(2\cdot2\cdot32)$$

This is a more general expression than (2·2·22) since the factorial function is not defined for noninteger values of its argument, whereas the gamma function is defined in this case.

The solution then becomes

$$u_1(x) = a_0 x^k \sum_{n=0}^{\infty} \frac{(-1)^n x^{2n} \Gamma(k+1)}{2^{2n} n! \Gamma(n+k+1)}$$

or, since $n! = \Gamma(n+1)$,

$$u_1(x) = a_0 x^k \Gamma(k+1) \sum_{n=0}^{\infty} \frac{(-1)^n x^{2n}}{2^{2n} \Gamma(n+1)\Gamma(n+k+1)}$$

If a_0 is arbitrarily replaced by $a_0'/2^k \Gamma(k+1)$, the solution can be written as

$$u_1(x) = a_0' J_k(x)$$

where

$$J_k(x) = \frac{x^k}{2^k} \sum_{n=0}^{\infty} \frac{(-1)^n x^{2n}}{2^{2n} \Gamma(n+1)\Gamma(n+k+1)} \qquad (2\cdot2\cdot33)$$

This is more general than (2·2·24) since it holds for all k, integer and noninteger, positive and negative.

An interesting relation can now be shown between J_k and J_{-k} when k = integer. The series for $J_{-k}(x)$ can be obtained simply by replacing k by $-k$ in the series for $J_k(x)$. Thus

$$J_{-k}(x) = \frac{x^{-k}}{2^{-k}} \sum_{n=0}^{\infty} \frac{(-1)^n x^{2n}}{2^{2n} \Gamma(n+1)\Gamma(n+1-k)}$$

For $n < k-1$, $\Gamma(n+1-k) = \infty$, and the term in the series is zero. Thus the terms in the series from $n = 0$ through and including $n = k-1$ are all zero. Therefore the series may be rewritten as

$$J_{-k}(x) = \frac{x^{-k}}{2^{-k}} \sum_{n=k}^{\infty} \frac{(-1)^n x^{2n}}{2^{2n} \Gamma(n+1)\Gamma(n+1-k)}$$

Letting $\ell = n - k$, $n = \ell + k$,

$$J_{-k}(x) = \frac{x^k}{2^k} \sum_{\ell=0}^{\infty} \frac{(-1)^{\ell+k} x^{2\ell}}{2^{2\ell} \Gamma(\ell+k+1)\Gamma(\ell+1)} = (-1)^k J_k(x)$$

Therefore J_{-k} and J_k are not independent solutions when k = integer. A similar relation holds for Y_k and Y_{-k}.

By substitution of $k = \pm\frac{1}{2}$ into (2·2·33), it can be shown that

$$J_{\frac{1}{2}}(x) = \sqrt{\frac{2}{\pi x}} \sin(x) \tag{2·2·34}$$

$$J_{-\frac{1}{2}}(x) = \sqrt{\frac{2}{\pi x}} \cos(x) \tag{2·2·35}$$

It is evident that these are independent since sine and cosine are independent. More generally, Bessel functions of order $k = n + \frac{1}{2}$ are related to functions called *spherical Bessel functions*.[1]

It can also be shown that the recurrence relations given in Section 2•2•4 also hold for noninteger values of k. Thus

$$J_{k-1}(x) + J_{k+1}(x) = \frac{2k}{x} J_k(x) \tag{2·2·36}$$

$$J_{k-1}(x) - J_{k+1}(x) = 2J_k'(x) \tag{2·2·37}$$

Similar relations can be found for the Y_k's. Some of the more frequently used recurrence relations are presented in [10] and in Appendix B.

---

[1] The spherical Bessel functions $j_n(x)$ and $y_n(x)$ are defined as

$$j_n(x) = \sqrt{\frac{\pi}{2x}} J_{n+\frac{1}{2}}(x) \quad \text{where } n = 0, \pm 1, \pm 2, \ldots$$

$$y_n(x) = \sqrt{\frac{\pi}{2x}} Y_{n+\frac{1}{2}}(x) \quad \text{where } n = 0, \pm 1, \pm 2, \ldots$$

The following relation can be shown to hold between these functions:

$$y_n(x) = (-1)^{n+1} j_{-n-1}(x) \qquad \text{for } n = 0, \pm 1, \pm 2, \ldots$$

For $n = 0$ the following is then true:

$$Y_{\frac{1}{2}}(x) = -J_{-\frac{1}{2}}(x)$$

and for $n = -1$,

$$Y_{-\frac{1}{2}}(x) = J_{\frac{1}{2}}(x)$$

Thus, $Y_{\frac{1}{2}}(x)$ and $J_{-\frac{1}{2}}(x)$ are linearly dependent as are $Y_{-\frac{1}{2}}(x)$ and $J_{\frac{1}{2}}(x)$. For a further discussion of spherical Bessel functions, see [10].

2•2•7 Integration of Bessel functions

Bessel functions may be integrated just as any other function is integrated. Some of the more common integrals are found in [10] and in Appendix B. A more complete discussion and tabulation can be found in [5].

An example of how one goes about deriving these relations might be helpful. Consider

$$\int x J_0(x)\,dx$$

The series expression for $J_0(x)$ [(2·2·33) with $k = 0$] may be substituted to give

$$\int x J_0(x)\,dx = \int x \sum_{n=0}^{\infty} \frac{(-1)^n x^{2n}}{2^{2n}(n!)^2}\,dx$$

The integration and the summation may usually[1] be interchanged in most problems that arise in engineering. The series can then be integrated term by term as follows:

$$\int x J_0(x)\,dx = \sum_{n=0}^{\infty} \frac{(-1)^n}{2^{2n}(n!)^2} \int x^{2n+1}\,dx = \sum_{n=0}^{\infty} \frac{(-1)^n x^{2n+2}}{2^{2n}(n!)^2(2n+2)}$$

$$= x \sum_{n=0}^{\infty} \frac{(-1)^n x^{2n+1}}{2^{2n+1} n!(n+1)!}$$

The resulting series is then recognized, by comparison to (2·2·33), to be $J_1(x)$. Thus

$$\int x J_0(x)\,dx = x J_1(x)$$

The engineer, of course, will not go through all these steps each time to integrate a Bessel function. Rather, an engineer relies on integral tables. As part of your background however, you should understand how the integrals might be obtained.

2•2•8 Asymptotic behavior of Bessel functions

One can often learn something about a new function by considering its asymptotic behavior as its argument becomes large. It is also handy to know about this as it is often much easier to compute values of the function for large arguments using the asymptotic expression rather than the exact expression. Consider Bessel's equation

$$x^2 u'' + xu' + (x^2 - k^2)u = 0$$

A solution to this has already been found to be $u = J_k(x)$. The behavior of $J_k(x)$ for large x can be indicated by letting

$$u = \frac{v}{\sqrt{x}} = x^{-\frac{1}{2}}v$$

Then

$$u' = x^{-\frac{1}{2}}v' + v(-\tfrac{1}{2})x^{-\frac{3}{2}}$$

$$u'' = x^{-\frac{1}{2}}v'' - \tfrac{1}{2}v'x^{-\frac{3}{2}} + \tfrac{1}{2}v\tfrac{3}{2}x^{-\frac{5}{2}} - \tfrac{1}{2}x^{-\frac{3}{2}}v'$$

Substituting u, u', and u'' into Bessel's equation gives

---

[1] This requires that the series be uniformly convergent. In most problems the engineer automatically assumes this to be true. Usually this is right.

$$x^{\frac{3}{2}}v'' - \frac{1}{2}v'x^{\frac{1}{2}} + \frac{1}{2}v\frac{3}{2}x^{-\frac{1}{2}} - \frac{1}{2}x^{\frac{1}{2}}v' + x^{\frac{1}{2}}v' + v(-\frac{1}{2})x^{-\frac{1}{2}} + x^{\frac{3}{2}}v - k^2x^{-\frac{1}{2}}v = 0$$

This reduces to

$$x^{\frac{3}{2}}v'' + v\left[\frac{3}{4}x^{-\frac{1}{2}} - \frac{1}{2}x^{-\frac{1}{2}} + x^{\frac{3}{2}} - k^2x^{-\frac{1}{2}}\right] = 0$$

$$x^{\frac{3}{2}}v'' + v\left[x^{\frac{3}{2}} + \frac{\frac{1}{4} - k^2}{x^{\frac{1}{2}}}\right] = 0$$

$$v'' + v\left[1 + \frac{\frac{1}{4} - k^2}{x^2}\right] = 0$$

Observe that, as x becomes large, the second term in parentheses becomes zero. Thus, for large x, the equation reduces to

$$v'' + v = 0 \tag{2·2·38}$$

This familiar equation has sines and cosines for its solution,

$$v = A\sin(x) + B\cos(x)$$

or equivalently,

$$v = C\cos(x - \phi)$$

where $C = \sqrt{A^2 + B^2}$ and $\phi = \tan^{-1}(A/B)$. Consequently, for large x,

$$u = \frac{v}{\sqrt{x}} = \frac{C}{\sqrt{x}}\cos(x - \phi)$$

This expression for u describes the behavior of $J_k(x)$ for large x. The analysis to find C and ϕ, which are functions of k, is complicated and will be omitted here. The result is

$$J_k(x) \approx \sqrt{\frac{2}{\pi x}}\cos(x - \phi_k) \tag{2·2·39}$$

where

$$\phi_k = (2k + 1)\frac{\pi}{4}$$

It can also be shown that

$$Y_k(x) \approx \sqrt{\frac{2}{\pi x}}\sin(x - \phi_k) \tag{2·2·40}$$

Observe that ϕ_k is zero for $k = -\frac{1}{2}$ and

$$J_{-\frac{1}{2}}(x) \approx \sqrt{\frac{2}{\pi x}}\cos(x)$$

$$Y_{-\frac{1}{2}}(x) \approx \sqrt{\frac{2}{\pi x}}\sin(x)$$

In fact, for this particular case, as can be seen from (2·2·35), it happens to be exact for all x. Once again, the similarity to sines and cosines is evident.

Additional asymptotic relations for Bessel functions are given in [10].

2•3 MODIFIED BESSEL FUNCTIONS

The Bessel equation considered in Section 2•2•1 often appears in a "modified" form. It is identical to (2·2·2) but with one sign changed to give

$$xu'' + u' - xu = 0 \qquad (2.3.1)$$

The functions obtained in the solution to (2·3·1) are called *modified Bessel functions*. Just as there were more complete forms of (2·2·2), such as (2·2·21) which gave rise to a family of ordinary Bessel functions, there are also more complete forms of (2·3·1) that give rise to a family of modified Bessel functions. These functions are discussed in the next few sections.

2•3•1 Series solution of a modified Bessel equation

If one begins with the method of Frobenius to solve (2·3·1), the two independent solutions can be shown to be

$$u_1(x) = \sum_{n=0}^{\infty} \frac{x^{2n}}{2^{2n}(n!)^2} \qquad (2.3.2)$$

and

$$u_2(x) = u_1(x)\ln(x) - \sum_{n=1}^{\infty} \frac{x^{2n}}{2^{2n}(n!)^2} \sum_{m=1}^{n} \frac{1}{m} \qquad (2.3.3)$$

Comparison of (2·3·2) with the corresponding solution (2·2·13) of the unmodified equation shows that the only difference is that the signs in the modified solution do not alternate. Thus the solution is called a *modified Bessel function* and is given the symbol $I_0(x)$ in place of $J_0(x)$. The first solution is then written as

$$u_1(x) = I_0(x) = \sum_{n=0}^{\infty} \frac{x^{2n}}{2^{2n}(n!)^2} \qquad (2.3.4)$$

The second solution (2·3·3) is also a modified form of the corresponding function $N_0(x)$. It is more common, however, to use a function $K_0(x)$ defined by $K_0(x) = -[\gamma - \ln(2)]I_0(x) - u_2(x)$ where γ is the Euler constant defined by (2·2·17). Thus the second solution is taken to be

$$K_0(x) = -\left[\gamma + \ln(\frac{x}{2})\right]I_0(x) + \sum_{n=1}^{\infty} \frac{x^{2n}}{2^{2n}(n!)^2} \sum_{m=1}^{n} \frac{1}{m} \qquad (2.3.5)$$

The complete solution to (2·3·1) is then given by

$$u(x) = AI_0(x) + BK_0(x) \qquad (2.3.6)$$

Again, if we had been more general, the solution to

$$xu'' + u' - m^2 xu = 0 \qquad (2.3.7)$$

would be given by

$$u(x) = AI_0(mx) + BK_0(mx) \qquad (2.3.8)$$

This can be verified by making a change in the independent variable from x to $\xi = mx$ in (2·3·7). This reduces (2·3·7) to (2·3·1), whose solution is given by (2·3·6).

The solution $I_0(x)$ is called a *modified Bessel function of the first kind of order zero* and $K_0(x)$ is called a *modified Bessel function of the second kind of order zero*. The next few sections discuss some of the more important properties of these functions.

Modified Bessel functions of nonzero order are found as the solution to

$$x^2u'' + xu' - (x^2 + k^2)u = 0 \tag{2.3.9}$$

where k will be considered an integer here. The solution is written as

$$u = AI_k(x) + BK_k(x) \tag{2.3.10}$$

where

$$I_k(x) = \sum_{n=0}^{\infty} \frac{x^{2n+k}}{2^{2n+k} n!(n+k)!} \tag{2.3.11}$$

and

$$K_k(x) = (-1)^{k+1}\left[\gamma + \ln(\frac{x}{2})\right]I_k(x) + \frac{1}{2}\sum_{n=0}^{k-1}\frac{(-1)^n(k-n-1)!x^{2n-k}}{2^{2n-k} n!}$$
$$+ \frac{(-1)^k}{2}\sum_{n=0}^{\infty}\frac{x^{2n+k}}{2^{2n+k} n!(n+k)!}\left(2\sum_{m=1}^{n}\frac{1}{m} + \sum_{m=1}^{k}\frac{1}{n+m}\right) \tag{2.3.12}$$

2•3•2 Behavior of modified Bessel functions

The behavior of these functions is shown in Figure 2•6. Observe that these functions do not oscillate as did $J_k(x)$ and $Y_k(x)$. The $K_k(x)$ become infinite as x approaches zero because of the logarithmic term just as the $Y_k(x)$ did.

The asymptotic behavior for large x can be evaluated just as was done for $J(x)$ in Section 2•2•8. Here, letting $u = v/\sqrt{x}$ reduces (2.3.9) to

$$v'' - v = 0$$

This has the solution

$$v = A'\exp(x) + B'\exp(-x)$$

so that

$$u = \frac{A'\exp(x)}{\sqrt{x}} + \frac{B'\exp(-x)}{\sqrt{x}} \approx AI_k(x) + BK_k(x)$$

Therefore it is not surprising to find, for large x, that

$$I_k(x) \approx \frac{\exp(x)}{\sqrt{2\pi x}} \tag{2.3.13}$$

and

$$K_k(x) \approx \sqrt{\frac{\pi}{2x}}\exp(-x) \tag{2.3.14}$$

Thus, whereas ordinary Bessel functions are related to the trigonometric functions, modified Bessel functions are related to exponential functions.

Asymptotic expressions are handy for numerical computations at large values of the argument, say, $x > 10$ or 20. More complete expressions can be found in [8, 10].

2•3•3 Modified Bessel function relations

As was the case with the ordinary Bessel functions, there are relationships among modified Bessel functions. For example, consider $I_0(x)$.

$$I_0(x) = \sum_{n=0}^{\infty}\frac{x^{2n}}{2^{2n}(n!)^2}$$

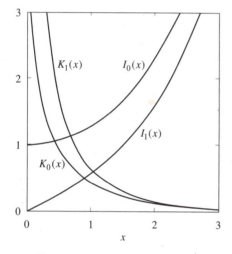

Figure 2•6 *Modified Bessel functions.*

Differentiation gives

$$I_0'(x) = \sum_{n=0}^{\infty} \frac{2n x^{2n-1}}{2^{2n}(n!)^2} = \sum_{n=1}^{\infty} \frac{x^{2n-1}}{2^{2n-1} n!(n-1)!}$$

Letting $\ell = n - 1$ gives

$$I_0'(x) = \sum_{\ell=0}^{\infty} \frac{x^{2\ell+1}}{2^{2\ell+1}(\ell+1)!\ell!}$$

The series thus formed is recognized as $I_1(x)$. Thus

$$I_0'(x) = I_1(x) \tag{2.3.15}$$

Additional recurrence relations can be found in [10] and in Appendix B.

The modified Bessel functions are also related to the ordinary Bessel functions through the use of complex arguments. Thus, from (2·2·15),

$$J_0(ix) = \sum_{n=0}^{\infty} \frac{(-1)^n (ix)^{2n}}{2^{2n}(n!)^2} = \sum_{n=0}^{\infty} \frac{(-1)^n (-1)^n x^{2n}}{2^{2n}(n!)^2} = \sum_{n=0}^{\infty} \frac{x^{2n}}{2^{2n}(n!)^2}$$

This series is recognized as $I_0(x)$. Thus

$$J_0(ix) = I_0(x) \tag{2.3.16}$$

2•4 THE GENERAL BESSEL EQUATION

In the previous sections you have been introduced to the family of Bessel equations (ordinary and modified) and the corresponding families of functions. As engineers it is convenient to have at our disposal a fast method of obtaining solutions to the various forms of Bessel's equation. This section presents a "cookbook" method of obtaining solutions.

The general Bessel equation is given in [4, 9] as

$$x^2 u'' + [(1 - 2A)x - 2Bx^2]u'$$
$$+ [C^2 D^2 x^{2C} + B^2 x^2 - B(1 - 2A)x + A^2 - C^2 n^2]u = 0 \qquad (2\cdot4\cdot1)$$

This has the solution

$$u(x) = x^A \exp(Bx)[c_1 J_n(Dx^C) + c_2 Y_n(Dx^C)] \qquad (2\cdot4\cdot2)$$

When you meet an equation you think might be a Bessel equation, first attempt to put it into the form shown in (2·4·1). The next step is to solve for A and B by comparing the coefficient of u' in your equation to (2·4·1). Then solve for C, D and n by comparing the coefficients of u in each equation and using the previously found values of A and B. If you cannot find A, B, C, D and n, you have either made a mistake or else your equation is not a Bessel equation.

Whenever D is found to be an imaginary number, J_n and Y_n should be replaced by I_n and K_n with the same argument except for omitting the $i = \sqrt{-1}$. This follows from the relationship between the ordinary Bessel functions of complex argument and the modified Bessel functions developed in Section 2•3•3.

Example To show how to make use of the general Bessel equation, consider the circular-fin problem discussed in Section 2•0. The differential equation (2·0·1) was found to be

$$\frac{d^2 t}{dr^2} + \frac{1}{r}\frac{dt}{dr} - \frac{2h}{k\delta}t = -\frac{2h}{k\delta}t_\infty \qquad (2\cdot4\cdot3)$$

The boundary conditions are that $t(r_i) = t_i$ and $t'(r_o) = 0$.

This is not a Bessel equation because of the nonzero right-hand side. To solve this equation we will assume that $t(r)$ is the sum of a particular solution $a(r)$ and a homogeneous solution $u(r)$ and write

$$t(r) = a(r) + u(r) \qquad (2\cdot4\cdot4)$$

Substituting (2·4·4) into (2·4·3),

$$\frac{d^2 a}{dr^2} + \frac{d^2 u}{dr^2} + \frac{1}{r}\left[\frac{da}{dr} + \frac{du}{dr}\right] - \frac{2h}{k\delta}[a + u] = -\frac{2h}{k\delta}t_\infty$$

To satisfy this differential equation we will take $u(r)$ to be the solution of the homogeneous equation

$$\frac{d^2 u}{dr^2} + \frac{1}{r}\frac{du}{dr} - \frac{2h}{k\delta}u = 0 \qquad (2\cdot4\cdot5)$$

This means that $a(r)$ must be a particular solution to

$$\frac{d^2 a}{dr^2} + \frac{1}{r}\frac{da}{dr} - \frac{2h}{k\delta}a = -\frac{2h}{k\delta}t_\infty \qquad (2\cdot4\cdot6)$$

Since the right-hand side of (2·4·6) is a constant we see, by inspection, that

$$a(r) = t_\infty \qquad (2\cdot4\cdot7)$$

To solve (2·4·5) we will put it into the standard form comparable to (2·4·1) by multiplying through by r^2. We will also define $M^2 = 2h/k\delta$ to give

$$r^2 \frac{d^2u}{dr^2} + r\frac{du}{dr} - M^2 r^2 u = 0 \qquad (2\cdot4\cdot8)$$

By comparing (2·4·8) to (2·4·1),

$1 - 2A = 1$ gives $A = 0$ $C^2 D^2 = -M^2$ gives $D = iM/C = iM$

$2B = 0$ gives $B = 0$ $C^2 n^2 = 0$ gives $n = 0$

$2C = 2$ gives $C = 1$

Therefore the solution to (2·4·8) is given by (2·4·2) as

$$u(r) = c_1 J_0(iMr) + c_2 Y_0(iMr) \qquad (2\cdot4\cdot9)$$

Since J_0 and Y_0 have an imaginary argument i, they can be related to I_0 and K_0. From Appendix B we find that $J_0(ix) = I_0(x)$ and $Y_0(ix) = iI_0(x) - (2/\pi)K_0(x)$. Thus (2·4·9) can be rewritten as

$$u(r) = c_1 I_0(Mr) + c_2 \left[iI_0(Mr) - \frac{2}{\pi}K_0(Mr) \right]$$

$$= (c_1 + ic_2)I_0(Mr) - \frac{2}{\pi}c_2 K_0(Mr)$$

or

$$u(r) = c_3 I_0(Mr) + c_4 K_0(Mr) \qquad (2\cdot4\cdot10)$$

Engineers usually skip over the intermediate steps shown above for going from (2·4·9) to (2·4·10). They are shown here to be sure you understand what is behind this change of functions.

The solution for $t(r)$ is obtained by substituting (2·4·7) and (2·4·10) into (2·4·4) to obtain

$$t(r) = t_\infty + c_3 I_0(Mr) + c_4 K_0(Mr) \qquad (2\cdot4\cdot11)$$

The constants c_3 and c_4 are evaluated by making the solution satisfy the boundary conditions. First, since $t(r_i) = t_i$,

$$t_\infty + c_3 I_0(Mr_i) + c_4 K_0(Mr_i) = t_i \qquad (2\cdot4\cdot12)$$

Next, we must evaluate the derivative of $t(r)$ to use in the second boundary condition. Differentiating (2·4·11),

$$\frac{dt}{dr} = c_3 \frac{d}{dr}I_0(Mr) + c_4 \frac{d}{dr}K_0(Mr)$$

$$= c_3 \frac{dI_0(Mr)}{d(Mr)}\frac{d(Mr)}{dr} + c_4 \frac{dK_0(Mr)}{d(Mr)}\frac{d(Mr)}{dr}$$

Now, from Appendix B,

$$I_0'(x) = \frac{d}{dx}I_0(x) = I_1(x) \qquad \text{and} \qquad K_0'(x) = \frac{d}{dx}K_0(x) = -K_1(x)$$

Thus

$$\frac{dI_0(Mr)}{d(Mr)} = I_1(Mr) \qquad \text{and} \qquad \frac{dK_0(Mr)}{d(Mr)} = -K_1(Mr)$$

The expression for the derivative then becomes

$$\frac{dt}{dr} = c_3 I_1(Mr)M - c_4 K_1(Mr)M$$

Since the second boundary condition is $t'(r_o) = 0$, the above yields

$$c_3 I_1(Mr_o) - c_4 K_1(Mr_o) = 0 \qquad (2\cdot4\cdot13)$$

Equations (2·4·12) and (2·4·13) must now be solved simultaneously for c_3 and c_4. This is most easily done by rewriting the equations as

$$I_0(Mr_i)c_3 + K_0(Mr_i)c_4 = t_i - t_\infty$$

$$I_1(Mr_o)c_3 - K_1(Mr_o)c_4 = 0$$

and then solving by determinants (Cramer's rule) to get

$$c_3 = \frac{\begin{vmatrix} t_i - t_\infty & K_0(Mr_i) \\ 0 & -K_1(Mr_o) \end{vmatrix}}{\begin{vmatrix} I_0(Mr_i) & K_0(Mr_i) \\ I_1(Mr_o) & -K_1(Mr_o) \end{vmatrix}} = \frac{-(t_i - t_\infty)K_1(Mr_o)}{-I_0(Mr_i)K_1(Mr_o) - I_1(Mr_o)K_0(Mr_i)}$$

$$c_4 = \frac{\begin{vmatrix} I_0(Mr_i) & t_i - t_\infty \\ I_1(Mr_o) & 0 \end{vmatrix}}{\begin{vmatrix} I_0(Mr_i) & K_0(Mr_i) \\ I_1(Mr_o) & -K_1(Mr_o) \end{vmatrix}} = \frac{-(t_i - t_\infty)I_1(Mr_o)}{-I_0(Mr_i)K_1(Mr_o) - I_1(Mr_o)K_0(Mr_i)}$$

The negative signs can all be canceled in the above and c_3 and c_4 substituted into (2·4·11) to yield

$$t(r) = t_\infty + (t_i - t_\infty)\frac{K_1(Mr_o)I_0(Mr) + I_1(Mr_o)K_0(Mr)}{I_0(Mr_i)K_1(Mr_o) + I_1(Mr_o)K_0(Mr_i)} \qquad (2\cdot4\cdot14)$$

This completes the solution of (2·4·3) using the general Bessel equation procedure. From here on, this will be the way to solve Bessel equations. As engineers we need not be concerned with series solutions each time we encounter a Bessel equation. The notion of a series solution should be valuable, however, in that

1. It should have helped your understanding of Bessel functions and

2. You may sometime need this method to solve equations that are not Bessel equations.

To complete this section, it might be well to give a numerical example to be sure that the Bessel function tables are being correctly used.

Numerical example   For the circular-fin problem discussed in Section 2•0 and solved above, determine the tip temperature if $M = 2.0$ in^{-1}, $r_i = 0.8$ in, $r_o = 1.0$ in, $t_i = 300$ F and $t_\infty = 100$ F.

At the tip, the normalized $r = r_o$. Thus (2·4·14) gives

$$t(r_o) = t_\infty + (t_i - t_\infty)\frac{K_1(Mr_o)I_0(Mr_o) + I_1(Mr_o)K_0(Mr_o)}{I_0(Mr_i)K_1(Mr_o) + I_1(Mr_o)K_0(Mr_i)}$$

$$= 100 + (300 - 100)\frac{K_1(2.0)I_0(2.0) + I_1(2.0)K_0(2.0)}{I_0(1.6)K_1(2.0) + I_1(2.0)K_0(1.6)}$$

Using a table of Bessel functions [8],

$$t(r_o) = 100 + (300 - 100)\frac{0.1399 \times 2.280 + 1.591 \times 0.1139}{1.750 \times 0.1399 + 1.591 \times 0.1880} = 283.9 \text{ F}$$

2•5 KELVIN FUNCTIONS

Another set of functions, closely related to the Bessel functions, are the *Kelvin functions*. These often appear in the analysis of oscillating heat-transfer systems. They are defined and briefly discussed in this section. For more detailed comments you should refer to any of several books discussing these functions [1, 2, 3].

The Kelvin functions $\text{ber}(x)$, $\text{bei}(x)$, $\text{ker}(x)$ and $\text{kei}(x)$ are best defined by their relation to the modified Bessel functions as follows:

$$I_0(x\sqrt{i}) = \text{ber}(x) + i\,\text{bei}(x) \qquad\qquad (2\cdot5\cdot1)$$

$$K_0(x\sqrt{i}) = \text{ker}(x) + i\,\text{kei}(x) \qquad\qquad (2\cdot5\cdot2)$$

The Kelvin functions are real functions that have been tabulated [8, 10]. Their series representations can be found by substituting $x\sqrt{i}$ into the series for $I_0(x)$ or $K_0(x)$ and separating the result into real and imaginary parts. For example, consider the series for $I_0(x)$:

$$I_0(x) = \sum_{n=0}^{\infty} \frac{x^{2n}}{2^{2n}(n!)^2}$$

Replacing x by $x\sqrt{i}$ gives

$$I_0(x\sqrt{i}) = \sum_{n=0}^{\infty} \frac{x^{2n}(\sqrt{i})^{2n}}{2^{2n}(n!)^2} = \sum_{n=0}^{\infty} \frac{x^{2n}i^n}{2^{2n}(n!)^2}$$

This series can be rewritten as two series, the first containing the even n terms in the above and the second containing the odd n terms. Thus

$$I_0(x\sqrt{i}) = \sum_{\ell=0}^{\infty} \frac{x^{4\ell}i^{2\ell}}{2^{4\ell}[(2\ell)!]^2} + \sum_{\ell=0}^{\infty} \frac{x^{2(2\ell+1)}i^{(2\ell+1)}}{2^{2(2\ell+1)}[(2\ell+1)!]^2}$$

Since $i^{2\ell} = (-1)^\ell$ and $i^{2\ell+1} = i(-1)^\ell$, the above reduces to

$$I_0(x\sqrt{i}) = \sum_{\ell=0}^{\infty} \frac{x^{4\ell}(-1)^\ell}{2^{4\ell}[(2\ell)!]^2} + i\sum_{\ell=0}^{\infty} \frac{x^{4\ell+2}(-1)^\ell}{2^{4\ell+2}[(2\ell+1)!]^2}$$

Thus, by comparison to $(2\cdot5\cdot1)$,

$$\text{ber}(x) = \sum_{\ell=0}^{\infty} \frac{(-1)^\ell x^{4\ell}}{2^{4\ell}[(2\ell)!]^2} \qquad \text{and} \qquad \text{bei}(x) = \sum_{\ell=0}^{\infty} \frac{(-1)^\ell x^{4\ell+2}}{2^{4\ell+2}[(2\ell+1)!]^2}$$

It can also be shown that

$$\text{ker}(x) = \frac{\pi}{4}\text{bei}(x) - \left[\gamma + \ln\left(\frac{x}{2}\right)\right]\text{ber}(x) + \sum_{n=1}^{\infty} \frac{(-1)^n x^{4n}}{2^{4n}[(2n)!]^2} \sum_{m=1}^{2n} \frac{1}{m}$$

and

$$\text{kei}(x) = -\frac{\pi}{4}\text{ber}(x) - \left[\gamma + \ln\left(\frac{x}{2}\right)\right]\text{bei}(x) - \sum_{n=1}^{\infty} \frac{(-1)^n x^{4n-2}}{2^{4n-2}[(2n-1)!]^2} \sum_{m=1}^{2n-1} \frac{1}{m}$$

These functions are pictured in Figure 2•7. Observe that $\text{ber}(x)$ and $\text{bei}(x)$ are each finite at $x = 0$ and oscillate with ever-increasing amplitude as x increases. This behavior is a cross between the ordinary Bessel functions which oscillate and the modified Bessel functions which increase in magnitude as x increases.

Another set of independent functions are the derivatives of $\text{ber}(x)$, $\text{bei}(x)$, $\text{ker}(x)$ and $\text{kei}(x)$. These are called $\text{ber}'(x)$, $\text{bei}'(x)$, $\text{ker}'(x)$ and $\text{kei}'(x)$, respectively, and have also been tabulated [8].

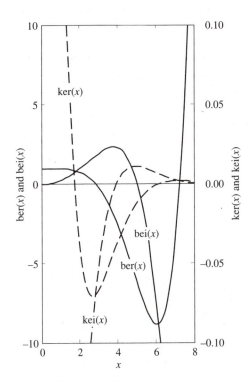

Figure 2•7 *Kelvin functions.*

SELECTED REFERENCES

1. Relton, F. E.: *Applied Bessel Functions*, Blackie & Son, Ltd., Glasgow, 1946.

2. Hildebrand, F. B.: *Advanced Calculus for Applications*, 2nd ed., Prentice-Hall, Inc., Englewood Cliffs, NJ, 1976.

3. Boyce, W. E. and R. C. DiPrima: *Elementary Differential Equations and Boundary Value Problems*, 4th ed., John Wiley & Sons, Inc., New York, 1986.

4. Wylie, C. R. and L. C. Barrett: *Advanced Engineering Mathematics*, 5th ed., McGraw-Hill Book Company, New York, 1982.

5. Luke, Y. L.: *Integrals of Bessel Functions*, McGraw-Hill Book Company, New York, 1962.

6. Kaplan, W.: *Ordinary Differential Equations*, Addison-Wesley Publishing Company, Inc., Reading, MA, 1958.

7. Piaggio, H. T. H.: *An Elementary Treatise on Differential Equations and Their Applications*, Bell & Sons, Ltd., London, 1954.

8. Flügge, W.: *Four-Place Tables of Transcendental Functions*, McGraw-Hill Book Company, New York, 1954.

9. Schneider, P. J.: *Conduction Heat Transfer*, Addison-Wesley Publishing Company, Inc., Reading, MA, 1955.

10. Abramowitz, M. and I. A. Stegun (eds.): *Handbook of Mathematical Functions*, Applied Mathematics Series 55, National Bureau of Standards, 1964. [Dover, 1965].

EXERCISES

2•1 Evaluate $\displaystyle\prod_{m=1}^{4}(2m+1)$.

2•2 Evaluate $\displaystyle\sum_{m=1}^{3}\frac{1}{2m+1}$.

2•3 Verify that (2·2·15) is a solution of (2·2·1) by direct substitution into (2·2·1).

2•4 Verify that (2·2·20) is the solution to (2·2·19) by making a change of independent variable from x to $\xi = mx$ in (2·2·19).

2•5 Verify that (2·2·24) is a solution of (2·2·21) by direct substitution into (2·2·21).

2•6 Verify (2·2·27).

2•7 Verify (2·2·28).

2•8 Show that $\dfrac{d}{dx}[x^k J_k(x)] = x^k J_{k-1}(x)$.

2•9 Consider the function

$$J_0(x;N) = \sum_{n=0}^{N}\frac{(-1)^n x^{2n}}{2^{2n}(n!)^2} \qquad \text{where } n \text{ is an integer}$$

(a) Evaluate $J_0(4;\infty)$ by whatever means you want.
(b) Plot a graph of $J_0(4;N)$ versus N for $N = 0$ to $N =$ whatever value is required to give $J_0(4;N) = J_0(4;\infty)$ to five decimal places. Assume $J_0(4;N)$ varies linearly between integer values of N.
(c) Why do you think this exercise was assigned?

2•10 Evaluate $J_0(1.7)$ and $J_0(3.2)$ from a table.

2•11 Evaluate $J_2(6.2)$ from (2·2·36) using tables to find $J_0(6.2)$ and $J_1(6.2)$.

2•12 Plot $J_1(x)$ for $0 \le x \le 10$.

2•13 Plot $J_2(x)$ for $0 \le x \le 10$.

2•14 Evaluate $\Gamma(3.31)$.

2•15 Evaluate $\Gamma(-1.3)$.

2•16 Show that, when k is an integer, (2·2·33) reduces to (2·2·24).

2•17 Verify (2·2·34) and (2·2·35).

2•18 Verify that (2·2·36) and (2·2·37) are valid for noninteger values of k.

2•19 Solve the modified Bessel equation

$$u'' + \frac{1}{x}u' - u = 0$$

by the method of Frobenius to obtain (2·3·2) and (2·3·3).

2•20 Verify (2·3·8) by defining a new independent variable $\xi = mx$ in (2·3·7).

2•21 Verify that the substitution of $u = v/\sqrt{x}$ into (2·3·9) reduces (2·3·9) to the form

$$v'' - v = 0 \qquad \text{for large } x$$

and that for large x,

$$u(x) = c_1\frac{\exp(x)}{\sqrt{x}} + c_2\frac{\exp(-x)}{\sqrt{x}}$$

2•22 Evaluate $I_0(3.7)$ and $K_1(6.5)$ from tables.

2•23 Evaluate $I_0(21.3)$ from (2·3·13) and from tables.

2•24 Using (2·4·1) and (2·4·2), determine the solution to

$$4x^2u'' + 8x^2u' + (4x^2 - x + 1)u = 0$$

2•25 The normalized differential equation for a particular circular fin of uniform thickness is

$$u'' + \frac{1}{r}u' - 4u = 0$$

For a specified temperature at the inside radius and zero end leak at the outer radius, the normalized boundary conditions might be

$$u(\tfrac{1}{2}) = 1 \qquad \text{and} \qquad u'(1) = 0$$

(a) What does the coefficient 4 represent physically?
(b) Determine the analytical solution of the problem.
(c) Plot the normalized temperature distribution.

2•26 For the circular fin discussed in the example in Section 2•4, evaluate the heat transfer from the fin. In addition to the numerical values given in Section 2•4, take $\delta = 0.1$ in and $k = 60$ Btu/hr-ft-F.

The mathematical relations in this chapter have been in nondimensional form. While this is sometimes handy, it is rarely ever necessary. The following exercises are given in dimensional form and are intended to be worked without normalization.

2•27 A spine protruding from a wall at temperature t_L has the shape of a circular cone. The radius of the cone base is R. The spine comes to a point at its tip and its length is L.

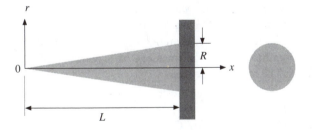

(a) From basic principles, derive the ordinary differential equation that describes the steady-state temperature distribution along the fin in the x direction. Assume the temperature is uniform in the r direction.
(b) Determine the solution to this equation.
(c) Determine an algebraic expression for the tip temperature.
(d) Determine an algebraic expression for the total heat-transfer rate from the spine.
(e) Determine the tip temperature [F] and the total heat-transfer rate [Btu/hr] for the following data:

$$h = 10 \text{ Btu/hr-ft}^2\text{-F} \qquad k = 117 \text{ Btu/hr-ft-F}$$
$$L = 2 \text{ in} \qquad\qquad\quad R = 0.125 \text{ in}$$
$$t_L = 200 \text{ F} \qquad\qquad\quad t_\infty = 70 \text{ F}$$

2•28 A straight triangular fin made of aluminum is shown below. The thickness of the fin increases linearly with x from zero at $x = 0$ to $2Y$ at $x = L$. The fin thickness $2Y$ at the base $x = L$ is small enough so that the temperature may be assumed uniform in the y direction. The heat transfer from the fin edges (one is shown cross hatched) may be neglected in comparison to that from the slanted faces of the fin. There is no temperature variation in the z direction.

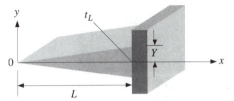

(a) From basic principles, derive the ordinary differential equation that describes the steady-state temperature distribution along the fin in the x direction.
(b) Obtain the solution to this equation.
(c) Obtain an algebraic expression for the tip temperature.
(d) Obtain an algebraic expression for the heat loss from the fin.
(e) Evaluate the tip temperature [F] and the energy leaving the fin [Btu/hr-ft] for the following data:

$$h = 10 \text{ Btu/hr-ft}^2\text{-F} \qquad k = 117 \text{ Btu/hr-ft-F}$$
$$L = 2 \text{ in} \qquad\qquad\quad Y = 0.125 \text{ in}$$
$$t_L = 200 \text{ F} \qquad\qquad\quad t_\infty = 70 \text{ F}$$

2•29 For the fins considered in exercises 2•27 and 2•28:

(a) Compare their tip temperatures.
(b) Compare their heat-transfer rates on the basis of equal base area.
(c) Compare their heat-transfer rates on the basis of equal convection surface area.
(d) Compare their heat-transfer rates on the basis of volume of fin material.
(e) Which method of comparison is best from an economic standpoint? Why?

2•30 A thin circular disk is held at 200 F (t_R) at its outer edge. There is 100-F (t_∞) air on both sides of the disk. The average heat-transfer coefficient h between the disk and the air is 10 Btu/hr-ft^2-F. The disk radius R is 4 in and its thickness δ is 0.06 in. The thermal conductivity k of the disk is 25 Btu/hr-ft-F. The governing differential equation is

$$\frac{d^2t}{dr^2} + \frac{1}{r}\frac{dt}{dr} - \frac{2h}{k\delta}t = -\frac{2h}{k\delta}t_\infty$$

Determine an algebraic expression for the temperature distribution and then evaluate

(a) The temperature at the center of the disk [F].
(b) The total heat-transfer rate from the disk [Btu/hr].

3

SEPARATION OF VARIABLES

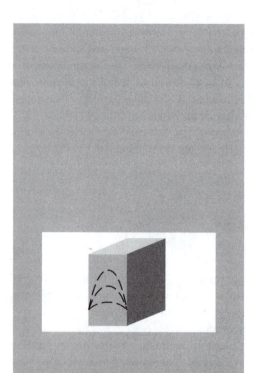

3•0 INTRODUCTION

So far, we have been concerned only with solutions to ordinary differential equations (one independent variable). This chapter will study the solution of partial differential equations (two or more independent variables) using the method of *separation of variables*. This is one of the most common methods available and is used in many more areas than heat transfer.

Separation of variables only works in special situations however. The method requires that:

(1) The partial differential equation must be both linear and homogeneous. A differential equation is *linear* if it contains no products of the dependent variable or its derivatives. That is, terms like t^4 or uu_x are not permitted. The partial differential equation describing laminar boundary-layer flow over a flat plate is a good example of a nonlinear equation. The equation is

$$\rho(uu_x + vu_y) = \mu u_{yy}$$

A differential equation is *homogeneous* if, when it is satisfied by t, it is also satisfied by Ct where C is an arbitrary constant ($C = 0$ is a special case). An example of a nonhomogeneous equation would be the one describing the temperature transient of a plane wall with energy generation. The equation is given by

$$kt_{xx} + g''' = \rho c t_\theta$$

To check homogeneity, replace t by Ct to get

$$kCt_{xx} + g''' = \rho c C t_\theta$$

or upon dividing by C,

$$kt_{xx} + \frac{1}{C}g''' = \rho c t_\theta$$

Since this is not identical to the original equation, the original equation is nonhomogeneous. If g''' were zero, the equation would then be homogeneous.

(2) The boundary conditions must be linear and "mostly" homogeneous. A boundary condition is *linear* if it contains no products of the dependent variable or its derivatives. Surface radiation is a good example of a nonlinear boundary condition. This boundary condition would be

$$-\left[k\frac{\partial T}{\partial x}\right]_{x=L} = \varepsilon\sigma(T^4 - T_\infty^4)$$

A boundary condition (different from an initial condition) is *homogeneous* if, when satisfied by t, it is also satisfied by Ct where C is an arbitrary constant. The following convective boundary condition is nonhomogeneous:

$$-\left[k\frac{\partial t}{\partial x}\right]_{x=L} = h(t - t_\infty)$$

The test is to substitute Ct in place of t to give

$$-\left[kC\frac{\partial t}{\partial x}\right]_{x=L} = h(Ct - t_\infty)$$

or, upon dividing by C,

$$-\left[k\frac{\partial t}{\partial x}\right]_{x=L} = h(t - \frac{1}{C}t_\infty)$$

This is not identical to the original boundary condition. Therefore the boundary condition is nonhomogeneous. If t_∞ were zero, the boundary condition would then be homogeneous.

For transients, *all* boundary conditions must be homogeneous. For two-dimensional, steady-state problems, both boundary conditions must be homogeneous in at least one of the two coordinate directions.

(3) The solid must have "simple" boundaries. That is, the boundaries of the solid must be along constant coordinate curves (*e.g.*, x = constant, r = constant, …). Plane walls, rectangles, right circular cylinders and spheres are acceptable shapes. Triangles and leaning or variable-diameter cylinders are not acceptable.

Figure 3•1 describes a fairly general two-dimensional conduction transient. The parameters k, g''', ρ, c, h and t_∞ are each assumed to be constant. This problem cannot be solved by separation of variables for three reasons: (1) the differential equation is not homogeneous since $g''' \neq 0$, (2) the boundary condition is not homogeneous since $t_\infty \neq 0$ and (3) the region does not have a simple shape.

To be able to use separation of variables to solve the problem described in Figure 3•1, the region would have to be rectangular and g''' and t_∞ both would have to be 0. This special case would rarely be of interest in practice. The region must be simple, but there are many simple solids in practice. The differential equation and its boundary conditions must always be linear. One can see that many interesting heat-transfer problems cannot be solved by separation of variables. Boundary layers and radiation problems are nonlinear. Nonzero values of specified temperatures, heat fluxes and generation give nonhomogeneous terms. Fortunately, there are ways to get around the necessity of having a homogeneous differential equation and boundary conditions.

The remainder of this chapter starts with relatively simple problems and builds up to more-complicated ones. Section 3•1 considers one-dimensional transients (two independent variables). Section 3•2 then considers two-dimensional problems. Two-dimensional, steady-state problems have two independent variables. Two-dimensional transients have three independent variables. Finally, Section 3•3 treats some supplementary problems to show how to use separation of variables for cylindrical and for spherical regions.

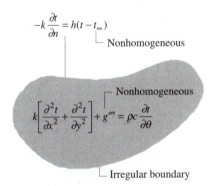

Figure 3•1 *Two-dimensional transient with non-homogeneous differential equation and boundary condition and an irregular boundary.*

3•1 ONE-DIMENSIONAL PROBLEMS

One-dimensional transients have two independent variables: position and time. Most of the mathematics needed in separation of variables is presented in this section. A plane wall is used in this section to illustrate three different nonhomogeneous boundary conditions (one at a time).

3•1•1 Plane-wall transient, specified-temperature boundary

To see what is involved in the method of separation of variables, let us consider a relatively simple engineering situation. Suppose a wall, initially having a known initial temperature distribution $t^{(0)}(x)$, suddenly has its surface temperature at $x = L$ changed to t_L. The surface at $x = 0$ is assumed to be adiabatic. The temperature-time history of the wall during the transient may be desired to see how long it takes the temperature at $x = 0$ to reach a specified value.

If it is assumed that the wall is large in all directions perpendicular to the x axis, the heat transfer may be taken to be one dimensional. For constant conductivity k the governing differential equation in this case is the heat equation given by

$$k\frac{\partial^2 t}{\partial x^2} = \rho c \frac{\partial t}{\partial \theta} \tag{3·1·1}$$

The boundary conditions that apply for $\theta > 0$ are given by

$$\left.\frac{\partial t}{\partial x}\right|_{x=0} = 0 \tag{3·1·2}$$

and

$$t(L,\theta) = t_L \tag{3·1·3}$$

The initial condition is given by

$$t(x,0) = t^{(0)}(x) \tag{3·1·4}$$

It is convenient to adopt a shorthand notation for partial derivatives by letting

$$t_x = \frac{\partial t}{\partial x}, \quad t_{xx} = \frac{\partial^2 t}{\partial x^2} \quad \text{and} \quad t_\theta = \frac{\partial t}{\partial \theta}$$

The mathematical description of this problem is pictured in Figure 3•2 and summarized in Table 3•1. We must satisfy these four equations. The partial differential equation (3·1·1) is homogeneous as is the boundary condition (3·1·2) at $x = 0$. The only nonhomogeneous term is the t_L in the boundary condition (3·1·3) at $x = L$. The initial condition (3·1·4) does not need to be homogeneous.

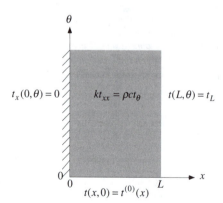

Figure 3•2 *Plane wall, initially at temperature $t^{(0)}(x)$, is suddenly exposed to a step change in surface temperature at $x = L$.*

Table 3•1 *Mathematical problem description for $t(x,\theta)$.*

Problem	Equation	Boundary/Initial Conditions	
$t(x,\theta)$	$kt_{xx} = \rho c t_\theta$		(3·1·1)
		$t_x(0,\theta) = 0$	(3·1·2)
		$t(L,\theta) = t_L$	(3·1·3)
		$t(x,0) = t^{(0)}(x)$	(3·1·4)

Since the problem for $t(x,\theta)$ is nonhomogeneous (due to the boundary condition at $x = L$ in this case), we have stated earlier (and will see later) that separation of variables cannot be applied directly to find $t(x,\theta)$. To get around this difficulty, we will assume $t(x,\theta)$ can be written as

$$t(x,\theta) = a(x) + b(\theta) + u(x,\theta) \qquad (3\cdot1\cdot5)$$

This is the most-general form that the solution can have. Since $a(x)$ and $b(\theta)$ are each functions of only one independent variable, they will be the solutions of *ordinary* differential equations. Nonhomogeneous terms are not a problem in finding solutions to ordinary differential equations. Since $u(x,\theta)$ is a function of two independent variables, it will be the solution of a *partial* differential equation. To use separation of variables to find $u(x,\theta)$ we must have a homogeneous partial differential equation and homogeneous boundary conditions for $u(x,\theta)$. We will therefore choose $a(x)$ and $b(\theta)$ so that the problem for $u(x,\theta)$ will be homogeneous.

Upon substituting the assumed form of the solution (3·1·5) into the partial differential equation (3·1·1) we obtain

$$k[a''(x) + u_{xx}] = \rho c[b'(\theta) + u_\theta]$$

This can be satisfied by taking $u(x,\theta)$ to satisfy the homogeneous partial differential equation

$$ku_{xx} = \rho c u_\theta \qquad (3\cdot1\cdot6)$$

This means that $a(x)$ and $b(\theta)$ must satisfy

$$ka''(x) = \rho c b'(\theta)$$

The only way that a function of x can equal a function of θ is if each function is equal to the same constant. Thus we will require that $a(x)$ satisfy the second-order ordinary differential equation

$$ka''(x) = c_1 \qquad (3\cdot1\cdot7)$$

and $b(\theta)$ satisfy the first-order ordinary differential equation

$$\rho c b'(\theta) = c_1 \qquad (3\cdot1\cdot8)$$

The value of c_1 that appears in (3·1·7) and (3·1·8) is arbitrary at this point. Since (3·1·7) is second order, it needs two boundary conditions. Equation (3·1·8) is first order in time so it needs one initial condition. The solution of these ordinary differential equations should not be troublesome.

Next, we must choose a boundary condition for $a(x)$ at $x = 0$ so that the boundary condition for $u(x,\theta)$ at $x = 0$ will be homogeneous. Substituting the assumed solution (3·1·5) into the boundary condition (3·1·2) at $x = 0$ we obtain

$$a'(0) + u_x(0,\theta) = 0$$

This can be satisfied by taking $u(x,\theta)$ to satisfy the homogeneous condition

$$u_x(0,\theta) = 0 \qquad (3\cdot1\cdot9)$$

Thus, $a(x)$ will have to satisfy

$$a'(0) = 0 \qquad (3\cdot1\cdot10)$$

To obtain a homogeneous boundary condition for $u(x,\theta)$ at $x = L$, we will substitute the general solution (3·1·5) into the boundary condition (3·1·3) at $x = L$ to obtain

$$a(L) + b(\theta) + u(L,\theta) = t_L$$

This can be satisfied by taking $u(x,\theta)$ to satisfy the homogeneous condition

$$u(L,\theta) = 0 \tag{3$\cdot$1$\cdot$11}$$

If we also take

$$a(L) = t_L \tag{3$\cdot$1$\cdot$12}$$

we must then take

$$b(\theta) = 0 \tag{3$\cdot$1$\cdot$13}$$

Now we must insist that the general solution (3$\cdot$1$\cdot$5) satisfy the initial condition (3$\cdot$1$\cdot$4). Substitution of (3$\cdot$1$\cdot$5) into (3$\cdot$1$\cdot$4) gives

$$t(x,0) = t^{(0)}(x) = a(x) + b(0) + u(x,0)$$

Upon solving for $u(x,0)$ and noting from (3$\cdot$1$\cdot$13) that $b(0) = 0$,

$$u(x,0) = t^{(0)}(x) - a(x) \tag{3$\cdot$1$\cdot$14}$$

We now have enough relations to specify the subproblems for $a(x)$, $b(\theta)$ and $u(x,\theta)$. These subproblems are summarized in Table 3•2. The problems for $a(x)$ and $b(\theta)$ are nonhomogeneous, but this does not matter since they obey ordinary differential equations. The problem for $u(x,\theta)$, where homogeneity is important, has been constructed to be homogeneous. Physical interpretations may be given to these three subproblems. If c_1 were negative, $a(x)$ would be the temperature distribution in a plane wall with generation that was adiabatic at $x = 0$ and t_L at $x = L$. The problem represented by $b(\theta)$ is a lumped-system transient with a constant heat input or generation with an initial temperature of zero. The problem for $u(x,\theta)$, pictured in Figure 3•3, is a plane-wall transient similar to the original problem for $t(x,\theta)$ except that the nonhomogeneous term is now zero and the initial condition has been modified.

Table 3•2 *Subproblems for $t(x,\theta) = a(x) + b(\theta) + u(x,\theta)$.*

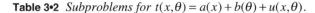

Subproblem	Equation	Boundary/Initial Conditions	
$a(x)$	$ka''(x) = c_1$		(3$\cdot$1$\cdot$7)
		$a'(0) = 0$	(3$\cdot$1$\cdot$10)
		$a(L) = t_L$	(3$\cdot$1$\cdot$12)
$b(\theta)$	$\rho c b'(\theta) = c_1$		(3$\cdot$1$\cdot$8)
		$b(\theta) = 0$	(3$\cdot$1$\cdot$13)
$u(x,\theta)$	$ku_{xx} = \rho c u_\theta$		(3$\cdot$1$\cdot$6)
		$u_x(0,\theta) = 0$	(3$\cdot$1$\cdot$9)
		$u(L,\theta) = 0$	(3$\cdot$1$\cdot$11)
		$u(x,0) = t^{(0)}(x) - a(x)$	(3$\cdot$1$\cdot$14)

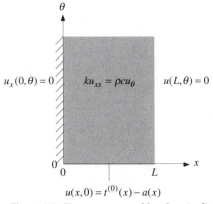

$$u(x,0) = t^{(0)}(x) - a(x)$$

Figure 3•3 *Homogeneous problem for $u(x,\theta)$ corresponding to Figure 3•2.*

Equation (3$\cdot$1$\cdot$13) is the solution for $b(\theta)$. Since $b(\theta)$ is a constant, $b'(\theta) = 0$ and from (3$\cdot$1$\cdot$8) we see that c_1 must be 0.

The solution of (3$\cdot$1$\cdot$7) for $c_1 = 0$ may be found by integrating once to obtain

$$a'(x) = a_1$$

From (3·1·10) we see that $a_1 = 0$. Integrating again for $a_1 = 0$,

$a(x) = a_2$

Since $a(x)$ is a constant, to satisfy (3·1·12) we must have

$a(x) = t_L$ (3·1·15)

Upon substituting (3·1·15) for $a(x)$ and (3·1·13) for $b(\theta)$ into (3·1·5),

$t(x,\theta) = t_L + u(x,\theta)$ (3·1·16)

Upon substituting (3·1·15) for $a(x)$ into (3·1·14),

$u(x,0) = t^{(0)}(x) - t_L$ (3·1·17)

For convenience the updated problems for $t(x,\theta)$ and $u(x,\theta)$ are summarized in Table 3•3. The updated problem that must be solved for $u(x,\theta)$ is depicted in Figure 3•4. This is the same problem as pictured in Figure 3•3 except that the $a(x)$ in the initial condition has now been found. The problem for $u(x,\theta)$ is both linear and homogeneous. From the physics of the problem, $u(x,\theta)$ starts at $u(x,0)$ and approaches 0 as steady state is attained.

Table 3•3 *Problems for $t(x,\theta)$ and $u(x,\theta)$.*

Problem	Equation	Boundary/Initial Conditions	
$t(x,\theta) = t_L + u(x,\theta)$			(3·1·16)
$u(x,\theta)$	$ku_{xx} = \rho c u_\theta$		(3·1·6)
		$u_x(0,\theta) = 0$	(3·1·9)
		$u(L,\theta) = 0$	(3·1·11)
		$u(x,0) = t^{(0)}(x) - t_L$	(3·1·17)

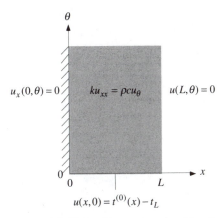

Figure 3•4 *Updated homogeneous problem for $u(x,\theta)$ corresponding to Figure 3•2.*

We must now learn to use separation of variables to solve for $u(x,\theta)$. Then $u(x,\theta)$ will be substituted into (3·1·16) to give $t(x,\theta)$.

The method of separation of variables starts by assuming that the solution for $u(x,\theta)$ can be written as a product of a function of x, $X(x)$, and a function of θ, $\Theta(\theta)$. Thus we will write

$u(x,\theta) = X(x)\Theta(\theta)$ (3·1·18)

Substitution of this assumed solution into the partial differential equation (3·1·6) gives

$kX''(x)\Theta(\theta) = \rho c X(x)\Theta'(\theta)$

where $X''(x)$ denotes the second derivative of $X(x)$ with respect to x and $\Theta'(\theta)$ denotes the first derivative of $\Theta(\theta)$ with respect to θ. Rearranging to separate the variables,

$$\frac{X''(x)}{X(x)} = \frac{\rho c}{k}\frac{\Theta'(\theta)}{\Theta(\theta)}$$ (3·1·19)

The only way that a function of x (left-hand side) can be equal to a function of θ (right-hand side) for all x and θ is if both sides are equal to the same constant. The choice of a positive constant (or zero) would give a solution that becomes infinite as time increases (or has no time dependence). Since this choice would not fit our physical knowledge that eventually the solution must approach 0, we must rule out a positive constant (or 0). It will be convenient to write this negative constant as $-\lambda^2$ where λ^2 is a positive number. Upon setting each side of (3·1·19) equal to $-\lambda^2$ and introducing thermal diffusivity, $\alpha = k/\rho c$, to simplify notation,

$$\frac{X''(x)}{X(x)} = -\lambda^2 \qquad \text{or} \qquad X''(x) + \lambda^2 X(x) = 0 \qquad (3\cdot1\cdot20)$$

and

$$\frac{1}{\alpha}\frac{\Theta'(\theta)}{\Theta(\theta)} = -\lambda^2 \qquad \text{or} \qquad \Theta'(\theta) + \lambda^2 \alpha \Theta(\theta) = 0 \qquad (3\cdot1\cdot21)$$

Equation (3·1·18) may next be substituted into the boundary condition (3·1·9) at $x = 0$ to obtain

$$X'(0)\Theta(\theta) = 0 \qquad (3\cdot1\cdot22)$$

This condition can be satisfied by taking $\Theta(\theta) = 0$, but substituting this into (3·1·18) would give $u(x,\theta) = 0$ for all time. This is not a helpful solution. Therefore, to satisfy the boundary condition at $x = 0$, we will take

$$X'(0) = 0 \qquad (3\cdot1\cdot23)$$

If the right-hand side of (3·1·22) had been anything other than 0 (*i.e.*, if the boundary condition at $x = 0$ had been nonhomogeneous) we could not have arrived at this result (3·1·23). For example, if (3·1·22) had been $X'(0)\Theta(\theta) = C$ we could only conclude that $\Theta(\theta) = C/X'(0)$. This says that $\Theta(\theta)$ must be a constant (*i.e.*, not a function of θ) and substitution into (3·1·18) would give $u(x,\theta)$ as a function only of x, not of θ.

Upon substituting (3·1·18) into the boundary condition (3·1·11) at $x = L$ gives

$$X(L)\Theta(\theta) = 0 \qquad \text{or} \qquad X(L) = 0 \qquad (3\cdot1\cdot24)$$

The reasoning leading to $X(L) = 0$ is similar to the reasoning that led to (3·1·23). Again, the boundary condition must be homogeneous.

The problem we must solve for $X(x)$ is called an *eigenproblem* and is summarized in Table 3•4. Equation (3·1·20) is a second order ordinary differential equation for the *eigenfunction*, $X(x)$, and its *eigenvalue*, λ. The boundary conditions are given by (3·1·23) and (3·1·24).

Table 3•4 *Eigenproblem for $u(x,\theta)$.*

Equation	Boundary Conditions	
$X''(x) + \lambda^2 X(x) = 0$		(3·1·20)
	$X'(0) = 0$	(3·1·23)
	$X(L) = 0$	(3·1·24)

The general solution to (3·1·20) is given by

$$X(x) = A\sin(\lambda x) + B\cos(\lambda x) \qquad (3\cdot1\cdot25)$$

We now must satisfy the boundary conditions (3·1·23) and (3·1·24). The first derivative of $X(x)$ is given by

$$X'(x) = A\lambda\cos(\lambda x) - B\lambda\sin(\lambda x)$$

Upon setting $X'(x)$ equal to 0 when $x = 0$ we obtain

$$X'(0) = A\lambda\cos(\lambda 0) - B\lambda\sin(\lambda 0) = A\lambda = 0$$

The condition that $A\lambda = 0$ may be satisfied by taking $\lambda = 0$, but then (3·1·25) would give

$$X(x) = A\sin(0x) + B\cos(0x) = B$$

This would eliminate x from the solution. This is not what we want. Thus, we will satisfy the boundary condition at $x = 0$ by taking $A = 0$. Upon setting $A = 0$ in (3·1·25) we obtain

$$X(x) = B\cos(\lambda x)$$

Upon substituting this result into the boundary condition (3·1·24) at $x = L$,

$$X(L) = B\cos(\lambda L) = 0$$

This may be satisfied by taking $B = 0$, but this would give $X(x) = 0$ which would eliminate the dependence of the solution on x. This is not what we want. Therefore we will satisfy the boundary condition at $x = L$ by taking

$$\cos(\lambda L) = 0$$

This relation is called an *eigencondition*. It restricts the values of λ. The cosine function is 0 at odd multiples of $\pi/2$. Therefore we must have

$$\lambda L = \frac{\pi}{2}, 3\frac{\pi}{2}, 5\frac{\pi}{2}, \cdots$$

Or, in general, the nth eigenvalue λ_n is given by

$$\lambda_n L = (2n-1)\frac{\pi}{2}$$

Thus the eigenfunctions are given by

$$X_n(x) = B_n \cos(\lambda_n x) \qquad \text{where} \qquad \lambda_n L = (2n-1)\frac{\pi}{2} \tag{3·1·26}$$

There are an infinite number of eigenvalues λ_n corresponding to $n = 1, 2, 3, \cdots$. There is an eigenfunction $X_n(x)$ corresponding to each eigenvalue λ_n.

The second separated equation (3·1·21) may now be written as

$$\Theta_n'(\theta) + \lambda_n^2 \alpha\Theta_n(\theta) = 0$$

There is a $\Theta_n(\theta)$ corresponding to each λ_n. The general solution to this ordinary differential equation may be written as

$$\Theta_n(\theta) = \exp(-\lambda_n^2 \alpha\theta) \tag{3·1·27}$$

An arbitrary constant coefficient could be included in $\Theta_n(\theta)$ but it won't be needed since we already have a constant B_n in $X_n(x)$.

Upon substituting (3·1·26) and (3·1·27) into (3·1·18) we may write

$$u_n(x,\theta) = B_n \cos(\lambda_n x)\exp(-\lambda_n^2 \alpha\theta)$$

where

$$\lambda_n L = (2n-1)\frac{\pi}{2} \qquad \text{and} \qquad n = 1, 2, 3, \cdots$$

There are an infinite number of functions $u_n(x,\theta)$. Each of these functions satisfy the partial differential equation (3·1·6) and the boundary conditions at $x = 0$ and at $x = L$ [(3·1·9) and (3·1·11)].

Due to the *linearity* and the *homogeneity* of the problem, if u_1 and u_2 *each* satisfy the partial differential equation [$ku_{xx} = \rho c u_\theta$] and the boundary conditions [$u_x(0,\theta) = 0$ and $u(L,\theta) = 0$], then

$u = c_1 u_1 + c_2 u_2$ where c_1 and c_2 are arbitrary constants

also satisfies the partial differential equation $[ku_{xx} = \rho c u_\theta]$ and the boundary conditions $[u_x(0,\theta) = 0$ and $u(L,\theta) = 0]$. This can be verified by direct substitution into the equation.

$$ku_{xx} = \rho c u_\theta$$

$$k(c_1 u_1 + c_2 u_2)_{xx} = \rho c (c_1 u_1 + c_2 u_2)_\theta$$

$$c_1 k u_{1xx} + c_2 k u_{2xx} = c_1 \rho c u_{1\theta} + c_2 \rho c u_{2\theta}$$

$$c_1(k u_{1xx} - \rho c u_{1\theta}) + c_2(k u_{2xx} - \rho c u_{2\theta}) = 0$$

Each of the parenthesized terms in the last equation is zero since u_1 and u_2 each satisfy $ku_{xx} = \rho c u_\theta$. The same type of substitution will show that u also satisfies the boundary conditions. The solutions u_1 and u_2 are called *partial solutions* since they satisfy the partial differential equation and its boundary conditions, but not the initial condition.

The method of separation of variables makes use of the fact that the addition of partial solutions is also another partial solution. By adding all of the partial solutions $u_n(x,\theta)$ we can get a solution which also satisfies the initial condition (3·1·17). Thus the solution will be expressed as

$$u(x,\theta) = \sum_{n=1}^{\infty} B_n \cos(\lambda_n x) \exp(-\lambda_n^2 \alpha \theta) \tag{3·1·28}$$

This expression for $u(x,\theta)$ satisfies the partial differential equation (3·1·6) and its boundary conditions (3·1·9) and (3·1·11). There are an infinite number of values of B_n still to be chosen. We want to choose the B_n so that $u(x,\theta)$ also satisfies the initial condition (3·1·17).

Upon setting $\theta = 0$ in (3·1·28) we obtain

$$u(x,0) = \sum_{n=1}^{\infty} B_n \cos(\lambda_n x)$$

This is a *Fourier series* representation of $u(x,0)$. To find a particular B_m we will multiply by $\cos(\lambda_m x)$ and integrate from $x = 0$ to $x = L$. This gives

$$\int_{x=0}^{L} u(x,0)\cos(\lambda_m x)\,dx = \sum_{n=1}^{\infty} B_n \int_{x=0}^{L} \cos(\lambda_n x)\cos(\lambda_m x)\,dx$$

If you look at your integral tables, you will find that

$$\int \cos(ax)\cos(bx)\,dx = \frac{\sin\{(a-b)x\}}{2(a-b)} + \frac{\sin\{(a+b)x\}}{2(a+b)} \qquad \text{for } a^2 \neq b^2$$

and

$$\int \cos^2(ax)\,dx = \frac{x}{2} + \frac{1}{4a}\sin(2ax)$$

Thus, for $n \neq m$ we may write

$$\int_{x=0}^{L} \cos(\lambda_n x)\cos(\lambda_m x)\,dx = \left[\frac{\sin\{(\lambda_n - \lambda_m)x\}}{2(\lambda_n - \lambda_m)} + \frac{\sin\{(\lambda_n + \lambda_m)x\}}{2(\lambda_n + \lambda_m)}\right]_{x=0}^{L}$$

$$= \frac{\sin(\lambda_n L - \lambda_m L)}{2(\lambda_n - \lambda_m)} + \frac{\sin(\lambda_n L + \lambda_m L)}{2(\lambda_n + \lambda_m)}$$

For $\lambda_n L = (2n-1)\pi/2$ and $\lambda_m L = (2m-1)\pi/2$ we can write

$$\lambda_n L - \lambda_m L = (2n - 1 - 2m + 1)\frac{\pi}{2} = (n - m)\pi$$

and

$$\lambda_n L + \lambda_m L = (2n - 1 + 2m - 1)\frac{\pi}{2} = (2n + 2m - 2)\frac{\pi}{2} = (n + m - 1)\pi$$

Since the sine of any integer multiple of π is 0,

$$\int_{x=0}^{L} \cos(\lambda_n x)\cos(\lambda_m x)\, dx = 0 \qquad \text{for } n \neq m$$

This means that the only term left in the series is the one with $n = m$. Thus,

$$\int_{x=0}^{L} u(x,0)\cos(\lambda_m x)\, dx = B_m \int_{x=0}^{L} \cos^2(\lambda_m x)\, dx$$

The integral on the right-hand side may be evaluated to give

$$\int_{x=0}^{L} \cos^2(\lambda_m x)\, dx = \left[\frac{x}{2} + \frac{1}{4\lambda_m}\sin(2\lambda_m x)\right]_{x=0}^{L} = \frac{L}{2} + \frac{1}{4\lambda_m}\sin(2\lambda_m L) = \frac{L}{2}$$

Therefore we may then solve for B_m as

$$B_m = \frac{2}{L}\int_{x=0}^{L} u(x,0)\cos(\lambda_m x)\, dx \qquad\qquad (3\cdot1\cdot29)$$

The function $u(x,0)$ is the initial condition for $u(x,\theta)$. In this problem, $u(x,0) = t^{(0)}(x) - t_L$ where $t^{(0)}(x)$ is the initial temperature distribution in the wall and t_L is the temperature at $x = L$ that is maintained for all $\theta > 0$. If we now take $t^{(0)}(x) = t^{(0)} = \text{constant}$,

$$B_m = \frac{2}{L}(t^{(0)} - t_L)\int_{x=0}^{L} \cos(\lambda_m x)\, dx = \frac{2}{L}(t^{(0)} - t_L)\frac{1}{\lambda_m}\sin(\lambda_m x)\Big|_{x=0}^{L}$$

$$= \frac{2}{\lambda_m L}(t^{(0)} - t_L)\sin(\lambda_m L)$$

Upon replacing m by n and substituting this expression for B_n back into the solution (3·1·28) for $u(x,\theta)$,

$$u(x,\theta) = \sum_{n=1}^{\infty} \frac{2}{\lambda_n L}(t^{(0)} - t_L)\sin(\lambda_n L)\cos(\lambda_n x)\exp(-\lambda_n^2 \alpha\theta)$$

where $\lambda_n L = (2n - 1)\pi/2$.

The solution for $t(x,\theta)$ is found by substituting $u(x,\theta)$ into (3·1·18) to obtain

$$t(x,\theta) = t_L + 2(t^{(0)} - t_L)\sum_{n=1}^{\infty} \frac{\sin(\lambda_n L)}{\lambda_n L}\cos(\lambda_n x)\exp(-\lambda_n^2 \alpha\theta) \qquad (3\cdot1\cdot30)$$

It is convenient to *normalize* this result so that t, x and θ appear in nondimensional groups. Thus we will write

$$\frac{t(x,\theta) - t_L}{(t^{(0)} - t_L)} = 2\sum_{n=1}^{\infty} \frac{\sin(\lambda_n L)}{\lambda_n L}\cos(\lambda_n L\frac{x}{L})\exp(-\lambda_n^2 L^2 \frac{\alpha\theta}{L^2})$$

Upon defining the following nondimensional variables:

$$\bar{x} = \frac{x}{L} \quad, \quad \bar{\theta} = \frac{\alpha\theta}{L^2} \quad \text{and} \quad \bar{t}(\bar{x},\bar{\theta}) = \frac{t(x,\theta) - t_L}{(t^{(0)} - t_L)}$$

we may write the solution as

$$\bar{t}(\bar{x},\bar{\theta}) = 2\sum_{n=1}^{\infty}\frac{\sin(\lambda_n L)}{(\lambda_n L)}\cos[(\lambda_n L)\bar{x}]\exp[-(\lambda_n L)^2\bar{\theta}]$$

Since $\lambda_n L = (2n-1)\pi/2$ and $\sin(\lambda_n L) = (-1)^{n+1}$ we can write

$$\bar{t}(\bar{x},\bar{\theta}) = \frac{4}{\pi}\sum_{n=1}^{\infty}\frac{(-1)^{n+1}}{(2n-1)}\cos[(2n-1)\frac{\pi}{2}\bar{x}]\exp[-(2n-1)^2\frac{\pi^2}{4}\bar{\theta}] \qquad (3\cdot1\cdot31)$$

We are interested in plotting $\bar{t}(\bar{x},\bar{\theta})$ as a function of $\bar{x}$ at various values of $\bar{\theta}$. At $\bar{\theta} = 0$, $\bar{t}$ should be 1. This is not readily apparent from the solution and will have to be discussed further. At $\bar{\theta} = 0$, each of the exponential terms in (3·1·31) is equal to 1 and the solution (3·1·31) reduces to

$$\bar{t}(\bar{x},0) = \frac{4}{\pi}\sum_{n=1}^{\infty}\frac{(-1)^{n+1}}{(2n-1)}\cos[(2n-1)\frac{\pi}{2}\bar{x}] \qquad (3\cdot1\cdot32)$$

Upon writing out the first four terms in the series,

$$\bar{t}(\bar{x},0) = \frac{4}{\pi}\cos(\frac{\pi}{2}\bar{x}) - \frac{4}{3\pi}\cos(\frac{3\pi}{2}\bar{x}) + \frac{4}{5\pi}\cos(\frac{5\pi}{2}\bar{x}) - \frac{4}{7\pi}\cos(\frac{7\pi}{2}\bar{x}) + \cdots$$

The first four terms are shown separately in Figure 3•5. Each term has a slope of 0 at $\bar{x} = 0$ and a value of 0 at $\bar{x} = 1$ as required by the boundary conditions. The terms have more oscillation and smaller amplitude as you go farther out in the series. The sum of the terms is also shown in Figure 3•5. Sum 1 includes only one term, Sum 2 includes the first two terms, etc. No matter how many terms are included in the sum, the sum always has a slope of 0 at $\bar{x} = 0$ and has a value of 0 at $\bar{x} = 1$ as required by the boundary conditions. Each sum oscillates about a value of 1 (the initial condition in this example). As more terms are added to the sum, the oscillations increase but the deviation from 1 decreases. In the limit, as the number of terms included in the sum increases, the solution will show tiny ripples about a value of 1 for $\bar{x} < 1$ and be 0 at $\bar{x} = 1$.

The convergence of this series solution at $\bar{\theta} = 0$ (3·1·32) is very slow. The amplitude of the oscillations in the nth term is given by

$$\text{Amplitude of term } n = \frac{4}{\pi}\frac{1}{(2n-1)}$$

It requires about 637 terms to reach an amplitude of 0.001. This is a lot of terms to evaluate (even if you have a computer). Rather than just using brute force and summing a large number of terms, look for special techniques to accelerate convergence. There are numerical tricks that can sometimes help.

Sometimes it is possible to use the helpful series summations listed in Appendix C to check the initial condition at one or two particular points. Upon setting $\bar{x} = 0$ in (3·1·32) we obtain

$$\bar{t}(0,0) = \frac{4}{\pi}\sum_{n=1}^{\infty}\frac{(-1)^{n+1}}{(2n-1)}$$

The sum of this alternating-sign series is given in Appendix C as $\pi/4$. Thus, $\bar{t}(0,0) = 1$ as required by the normalized initial condition. Upon setting $\bar{x} = 0.5$ in (3·1·32) we obtain

$$\bar{t}(0.5,0) = \frac{4}{\pi}\sum_{n=1}^{\infty}\frac{(-1)^{n+1}}{(2n-1)}\cos[(2n-1)\pi/4]$$

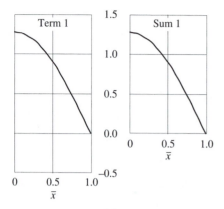

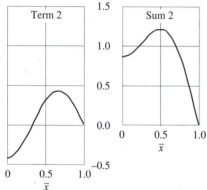

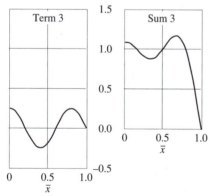

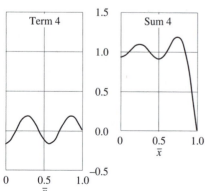

Figure 3•5 *Convergence of (3·1·31) at $\bar{\theta} = 0$ to $\bar{t}(\bar{x},0) = 1$.*

The sum of this series is also given in Appendix C as $\pi/4$. Therefore, $\bar{t}(0.5,0) = 1$ as required by the normalized initial condition. By verifying the initial condition at a couple points, we have a little more confidence that the solution is correct.

From Figure 3•5 it appears that the convergence to the initial condition $\bar{t}(\bar{x},0) = 1$ is worst when $\bar{x}$ is close to 1. The reason for this is that the initial condition gives $\bar{t}(1,0) = 1$ whereas the boundary condition gives $\bar{t}(1,0) = 0$. This discontinuity in temperature at $\bar{x} = 1$ is very hard to obtain physically and the mathematics also has a difficult time representing it. When series convergence is very slow (*e.g.*, $\bar{x}$ close to 1 and $\bar{\theta}$ near 0 in this problem), it may be possible to change the physical model to obtain a solution that is easier to evaluate.[1]

As $\bar{\theta}$ moves away from 0, the exponentials in the series (3·1·31) help convergence. As $\bar{\theta}$ becomes infinite, each of the exponential terms become 0 and $\bar{t}$ approaches 0 as expected.

In this problem $\bar{\theta}$ ranges from 0 (where $\bar{t} = 1$) to infinity (where $\bar{t} = 0$). We would also like to obtain the solution for some intermediate values of $\bar{\theta}$. One of the advantages of expressing the solution in normalized form is that a value of $\bar{\theta} = 1$ can then be expected to be a reasonable value to consider. For $\bar{\theta} = 1$ the solution (3·1·31) becomes

$$\bar{t}(\bar{x},1) = 2\sum_{n=1}^{\infty} \frac{(-1)^{n+1}}{(\lambda_n L)} \cos[(\lambda_n L)\bar{x}]\exp[-(\lambda_n L)^2]$$

Upon substituting $\lambda_n L = (2n-1)\pi/2$ we obtain

$$\bar{t}(\bar{x},1) = \frac{4}{\pi}\sum_{n=1}^{\infty} \frac{(-1)^{n+1}}{(2n-1)} \cos[(2n-1)\frac{\pi}{2}\bar{x}]\exp[-(2n-1)^2\frac{\pi^2}{4}]$$

Upon writing out the first two terms of the series,

$$\bar{t}(\bar{x},1) = \frac{4}{\pi}\cos(\frac{\pi}{2}\bar{x})\exp(-\frac{\pi^2}{4}) - \frac{4}{3\pi}\cos(\frac{3\pi}{2}\bar{x})\exp(-9\frac{\pi^2}{4}) + \cdots$$

or,

$$\bar{t}(\bar{x},1) = 0.1080 \times \cos(\frac{\pi}{2}\bar{x}) - 9.629 \times 10^{-11} \times \cos(\frac{3\pi}{2}\bar{x}) + \cdots$$

Since the coefficient of the second term is so small, only one term in the series is needed to obtain a satisfactory answer.

For $\bar{\theta} = 0.1$ (an order of magnitude smaller than $\bar{\theta} = 1$), the first four terms in the series solution (3·1·31) give

$$\bar{t}(\bar{x},0.1) = 0.9948 \times \cos(\frac{\pi}{2}\bar{x}) - 4.606 \times 10^{-2} \times \cos(\frac{3\pi}{2}\bar{x})$$

$$+5.333 \times 10^{-4} \times \cos(\frac{5\pi}{2}\bar{x}) - 1.021 \times 10^{-6} \times \cos(\frac{7\pi}{2}\bar{x}) + \cdots$$

At $\bar{\theta} = 0.1$ the coefficients are not decreasing as rapidly as for $\bar{\theta} = 1$. For four significant figures, you need about three terms in the series rather than only one. The smaller the time, the more terms in the series are required to obtain a given accuracy.

Figure 3•6 shows the solution for $\bar{\theta} = 0.1$ and 1.0. For $\bar{\theta} = 10$ (an order of magnitude more than $\bar{\theta} = 1$), the transient is over for all practical purposes.

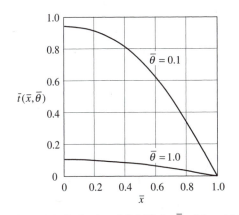

Figure 3•6 *Evaluation of (3·1·31) for $\bar{\theta} = 0.1$ and 1.0.*

---

[1] Later we will see that the semi-infinite solid can be used to model this problem for short times near $\bar{x} = 1$.

The heat flux in the wall may be computed from the conduction rate equation given by

$$q_x''(x,\theta) = -kt_x(x,\theta)$$

Upon differentiating the temperature distribution (3·1·30) and substituting into the rate equation,

$$q_x''(x,\theta) = -k\left[-2(t^{(0)} - t_L)\sum_{n=1}^{\infty}\frac{\sin(\lambda_n L)}{\lambda_n L}\lambda_n \sin(\lambda_n x)\exp(-\lambda_n^2\alpha\theta)\right]$$

This simplifies to give

$$q_x''(x,\theta) = \frac{2k}{L}(t^{(0)} - t_L)\sum_{n=1}^{\infty}\sin(\lambda_n L)\sin(\lambda_n x)\exp(-\lambda_n^2\alpha\theta) \qquad (3·1·33)$$

The convergence problems we noted for $t(x,\theta)$ are worse for $q_x''(x,\theta)$. There is no λ_n in the denominator to help convergence. Convergence is very slow until θ increases enough so that the exponentials help. At $x = 0$ the heat flux given by (3·1·33) is 0 since $\sin(\lambda_n 0) = 0$. This agrees with the boundary condition at $x = 0$. At $x = L$ the sine terms are both +1 or both −1 so their product is always +1 and the heat flux (3·1·33) reduces to

$$q_x''(L,\theta) = \frac{2k}{L}(t^{(0)} - t_L)\sum_{n=1}^{\infty}\exp(-\lambda_n^2\alpha\theta)$$

Convergence comes only from the exponentials. For θ near 0 the exponentials are close to 1. Many terms are required for convergence. Of course, at $\theta = 0$ the heat flux is infinite and an infinite number of terms are required. For short times and near the surface at $x = L$ the semi-infinite solid model and solution discussed in Chapter 6 are recommended.

Review A linear, *nonhomogeneous* problem for $t(x,\theta)$ was assumed to have a solution given by (3·1·5) as

$$t(x,\theta) = a(x) + b(\theta) + u(x,\theta)$$

The functions $a(x)$ and $b(\theta)$ were chosen so that the problem for $u(x,\theta)$ was linear and *homogeneous*. It was then assumed that the variables in $u(x,\theta)$ could be separated by writing (3·1·18)

$$u(x,\theta) = X(x)\Theta(\theta)$$

Substitution of this assumed form of the solution into the partial differential equation for $u(x,\theta)$ led to two ordinary differential equations to solve:

$$X_n''(x) + \lambda_n^2 X_n(x) = 0 \qquad \text{and} \qquad \Theta_n'(\theta) + \lambda_n^2\alpha\Theta_n(\theta) = 0$$

There are an infinite number of each of these equations corresponding to $n = 1, 2, 3, \ldots$. Each eigenfunction $X_n(x)$ involves the trigonometric functions sine and cosine (both oscillatory) and satisfies the homogeneous boundary conditions for $u(x,\theta)$. The solution for $\Theta_n(\theta)$ is $\exp(-\lambda_n^2\alpha\theta)$ which decays to 0 as θ increases.[1]

Due to the linear, homogeneous nature of the problem for $u(x,\theta)$, each of the infinite number of "separated product solutions" $X_n(x)\Theta_n(\theta)$ may be multiplied by a constant C_n and added together to obtain a more general solution. Thus, $u(x,\theta)$ can be written as

---

[1] The steady-state solution of a homogeneous problem with a specified-temperature or a convective boundary condition will always be 0.

$$u(x,\theta) = \sum_{n=1}^{\infty} C_n X_n(x)\Theta_n(\theta) \tag{3.1.34}$$

This solution satisfies the partial differential equation for $u(x,\theta)$ and its boundary conditions and has the correct steady-state solution $u(x,\infty) = 0$. The C_n are *Fourier coefficients* that are found by forcing the solution to also satisfy the initial condition for $u(x,\theta)$. Since $\Theta_n(0) = 1$ we may write

$$u(x,0) = \sum_{n=1}^{\infty} C_n X_n(x)$$

We can find a particular C_m by multiplying both sides by the eigenfunction $X_m(x)$ and integrating between the two values of x where the boundary conditions are given. This gives

$$\int_{x=0}^{L} u(x,0)X_m(x)\,dx = \sum_{n=1}^{\infty} C_n \int_{x=0}^{L} X_n(x)X_m(x)\,dx \tag{3.1.35}$$

The eigenfunctions $X_m(x)$ and $X_n(x)$ have the property[1] that the integral on the right-hand side is 0 for $n \neq m$ and is not 0 for $n = m$. The only term in the sum that will remain after integrating is the term for $n = m$. Thus,

$$\int_{x=0}^{L} u(x,0)X_m(x)\,dx = C_m \int_{x=0}^{L} X_m^2(x)\,dx$$

The value of C_m is then given by

$$C_m = \frac{\int_{x=0}^{L} u(x,0)X_m(x)\,dx}{\int_{x=0}^{L} X_m^2(x)\,dx}$$

These are the Fourier coefficients needed for (3.1.34).

3•1•2 Orthogonality X

Since *orthogonal functions* are essential to the technique of separation of variables, a few additional comments should be made to provide better understanding. In the previous section we obtained a function $X_n(x)$ which satisfied the differential equation

$$X_n'' + \lambda_n^2 X_n = 0$$

Another of these functions $X_m(x)$ would satisfy the differential equation

$$X_m'' + \lambda_m^2 X_m = 0$$

If we multiply the differential equation for $X_n(x)$ by $X_m(x)$ and multiply the differential equation for $X_m(x)$ by $X_n(x)$ and then subtract the two equations we obtain

$$X_m X_n'' - X_m'' X_n + (\lambda_n^2 - \lambda_m^2)X_m X_n = 0$$

The first two terms can be rewritten to rearrange the equation to give[2]

$$\frac{d}{dx}(X_m X_n' - X_m' X_n) = (\lambda_m^2 - \lambda_n^2)X_m X_n$$

---

[1] The eigenfunctions are *orthogonal* functions. Orthogonal functions are discussed in the next section.

[2] This can be verified by differentiating $(X_m X_n' - X_m' X_n)$ as follows:

$$\frac{d}{dx}(X_m X_n' - X_m' X_n) = X_m X_n'' + X_n' X_m' - X_m' X_n' - X_m'' X_n = X_m X_n'' - X_m'' X_n$$

Upon multiplying by dx and integrating from $x = a$ to $x = b$ where the boundary conditions are given,

$$\int_{x=a}^{b} d(X_m X'_n - X'_m X_n) = (\lambda_m^2 - \lambda_n^2) \int_{x=a}^{b} X_m X_n \, dx$$

Substitution of limits on the left-hand side gives

$$[X_m(b)X'_n(b) - X'_m(b)X_n(b)] - [X_m(a)X'_n(a) - X'_m(a)X_n(a)]$$

$$= (\lambda_m^2 - \lambda_n^2) \int_{x=a}^{b} X_m X_n \, dx$$

The integrated term at $x = a$ will be 0 for any one of the following boundary conditions:

$$X(a) = 0 \quad , \quad X'(a) = 0 \quad \text{or} \quad X'(a) = CX(a)$$

Similarly, the integrated term at $x = b$ will be 0 for any one of the following boundary conditions:

$$X(b) = 0 \quad , \quad X'(b) = 0 \quad \text{or} \quad X'(b) = CX(b)$$

These boundary conditions on the eigenfunctions are homogeneous. They result from the thermal boundary conditions shown in Table 3•5.

Table 3•5 *Boundary conditions for orthogonal eigenfunctions.*

Thermal condition	Eigenfunction condition
Specified temperature	$X = 0$
Specified heat flux	$X' = 0$
Convection	$X' = CX$

With appropriate thermal boundary conditions the integrated terms will be 0 and we can write

$$0 = (\lambda_m^2 - \lambda_n^2) \int_{x=a}^{b} X_m X_n \, dx$$

There are two possibilities

$$\int_{x=a}^{b} X_m X_n \, dx = 0 \qquad\qquad m \neq n \qquad\qquad (3\cdot1\cdot36)$$

or

$$\int_{x=a}^{b} X_m X_n \, dx = \int_{x=a}^{b} X_m^2 \, dx > 0 \qquad m = n$$

When these relations are true we say that X_m and X_n are orthogonal to one another. The eigenfunctions are orthogonal functions.[1] The fact that the integral in (3•1•36) is 0 was used to select one term in the infinite series in (3•1•35) to evaluate its Fourier coefficient.

---

[1] Two functions, $p(x)$ and $q(x)$, which are piecewise continuous for $a \leq x \leq b$, are orthogonal in this interval if

$$\int_{x=a}^{b} p(x)q(x) \, dx = 0$$

3•1•3 Plane-wall transient, initial conditions

The rate of convergence of the Fourier series representation of the initial condition depends upon the initial temperature profile and on the boundary conditions. Let us again consider a plane wall, insulated at $x = 0$, with a specified temperature at $x = L$ and having an initial temperature of $t^{(0)}(x)$ as pictured in Figure 3•7. This problem was considered in Section 3•1•1.

The nonhomogeneous boundary condition at $x = L$ can be handled by assuming the solution to have the form

$$t(x,\theta) = a(x) + b(\theta) + u(x,\theta)$$

Substituting this assumed form of the solution into the partial differential equation and its boundary and initial conditions gives the homogeneous problem for $u(x,\theta)$ stated in Table 3•6 and pictured in Figure 3•8.

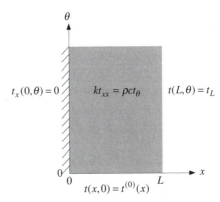

Figure 3•7 *Plane wall, initially at temperature $t^{(0)}(x)$, is suddenly exposed to a step change in surface temperature at $x = L$.*

Table 3•6 *Problems for $t(x,\theta)$ and $u(x,\theta)$.*

Problem	Equation	Boundary/Initial Conditions	
$t(x,\theta) = t_L + u(x,\theta)$			(3·1·16)
$u(x,\theta)$	$ku_{xx} = \rho c u_\theta$		(3·1·6)
		$u_x(0,\theta) = 0$	(3·1·9)
		$u(L,\theta) = 0$	(3·1·11)
		$u(x,0) = t^{(0)}(x) - t_L$	(3·1·17)

The separation of variables procedure found $u(x,\theta)$ given by (3·1·28) as

$$u(x,\theta) = \sum_{n=1}^{\infty} B_n \cos(\lambda_n x) \exp(-\lambda_n^2 \alpha \theta) \tag{3·1·28}$$

where $\lambda_n L = (2n-1)\pi/2$ and B_n is given by (3·1·29) as

$$B_m = \frac{2}{L} \int_{x=0}^{L} u(x,0) \cos(\lambda_m x)\, dx \tag{3·1·29}$$

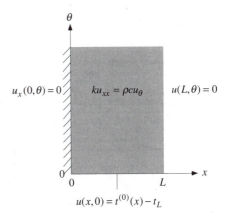

Figure 3•8 *Homogeneous problem for $u(x,\theta)$ corresponding to Figure 3•7.*

Up to this point the solution has been exactly the same as shown in Section 3•1•1. Rather than considering a uniform initial temperature profile as in Section 3•1•1, let us consider two other initial conditions as examples.

Example 1 Consider an initial temperature that varies linearly from t_0 at $x = 0$ to t_L at $x = L$ as given by

$$t^{(0)}(x) = t_0 + \frac{t_L - t_0}{L} x \tag{3·1·37}$$

Two things should be pointed out about this initial condition relative to the boundary conditions. At $x = 0$ the gradient from (3·1·37) is $(t_L - t_0)/L$. Thus at $\theta = 0$ there is a step in gradient from $(t_L - t_0)/L$ to 0. At $x = L$ the initial temperature from (3·1·37) is t_L. This is the same as the boundary condition at $x = L$. There is no step in the boundary temperature at $\theta = 0$. Discontinuities in temperature and temperature gradient are related to series convergence.

The initial condition for $u(x,\theta)$ is found by substituting (3·1·39) into (3·1·17) to obtain

$$u(x,0) = t_0 + \frac{t_L - t_0}{L} x - t_L = \frac{t_0 - t_L}{L}(L - x)$$

Upon substituting $u(x,0)$ into (3·1·29) and integrating we find that

$$B_m = \frac{2(t_0 - t_L)}{(\lambda L)^2}$$

Substituting these coefficients into (3·1·28) and then substituting $u(x,\theta)$ into (3·1·16) we obtain

$$t(x,\theta) = t_L + 2(t_0 - t_L)\sum_{n=1}^{\infty}\frac{1}{(\lambda_n L)^2}\cos(\lambda_n x)\exp(-\lambda_n^2 \alpha\theta)$$

Upon defining nondimensional variables

$$\bar{x} = \frac{x}{L} \quad , \quad \bar{\theta} = \frac{\alpha\theta}{L^2} \quad \text{and} \quad \bar{t}(\bar{x},\bar{\theta}) = \frac{t(x,\theta) - t_L}{(t_0 - t_L)}$$

and substituting $\lambda_n L = (2n-1)\pi/2$, we may write the solution as

$$\bar{t}(\bar{x},\bar{\theta}) = \frac{8}{\pi^2}\sum_{n=1}^{\infty}\frac{1}{(2n-1)^2}\cos[(2n-1)\frac{\pi}{2}\bar{x}]\exp[-(2n-1)^2\frac{\pi^2}{4}\bar{\theta}] \qquad (3·1·38)$$

The first four terms of $\bar{t}(\bar{x},0)$ are shown separately in Figure 3•9. Note that the amplitudes of these terms decrease more rapidly than the amplitudes shown in Figure 3•5 for a uniform initial temperature profile. The sum of the terms is also shown in Figure 3•9. Convergence toward the linear initial temperature profile is much better than for the uniform temperature case shown in Figure 3•5. The improved convergence shown in Figure 3•9 relative to Figure 3•5 is because (3·1·38) has $(2n-1)^2$ in its denominator whereas (3·1·31) had only $(2n-1)$ in its denominator. Thus, the amplitude of the terms in (3·1·38) decreases much faster than the amplitude in (3·1·31) as n increases.

One check of this solution (3·1·38) is to evaluate it for $\bar{\theta} = 0$ at the boundaries. Upon setting $\bar{\theta} = 0$ in (3·1·38)

$$\bar{t}(\bar{x},0) = \frac{8}{\pi^2}\sum_{n=1}^{\infty}\frac{1}{(2n-1)^2}\cos[(2n-1)\frac{\pi}{2}\bar{x}]$$

At $\bar{x} = 0$ the cosine is always 1 and the solution reduces to

$$\bar{t}(0,0) = \frac{8}{\pi^2}\sum_{n=1}^{\infty}\frac{1}{(2n-1)^2}$$

Appendix C contains several helpful series summations. From Appendix C we see that the sum of this series is $\pi^2/8$. Thus, $\bar{t}(0,0) = 1$. This agrees with the normalized initial condition. At the other boundary, $\bar{x} = 1$, the cosine is always 0 (the eigencondition) and $\bar{t}(1,0) = 0$. This agrees with the normalized initial condition. Sometimes it is helpful to evaluate the initial condition at an internal position in the wall. For $\bar{x} = 0.5$ the solution gives

$$\bar{t}(0.5,0) = \frac{8}{\pi^2}\sum_{n=1}^{\infty}\frac{1}{(2n-1)^2}\cos[(2n-1)\pi/4]$$

From Appendix C, this series sums to $\pi^2/16$. Thus, $\bar{t}(0.5,0) = 0.5$. This also agrees with the normalized initial condition.

Example 2 In this example we will consider an initial temperature profile that varies parabolically from t_0 at $x = 0$ to t_L at $x = L$ as given by

$$t^{(0)}(x) = t_0 + \frac{t_L - t_0}{L^2}x^2$$

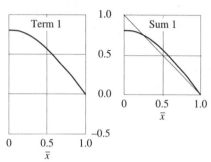
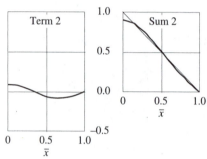
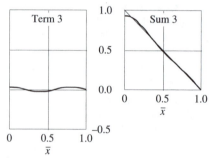
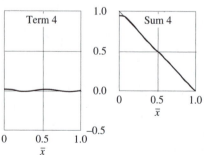

Figure 3•9 *Convergence of (3·1·38) at $\bar{\theta} = 0$ to $\bar{t}(\bar{x},0) = 1 - \bar{x}$.*

Note that at $x = 0$ the initial temperature gradient is 0 and at $x = L$ the initial temperature is t_L. There is no step in temperature or temperature gradient in this example. This initial profile might be the result of steady-state uniform energy generation within the wall prior to time 0 with an adiabatic face at $x = 0$ and a temperature of t_L at $x = L$. The solution for temperature in normalized form is given by

$$\bar{t}(\bar{x},\bar{\theta}) = \frac{32}{\pi^3} \sum_{n=1}^{\infty} \frac{(-1)^{n+1}}{(2n-1)^3} \cos[(2n-1)\frac{\pi}{2}\bar{x}] \exp[-(2n-1)^2 \frac{\pi^2}{4}\bar{\theta}] \qquad (3\cdot1\cdot39)$$

The first term of (3·1·39) for $\bar{\theta} = 0$ is shown in Figure 3•10 compared to the exact initial condition. Convergence will be much better than in Example 1 since $(2n-1)^3$ appears in the denominator rather than $(2n-1)^2$.

We may check the solution (3·1·39) for $\bar{\theta} = 0$ at three values of $\bar{x}$. Upon setting $\bar{\theta} = 0$,

$$\bar{t}(\bar{x},0) = \frac{32}{\pi^3} \sum_{n=1}^{\infty} \frac{(-1)^{n+1}}{(2n-1)^3} \cos[(2n-1)\frac{\pi}{2}\bar{x}]$$

At $\bar{x} = 0$ the cosine is always 1 and the solution reduces to

$$\bar{t}(0,0) = \frac{32}{\pi^3} \sum_{n=1}^{\infty} \frac{(-1)^{n+1}}{(2n-1)^3}$$

Appendix C gives the sum of this series as $\pi^3/32$. Therefore, $\bar{t}(0,0) = 1$. For $\bar{x} = 0.5$ the solution reduces to

$$\bar{t}(0.5,0) = \frac{32}{\pi^3} \sum_{n=1}^{\infty} \frac{(-1)^{n+1}}{(2n-1)^3} \cos[(2n-1)\pi/4]$$

From Appendix C the sum of this infinite series is $3\pi^3/128$. This gives $\bar{t}(0.5,0) = 96/128 = 0.75$. At $\bar{x} = 1$, the cosine is always 0 (from the eigencondition) and $\bar{t}(1,0) = 0$. Each of these results agree with the normalized initial condition.

There is some rhyme and reason regarding the rates of convergence shown by these two examples and the one considered in Section 3•1•1. In Section 3•1•1 there was a step change in the temperature (at $x = L$) and the series converged slowly as $1/(2n-1)$. In Example 1 in this section there was only a step in temperature gradient (at $x = 0$) and the series converged faster as $1/(2n-1)^2$. In Example 2 there are no steps in either temperature or in temperature gradient and convergence is still faster as $1/(2n-1)^3$. A step in temperature is the most difficult to achieve physically and gives the slowest rate of convergence. A step change in the gradient is easier to achieve physically and gives better convergence. The best convergence is when there are no steps in either temperature or temperature gradient. You can check your analytical solution to see if its rate of convergence is what you expect based on the initial and boundary conditions of the problem. Table 3•7 shows what you should expect.

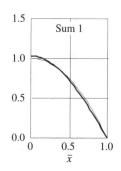

Figure 3•10 *First term of (3·1·39) for $\bar{\theta} = 0$ compared to $\bar{t}(\bar{x},0) = 1 - \bar{x}^2$.*

Table 3•7 *Rate of series convergence.*

For a step in	Series converges as
Temperature	$1/n$
Gradient	$1/n^2$
None	$1/n^3$

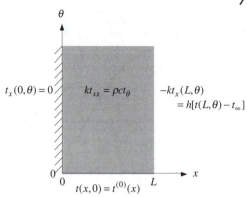

Figure 3•11 *Plane wall, initially at temperature $t^{(0)}(x)$, suddenly exposed to convection at $x = L$.*

3•1•4 Plane-wall transient, convection boundary

Let us consider the same plane wall as discussed in Section 3•1•1 except that now the specified-temperature boundary condition at $x = L$ will be replaced by a convection boundary condition. A mathematical description of this problem may be pictured as shown in Figure 3•11. This problem is nonhomogeneous if, as is usually the case, the ambient temperature t_∞ is not equal to 0.

To handle this nonhomogeneous problem we will again assume that the solution for $t(x,\theta)$ can be written as

$$t(x,\theta) = a(x) + b(\theta) + u(x,\theta) \tag{3.1.40}$$

Substituting this assumed form of the solution into the partial differential equation gives

$$k[a''(x) + u_{xx}] = \rho c[b'(\theta) + u_\theta]$$

This can be satisfied by taking $u(x,\theta)$ to satisfy the homogeneous partial differential equation

$$ku_{xx} = \rho c u_\theta \tag{3.1.41}$$

This means that $a(x)$ and $b(\theta)$ must satisfy

$$ka''(x) = \rho c b'(\theta)$$

The only way that a function of x can equal a function of θ is if each function is equal to the same constant. Thus we will require that $a(x)$ satisfy the second-order ordinary differential equation

$$ka''(x) = c_1 \tag{3.1.42}$$

and $b(\theta)$ satisfy the first-order ordinary differential equation

$$\rho c b'(\theta) = c_1 \tag{3.1.43}$$

Substituting (3.1.40) into the boundary condition at $x = 0$ gives

$$a'(0) + u_x(0,\theta) = 0$$

This can be satisfied by taking $u(x,\theta)$ to obey the homogeneous condition

$$u_x(0,\theta) = 0 \tag{3.1.44}$$

Thus, $a(x)$ will have to satisfy

$$a'(0) = 0 \tag{3.1.45}$$

To obtain a homogeneous boundary condition for $u(x,\theta)$ at $x = L$, we will substitute (3.1.40) into the boundary condition at $x = L$ to obtain

$$-k[a'(L) + u_x(L,\theta)] = h[a(L) + b(\theta) + u(L,\theta) - t_\infty]$$

This can be satisfied by taking $u(x,\theta)$ to obey the homogeneous condition

$$-ku_x(L,\theta) = hu(L,\theta) \tag{3.1.46}$$

We will also take

$$-ka'(L) = h[a(L) - t_\infty] \tag{3.1.47}$$

Therefore we must take

$$b(\theta) = 0 \tag{3.1.48}$$

The initial condition for $u(x,\theta)$ is found by substituting (3.1.40) into the initial condition for $t(x,\theta)$ to obtain

$$t(x,0) = t^{(0)}(x) = a(x) + b(0) + u(x,0)$$

Upon solving for $u(x,0)$ and noting from (3·1·48) that $b(0) = 0$,

$$u(x,0) = t^{(0)}(x) - a(x) \qquad\qquad (3·1·49)$$

We now have enough relations to specify the subproblems for $a(x)$, $b(\theta)$ and $u(x,\theta)$. These three subproblems are summarized in Table 3•8. The problem for $u(x,\theta)$ is pictured in Figure 3•12. The problem for $a(x)$ is nonhomogeneous, but this does not matter since $a(x)$ obeys an ordinary differential equation. The problem for $u(x,\theta)$, where homogeneity is important, has been constructed to be homogeneous.

Table 3•8 *Subproblems for $t(x,\theta) = a(x) + b(\theta) + u(x,\theta)$.*

Subproblem	Equation	Boundary/Initial Conditions	
$a(x)$	$ka''(x) = c_1$		(3·1·42)
		$a'(0) = 0$	(3·1·45)
		$-ka'(L) = h[a(L) - t_\infty]$	(3·1·47)
$b(\theta)$	$\rho c b'(\theta) = c_1$		(3·1·43)
		$b(\theta) = 0$	(3·1·48)
$u(x,\theta)$	$ku_{xx} = \rho c u_\theta$		(3·1·41)
		$u_x(0,\theta) = 0$	(3·1·44)
		$-ku_x(L,\theta) = hu(L,\theta)$	(3·1·46)
		$u(x,0) = t^{(0)}(x) - a(x)$	(3·1·49)

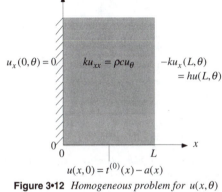

Figure 3•12 *Homogeneous problem for $u(x,\theta)$ corresponding to Figure 3•11.*

Equation (3·1·48) is the solution for $b(\theta)$. Since $b(\theta)$ is a constant, $b'(\theta) = 0$ and from (3·1·43) we see that c_1 must be 0.

The solution of (3·1·42) for $c_1 = 0$ may be found by integrating once to obtain

$$a'(x) = a_1$$

From (3·1·45) we see that $a_1 = 0$. Integrating again for $a_1 = 0$,

$$a(x) = a_2$$

Substituting this equation for a_1 into the boundary condition (3·1·47) at $x = L$ gives

$$-k \cdot 0 = h[a_2 - t_\infty]$$

Thus $a_2 = t_\infty$ and $a(x)$ is given by

$$a(x) = t_\infty \qquad\qquad (3·1·50)$$

Substituting (3·1·50) for $a(x)$ and (3·1·48) for $b(\theta)$ into (3·1·40),

$$t(x,\theta) = t_\infty + u(x,\theta) \qquad\qquad (3·1·51)$$

Upon substituting (3·1·50) for $a(x)$ into (3·1·49),

$$u(x,0) = t^{(0)}(x) - t_\infty \qquad\qquad (3·1·52)$$

For convenience the problems for $t(x,\theta)$ and $u(x,\theta)$ are summarized in Table 3•9. The problem that must be solved for $u(x,\theta)$ is depicted in Figure 3•13. The variable $u(x,\theta)$ may be thought of as the transient temperature distribution in a plane wall (no generation), initially at temperature $u(x,0)$, whose surface at $x = L$ is suddenly exposed to convection with an ambient at a temperature of 0. The surface at $x = 0$ is adiabatic. This is the same as the original problem for $t(x,\theta)$ except that the nonhomogeneous boundary condition $-kt_x(L,\theta) = h[t(x,\theta)-t_\infty]$ has been replaced by a homogeneous one $-ku_x(L,\theta) = hu(L,\theta)$ and the initial temperature distribution $t(x,0) = t^{(0)}(x)$ has been modified to $u(x,0) = t^{(0)}(x)-t_\infty$. Note that the problem for $u(x,\theta)$ is both linear and homogeneous. From the physics of the problem, $u(x,\theta)$ starts at $u(x,0)$ and approaches 0 as steady state is attained.

Table 3•9 *Problems for $t(x,\theta)$ and $u(x,\theta)$.*

Problem	Equation	Boundary/Initial Conditions	
$t(x,\theta) = t_\infty + u(x,\theta)$			(3·1·51)
$u(x,\theta)$	$ku_{xx} = \rho c u_\theta$		(3·1·41)
		$u_x(0,\theta) = 0$	(3·1·44)
		$-ku_x(L,\theta) = hu(L,\theta)$	(3·1·46)
		$u(x,0) = t^{(0)}(x) - t_\infty$	(3·1·52)

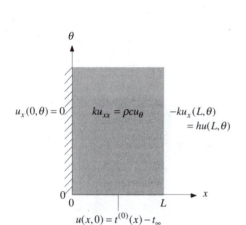

Figure 3•13 *Updated homogeneous problem for $u(x,\theta)$ corresponding to Figure 3•11.*

The method of separation of variables starts by assuming that the solution for $u(x,\theta)$ can be written as a product of a function of x, $X(x)$, and a function of θ, $\Theta(\theta)$. Thus we will write

$$u(x,\theta) = X(x)\Theta(\theta) \tag{3·1·53}$$

Substitution of this assumed solution into the partial differential equation (3·1·41) gives

$$kX''(x)\Theta(\theta) = \rho c X(x)\Theta'(\theta)$$

Upon rearranging to separate the variables,

$$\frac{X''(x)}{X(x)} = \frac{\rho c}{k}\frac{\Theta'(\theta)}{\Theta(\theta)} \tag{3·1·54}$$

Upon setting each side of (3·1·54) equal to $-\lambda^2$, and introducing the thermal diffusivity $\alpha = k/\rho c$ to simplify notation,

$$X''(x) + \lambda^2 X(x) = 0 \tag{3·1·55}$$

and

$$\Theta'(\theta) + \lambda^2\alpha\Theta(\theta) = 0 \tag{3·1·56}$$

Equation (3·1·53) may next be substituted into the boundary condition (3·1·44) at $x = 0$ to obtain

$$X'(0)\Theta(\theta) = 0$$

Since setting $\Theta(\theta) = 0$ is not be a helpful way to satisfy this condition, we will take

$$X'(0) = 0 \tag{3·1·57}$$

Substituting (3·1·53) into the boundary condition at $x = L$ (3·1·46) gives

$$-kX'(L)\Theta(\theta) = hX(L)\Theta(\theta)$$

The $\Theta(\theta)$ cancels and we must have

$$-kX'(L) = hX(L) \tag{3.1.58}$$

The eigenproblem for $X(x)$ is given by (3.1.55), (3.1.57) and (3.1.58).

The general solution to (3.1.55) is given by

$$X(x) = A\sin(\lambda x) + B\cos(\lambda x) \tag{3.1.59}$$

Taking the first derivative and substituting into the boundary condition at $x = 0$ (3.1.57) we obtain

$$X'(0) = A\lambda\cos(\lambda 0) - B\lambda\sin(\lambda 0) = A\lambda = 0$$

This will be satisfied by taking $A = 0$. Substitution of (3.1.59), with $A = 0$, into the boundary condition at $x = L$ (3.1.58) gives

$$-k[-B\lambda\sin(\lambda L)] = h[B\cos(\lambda L)]$$

Upon canceling B, multiplying both sides by L/k to give nondimensional groups λL and $hL/k = H$ and rearranging we obtain

$$\lambda L\sin(\lambda L) - H\cos(\lambda L) = 0 \tag{3.1.60}$$

The nondimensional group $H = hL/k$ is called the *Biot number*. This eigencondition (3.1.60) must be solved to find the eigenvalues λ.

Figure 3•14 shows $f(\lambda L) = \lambda L\sin(\lambda L) - H\cos(\lambda L)$ as a function of λL for $H = 1$. The acceptable values (the eigenvalues) of λL occur where $f(\lambda L) = 0$. There are an infinite number of values of λL which satisfy (3.1.60) since there are an infinite number of zeros of $f(\lambda L)$. The first four zeros are indicated in Figure 3•14. The first six eigenvalues are tabulated in Appendix D for a number of values of H and are listed in Table 3•10 for $H = 1$. As n increases $\lambda_n L$ increases and $\sin(\lambda_n L)$ must approach 0. Therefore the $\lambda_n L$ approach multiples of π. For large n, $\lambda_n L \approx (n-1)\pi$. For $n = 6$ the error is only 0.40 percent.

Table 3•10 *First six roots of* $\lambda L\sin(\lambda L) - \cos(\lambda L) = 0$.

n	$\lambda_n L$	$(n-1)\pi$	% Deviation
1	0.8603	0.0000	∞
2	3.4256	3.1416	9.04
3	6.4373	6.2832	2.45
4	9.5293	9.4248	1.11
5	12.6453	12.5664	0.63
6	15.7713	15.7080	0.40

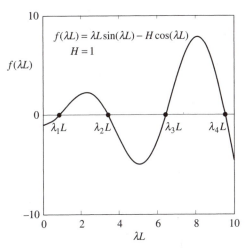

Figure 3•14 *Graph of* $f(\lambda L) = \lambda L\sin(\lambda L) - H\cos(\lambda L)$ *versus* λL *showing the first four values of* λL *where* $f(\lambda L) = 0$.

As the heat-transfer coefficient increases for the surface at $x = L$, the boundary condition at $x = L$ approaches a step change in temperature. From (3.1.60) we see that for large H, $\cos(\lambda_n L)$ must approach 0 so that $\lambda_n L$ will approach $(2n-1)\pi/2$. This agrees with the values in Table D•1. These are the eigenvalues that were obtained in Section 3•1•1 for a step change in surface temperature at $x = L$. Eigenvalues for $H > 100$ are not listed in Table D•1 because an engineer would most likely assume a step in surface temperature rather than a convective surface if $H > 100$.

The first separated equation (3.1.55) has a solution given by

$$X_n(x) = B_n\cos(\lambda_n x)$$

The second separated equation (3.1.56) has the solution

$$\Theta_n(\theta) = \exp(-\lambda_n^2\alpha\theta)$$

The eigenvalues λ_n in each of these equations are obtained from the eigencondition (3·1·60). We will substitute these functions into (3·1·53) to obtain the partial solution $u_n(x,\theta)$ given by

$$u_n(x,\theta) = B_n \cos(\lambda_n x)\exp(-\lambda_n^2 \alpha\theta)$$

This partial solution satisfies the partial differential equation (3·1·41) for $u(x,\theta)$ and its boundary conditions, (3·1·44) and (3·1·46), for $n = 1, 2, \cdots$. To satisfy the initial condition (3·1·52) we will sum up these partial solutions to obtain the more-general solution

$$u(x,\theta) = \sum_{n=1}^{\infty} B_n \cos(\lambda_n x)\exp(-\lambda_n^2 \alpha\theta) \tag{3·1·61}$$

where the λ_n must satisfy $\lambda_n L \tan(\lambda_n L) = H$. Equation (3·1·61) also satisfies the partial differential equation and its boundary conditions.

To satisfy the initial condition we will set $\theta = 0$ to obtain

$$u(x,0) = \sum_{n=1}^{\infty} B_n \cos(\lambda_n x)$$

To find a particular B_m we will multiply by $\cos(\lambda_m x)$ and integrate from $x = 0$ to $x = L$. This gives

$$\int_{x=0}^{L} u(x,0)\cos(\lambda_m x)\,dx = \sum_{n=1}^{\infty} B_n \int_{x=0}^{L} \cos(\lambda_n x)\cos(\lambda_m x)\,dx \tag{3·1·62}$$

The integral on the right-hand side may be looked up in your integral tables. For $n \neq m$, after substituting limits,

$$\int_{x=0}^{L} \cos(\lambda_n x)\cos(\lambda_m x)\,dx = \frac{\sin(\lambda_n L - \lambda_m L)}{2(\lambda_n - \lambda_m)} + \frac{\sin(\lambda_n L + \lambda_m L)}{2(\lambda_n + \lambda_m)}$$

This result may be simplified by using the trigonometric identities for sums and differences of angles. We may write

$$\int_{x=0}^{L} \cos(\lambda_n x)\cos(\lambda_m x)\,dx = \frac{\sin(\lambda_n L)\cos(\lambda_m L) - \cos(\lambda_n L)\sin(\lambda_m L)}{2(\lambda_n - \lambda_m)}$$
$$+ \frac{\sin(\lambda_n L)\cos(\lambda_m L) + \cos(\lambda_n L)\sin(\lambda_m L)}{2(\lambda_n + \lambda_m)}$$

From the eigencondition (3·1·60) we may write

$$\sin(\lambda L) = \frac{H\cos(\lambda L)}{\lambda L} \tag{3·1·63}$$

Upon using this expression to replace $\sin(\lambda_n L)$ and $\sin(\lambda_m L)$ we obtain

$$\int_{x=0}^{L} \cos(\lambda_n x)\cos(\lambda_m x)\,dx = \frac{\dfrac{H\cos(\lambda_n L)}{\lambda_n L}\cos(\lambda_m L) - \cos(\lambda_n L)\dfrac{H\cos(\lambda_m L)}{\lambda_m L}}{2(\lambda_n - \lambda_m)}$$
$$+ \frac{\dfrac{H\cos(\lambda_n L)}{\lambda_n L}\cos(\lambda_m L) + \cos(\lambda_n L)\dfrac{H\cos(\lambda_m L)}{\lambda_m L}}{2(\lambda_n + \lambda_m)}$$

After several more algebraic steps this integral turns out to be 0. From the discussion of orthogonality in Section 3•1•2, this was expected.

Since each of the integrals on the right-hand side of (3·1·62) is 0 when $n \neq m$, only the term for $n = m$ will remain. Thus we may write

$$\int_{x=0}^{L} u(x,0)\cos(\lambda_m x)\,dx = B_m \int_{x=0}^{L} \cos^2(\lambda_m x)\,dx \qquad (3\cdot1\cdot64)$$

The remaining integral on the right-hand side may be evaluated as

$$\int_{x=0}^{L} \cos^2(\lambda_m x)\,dx = \left[\frac{x}{2} + \frac{1}{4\lambda_m}\sin(2\lambda_m x)\right]_{x=0}^{L} = \frac{L}{2} + \frac{1}{4\lambda_m}\sin(2\lambda_m L)$$

Upon using the trigonometric relation for multiple angles we may write

$$\int_{x=0}^{L} \cos^2(\lambda_m x)\,dx = \frac{L}{2} + \frac{1}{4\lambda_m}2\sin(\lambda_m L)\cos(\lambda_m L)$$

Equation (3·1·63) may be used to replace $\sin(\lambda_m L)$ to give

$$\int_{x=0}^{L} \cos^2(\lambda_m x)\,dx = \frac{L}{2} + \frac{1}{2\lambda_m}\frac{H\cos(\lambda_m L)}{\lambda_m L}\cos(\lambda_m L)$$

Upon factoring $2\lambda_m^2 L$ out of the denominator we obtain

$$\int_{x=0}^{L} \cos^2(\lambda_m x)\,dx = \frac{1}{2\lambda_m^2 L}[(\lambda_m L)^2 + H\cos^2(\lambda_m L)] \qquad (3\cdot1\cdot65)$$

For a uniform initial temperature $t^{(0)}$ the initial condition for $u(x,\theta)$ is given by (3·1·52) as $u(x,0) = t^{(0)} - t_\infty$. Thus the integral on the left-hand side of (3·1·64) becomes

$$\int_{x=0}^{L} u(x,0)\cos(\lambda_m x)\,dx = (t^{(0)} - t_\infty)\int_{x=0}^{L}\cos(\lambda_m x)\,dx$$

$$= (t^{(0)} - t_\infty)\frac{1}{\lambda_m}\sin(\lambda_m x)\Big|_{x=0}^{L}$$

$$= (t^{(0)} - t_\infty)\frac{1}{\lambda_m}\sin(\lambda_m L)$$

Upon substituting this result and (3·1·65) into (3·1·64) we obtain

$$(t^{(0)} - t_\infty)\frac{1}{\lambda_m}\sin(\lambda_m L) = B_m\frac{1}{2\lambda_m^2 L}[(\lambda_m L)^2 + H\cos^2(\lambda_m L)]$$

This may be rearranged to solve for B_m as

$$B_m = \frac{2(t^{(0)} - t_\infty)(\lambda_m L)\sin(\lambda_m L)}{[(\lambda_m L)^2 + H\cos^2(\lambda_m L)]}$$

Equation (3·1·63) may be used to replace $(\lambda_m L)\sin(\lambda_m L)$ in the numerator to give

$$B_m = \frac{2(t^{(0)} - t_\infty)H\cos(\lambda_m L)}{(\lambda_m L)^2 + H\cos^2(\lambda_m L)}$$

These coefficients may now be substituted into (3·1·61) to obtain $u(x,\theta)$ which, in turn, may be substituted into (3·1·51) to give

$$t(x,\theta) = t_\infty + \sum_{n=1}^{\infty}\frac{2(t^{(0)} - t_\infty)H\cos(\lambda_n L)}{(\lambda_n L)^2 + H\cos^2(\lambda_n L)}\cos(\lambda_n x)\exp(-\lambda_n^2\alpha\theta) \qquad (3\cdot1\cdot66)$$

Upon defining the following nondimensional variables:

$$\bar{x} = \frac{x}{L} \quad , \quad \bar{\theta} = \frac{\alpha\theta}{L^2} \quad \text{and} \quad \bar{t}(\bar{x},\bar{\theta}) = \frac{t(x,\theta) - t_\infty}{(t^{(0)} - t_\infty)}$$

we may write the solution (3·1·66) as

$$\bar{t}(\bar{x},\bar{\theta}) = \sum_{n=1}^{\infty} \frac{2H\cos(\lambda_n L)}{(\lambda_n L)^2 + H\cos^2(\lambda_n L)}\cos[(\lambda_n L)\bar{x}]\exp[-(\lambda_n L)^2\bar{\theta}] \qquad (3\cdot1\cdot67)$$

The values of $\lambda_n L$ come from the eigencondition (3·1·60) and depend on the Biot number, H. Since there is no step change (discontinuity) in temperature in this problem, only a step in the temperature gradient at $x = L$, the solution (3·1·67) converges as $1/(\lambda_m L)^2$ once $\lambda_n L$ gets large enough to make $H\cos^2(\lambda_n L)$ in the denominator unimportant. This is much better convergence than was obtained in Section 3•1•1 for a step in surface temperature. As $\bar{\theta}$ increases, the exponential term will help the series to converge.

The value of H can range between 0 (adiabatic surface) and infinity (step change in surface temperature). A reasonable intermediate value to consider is $H = 1$. The value of $\bar{\theta}$ runs from 0 (initial state) to infinity (steady state). A reasonable value to consider is $\bar{\theta} = 1.0$. The solution for $H = 1$, $\bar{\theta} = 1.0$ is shown in Figure 3•15. The solution for $\bar{\theta} = 0.1$ (an order of magnitude less than $\bar{\theta} = 1.0$) is also shown. The solution for $\bar{\theta} = 10$ (an order of magnitude more than $\bar{\theta} = 1.0$ is so close to steady state $\bar{t}(\bar{x},\infty) = 0$ that it is not shown.

It is also of interest to look at (3·1·67) for an intermediate time ($\bar{\theta} = 1.0$) for a range of H. Values of $H = 1$ (an intermediate value), $H = 0.1$ (an order of magnitude less than $H = 1$) and $H = 10$ (an order of magnitude more than $H = 1$) are shown in Figure 3•16. The curve for $H = 1$ is the same curve shown in Figure 3•15. A value of $H = hL/k = 0.1$ (small) could be due to a small heat-transfer coefficient (the temperature has not responded much by this time) or a large thermal conductivity (the internal temperature is relatively uniform at this time) or a combination of both. For values of $H \leq 0.1$ an engineer often assumes that a lumped-parameter analysis (uniform internal temperature) is adequate. A value of $H = hL/k = 10$ (large) could be due to a large heat-transfer coefficient (the temperature has almost reached steady state at this time) or a small conductivity (there is a significant temperature variation within the wall).

Limiting case (small Biot number) It is interesting to look at the solution (3·1·67) for the limiting case of $H =$ small (but not 0). From (3·1·60), as H becomes small, $\sin(\lambda L)$ approaches 0, the values of $\lambda_n L$ thus approach 0, π, 2π, ... or $(n-1)\pi$ and the values of $\cos(\lambda_n L)$ approach ±1. Since $\lambda_1 L$ and H both approach 0, the coefficient of the first term in (3·1·67) approaches $0/0$ and must be studied further. All of the other coefficients in (3·1·67) approach 0 due to the H in the numerator. Only the first term in the solution remains and the solution reduces to

$$\bar{t}(\bar{x},\bar{\theta}) = \frac{2H\cos(\lambda_1 L)}{(\lambda_1 L)^2 + H\cos^2(\lambda_1 L)}\cos[(\lambda_1 L)\bar{x}]\exp[-(\lambda_1 L)^2\bar{\theta}] \qquad (3\cdot1\cdot68)$$

The value of $\lambda_1 L$ is related to H by the eigencondition (3·1·60). That is,

$$\lambda_1 L\sin(\lambda_1 L) = H\cos(\lambda_1 L)$$

For small $\lambda_1 L$ we may replace $\sin(\lambda_1 L)$ and $\cos(\lambda_1 L)$ by their power series to obtain

$$\lambda_1 L\left[\lambda_1 L - \frac{(\lambda_1 L)^3}{3!} + \frac{(\lambda_1 L)^5}{5!} - \cdots\right] = H\left[1 - \frac{(\lambda_1 L)^2}{2!} + \frac{(\lambda_1 L)^4}{4!} - \cdots\right]$$

For small $\lambda_1 L$ we will neglect all but the first term in each of the two power series to obtain $(\lambda_1 L)^2 = H$ or $\lambda_1 L = \sqrt{H}$. Substituting this result into (3·1·68),

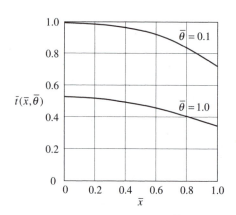

Figure 3•15 *Evaluation of (3·1·67) at $\bar{\theta} = 0.1$ and 1.0 for $H = 1$.*

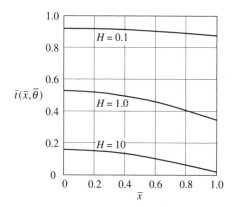

Figure 3•16 *Evaluation of (3·1·67) at $\bar{\theta} = 1.0$ for $H = 0.1$, 1.0 and 10.*

$$\bar{t}(\bar{x},\bar{\theta}) = \frac{2H\cos(\sqrt{H})}{H + H\cos^2(\sqrt{H})}\cos[(\sqrt{H})\bar{x}]\exp[-H\bar{\theta}]$$

As H becomes small, $\cos(\sqrt{H})$ approaches 1 and the solution approaches

$$\bar{t}(\bar{x},\bar{\theta}) = \cos[(\sqrt{H})\bar{x}]\exp[-H\bar{\theta}]$$

Since $\bar{x}$ is never greater than 1, $\cos[(\sqrt{H})\bar{x}]$ also approaches 1 as H becomes small. Thus, the dependence on position drops out and we obtain

$$\bar{t}(\bar{\theta}) = \exp[-H\bar{\theta}]$$

Upon substituting $H = hL/k$ and $\bar{\theta} = \alpha\theta/L^2$ we may then write

$$\bar{t}(\theta) = \exp\left[-\frac{hL}{k}\frac{\alpha\theta}{L^2}\right] = \exp\left[-\frac{hL}{k}\frac{k\theta}{\rho c L^2}\right] = \exp\left[-\frac{h}{\rho c L}\theta\right]$$

Upon multiplying the numerator and denominator of the exponential by A and noting that $AL = V$ and $\bar{t} = (t - t_\infty)/(t^{(0)} - t_\infty)$, we see that

$$t(\theta) = t_\infty + (t^{(0)} - t_\infty)\exp\left[-\frac{hA}{\rho V c}\theta\right]$$

This is same as the lumped-parameter solution given by (1·2·31).

Limiting case (large Biot number) Now let us look at the solution (3·1·67) for H becoming large. From (3·1·60), as H becomes infinite, $\cos(\lambda_n L)$ approaches 0 and thus the values of $\lambda_n L$ approach $\frac{\pi}{2}$, $\frac{3\pi}{2}$, $\frac{5\pi}{2}$, ... or $(2n-1)\pi/2$. The product $H\cos(\lambda_n L)$ approaches $\infty \times 0$, but we can use the eigencondition (3·1·60) to evaluate this indeterminate product as

$$H\cos(\lambda_n L) = \lambda_n L \sin(\lambda_n L)$$

Upon substituting this into (3·1·67) we obtain

$$\bar{t}(\bar{x},\bar{\theta}) = \sum_{n=1}^{\infty}\frac{2\lambda_n L\sin(\lambda_n L)}{(\lambda_n L)^2 + \lambda_n L\sin(\lambda_n L)\cos(\lambda_n L)}\cos[(\lambda_n L)\bar{x}]\exp[-(\lambda_n L)^2\bar{\theta}]$$

We may cancel $\lambda_n L$ from numerator and denominator. Since $\cos(\lambda_n L) = 0$ and $\sin(\lambda_n L) = (-1)^{n+1}$ the solution reduces to

$$\bar{t}(\bar{x},\bar{\theta}) = \sum_{n=1}^{\infty}\frac{2(-1)^{n+1}}{(\lambda_n L)}\cos[(\lambda_n L)\bar{x}]\exp[-(\lambda_n L)^2\bar{\theta}]$$

Finally, since $\lambda_n L = (2n-1)\pi/2$ we may write

$$\bar{t}(\bar{x},\bar{\theta}) = \frac{4}{\pi}\sum_{n=1}^{\infty}\frac{(-1)^{n+1}}{(2n-1)}\cos[(2n-1)\frac{\pi}{2}\bar{x}]\exp[-(2n-1)^2\frac{\pi^2}{4}\bar{\theta}]$$

The Biot number has dropped out of the solution and the result is identical to the specified-temperature solution (3·1·31).

Charts For values of $\bar{\theta} \geq 0.2$ the exponential in (3·1·67) provides enough convergence so that only one term in the series is needed for most purposes. This also means that $\bar{t}(0,\bar{\theta})$ varies exponentially with $\bar{\theta}$ and a plot of the logarithm of $\bar{t}(0,\bar{\theta})$ versus $\bar{\theta}$ will be a straight line. Such plots have been prepared for plane walls (and for infinitely long cylinders and for spheres) for H ranging from 0.01 (lumped approximation valid) to ∞ (step change in surface temperature). These charts (and companion charts for temperature distribution and total internal energy change) are not included in this book since they are presented in most undergraduate heat-transfer textbooks.

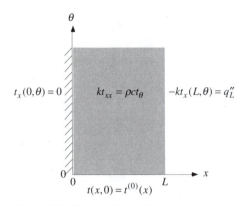

Figure 3•17 *Plane wall, initially at temperature $t^{(0)}(x)$, suddenly exposed to a nonzero heat flux q''_L at $x = L$.*

3•1•5 Plane-wall transient, nonzero-heat-flux boundary

Let us consider the same plane wall as discussed in Sections 3•1•1 and 3•1•4 except that now the boundary condition at $x = L$ will be a specified nonzero heat flux q''_L out of the wall. A mathematical description of this problem is shown in Figure 3•17. This problem is nonhomogeneous because of the nonzero-heat-flux boundary condition.

To handle this nonhomogeneous problem we will again assume that the solution for $t(x,\theta)$ can be written as

$$t(x,\theta) = a(x) + b(\theta) + u(x,\theta) \tag{3.1.69}$$

Substituting this assumed solution into the partial differential equation,

$$k[a''(x) + u_{xx}] = \rho c[b'(\theta) + u_\theta]$$

This can be satisfied by taking $u(x,\theta)$ to satisfy the homogeneous partial differential equation

$$ku_{xx} = \rho c u_\theta \tag{3.1.70}$$

This means that $a(x)$ and $b(\theta)$ must satisfy

$$ka''(x) = \rho c b'(\theta)$$

The only way that a function of x can equal a function of θ is if each function is equal to the same constant. Thus we will require that $a(x)$ satisfy the second-order ordinary differential equation

$$ka''(x) = c_1 \tag{3.1.71}$$

and $b(\theta)$ satisfy the first-order ordinary differential equation

$$\rho c b'(\theta) = c_1 \tag{3.1.72}$$

The value of c_1 in (3.1.71) and (3.1.72) is arbitrary at this point. Since (3.1.71) is second order, it will require two boundary conditions. Since (3.1.72) is first order in time, it will need one initial condition.

Next, we must choose a boundary condition for $a(x)$ at $x = 0$ so that the boundary condition for $u(x,\theta)$ at $x = 0$ will be homogeneous. Substituting the assumed solution (3.1.69) into the boundary condition at $x = 0$ gives

$$a'(0) + u_x(0,\theta) = 0$$

This can be satisfied by taking $u(x,\theta)$ to obey the homogeneous condition

$$u_x(0,\theta) = 0 \tag{3.1.73}$$

which means $a(x)$ will have to satisfy

$$a'(0) = 0 \tag{3.1.74}$$

To obtain a homogeneous boundary condition for $u(x,\theta)$ at $x = L$, we will substitute the general solution (3.1.69) into the boundary condition at $x = L$ to obtain

$$-k[a'(L) + u_x(L,\theta)] = q''_L$$

This can be satisfied by taking $u(x,\theta)$ to obey the homogeneous condition

$$u_x(L,\theta) = 0 \tag{3.1.75}$$

which means that $a(x)$ must satisfy the boundary condition

$$-ka'(L) = q''_L \tag{3.1.76}$$

We now have two boundary conditions for (3.1.70) and two for (3.1.71).

To obtain the initial condition for $u(x,\theta)$ we will substitute equation (3·1·69) into the initial condition for $t(x,\theta)$. This gives

$$t(x,0) = t^{(0)}(x) = a(x) + b(0) + u(x,0)$$

Upon solving for $u(x,0)$,

$$u(x,0) = t^{(0)}(x) - a(x) - b(0) \qquad (3·1·77)$$

We now have enough relations to determine $a(x)$, $b(\theta)$ and $u(x,\theta)$. The three subproblems we must solve are summarized in Table 3•11. The subproblem for $a(x)$ must satisfy an ordinary differential equation (3·1·71). Boundary conditions for an ordinary differential equation can be nonhomogeneous. The function $u(x,\theta)$ satisfies a homogeneous partial differential equation (3·1·70) with homogeneous boundary conditions (3·1·73) and (3·1·75). A homogeneous problem (partial differential equation and both boundary conditions) for $u(x,\theta)$ is essential for separation of variables. The problem for $u(x,\theta)$ is shown in Figure 3•18.

Table 3•11 *Subproblems for* $t(x,\theta) = a(x) + b(\theta) + u(x,\theta)$.

Subproblem	Equation	Boundary/Initial Conditions	
$a(x)$	$ka''(x) = c_1$		(3·1·71)
		$a'(0) = 0$	(3·1·74)
		$-ka'(L) = q_L''$	(3·1·76)
$b(\theta)$	$\rho c b'(\theta) = c_1$		(3·1·72)
$u(x,\theta)$	$ku_{xx} = \rho c u_\theta$		(3·1·70)
		$u_x(0,\theta) = 0$	(3·1·73)
		$u_x(L,\theta) = 0$	(3·1·75)
		$u(x,0) = t^{(0)}(x) - a(x) - b(0)$	(3·1·77)

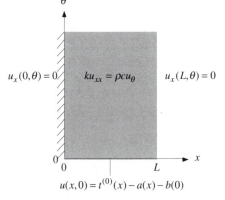

Figure 3•18 *Homogeneous problem for* $u(x,\theta)$ *corresponding to Figure 3•17.*

To obtain $a(x)$ we will integrate (3·1·71) once to obtain

$$a'(x) = \frac{c_1}{k}x + a_1 \qquad (3·1·78)$$

where a_1 is an arbitrary constant of integration. To satisfy (3·1·74) we see that we must have $a_1 = 0$. Upon setting $a_1 = 0$ and $x = L$ in (3·1·78) and then substituting into (3·1·76) gives

$$-k[\frac{c_1}{k}L] = q_L''$$

From this we can solve for c_1 as

$$c_1 = -\frac{q_L''}{L} \qquad (3·1·79)$$

Substituting (3·1·79) and $a_1 = 0$ into (3·1·78) gives

$$a'(x) = -\frac{q_L''}{kL}x$$

Upon integrating once more

$$a(x) = -\frac{q_L''}{kL}\frac{x^2}{2} + a_2$$

This relation satisfies (3·1·71), (3·1·74) and (3·1·76) for any value of a_2. The easiest thing to do is take $a_2 = 0$ to give

$$a(x) = -\frac{q_L''}{kL}\frac{x^2}{2} \tag{3·1·80}$$

To find $b(\theta)$, substitute (3·1·79) into (3·1·72) and integrate to obtain

$$b(\theta) = -\frac{q_L''}{\rho cL}\theta + b_1$$

Since this result satisfies (3·1·72) for any b_1 we will arbitrarily take $b_1 = 0$ to give

$$b(\theta) = -\frac{q_L''}{\rho cL}\theta \tag{3·1·81}$$

Substituting (3·1·80) and (3·1·81) into (3·1·69) gives

$$t(x,\theta) = -\frac{q_L''}{kL}\frac{x^2}{2} - \frac{q_L''}{\rho cL}\theta + u(x,\theta) \tag{3·1·82}$$

Substituting (3·1·80) and $b(0) = 0$ from (3·1·81) into (3·1·77) gives

$$u(x,0) = t^{(0)}(x) + \frac{q_L''}{kL}\frac{x^2}{2} \tag{3·1·83}$$

For convenience the problems for $t(x,\theta)$ and $u(x,\theta)$ are summarized in Table 3•12. The updated problem that must be solved for $u(x,\theta)$ is pictured in Figure 3•19. The variable $u(x,\theta)$ may be thought of as the transient temperature distribution in a plane wall (no generation), initially at a nonuniform temperature $u(x,0)$, whose surfaces at $x = 0$ and $x = L$ are both adiabatic. Since there is no generation and both surfaces are adiabatic, the total energy in this wall will remain constant. From the physics of the problem, $u(x,\theta)$ starts at $u(x,0)$ and should approach an average uniform value (with the same total energy as at the initial state) as steady state is attained.

Table 3•12 *Problems for $t(x,\theta)$ and $u(x,\theta)$.*

Problem	Equation	Boundary/Initial Conditions	
$t(x,\theta) = -\dfrac{q_L''}{kL}\dfrac{x^2}{2} - \dfrac{q_L''}{\rho cL}\theta + u(x,\theta)$			(3·1·82)
$u(x,\theta)$	$ku_{xx} = \rho cu_\theta$		(3·1·70)
		$u_x(0,\theta) = 0$	(3·1·73)
		$u_x(L,\theta) = 0$	(3·1·75)
		$u(x,0) = t^{(0)}(x) + \dfrac{q_L''}{kL}\dfrac{x^2}{2}$	(3·1·83)

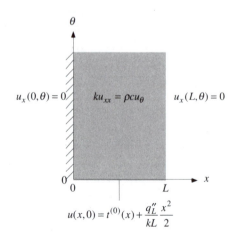

Figure 3•19 *Updated homogeneous problem for* $u(x,\theta)$ *corresponding to Figure 3•17.*

For the method of separation of variables we will assume that

$$u(x,\theta) = X(x)\Theta(\theta) \tag{3·1·84}$$

Substituting this assumed solution (3·1·84) into (3·1·70) and then separating variables gives

$$\frac{X''(x)}{X(x)} = \frac{\rho c}{k}\frac{\Theta'(\theta)}{\Theta(\theta)} \tag{3.1.85}$$

Upon setting each side of (3·1·85) equal to $-\lambda^2$, and introducing the thermal diffusivity $\alpha = k/\rho c$ to simplify notation,

$$X''(x) + \lambda^2 X(x) = 0 \tag{3.1.86}$$

and

$$\Theta'(\theta) + \lambda^2 \alpha \Theta(\theta) = 0 \tag{3.1.87}$$

Equation (3·1·84) may next be substituted into the boundary condition (3·1·73) at $x = 0$ to obtain

$$X'(0)\Theta(\theta) = 0$$

Since setting $\Theta(\theta) = 0$ is not be a helpful way to satisfy this condition, we will take

$$X'(0) = 0 \tag{3.1.88}$$

Substituting (3·1·84) into the boundary condition at $x = L$ (3·1·75) gives

$$X'(L)\Theta(\theta) = 0$$

Since $\Theta(\theta) = 0$ would not be helpful we will take

$$X'(L) = 0 \tag{3.1.89}$$

The eigenproblem for $X(x)$ is given by (3·1·86), (3·1·88) and (3·1·89).

The general solution to (3·1·86) is given by

$$X(x) = A\sin(\lambda x) + B\cos(\lambda x) \tag{3.1.90}$$

Upon taking the first derivative and substituting into the boundary condition at $x = 0$ (3·1·88) we obtain

$$X'(0) = A\lambda\cos(\lambda 0) - B\lambda\sin(\lambda 0) = A\lambda = 0$$

This will be satisfied by taking $A = 0$. Substitution of the first derivative of (3·1·90), with $A = 0$, into the boundary condition at $x = L$ (3·1·89) gives

$$-B\lambda\sin(\lambda L) = 0$$

We will satisfy this condition by taking $\sin(\lambda L) = 0$. The values of λL that satisfy this eigencondition are $\lambda L = 0$, π, 2π, …. The nth eigenfunction and corresponding eigenvalue are given by

$$X_n(x) = B_n\cos(\lambda_n x) \quad \text{where} \quad \lambda_n L = n\pi \quad \text{for} \quad n = 0, 1, 2, \cdots \tag{3.1.91}$$

The solution to (3·1·87) is given by

$$\Theta_n(\theta) = \exp(-\lambda_n^2 \alpha\theta) \tag{3.1.92}$$

Substituting (3·1·91) and (3·1·92) into (3·1·84) gives the partial solution

$$u_n(x,\theta) = B_n\cos(\lambda_n x)\exp(-\lambda_n^2 \alpha\theta)$$

This partial solution satisfies (3·1·70) and its boundary conditions, (3·1·73) and (3·1·75), for $n = 0, 1, 2, \cdots$. To satisfy the initial condition (3·1·83) we will sum up these partial solutions to obtain the more-general solution

$$u(x,\theta) = \sum_{n=0}^{\infty} B_n\cos(\lambda_n x)\exp(-\lambda_n^2 \alpha\theta) \tag{3.1.93}$$

Since $\lambda_n L = n\pi$, we see that $n = 0$ is a special case since $\lambda_0 = 0$ will give $\cos(\lambda_0 x) = 1$ and $\exp(-\lambda_0^2 \alpha\theta) = 1$ for any x and any θ. We will therefore separate the first term from the others and write (3·1·93) as

$$u(x,\theta) = B_0 + \sum_{n=1}^{\infty} B_n \cos(\lambda_n x) \exp(-\lambda_n^2 \alpha\theta) \tag{3·1·94}$$

We will evaluate the Fourier coefficients B_0 and B_n by making (3·1·94) match the initial condition (3·1·83). Upon setting $\theta = 0$ in (3·1·94),

$$u(x,0) = B_0 + \sum_{n=1}^{\infty} B_n \cos(\lambda_n x) \tag{3·1·95}$$

To find B_0, we will multiply (3·1·95) by $\cos(\lambda_0 x)\,dx = \cos(0x)\,dx = dx$ and integrate to obtain

$$\int_{x=0}^{L} u(x,0)\,dx = \int_{x=0}^{L} B_0 \,dx + \sum_{n=1}^{\infty} B_n \int_{x=0}^{L} \cos(\lambda_n x)\,dx$$

The integral in the summation is given by

$$\int_{x=0}^{L} \cos(\lambda_n x)\,dx = \frac{1}{\lambda_n} \sin(\lambda_n x)\Big|_{x=0}^{L} = \frac{1}{\lambda_n}\sin(\lambda_n L) = \frac{1}{\lambda_n}\sin(n\pi) = 0$$

Thus,

$$\int_{x=0}^{L} u(x,0)\,dx = B_0 L$$

Upon solving for B_0,

$$B_0 = \frac{1}{L}\int_{x=0}^{L} u(x,0)\,dx \tag{3·1·96}$$

This is the average value of the initial condition $u(x,0)$.

To find a B_m in (3·1·94) we will multiply (3·1·95) by $\cos(\lambda_m x)\,dx$ and integrate to obtain

$$\int_{x=0}^{L} u(x,0)\cos(\lambda_m x)\,dx = B_0 \int_{x=0}^{L} \cos(\lambda_m x)\,dx$$

$$+ \sum_{n=1}^{\infty} B_n \int_{x=0}^{L} \cos(\lambda_n x)\cos(\lambda_m x)\,dx$$

As we have just seen, the integral after B_0 is 0. Thus,

$$\int_{x=0}^{L} u(x,0)\cos(\lambda_m x)\,dx = \sum_{n=1}^{\infty} B_n \int_{x=0}^{L} \cos(\lambda_n x)\cos(\lambda_m x)\,dx \tag{3·1·97}$$

As we saw in Section 3•1•4, for $n \neq m$,

$$\int_{x=0}^{L} \cos(\lambda_n x)\cos(\lambda_m x)\,dx = \frac{\sin(\lambda_n L)\cos(\lambda_m L) - \cos(\lambda_n L)\sin(\lambda_m L)}{2(\lambda_n - \lambda_m)}$$

$$+ \frac{\sin(\lambda_n L)\cos(\lambda_m L) + \cos(\lambda_n L)\sin(\lambda_m L)}{2(\lambda_n + \lambda_m)}$$

Since the eigencondition in this problem is $\sin(\lambda_n L) = 0$, this integral becomes

$$\int_{x=0}^{L} \cos(\lambda_n x)\cos(\lambda_m x)\,dx = 0 \qquad n \neq m$$

Therefore each of the integrals on the right-hand side of (3·1·97) is 0 except when $n = m$. There is only one nonzero term on the right-hand side of (3·1·97) which then simplifies to give

$$\int_{x=0}^{L} u(x,0)\cos(\lambda_m x)\,dx = B_m \int_{x=0}^{L} \cos^2(\lambda_m x)\,dx$$

The remaining integral on the right-hand side may be evaluated as

$$\int_{x=0}^{L} \cos^2(\lambda_m x)\,dx = \left[\frac{x}{2} + \frac{1}{4\lambda_m}\sin(2\lambda_m x)\right]_{x=0}^{L} = \frac{L}{2} + \frac{1}{4\lambda_m}\sin(2\lambda_m L) = \frac{L}{2}$$

We may then solve for B_m as

$$B_m = \frac{2}{L}\int_{x=0}^{L} u(x,0)\cos(\lambda_m x)\,dx \qquad (3\cdot1\cdot98)$$

The initial condition for $u(x,\theta)$ is given by (3·1·83). If the initial condition for $t(x,\theta)$ is $t^{(0)}(x) = t^{(0)}$ where $t^{(0)}$ is a constant, then (3·1·83) may be substituted into (3·1·96) to give

$$B_0 = \frac{1}{L}\int_{x=0}^{L}\left[t^{(0)} + \frac{q_L''}{kL}\frac{x^2}{2}\right]dx = t^{(0)} + \frac{q_L'' L}{6k}$$

and into (3·1·98) to give, after using the eigencondition $\sin(\lambda_m L) = 0$ to simplify,

$$B_m = \frac{2}{L}\int_{x=0}^{L}\left[t^{(0)} + \frac{q_L''}{kL}\frac{x^2}{2}\right]\cos(\lambda_m x)\,dx = 2\frac{q_L'' L}{k}\frac{\cos(\lambda_n L)}{(\lambda_n L)^2}$$

These results for B_0 and B_m may be substituted into (3·1·94) to give $u(x,\theta)$ as

$$u(x,\theta) = t^{(0)} + \frac{q_L'' L}{6k} + 2\frac{q_L'' L}{k}\sum_{n=1}^{\infty}\frac{\cos(\lambda_n L)}{(\lambda_n L)^2}\cos(\lambda_n x)\exp(-\lambda_n^2\alpha\theta)$$

which may then be substituted into (3·1·82) to give $t(x,\theta)$ as

$$t(x,\theta) = -\frac{q_L''}{kL}\frac{x^2}{2} - \frac{q_L''}{\rho c L}\theta + t^{(0)} + \frac{q_L'' L}{6k}$$

$$+ 2\frac{q_L'' L}{k}\sum_{n=1}^{\infty}\frac{\cos(\lambda_n L)}{(\lambda_n L)^2}\cos(\lambda_n x)\exp(-\lambda_n^2\alpha\theta) \qquad (3\cdot1\cdot99)$$

where $\lambda_n L = n\pi$.

Four features should be noted about this solution. First, due to the second term on the right-hand side, there is no steady-state solution. The temperature decreases linearly toward $-\infty$ as θ increases. This is to be expected since energy at a rate of q_L'' is continually being removed from the wall at $x = L$. Second, the series has good convergence due to the $(\lambda_n L)^2$ in the denominator. This is the expected convergence since this problem has a step in temperature gradient, not temperature. Third, as θ increases the series decays to 0 due to the exponential term. The solution then reduces to

$$t(x,\theta) = -\frac{q_L''}{kL}\frac{x^2}{2} - \frac{q_L''}{\rho c L}\theta + t^{(0)} + \frac{q_L'' L}{6k} \qquad \text{[for } \theta \text{ large]}$$

At any θ the temperature varies parabolically with x. The temperature profile at one time is similar to the temperature profile at any other time. They only differ by a constant. Fourth, the maximum temperature occurs at

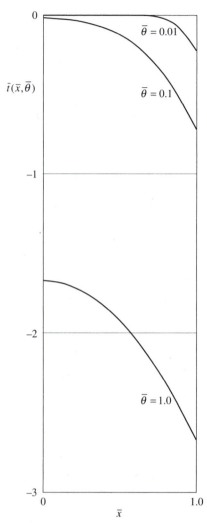

Figure 3•20 *Evaluation of (3·1·100) at $\bar{\theta}$ = 0.01, 0.1 and 1.0.*

$x = 0$ and the minimum temperature occurs at $x = L$. The difference between these two temperatures increases as θ increases. The maximum difference is $q_L'' L / 2k$.

Before numerically evaluating (3·1·99) it is wise to normalize. Upon defining the following nondimensional variables:

$$\bar{x} = \frac{x}{L} \quad , \quad \bar{\theta} = \frac{\alpha \theta}{L^2} \quad \text{and} \quad \bar{t}(\bar{x}, \bar{\theta}) = \frac{t(x,\theta) - t^{(0)}}{q_L'' L / 2k}$$

and substituting $\lambda_n L = n\pi$ and $\cos(n\pi) = (-1)^n$ the solution becomes

$$\bar{t}(\bar{x}, \bar{\theta}) = \frac{1}{3} - \bar{x}^2 - 2\bar{\theta} + \frac{4}{\pi^2} \sum_{n=1}^{\infty} \frac{(-1)^n}{n^2} \cos(n\pi\bar{x}) \exp(-n^2\pi^2\bar{\theta}) \qquad (3\cdot1\cdot100)$$

This solution can be checked at $\bar{\theta} = 0$ for $\bar{x} = 0$, 0.5 and 1.0 using the helpful series summations in Appendix C.

Figure 3•20 shows the temperature profile (3·1·100) at three different times. At $\bar{\theta} = 0.01$ the temperatures near $\bar{x} = 1$ ($x = L$) have begun to decrease, but the side of the wall near $\bar{x} = 0$ has not yet started to respond. At $\bar{\theta} = 0.1$ the surface at $\bar{x} = 0$ has just started to decrease. By $\bar{\theta} = 1$ all of the terms in the summation are negligible and the temperature profile has become a "fully established" parabola. As $\bar{\theta}$ continues to increase the temperature continues to decrease, but maintains its parabolic shape since the temperature at each value of $\bar{x}$ decreases at the same rate.

3•1•6 Summary

Separation of variables will only work when the problem is homogeneous. The solution of one-dimensional, transient problems with nonhomogeneous terms that are constant, independent of time, starts by assuming the solution may be written as

$$t(x,\theta) = a(x) + b(\theta) + u(x,\theta)$$

The functions $a(x)$ and $b(\theta)$ are chosen so that the problem for $u(x,\theta)$ will be homogeneous. The solution for $u(x,\theta)$ will always start from some nonzero initial condition $u(x,0)$ and decay to a constant, uniform value (most often = 0) as time goes on.

The function $b(\theta)$ is only needed when there is a net overall energy unbalance for the solid and there is no solid (or ambient) temperature that has been specified. In this case there is no steady-state solution and $t(x,\theta)$ will increase or decrease linearly with time once the starting transient $u(x,\theta)$ has decayed to a constant, uniform value. In this run-away situation, the function $b(\theta)$ will vary linearly with θ.

When the problem is not a run-away, $a(x)$ is the steady-state solution for $t(x,\theta)$. For run-away problems $a(x)$ will give the "fully-established" shape of the temperature profile that increases or decreases with time at a constant rate.

When the nonhomogeneous terms are functions of time, the method of solution we have been discussing will rarely work. Problems with time-dependent nonhomogeneous terms will be treated in the next section and again in Section 4•2.

3•1•7 Time-dependent nonhomogeneous terms

The technique of assuming $t(x,\theta) = a(x) + b(\theta) + u(x,\theta)$ works for steady boundary conditions and steady generation. Problems with time-dependent nonhomogeneous terms require other methods. In this section we will study *variation of parameters*, a method related to separation of variables. Chapter 4 will discuss a method that uses Duhamel's theorem.

The method of variation of parameters consists of the following steps:

Variation of Parameters

1. Set up a problem corresponding to the nonhomogeneous problem but with all the nonhomogeneous terms set equal to zero.

2. Determine the eigenvalues, eigenfunctions and eigencondition for the "corresponding homogeneous problem."

3. Construct a solution to the nonhomogeneous problem of the form

$$t(x,\theta) = \sum_{n=1}^{\infty} A_n(\theta)\Phi_n(x)$$

    where the $\Phi_n(x)$ are the eigenfunctions you have obtained above from the corresponding homogeneous problem.

4. Evaluate the time-dependent Fourier coefficients $A_n(\theta)$ in the usual manner making use of the orthogonality of the $\Phi_n(x)$. That is,

$$\int_{x=0}^{L} t(x,\theta)\Phi_m(x)\,dx = \sum_{n=1}^{\infty} A_n(\theta)\int_{x=0}^{L}\Phi_n(x)\Phi_m(x)\,dx$$

    The integral in the summation is 0 for $n \neq m$ and equal to a constant C for $n = m$. Thus only term m will remain and $A_m(\theta)$ can be found as

$$A_m(\theta) = \frac{1}{C}\int_{x=0}^{L} t(x,\theta)\Phi_m(x)\,dx$$

    However, note that in the integral for $A_m(\theta)$, $t(x,\theta)$ is still unknown.

5. Set up an ordinary differential equation and initial condition for $A_m(\theta)$.

6. Solve for $A_m(\theta)$.

7. Complete the solution by substituting the result for $A_m(\theta)$ into

$$t(x,\theta) = \sum_{n=1}^{\infty} A_n(\theta)\Phi_n(x)$$

Example 1 Suppose a wall, initially having an initial temperature distribution $t^{(0)}(x)$, has a surface temperature at $x = L$ that varies as specified by $t_L(\theta)$. The surface at $x = 0$ is assumed to be adiabatic. The mathematical problem is given in Table 3•13 and shown in Figure 3•21.

Table 3•13 *Mathematical problem description for* $t(x,\theta)$.

Problem	Equation	Boundary/Initial Conditions	
$t(x,\theta)$	$kt_{xx} = \rho c t_\theta$		(3·1·101)
		$t_x(0,\theta) = 0$	(3·1·102)
		$t(L,\theta) = t_L(\theta)$	(3·1·103)
		$t(x,0) = t^{(0)}(x)$	(3·1·104)

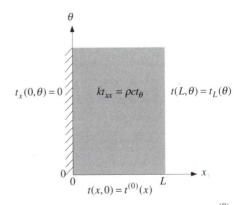

Figure 3•21 *Plane wall, initially at temperature* $t^{(0)}(x)$, *is suddenly exposed to a time-dependent surface temperature at* $x = L$.

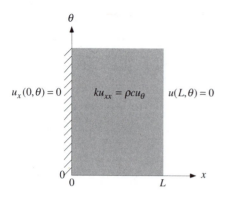

Figure 3•22 *Corresponding homogeneous problem for Figure 3•21.*

The first step is to obtain the eigenvalues and eigenfunctions of the "corresponding homogeneous problem." The corresponding homogeneous problem is obtained by setting the nonhomogeneous terms equal to 0 in the problem for $t(x,\theta)$. Separation of variables can then be used to set up the eigenproblem which can be solved for the eigenvalues and eigenfunctions. The corresponding homogeneous problem is stated in Table 3•14 and shown in Figure 3•22. An initial condition for $u(x,\theta)$ is not given since this problem will only be solved up to the point where the eigenproblem has been solved. Therefore an initial condition is not needed.

Table 3•14 *Corresponding homogeneous problem for $t(x,\theta)$.*

Problem	Equation	Boundary/Initial Conditions	
$u(x,\theta)$	$ku_{xx} = \rho c u_\theta$		(3·1·6)
		$u_x(0,\theta) = 0$	(3·1·9)
		$u(L,\theta) = 0$	(3·1·11)

The homogeneous problem in Table 3•14 was solved in Section 3•1•1. The eigenfunctions are $\cos(\lambda_n x)$ and the eigenvalues are $\lambda_n L = (2n-1)\pi/2$. This is all that we need from the corresponding homogeneous problem.

We will now return to the nonhomogeneous problem for $t(x,\theta)$ and assume the solution can be written as

$$t(x,\theta) = \sum_{n=1}^{\infty} A_n(\theta)\cos(\lambda_n x) \tag{3·1·105}$$

where the eigencondition, $\cos(\lambda_m L) = 0$, gives $\lambda_n L = (2n-1)\pi/2$.

We can now use the orthogonality of the eigenfunctions to evaluate the $A_n(\theta)$ in the usual way. Upon multiplying both sides by $\cos(\lambda_m x)\,dx$ and integrating from $x = 0$ to $x = L$,

$$\int_{x=0}^{L} t(x,\theta)\cos(\lambda_m x)\,dx = \sum_{n=1}^{\infty} A_n(\theta)\int_{x=0}^{L}\cos(\lambda_n x)\cos(\lambda_m x)\,dx$$

The integral on the right-hand side is 0 for $n \neq m$ and is equal to $L/2$ for $n = m$. Therefore only one term in the summation will remain and we may solve for $A_m(\theta)$ and write

$$A_m(\theta) = \frac{2}{L}\int_{x=0}^{L} t(x,\theta)\cos(\lambda_m x)\,dx \tag{3·1·106}$$

The next step is to construct an ordinary differential equation for $A_m(\theta)$. Upon differentiating (3·1·106) with respect to θ,

$$\frac{dA_m}{d\theta} = \frac{2}{L}\int_{x=0}^{L} t_\theta(x,\theta)\cos(\lambda_m x)\,dx$$

From the partial differential equation (3·1·101) for $t(x,\theta)$ we may write $t_\theta(x,\theta) = \alpha t_{xx}(x,\theta)$. Thus,

$$\frac{dA_m}{d\theta} = \frac{2\alpha}{L}\int_{x=0}^{L} t_{xx}(x,\theta)\cos(\lambda_m x)\,dx$$

The right-hand side may be rearranged to prepare for integrating by parts. We may write

$$\frac{dA_m}{d\theta} = \frac{2\alpha}{L} \int_{x=0}^{L} \cos(\lambda_m x) \frac{\partial t_x}{\partial x} dx \tag{3·1·107}$$

To integrate by parts we will let

$$v = \cos(\lambda_m x) \qquad \text{and} \qquad du = \frac{\partial t_x}{\partial x} dx$$

Then

$$dv = -\lambda_m \sin(\lambda_m x)\, dx \qquad \text{and} \qquad u = t_x$$

Integrating the integral in (3·1·107) by parts then gives

$$\frac{dA_m}{d\theta} = \frac{2\alpha}{L} \left\{ [t_x(x,\theta)\cos(\lambda_m x)]_{x=0}^{L} - \int_{x=0}^{L} t_x[-\lambda_m \sin(\lambda_m x)]dx \right\} \tag{3·1·108}$$

The limits can now be substituted into the integrated term. At the upper limit we need to know $t_x(L,\theta)$. Observe that $t_x(L,\theta)$ is not given in this problem; it is unknown. Fortunately, $\cos(\lambda_m L) = 0$ was the eigencondition for this problem. Thus $t_x(L,\theta)$ drops out of the problem. This "fortuitous" event (*i.e.*, the elimination of an unknown function at the boundary) has been designed to happen by cleverly constructing the assumed solution (3·1·105) for $t(x,\theta)$ to contain the eigenfunctions of the corresponding homogeneous problem. At the lower limit, $x = 0$, we see that $\cos(0) = 1$ and from (3·1·102) we see that $t_x(0,\theta) = 0$. Therefore the integrated term is 0 at both limits. Thus we may simplify and rearrange (3·1·108) to give

$$\frac{dA_m}{d\theta} = \frac{2\alpha\lambda_m}{L} \int_{x=0}^{L} \sin(\lambda_m x) t_x\, dx$$

Since things are going smoothly, we will integrate once more by parts to obtain

$$\frac{dA_m}{d\theta} = \frac{2\alpha\lambda_m}{L} \left\{ [t(x,\theta)\sin(\lambda_m x)]_{x=0}^{L} - \int_{x=0}^{L} t(x,\theta)\lambda_m \cos(\lambda_m x)dx \right\}$$

Upon substituting limits for the integrated term and noting that $\sin(0) = 0$ we obtain

$$\frac{dA_m}{d\theta} = \frac{2\alpha\lambda_m}{L} \left\{ t(L,\theta)\sin(\lambda_m L) - \lambda_m \int_{x=0}^{L} t(x,\theta)\cos(\lambda_m x)dx \right\} \tag{3·1·109}$$

The function $t(L,\theta) = t_L(\theta)$ is given as a boundary condition (3·1·103). From (3·1·106) we see that the integral on the right-hand side is equal to $LA_m(\theta)/2$. Thus, (3·1·109) becomes

$$\frac{dA_m}{d\theta} = \frac{2\alpha\lambda_m}{L} \left\{ t_L(\theta)\sin(\lambda_m L) - \frac{\lambda_m L}{2} A_m(\theta) \right\}$$

Upon removing the braces and rearranging this result,

$$\frac{dA_m}{d\theta} + \lambda_m^2 \alpha A_m = \frac{2\alpha\lambda_m}{L} \sin(\lambda_m L) t_L(\theta) \tag{3·1·110}$$

This is a first-order ordinary differential equation for $A_m = A_m(\theta)$. It requires an initial condition. From the definition of A_m given by (3·1·106),

$$A_m(0) = \frac{2}{L} \int_{x=0}^{L} t(x,0)\cos(\lambda_m x)dx \tag{3·1·111}$$

Since the initial condition (3·1·104) for temperature is known, we can evaluate this integral to determine $A_m(0)$.

The next step is to solve (3·1·110). We will use an integrating factor as we did in Section 1•2•5 to solve (1·2·22). The integrating factor is $\exp(\lambda_m^2 \alpha \theta)$. Upon multiplying (3·1·110) by the integrating factor,

$$\exp(\lambda_m^2 \alpha \theta)\frac{dA_m}{d\theta} + \lambda_m^2 \alpha \exp(\lambda_m^2 \alpha \theta)A_m = \frac{2\alpha\lambda_m}{L}\sin(\lambda_m L)t_L(\theta)\exp(\lambda_m^2 \alpha \theta)$$

The left-hand side may be written as an exact derivative to give

$$\frac{d}{d\theta}\Big[\exp(\lambda_m^2 \alpha \theta)A_m\Big] = \frac{2\alpha\lambda_m}{L}\sin(\lambda_m L)t_L(\theta)\exp(\lambda_m^2 \alpha \theta)$$

Next we will multiply by $d\theta$ and integrate from 0 to τ to obtain

$$\int_{\theta=0}^{\tau}d\Big[\exp(\lambda_m^2 \alpha \theta)A_m\Big] = \frac{2\alpha\lambda_m}{L}\sin(\lambda_m L)\int_{\theta=0}^{\tau}t_L(\theta)\exp(\lambda_m^2 \alpha \theta)\,d\theta$$

Integrating the left-hand side and substituting limits,

$$\exp(\lambda_m^2 \alpha \tau)A_m(\tau) - A_m(0) = \frac{2\alpha\lambda_m}{L}\sin(\lambda_m L)\int_{\theta=0}^{\tau}t_L(\theta)\exp(\lambda_m^2 \alpha \theta)\,d\theta$$

Upon solving for $A_m(\tau)$,

$$A_m(\tau) = A_m(0)\exp(-\lambda_m^2 \alpha \tau)$$

$$+ \frac{2\alpha\lambda_m}{L}\sin(\lambda_m L)\exp(-\lambda_m^2 \alpha \tau)\int_{\theta=0}^{\tau}t_L(\theta)\exp(\lambda_m^2 \alpha \theta)\,d\theta$$

Since $t_L(\theta)$ is known, we may evaluate $A_m(\tau)$. Once we have found $A_m(\tau)$, the solution is given by replacing θ by τ and n by m in (3·1·105) and then substituting $A_m(\tau)$ to obtain

$$t(x,\tau) = \sum_{m=1}^{\infty}A_m(0)\exp(-\lambda_m^2 \alpha \tau)\cos(\lambda_m x) + \frac{2\alpha}{L}\sum_{m=1}^{\infty}\lambda_m \sin(\lambda_m L)\exp(-\lambda_m^2 \alpha \tau)\cos(\lambda_m x)\int_{\theta=0}^{\tau}t_L(\theta)\exp(\lambda_m^2 \alpha \theta)\,d\theta$$

$$(3·1·112)$$

where $\lambda_n L = (2n-1)\pi/2$ and $A_m(0)$ is obtained by substituting the initial condition $t(x,0) = t^{(0)}(x)$ into (3·1·111).

There are two unsettling concerns with the form of (3·1·112):

1. The second series is slow to converge due to λ_m in the numerator.
2. At $x = L$, $\cos(\lambda_m L) = 0$ and thus it is hard to believe that the boundary condition $t(L,\tau) = t_L(\tau)$ is satisfied.

These concerns are inherent in variation of parameters because we solved the nonhomogeneous problem using eigenfunctions of a homogeneous problem. We can overcome these concerns by integrating by parts to obtain

$$\int_{\theta=0}^{\tau}t_L(\theta)\exp(\lambda_m^2 \alpha \theta)\,d\theta = \left[t_L(\theta)\frac{1}{\lambda_m^2 \alpha}\exp(\lambda_m^2 \alpha \theta)\right]_{\theta=0}^{\tau}$$

$$- \int_{\theta=0}^{\tau}\frac{1}{\lambda_m^2 \alpha}\exp(\lambda_m^2 \alpha \theta)\dot{t}_L(\theta)\,d\theta$$

where $\dot{t}_L(\theta)$ denotes the time derivative of $t_L(\theta)$. Upon factoring out $\lambda_m^2 \alpha$ and substituting limits for the integrated term,

$$\int_{\theta=0}^{\tau}t_L(\theta)\exp(\lambda_m^2 \alpha \theta)\,d\theta = \frac{1}{\lambda_m^2 \alpha}\Big[t_L(\tau)\exp(\lambda_m^2 \alpha \tau) - t_L(0)$$

$$- \int_{\theta=0}^{\tau}\exp(\lambda_m^2 \alpha \theta)\dot{t}_L(\theta)\,d\theta\Big]$$

This integral may be substituted back into (3·1·112) to obtain

$$t(x,\tau) = \sum_{m=1}^{\infty} A_m(0)\exp(-\lambda_m^2\alpha\tau)\cos(\lambda_m x)$$

$$+\frac{2\alpha}{L}\sum_{m=1}^{\infty}\lambda_m\sin(\lambda_m L)\exp(-\lambda_m^2\alpha\tau)\cos(\lambda_m x)\frac{1}{\lambda_m^2\alpha}\left[t_L(\tau)\exp(\lambda_m^2\alpha\tau)-t_L(0)-\int_{\theta=0}^{\tau}\exp(\lambda_m^2\alpha\theta)\dot{t}_L(\theta)d\theta\right]$$

Upon simplifying this result and writing the second sum in three parts,

$$t(x,\tau) = \sum_{m=1}^{\infty} A_m(0)\exp(-\lambda_m^2\alpha\tau)\cos(\lambda_m x)+2t_L(\tau)\sum_{m=1}^{\infty}\frac{\sin(\lambda_m L)}{\lambda_m L}\cos(\lambda_m x)$$

$$-2t_L(0)\sum_{m=1}^{\infty}\frac{\sin(\lambda_m L)}{\lambda_m L}\cos(\lambda_m x)\exp(-\lambda_m^2\alpha\tau)$$

$$-2\sum_{m=1}^{\infty}\frac{\sin(\lambda_m L)}{\lambda_m L}\cos(\lambda_m x)\exp(-\lambda_m^2\alpha\tau)\int_{\theta=0}^{\tau}\exp(\lambda_m^2\alpha\theta)\dot{t}_L(\theta)d\theta$$

$$(3\cdot1\cdot113)$$

Convergence has improved since the λ_m in the numerator has canceled out. Three of the four series have a decaying exponential that helps convergence. The sum without an exponential is a slowly convergent series. Fortunately we can obtain the sum of this series without having to actually add a large number of terms. If $t_L(\tau)$ were equal to t_L (a constant) the steady-state solution of this problem would be $t(x,\infty) = t_L$. For steady state, each of the decaying exponentials in (3·1·113) is 0 and (3·1·113) reduces to

$$t(x,\infty) = 2t_L\sum_{m=1}^{\infty}\frac{\sin(\lambda_m L)}{\lambda_m L}\cos(\lambda_m x)$$

Since $t(x,\infty) = t_L$ we see that sum of this series must be $1/2$ for all x. Since this series is only a function of x, its sum does not depend upon the clever analysis made to arrive at this conclusion.[1] We can therefore substitute this result into (3·1·113) to obtain

$$t(x,\tau) = \sum_{m=1}^{\infty} A_m(0)\exp(-\lambda_m^2\alpha\tau)\cos(\lambda_m x)+t_L(\tau)$$

$$-2t_L(0)\sum_{m=1}^{\infty}\frac{\sin(\lambda_m L)}{\lambda_m L}\cos(\lambda_m x)\exp(-\lambda_m^2\alpha\tau)$$

$$-2\sum_{m=1}^{\infty}\frac{\sin(\lambda_m L)}{\lambda_m L}\cos(\lambda_m x)\exp(-\lambda_m^2\alpha\tau)\int_{\theta=0}^{\tau}\exp(\lambda_m^2\alpha\theta)\dot{t}_L(\theta)d\theta$$

$$(3\cdot1\cdot114)$$

The series with the slowest convergence has been eliminated. At $x = L$ the cosine terms are each 0 as required by the eigencondition $\cos(\lambda_m L) = 0$. Thus at $x = L$, (3·1·114) reduces to $t(L,\tau) = t_L(\tau)$ as required by the boundary condition (3·1·103).

The original form of the solution (3·1·112) required the complete past history of $t_L(\theta)$ from $\theta = 0$ to time τ. The revised solution (3·1·114) requires $t_L(\theta)$ only at $\theta = \tau$, but needs the complete past history of the time derivative $\dot{t}_L(\theta)$ from $\theta = 0$ to time τ.

---

[1] We could also reach this conclusion by looking at the initial condition for the problem discussed in Section 3•1•1.

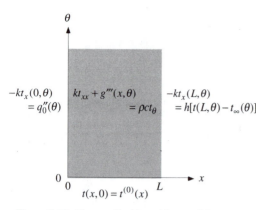

Figure 3·23 *Plane wall with position- and time-dependent generation $g'''(x,\theta)$, initially at temperature $t^{(0)}(x)$, is suddenly exposed to a time-dependent heat flux $q_0''(\theta)$ at $x = 0$ and a time-dependent ambient temperature $t_\infty(\theta)$ at $x = L$.*

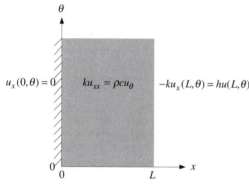

Figure 3·24 *Corresponding homogeneous problem for Figure 3·23.*

Example 2 The plane wall shown in Figure 3·23 has a specified position- and time-dependent generation rate $g'''(x,\theta)$. At $x = 0$ there is a specified time-dependent heat-flux input $q_0''(\theta)$. At $x = L$ the wall convects to an ambient that has a specified ambient temperature $t_\infty(\theta)$. The initial temperature distribution in the wall is $t^{(0)}(x)$. This example has three time-dependent nonhomogeneous terms. The mathematical problem is given in Table 3·15.

Table 3·15 *Mathematical problem description for $t(x,\theta)$.*

Problem	Equation	Boundary/Initial Conditions	
$t(x,\theta)$	$kt_{xx} + g'''(x,\theta) = \rho ct_\theta$		(3·1·115)
		$-kt_x(0,\theta) = q_0''(\theta)$	(3·1·116)
		$-kt_x(L,\theta) = h[t(x,\theta) - t_\infty(\theta)]$	(3·1·117)
		$t(x,0) = t^{(0)}(x)$	(3·1·118)

The first step is to obtain the "corresponding homogeneous problem" by setting the nonhomogeneous terms equal to 0 in the problem for $t(x,\theta)$. The corresponding homogeneous problem is stated in Table 3·16 and pictured in Figure 3·24. An initial condition for $u(x,\theta)$ is not given since this problem will only be solved up to the point where the eigenproblem has been solved. Therefore an initial condition is not needed.

Table 3·16 *Corresponding homogeneous problem for $t(x,\theta)$.*

Problem	Equation	Boundary/Initial Conditions	
$u(x,\theta)$	$ku_{xx} = \rho cu_\theta$		(3·1·115)
		$u_x(0,\theta) = 0$	(3·1·116)
		$-ku_x(L,\theta) = hu(x,\theta)$	(3·1·117)

The homogeneous problem in Table 3·16 was solved in Section 3·1·4. The eigenfunctions are $\cos(\lambda_n x)$ and the eigenvalues are the solutions of

$$\lambda_n L\sin(\lambda_n L) - H\cos(\lambda_n L) = 0 \qquad \text{where } H = hL/k \qquad (3·1·118)$$

This is all that we need from the corresponding homogeneous problem.

We will now return to the nonhomogeneous problem for $t(x,\theta)$ and assume the solution can be written as

$$t(x,\theta) = \sum_{n=1}^{\infty} A_n(\theta)\cos(\lambda_n x) \qquad (3·1·119)$$

where the eigenvalues are the solutions of $\lambda_n L\tan(\lambda_n L) = H$.

Upon multiplying both sides by $\cos(\lambda_m x)\,dx$ and integrating from $x = 0$ to $x = L$,

$$\int_{x=0}^{L} t(x,\theta)\cos(\lambda_m x)\,dx = \sum_{n=1}^{\infty} A_n(\theta)\int_{x=0}^{L}\cos(\lambda_n x)\cos(\lambda_m x)\,dx$$

The integral on the right-hand side is 0 for $n \neq m$ and is given by (3·1·65) for $n = m$. Therefore only one term in the summation will remain and we may write

$$\int_{x=0}^{L} t(x,\theta)\cos(\lambda_m x)\,dx = A_m(\theta)\frac{(\lambda_m L)^2 + H\cos^2(\lambda_m L)}{2\lambda_m^2 L} \tag{3·1·120}$$

Multiplying both sides by $2/L$ and defining $B_m(\theta)$ as

$$B_m(\theta) = A_m(\theta)\frac{(\lambda_m L)^2 + H\cos^2(\lambda_m L)}{(\lambda_m L)^2} \tag{3·1·121}$$

we may then write (3·1·120) as

$$B_m(\theta) = \frac{2}{L}\int_{x=0}^{L} t(x,\theta)\cos(\lambda_m x)\,dx \tag{3·1·122}$$

The next step is to construct an ordinary differential equation for $B_m(\theta)$. Differentiating (3·1·122) with respect to θ and then multiplying and dividing by ρc,

$$\frac{dB_m}{d\theta} = \frac{2}{\rho c L}\int_{x=0}^{L}\rho c\, t_\theta(x,\theta)\cos(\lambda_m x)\,dx$$

From the partial differential equation (3·1·115) for $t(x,\theta)$ we may replace $\rho c\, t_\theta(x,\theta)$ by $k t_{xx}(x,\theta) + g'''(x,\theta)$. Thus,

$$\frac{dB_m}{d\theta} = \frac{2}{\rho c L}\int_{x=0}^{L}\left[k t_{xx}(x,\theta) + g'''(x,\theta)\right]\cos(\lambda_m x)\,dx$$

Removing the brackets and writing the right-hand side as two integrals,

$$\frac{dB_m}{d\theta} = \frac{2\alpha}{L}\int_{x=0}^{L} t_{xx}(x,\theta)\cos(\lambda_m x)\,dx + \frac{2}{\rho c L}\int_{x=0}^{L} g'''(x,\theta)\cos(\lambda_m x)\,dx$$

Since $g'''(x,\theta)$ is a specified function the second integral on the right can be evaluated. The first integral will be integrated by parts to give

$$\frac{dB_m}{d\theta} = \frac{2\alpha}{L}\left\{\left[t_x(x,\theta)\cos(\lambda_m x)\right]_{x=0}^{L} - \int_{x=0}^{L} t_x[-\lambda_m \sin(\lambda_m x)]\,dx\right\}$$

$$+ \frac{2}{\rho c L}\int_{x=0}^{L} g'''(x,\theta)\cos(\lambda_m x)\,dx$$

The limits can now be substituted into the integrated term to give

$$\frac{dB_m}{d\theta} = \frac{2\alpha}{L}\left\{t_x(L,\theta)\cos(\lambda_m L) - t_x(0,\theta) + \lambda_m \int_{x=0}^{L} t_x \sin(\lambda_m x)\,dx\right\}$$

$$+ \frac{2}{\rho c L}\int_{x=0}^{L} g'''(x,\theta)\cos(\lambda_m x)\,dx$$

In Example 1 both of the integrated terms were 0. Here, neither term is 0. From the boundary condition (3·1·117) at $x = L$, we will replace $t_x(L,\theta)$ by $-h[t(x,\theta) - t_\infty(\theta)]/k$. Using the boundary condition (3·1·116) at $x = 0$, we will replace $t_x(0,\theta)$ by $-q_0''(\theta)/k$. Upon making these substitutions and also integrating the first integral by parts once again,

$$\frac{dB_m}{d\theta} = \frac{2\alpha}{L}\left\{-\frac{h}{k}[t(L,\theta) - t_\infty(\theta)]\cos(\lambda_m L) + \frac{1}{k}q_0''(\theta)\right\}$$

$$+ \frac{2\alpha}{L}\lambda_m\left\{\left[t(x,\theta)\sin(\lambda_m x)\right]_{x=0}^{L} - \int_{x=0}^{L} t(x,\theta)\lambda_m \cos(\lambda_m x)\,dx\right\}$$

$$+ \frac{2}{\rho c L}\int_{x=0}^{L} g'''(x,\theta)\cos(\lambda_m x)\,dx$$

Substituting limits on the integrated term and rearranging,

$$\frac{dB_m}{d\theta} = -\frac{2h}{\rho cL}[t(L,\theta) - t_\infty(\theta)]\cos(\lambda_m L) + \frac{2}{\rho cL}q_0''(\theta)$$

$$+ \frac{2\alpha}{L}\lambda_m t(L,\theta)\sin(\lambda_m L) - \frac{2\alpha}{L}\lambda_m^2 \int_{x=0}^{L} t(x,\theta)\cos(\lambda_m x)\,dx$$

$$+ \frac{2}{\rho cL}\int_{x=0}^{L} g'''(x,\theta)\cos(\lambda_m x)\,dx$$

The two terms containing $t(L,\theta)$ may be grouped together. From (3·1·122) we see that the first integral on the right-hand side is equal to $LB_m(\theta)/2$. We may then write

$$\frac{dB_m}{d\theta} = t(L,\theta)\left[-\frac{2h}{\rho cL}\cos(\lambda_m L) + \frac{2\alpha}{L}\lambda_m \sin(\lambda_m L)\right]$$

$$+ \frac{2h}{\rho cL}t_\infty(\theta)\cos(\lambda_m L) + \frac{2}{\rho cL}q_0''(\theta) - \frac{2\alpha}{L}\lambda_m^2 \frac{LB_m(\theta)}{2}$$

$$+ \frac{2}{\rho cL}\int_{x=0}^{L} g'''(x,\theta)\cos(\lambda_m x)\,dx$$

Next we will factor $2\alpha/L^2$ out of the brackets and move the term with $B_m(\theta)$ to the left-hand side of the equation to give

$$\frac{dB_m}{d\theta} + \lambda_m^2 \alpha B_m(\theta) = \frac{2\alpha}{L^2}t(L,\theta)\left[-\frac{hL}{k}\cos(\lambda_m L) + \lambda_m L\sin(\lambda_m L)\right]$$

$$+ \frac{2h}{\rho cL}t_\infty(\theta)\cos(\lambda_m L) + \frac{2}{\rho cL}q_0''(\theta)$$

$$+ \frac{2}{\rho cL}\int_{x=0}^{L} g'''(x,\theta)\cos(\lambda_m x)\,dx$$

From the eigencondition (3·1·118) we see that the term in brackets is 0. The unknown temperature $t(L,\theta)$ drops out of the equation! Nice things should always happen like this in variation of parameters. If they don't you have probably made a mistake. The differential equation then reduces to give

$$\frac{dB_m}{d\theta} + \lambda_m^2 \alpha B_m(\theta) = \frac{2h}{\rho cL}t_\infty(\theta)\cos(\lambda_m L) + \frac{2}{\rho cL}q_0''(\theta)$$

$$+ \frac{2}{\rho cL}\int_{x=0}^{L} g'''(x,\theta)\cos(\lambda_m x)\,dx$$

We have arrived at a nonhomogeneous ordinary differential equation for $B_m(\theta)$. The right-hand side is a function of θ that can be evaluated since the time-dependent nonhomogeneous terms, $t_\infty(\theta)$, $q_0''(\theta)$ and $g'''(x,\theta)$, are all specified. The initial condition for $B_m(\theta)$ can be found from (3·1·122) as

$$B_m(0) = \frac{2}{L}\int_{x=0}^{L} t(x,0)\cos(\lambda_m x)\,dx = \frac{2}{L}\int_{x=0}^{L} t^{(0)}(x)\cos(\lambda_m x)\,dx$$

The solution for $B_m(\theta)$ can then be found and substituted into (3·1·121) to obtain $A_m(\theta)$ which, in turn, can then be substituted into (3·1·119) to give the solution for $t(x,\theta)$.

3•2 TWO-DIMENSIONAL PROBLEMS

The solution of two-dimensional problems using separation of variables requires that the region have a relatively simple shape. The boundaries of the region must lie along constant coordinate lines. In the rectangular coordinate system, we are restricted to squares and rectangles. The treatment of irregular two-dimensional regions is best left to numerical computer methods.[1]

The region of interest in sections 3•2•1 and 3•2•2 will be square. Rather than treating a variety of boundary conditions as we did in Section 3•1, these two sections are only intended to illustrate how the techniques of Section 3•1 can be applied to two-dimensional problems (if the region has a simple shape). Section 3•2•1 will consider a steady-state example. Section 3•2•2 will treat a transient problem.

3•2•1 Square region, steady-state energy generation

Consider the square region shown in Figure 3•25. Since the partial differential equation is second order in x and second order in y, two x-boundary conditions and two y-boundary conditions are needed. In this problem the differential equation is nonhomogeneous due to the generation term. The boundary conditions along $x = L$ and along $y = L$ are nonhomogeneous since, in general, $t_L \neq 0$. The mathematical description of the problem is given in Table 3•17.

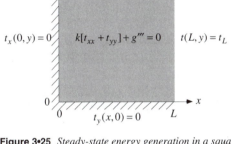

Figure 3•25 *Steady-state energy generation in a square region.*

Table 3•17 *Mathematical problem description for $t(x,y)$.*

Problem	Equation	Boundary Conditions	
$t(x,y)$	$k[t_{xx} + t_{yy}] + g''' = 0$		(3·2·1)
		$t_x(0,y) = 0$	(3·2·2)
		$t(L,y) = t_L$	(3·2·3)
		$t_y(x,0) = 0$	(3·2·4)
		$t(x,L) = t_L$	(3·2·5)

As we have seen earlier in the chapter, nonhomogeneous terms also cause difficulty in using the method of separation of variables for two-dimensional problems. The partial differential equation must be homogeneous. In two-dimensional, steady-state problems either both x-boundary conditions must be homogeneous or both y-boundary conditions. We will provide the necessary homogeneity by assuming the solution can be written as

$$t(x,y) = a(x) + b(y) + u(x,y) \qquad (3 \cdot 2 \cdot 6)$$

A solution of this form will be helpful in only a few cases. Not all boundary conditions can be accommodated. The idea is to try it and see if it works. If it doesn't you may be out of luck as far as obtaining a solution by separation of variables.

Upon substituting (3·2·6) into the partial differential equation (3·2·1),

$$k[a''(x) + u_{xx} + b''(y) + u_{yy}] + g''' = 0$$

To obtain a homogeneous equation for $u(x,y)$ we will take

---

[1] The finite-element method is developed in Chapter 9 for two-dimensional problems.

$$u_{xx} + u_{yy} = 0 \qquad (3\cdot2\cdot7)$$

This means that

$$k[a''(x) + b''(y)] + g''' = 0$$

For constant g''' we can satisfy this equation by taking

$$ka''(x) + g''' = 0 \qquad (3\cdot2\cdot8)$$

and

$$b''(y) = 0 \qquad (3\cdot2\cdot9)$$

Substituting (3·2·6) into the boundary condition (3·2·2) along $x = 0$ gives

$$a'(0) + u_x(0, y) = 0$$

We can obtain a homogeneous boundary condition for $u(x, y)$ by taking

$$u_x(0, y) = 0 \qquad (3\cdot2\cdot10)$$

This means that we must also take

$$a'(0) = 0 \qquad (3\cdot2\cdot11)$$

Substituting (3·2·6) into the boundary condition (3·2·3) along $x = L$ gives

$$a(L) + b(y) + u(L, y) = t_L$$

We can obtain a homogeneous boundary condition for $u(x, y)$ by taking

$$u(L, y) = 0 \qquad (3\cdot2\cdot12)$$

This means that we must also satisfy

$$a(L) + b(y) = t_L$$

We see from this that $b(y)$ must be a constant. This will also satisfy (3·2·9). The easiest thing to do is to take

$$a(L) = t_L \qquad (3\cdot2\cdot13)$$

and

$$b(y) = 0 \qquad (3\cdot2\cdot14)$$

Substituting (3·2·6) into the boundary condition (3·2·4) along $y = 0$ gives

$$b'(0) + u_y(x, 0) = 0$$

Since we have taken $b(y) = 0$ in (3·2·14), we see that $b'(0) = 0$ and thus,

$$u_y(x, 0) = 0 \qquad (3\cdot2\cdot15)$$

Substituting (3·2·6) into the boundary condition (3·2·5) along $y = L$ gives

$$a(x) + b(L) + u(x, L) = t_L$$

Upon solving for $u(x, L)$ and setting $b(L) = 0$ from (3·2·14),

$$u(x, L) = t_L - a(x) \qquad (3\cdot2\cdot16)$$

The subproblems for $a(x)$, $b(y)$ and $u(x, y)$ are summarized in Table 3•18. In the problem for $u(x, y)$, the partial differential equation (3·2·7) and the two x-boundary conditions (3·2·10) and (3·2·12) are homogeneous.

Since the two x-boundary conditions are homogeneous, neither of the y-boundary conditions needs to be homogeneous (although one is). The "partially homogeneous" problem for $u(x,y)$ is pictured in Figure 3•26.

Table 3•18 *Subproblems for* $t(x,y) = a(x) + b(y) + u(x,y)$.

Subproblem	Equation	Boundary Conditions	
$a(x)$	$ka''(x) + g''' = 0$		(3·2·8)
		$a'(0) = 0$	(3·2·11)
		$a(L) = t_L$	(3·2·13)
$b(y)$	$b''(y) = 0$		(3·2·9)
		$b(y) = 0$	(3·2·14)
$u(x,y)$	$u_{xx} + u_{yy} = 0$		(3·2·7)
		$u_x(0,y) = 0$	(3·2·10)
		$u(L,y) = 0$	(3·2·12)
		$u_y(x,0) = 0$	(3·2·15)
		$u(x,L) = t_L - a(x)$	(3·2·16)

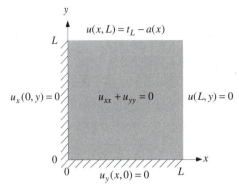

Figure 3•26 *Partially homogeneous problem for* $u(x,y)$ *corresponding to Figure 3•25.*

The solution to (3·2·8) may be obtained by integrating twice and using (3·2·11) and (3·2·13) as boundary conditions. The result is given by

$$a(x) = t_L + \frac{g'''}{2k}(L^2 - x^2) \tag{3·2·17}$$

This result may be substituted into (3·2·16) to give

$$u(x,L) = t_L - t_L - \frac{g'''}{2k}(L^2 - x^2) = \frac{g'''}{2k}(x^2 - L^2) \tag{3·2·18}$$

Equation (3·2·14) is the solution for $b(y)$ and it satisfies (3·2·9). Equations (3·2·17) and (3·2·14) can be substituted into (3·2·6) to obtain

$$t(x,y) = t_L + \frac{g'''}{2k}(L^2 - x^2) + u(x,y) \tag{3·2·19}$$

For convenience the problems for $t(x,y)$ and $u(x,y)$ are summarized in Table 3•19. The updated partially homogeneous problem for $u(x,y)$ is now shown in Figure 3•27.

Table 3•19 *Subproblems for* $t(x,y)$ *and* $u(x,y)$.

Subproblem	Equation	Boundary Conditions	
$t(x,y) = t_L + \frac{g'''}{2k}(L^2 - x^2) + u(x,y)$			(3·2·19)
$u(x,y)$	$u_{xx} + u_{yy} = 0$		(3·2·7)
		$u_x(0,y) = 0$	(3·2·10)
		$u(L,y) = 0$	(3·2·12)
		$u_y(x,0) = 0$	(3·2·15)
		$u(x,L) = \frac{g'''}{2k}(x^2 - L^2)$	(3·2·18)

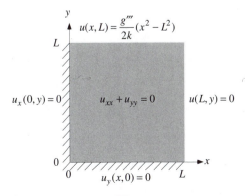

Figure 3•27 *Updated partially homogeneous problem for* $u(x,y)$ *corresponding to Figure 3•25.*

Separation of variables will now be used to find $u(x, y)$. The method starts by assuming that the variables can be separated and the solution can be written as

$$u(x, y) = X(x)Y(y) \tag{3.2.20}$$

Substitution of this assumed solution into the partial differential equation (3.2.7) gives

$$X''(x)Y(y) + X(x)Y''(y) = 0$$

Upon rearranging to separate the variables,

$$\frac{X''(x)}{X(x)} = -\frac{Y''(y)}{Y(y)}$$

The only way that a function of x (left-hand side) can be equal to a function of y (right-hand side) for all x and y is if both sides are equal to the same constant. Since the x-boundary conditions are both homogeneous, we will choose the constant so that the solution for $X(x)$ will be oscillatory. This will give an infinite number of eigenvalues and eigenfunctions that can be used to satisfy the nonhomogeneous boundary conditions in the y direction. Thus, we will take the constant to be $-\lambda^2$ to give

$$\frac{X''(x)}{X(x)} = -\lambda^2 \qquad \text{or} \qquad X''(x) + \lambda^2 X(x) = 0 \tag{3.2.21}$$

and

$$-\frac{Y''(y)}{Y(y)} = -\lambda^2 \qquad \text{or} \qquad Y''(y) - \lambda^2 Y(y) = 0 \tag{3.2.22}$$

Upon substituting (3.2.20) into the boundary condition (3.2.10) along $x = 0$ we obtain

$$X'(0)Y(y) = 0$$

To satisfy this condition we will take

$$X'(0) = 0 \tag{3.2.23}$$

Substituting (3.2.20) into the boundary condition (3.2.12) along $x = L$,

$$X(L)Y(y) = 0$$

We will satisfy this relation by taking

$$X(L) = 0 \tag{3.2.24}$$

The eigenproblem for $X(x)$ is given by (3.2.21), (3.2.23) and (3.2.24). The general solution to (3.2.21) is given by

$$X(x) = A\sin(\lambda x) + B\cos(\lambda x)$$

To satisfy (3.2.23) we must take $A = 0$. To satisfy (3.2.24) requires that $\cos(\lambda L) = 0$ which gives $\lambda L =$ odd multiples of $\pi/2$. Therefore the eigenfunctions are given by

$$X_n(x) = B_n \cos(\lambda_n L) \qquad \text{where} \qquad \lambda_n L = (2n-1)\frac{\pi}{2} \tag{3.2.25}$$

The general solution to (3.2.22) is given by

$$Y_n(y) = C_n \sinh(\lambda_n y) + D_n \cosh(\lambda_n y) \tag{3.2.26}$$

Equations (3.2.25) and (3.2.26) may be substituted into (3.2.20) to give

$$u_n(x,y) = \cos(\lambda_n x)\left[C_n \sinh(\lambda_n y) + D_n \cosh(\lambda_n y)\right]$$

where the B_n has been "absorbed" into the C_n and D_n. This solution satisfies the partial differential equation (3·2·7) and the x-boundary conditions (3·2·10) and (3·2·12). A more general solution may be obtained by summing over all of the eigenvalues to give

$$u(x,y) = \sum_{n=1}^{\infty} \cos(\lambda_n x)\left[C_n \sinh(\lambda_n y) + D_n \cosh(\lambda_n y)\right] \tag{3·2·27}$$

The C_n and D_n will be chosen to make (3·2·27) satisfy the y-boundary conditions (3·2·15) and (3·2·18).

Upon differentiating (3·2·27) with respect to y

$$u_y(x,y) = \sum_{n=1}^{\infty} \cos(\lambda_n x)\left[C_n \lambda_n \cosh(\lambda_n y) + D_n \lambda_n \sinh(\lambda_n y)\right]$$

and then setting $y = 0$ and substituting into (3·2·15),

$$\sum_{n=1}^{\infty} \cos(\lambda_n x) C_n \lambda_n = 0$$

The only way this can be satisfied for all x is to take each $C_n = 0$. We could also arrive at this conclusion by multiplying both sides by $\cos(\lambda_m x)\,dx$ and then integrating from $x = 0$ to $x = L$. Due to orthogonality, only the term for $n = m$ would remain in the sum. The right-hand side is 0. Thus, $C_n = 0$. Therefore (3·2·27) reduces to

$$u(x,y) = \sum_{n=1}^{\infty} \cos(\lambda_n x) D_n \cosh(\lambda_n y) \tag{3·2·28}$$

Upon setting $y = L$ and substituting (3·2·28) into (3·2·18),

$$\sum_{n=1}^{\infty} \cos(\lambda_n x) D_n \cosh(\lambda_n L) = \frac{g'''}{2k}(x^2 - L^2)$$

To find D_m we will multiply both sides by $\cos(\lambda_m x)\,dx$ and then integrate from $x = 0$ to $x = L$. This gives

$$\int_{x=0}^{L} \sum_{n=1}^{\infty} \cos(\lambda_n x) D_n \cosh(\lambda_n L)\cos(\lambda_m x)\,dx = \int_{x=0}^{L} \frac{g'''}{2k}(x^2 - L^2)\cos(\lambda_m x)\,dx$$

From orthogonality, only the term for $n = m$ will remain in the sum. Then, upon also factoring $D_m \cosh(\lambda_m L)$ out of the integral on the left-hand side,

$$D_m \cosh(\lambda_m L)\int_{x=0}^{L} \cos^2(\lambda_m x)\,dx = \int_{x=0}^{L} \frac{g'''}{2k}(x^2 - L^2)\cos(\lambda_m x)\,dx$$

Evaluating these integrals and then solving for D_m gives

$$D_m = 2\frac{g''' L^2}{k}\frac{(-1)^m}{(\lambda_m L)^3 \cosh(\lambda_m L)}$$

These Fourier coefficients may be substituted into (3·2·28) to give $u(x,y)$ which may then be substituted into (3·2·19) to give

$$t(x,y) = t_L + \frac{g'''}{2k}(L^2 - x^2) + 2\frac{g''' L^2}{k}\sum_{n=1}^{\infty}\frac{(-1)^n \cos(\lambda_n x)\cosh(\lambda_n y)}{(\lambda_n L)^3 \cosh(\lambda_n L)} \tag{3·2·29}$$

The heat flux in the x direction may be found from the rate equation as

$$q_x''(x,y) = -kt_x(x,y)$$

Differentiating (3·2·29) with respect to x gives

$$t_x(x,y) = -\frac{g'''}{2k}2x + 2\frac{g'''L^2}{k}\sum_{n=1}^{\infty}\frac{(-1)^{n+1}\lambda_n\sin(\lambda_n x)\cosh(\lambda_n y)}{(\lambda_n L)^3\cosh(\lambda_n L)}$$

Simplifying and multiplying this gradient by $-k$ to obtain the heat flux,

$$q_x''(x,y) = g'''x + 2g'''L\sum_{n=1}^{\infty}\frac{(-1)^n\sin(\lambda_n x)\cosh(\lambda_n y)}{(\lambda_n L)^2\cosh(\lambda_n L)} \qquad (3\cdot2\cdot30)$$

Note that the series convergence in (3·2·30) is not as good as the series in (3·2·29) since (3·2·30) only has λ_n^2 in the denominator whereas (3·2·29) has λ_n^3 in the denominator. Near $y = 0$ the $\cosh(\lambda_n L)$ in the denominator helps both (3·2·29) and (3·2·30) to converge. Along $y = L$ the hyperbolic functions cancel and $\cosh(\lambda_n L)$ in the denominator is no longer present to help convergence.

The heat flux in the y direction may be found from the rate equation as

$$q_y''(x,y) = -kt_y(x,y)$$

Differentiating (3·2·29) with respect to y gives

$$t_y(x,y) = 2\frac{g'''L^2}{k}\sum_{n=1}^{\infty}\frac{(-1)^n\cos(\lambda_n x)\lambda_n\sinh(\lambda_n y)}{(\lambda_n L)^3\cosh(\lambda_n L)}$$

Simplifying and multiplying this gradient by $-k$ to obtain the heat flux,

$$q_y''(x,y) = 2g'''L\sum_{n=1}^{\infty}\frac{(-1)^{n+1}\cos(\lambda_n x)\sinh(\lambda_n y)}{(\lambda_n L)^2\cosh(\lambda_n L)} \qquad (3\cdot2\cdot31)$$

Before making numerical calculations it is convenient to first normalize (3·2·29), (3·2·30) and (3·2·31). We will define

$$\bar{x} = \frac{x}{L} \quad , \quad \bar{y} = \frac{y}{L} \quad , \quad \bar{t}(\bar{x},\bar{y}) = \frac{t(x,y)-t_L}{g'''L^2/k} \quad \text{and} \quad \bar{q}''(\bar{x},\bar{y}) = \frac{q''(x,y)}{g'''L}$$

With these definitions, temperature (3·2·29) becomes

$$\bar{t}(\bar{x},\bar{y}) = \frac{1}{2}(1-\bar{x}^2) + 2\sum_{n=1}^{\infty}\frac{(-1)^n\cos[(\lambda_n L)\bar{x}]\cosh[(\lambda_n L)\bar{y}]}{(\lambda_n L)^3\cosh(\lambda_n L)} \qquad (3\cdot2\cdot32)$$

The two heat fluxes (3·2·30) and (3·2·31) may be written as

$$\bar{q}_x''(\bar{x},\bar{y}) = \bar{x} + 2\sum_{n=1}^{\infty}\frac{(-1)^n\sin[(\lambda_n L)\bar{x}]\cosh[(\lambda_n L)\bar{y}]}{(\lambda_n L)^2\cosh(\lambda_n L)} \qquad (3\cdot2\cdot33)$$

$$\bar{q}_y''(\bar{x},\bar{y}) = 2\sum_{n=1}^{\infty}\frac{(-1)^{n+1}\cos[(\lambda_n L)\bar{x}]\sinh[(\lambda_n L)\bar{y}]}{(\lambda_n L)^2\cosh(\lambda_n L)} \qquad (3\cdot2\cdot34)$$

Equations (3·2·32), (3·2·33) and (3·2·34) have been evaluated at several points. The results are given in Table 3•20.

A graph of the solution can be helpful. Curves of $\bar{t}(\bar{x},\bar{y}) = 0.0$, 0.1 and 0.2 (temperature contours) are shown in Figure 3•28. The quantities $\bar{q}_x''$ and $\bar{q}_y''$ are the x and y components of the nondimensional heat-flux vector. The vector sums of these heat-flux components are shown for several points lying on the two temperature contours are also shown in Figure 3•28. The heat-flux arrows are perpendicular to the temperature contours and their lengths are proportional to their magnitudes.

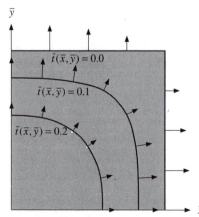

Figure 3•28 *Temperature contours and heat-flux vectors from (3·2·32), (3·2·33) and (3·2·34).*

Table 3•20 *Evaluation of (3·2·32), (3·2·33) and (3·2·34) at 15 locations.*

$\bar{x}$	$\bar{y}$	$\bar{t}(\bar{x},\bar{y})$	$\bar{q}_x''(\bar{x},\bar{y})$	$\bar{q}_y''(\bar{x},\bar{y})$
0.00	0.00	0.2947	0.0000	0.0000
0.25	0.00	0.2789	0.1278	0.0000
0.50	0.00	0.2293	0.2727	0.0000
0.75	0.00	0.1397	0.4509	0.0000
1.00	0.00	0.0000	0.6753	0.0000
0.25	0.25	0.2641	0.1193	0.1193
0.50	0.25	0.2178	0.2558	0.0936
0.75	0.25	0.1333	0.4272	0.0521
1.00	0.25	0.0000	0.6488	0.0000
0.50	0.50	0.1811	0.2039	0.2039
0.75	0.50	0.1127	0.3517	0.1158
1.00	0.50	0.0000	0.5628	0.0000
0.75	0.75	0.0728	0.2117	0.2117
1.00	0.75	0.0000	0.3918	0.0000
1.00	1.00	0.0000	0.0000	0.0000

Another check on the solution is to calculate the total heat flow per unit depth q' across the boundary along $x = L$. This may be found by setting $x = L$ in (3·2·30) and then integrating from $y = 0$ to $y = L$. This gives

$$q'\big|_{x=L} = \int_{y=0}^{L} q_x''(L,y)\,dy$$

$$= \int_{y=0}^{L}\left[g'''L + 2g'''L\sum_{n=1}^{\infty}\frac{(-1)^n \sin(\lambda_n L)\cosh(\lambda_n y)}{(\lambda_n L)^2 \cosh(\lambda_n L)}\right]dy$$

$$= g'''L^2 + 2g'''L\sum_{n=1}^{\infty}\frac{(-1)^n \sin(\lambda_n L)}{(\lambda_n L)^2 \cosh(\lambda_n L)}\int_{y=0}^{L}\cosh(\lambda_n y)\,dy$$

Upon carrying out the remaining integral and recalling that $\sin(\lambda_n L) = (-1)^{n+1}$ for $\lambda_n L = (2n-1)\pi/2$,

$$q'\big|_{x=L} = g'''L^2 - 2g'''L\sum_{n=1}^{\infty}\frac{1}{(\lambda_n L)^2 \cosh(\lambda_n L)}\frac{\sinh(\lambda_n L)}{\lambda_n}$$

Multiplication of the numerator and denominator by L and rearranging,

$$q'\big|_{x=L} = g'''L^2 - 2g'''L^2\sum_{n=1}^{\infty}\frac{\tanh(\lambda_n L)}{(\lambda_n L)^3}$$

If enough terms (*e.g.*, 18) in the series are considered, we see that the series sums to $1/4$. Thus,

$$q'\big|_{x=L} = g'''L^2 - 2g'''L^2\frac{1}{4} = \frac{g'''L^2}{2}$$

Since $g'''L^2$ is the total energy generated in the square, we see that half of the energy leaves through the boundary along $x = L$. This is expected due to the symmetry of the problem.

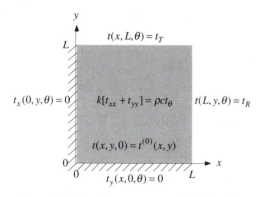

Figure 3•29 *Transient temperature distribution in a square region.*

3•2•2 Square region, transient problem

Two-dimensional transients have three independent variables, two position variables and time. The analytical solution of these problems is more complicated than previous solutions in this chapter which have involved only two independent variables. Separation of variables can still be used if the shape of the region and the boundary conditions are simple enough.

Consider the problem described in Figure 3•29. A square region is initially at $t^{(0)}(x,y)$. Then at $\theta = 0$, the boundaries along $x = 0$ and $y = 0$ are insulated, the boundary along $x = L$ has its temperature stepped to t_R and the boundary temperature along $y = L$ is stepped to t_T. The mathematical description of the problem is given in Table 3•21.

Table 3•21 *Mathematical problem description for $t(x,y,\theta)$.*

Problem	Equation	Boundary/Initial Conditions	
$t(x,y,\theta)$	$k[t_{xx} + t_{yy}] = \rho c t_\theta$		(3·2·35)
		$t_x(0,y,\theta) = 0$	(3·2·36)
		$t(L,y,\theta) = t_R$	(3·2·37)
		$t_y(x,0,\theta) = 0$	(3·2·38)
		$t(x,L,\theta) = t_T$	(3·2·39)
		$t(x,y,0) = t^{(0)}(x,y)$	(3·2·40)

In transient problems, all boundary conditions must be homogeneous. Since the boundary conditions along $x = L$ and along $y = L$ are not homogeneous, separation of variables will not work directly. To obtain the necessary homogeneous problem we will assume the solution can be written as

$$t(x,y,\theta) = a(x) + b(y) + u(x,y) + v(x,y,\theta) \tag{3.2.41}$$

We will try to construct a solution for $t(x,y,\theta)$ having this form. The problems for $a(x)$ and $b(y)$ do not need to be homogeneous. The steady-state problem for $u(x,y)$ must have homogeneous boundary conditions in one coordinate direction. In this example we will construct homogeneous boundary conditions for $u(x,y)$ in the x direction. The transient problem for $v(x,y,\theta)$ must have homogeneous boundary conditions in both coordinate directions. If (3·2·41) does not work we may be out of luck using separation of variables. A function $c(\theta)$ was not included in (3·2·41) since such a function is only needed when a run-away solution is possible. In this example the temperatures are tied to t_R and t_T which means there is no chance of a run-away solution.

Substitution of (3·2·41) into (3·2·35) gives

$$k[a''(x) + u_{xx} + v_{xx} + b''(y) + u_{yy} + v_{yy}] = \rho c v_\theta$$

To obtain homogeneous differential equations for $u(x,y)$ and $v(x,y,\theta)$ we will take

$$u_{xx} + u_{yy} = 0 \tag{3.2.42}$$

and

$$k[v_{xx} + v_{yy}] = \rho c v_\theta \tag{3.2.43}$$

This means that

$$a''(x) + b''(y) = 0$$

The only way this can be satisfied is to take

$$a''(x) = c_1 \tag{3·2·44}$$

and

$$b''(y) = -c_1 \tag{3·2·45}$$

Substituting (3·2·41) into the boundary condition (3·2·36) along $x = 0$ gives

$$a'(0) + u_x(0,y) + v_x(0,y,\theta) = 0$$

We want to have homogeneous x-boundary conditions for both $u(x,y)$ and $v(x,y,\theta)$. Thus we will take

$$u_x(0,y) = 0 \tag{3·2·46}$$

$$v_x(0,y,\theta) = 0 \tag{3·2·47}$$

This means that we must also have

$$a'(0) = 0 \tag{3·2·48}$$

Substitution of (3·2·41) into the boundary condition (3·2·37) along $x = L$ gives

$$a(L) + b(y) + u(L,y) + v(L,y,\theta) = t_R$$

We want to have homogeneous x-boundary conditions for both $u(x,y)$ and $v(x,y,\theta)$. Thus we will take

$$u(L,y) = 0 \tag{3·2·49}$$

$$v(L,y,\theta) = 0 \tag{3·2·50}$$

This means that we must also have

$$a(L) + b(y) = t_R$$

One way to satisfy this relation is to take

$$a(L) = 0 \tag{3·2·51}$$

and

$$b(y) = t_R \tag{3·2·52}$$

Substituting (3·2·41) into the boundary condition (3·2·38) along $y = 0$ gives

$$b'(0) + u_y(x,0) + v_y(x,0,\theta) = 0$$

We want to have a homogeneous y-boundary condition for $v(x,y,\theta)$, but we do not care about $u(x,y)$ here. Thus we will take

$$v_y(x,0,\theta) = 0 \tag{3·2·53}$$

This means that we must also have

$$u_y(x,0) = -b'(0) \tag{3·2·54}$$

Substituting (3·2·41) into the boundary condition (3·2·39) along $y = L$ gives

$$a(x) + b(L) + u(x,L) + v(x,L,\theta) = t_T$$

To have a homogeneous boundary condition for $v(x,y,\theta)$ we will take

$$v(x, L, \theta) = 0 \tag{3.2.55}$$

This means that we must also have

$$u(x, L) = t_T - a(x) - b(L) \tag{3.2.56}$$

Since the x-boundary conditions for $u(x, y)$ were homogeneous, it is not necessary that the y-boundary conditions be homogeneous.

Substitution of (3.2.41) into the initial condition (3.2.40) gives

$$a(x) + b(y) + u(x, y) + v(x, y, 0) = t^{(0)}(x, y)$$

This may be solved for the initial condition for $v(x, y, \theta)$ as

$$v(x, y, 0) = t^{(0)}(x, y) - u(x, y) - a(x) - b(y) \tag{3.2.57}$$

The subproblems for $a(x)$, $b(y)$, $u(x, y)$ and $v(x, y, \theta)$ are summarized in Table 3•22. The functions $a(x)$ and $b(y)$ satisfy nonhomogeneous ordinary differential equations that can have nonhomogeneous boundary conditions. The functions $a(x)$ and $b(y)$ allow the x-boundary conditions for $u(x, y)$ and $v(x, y, \theta)$ to be homogeneous. The y-boundary conditions for $u(x, y)$ can be nonhomogeneous since $u(x, y)$ is a steady problem. The nonhomogeneous y-boundary conditions for $u(x, y)$ are chosen so that the y-boundary conditions for $v(x, y, \theta)$ are homogeneous (as required to use separation of variables). The partially homogeneous problem for $u(x, y)$ is shown in Figure 3•30. The completely homogeneous problem for $v(x, y, \theta)$ is shown in Figure 3•31.

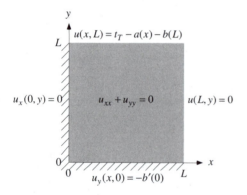

Figure 3•30 *Partially homogeneous steady-state problem for $u(x, y)$ corresponding to Figure 3•29.*

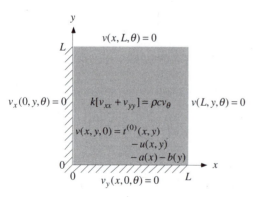

Figure 3•31 *Completely homogeneous transient problem for $v(x, y, \theta)$ corresponding to Figure 3•29.*

Table 3•22 *Subproblems for $t(x, y, \theta) = a(x) + b(y) + u(x, y) + v(x, y, \theta)$.*

Subproblem	Equation	Boundary/Initial Conditions	
$a(x)$	$a''(x) = c_1$		(3.2.44)
		$a'(0) = 0$	(3.2.48)
		$a(L) = 0$	(3.2.51)
$b(y)$	$b''(y) = -c_1$		(3.2.45)
		$b(y) = t_R$	(3.2.52)
$u(x, y)$	$u_{xx} + u_{yy} = 0$		(3.2.42)
		$u_x(0, y) = 0$	(3.2.46)
		$u(L, y) = 0$	(3.2.49)
		$u_y(x, 0) = -b'(0)$	(3.2.54)
		$u(x, L) = t_T - a(x) - b(L)$	(3.2.56)
$v(x, y, \theta)$	$k[v_{xx} + v_{yy}] = \rho c v_\theta$		(3.2.43)
		$v_x(0, y, \theta) = 0$	(3.2.47)
		$v(L, y, \theta) = 0$	(3.2.50)
		$v_y(x, 0, \theta) = 0$	(3.2.53)
		$v(x, L, \theta) = 0$	(3.2.55)
		$v(x, y, 0) = t^{(0)}(x, y) - u(x, y)$ $- a(x) - b(y)$	(3.2.57)

Let us first consider the problem for $b(y)$. Integration of (3.2.45) gives

$$b(y) = -c_1 \frac{y^2}{2} + b_1 y + b_2$$

Thus, $b(y)$ can at most be quadratic in y. The other restriction on $b(y)$ is given by (3·2·52). By comparison, $c_1 = 0$, $b_1 = 0$, $b_2 = t_R$ and the solution for $b(y)$ is given by (3·2·52).

Upon setting $c_1 = 0$ in (3·2·44), integrating twice and then applying the boundary conditions (3·2·48) and (3·2·51) we find that

$$a(x) = 0 \tag{3·2·58}$$

Substituting (3·2·58) and (3·2·52) into (3·2·41) gives

$$t(x, y, \theta) = t_R + u(x, y) + v(x, y, \theta) \tag{3·2·59}$$

Substituting (3·2·58) and (3·2·52) into the y-boundary conditions (3·2·54) and (3·2·56) for $u(x, y)$ gives

$$u_y(x, 0) = 0 \tag{3·2·60}$$

and

$$u(x, L) = t_T - t_R \tag{3·2·61}$$

Substituting (3·2·58) and (3·2·52) into the initial condition (3·2·57) for $v(x, y, \theta)$ gives

$$v(x, y, 0) = t^{(0)}(x, y) - u(x, y) - t_R \tag{3·2·62}$$

The updated problems for $t(x, y, \theta)$, $u(x, y)$ and $v(x, y, \theta)$ are summarized in Table 3•23. The updated sketches for $u(x, y)$ and $v(x, y, \theta)$ are shown in figures 3•32 and 3•33, respectively.

Table 3•23 *Subproblems for* $t(x, y, \theta)$, $u(x, y)$ *and* $v(x, y, \theta)$.

Subproblem	Equation	Boundary/Initial Conditions	
$t(x, y, \theta) = t_R + u(x, y) + v(x, y, \theta)$			(3·2·59)
$u(x, y)$	$u_{xx} + u_{yy} = 0$		(3·2·42)
		$u_x(0, y) = 0$	(3·2·46)
		$u(L, y) = 0$	(3·2·49)
		$u_y(x, 0) = 0$	(3·2·60)
		$u(x, L) = t_T - t_R$	(3·2·61)
$v(x, y, \theta)$	$k[v_{xx} + v_{yy}] = \rho c v_\theta$		(3·2·43)
		$v_x(0, y, \theta) = 0$	(3·2·47)
		$v(L, y, \theta) = 0$	(3·2·50)
		$v_y(x, 0, \theta) = 0$	(3·2·53)
		$v(x, L, \theta) = 0$	(3·2·55)
		$v(x, y, 0) = t^{(0)}(x, y) - u(x, y) - t_R$	(3·2·62)

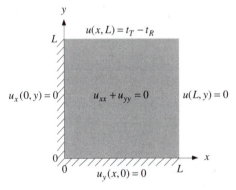

Figure 3•32 *Partially homogeneous steady-state problem for* $u(x, y)$ *corresponding to Figure 3•29.*

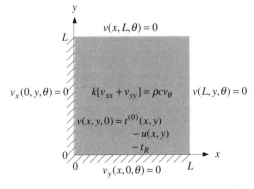

Figure 3•33 *Completely homogeneous transient problem for* $v(x, y, \theta)$ *corresponding to Figure 3•29.*

The next step in the solution is to solve for $u(x, y)$ using separation of variables. We will assume that

$$u(x, y) = X(x)Y(y) \tag{3·2·63}$$

Substitution of this assumed solution into the partial differential equation (3·2·42) gives

$$X''(x)Y(y) + X(x)Y''(y) = 0$$

Upon rearranging to separate the variables,

$$\frac{X''(x)}{X(x)} = -\frac{Y''(y)}{Y(y)}$$

Upon setting each side equal to $-\lambda^2$,

$$X''(x) + \lambda^2 X(x) = 0 \qquad\qquad (3\cdot2\cdot64)$$

and

$$Y''(y) - \lambda^2 Y(y) = 0 \qquad\qquad (3\cdot2\cdot65)$$

Upon substituting (3·2·63) into the boundary condition (3·2·46) along $x = 0$ we obtain

$$X'(0)Y(y) = 0$$

To satisfy this condition we will take

$$X'(0) = 0 \qquad\qquad (3\cdot2\cdot66)$$

Substituting (3·2·63) into the boundary condition (3·2·49) along $x = L$,

$$X(L)Y(y) = 0$$

We will satisfy this relation by taking

$$X(L) = 0 \qquad\qquad (3\cdot2\cdot67)$$

The eigenproblem for $X(x)$ is given by (3·2·64), (3·2·66) and (3·2·67). The general solution to (3·2·64) is given by

$$X(x) = A\sin(\lambda x) + B\cos(\lambda x)$$

To satisfy (3·2·66) we must take $A = 0$. To satisfy (3·2·67) requires that $\cos(\lambda L) = 0$ which gives $\lambda L =$ odd multiples of $\pi/2$. Therefore the eigenfunctions are given by

$$X_n(x) = B_n \cos(\lambda_n L) \qquad \text{where} \qquad \lambda_n L = (2n-1)\pi/2 \qquad (3\cdot2\cdot68)$$

The general solution to (3·2·65) is given by

$$Y_n(y) = C_n \sinh(\lambda_n y) + D_n \cosh(\lambda_n y) \qquad\qquad (3\cdot2\cdot69)$$

Equations (3·2·68) and (3·2·69) may be substituted into (3·2·63) to give

$$u_n(x,y) = \cos(\lambda_n x)\left[C_n \sinh(\lambda_n y) + D_n \cosh(\lambda_n y)\right]$$

where the B_n has been "absorbed" into the C_n and D_n. This solution satisfies the partial differential equation (3·2·42) and the x-boundary conditions (3·2·46) and (3·2·49). A more general solution may be obtained by summing over all of the eigenvalues to give

$$u(x,y) = \sum_{n=1}^{\infty} \cos(\lambda_n x)\left[C_n \sinh(\lambda_n y) + D_n \cosh(\lambda_n y)\right] \qquad (3\cdot2\cdot70)$$

The C_n and D_n will be chosen to make (3·2·70) satisfy the y-boundary conditions (3·2·60) and (3·2·61).

Upon differentiating (3·2·70) with respect to y

$$u_y(x,y) = \sum_{n=1}^{\infty} \cos(\lambda_n x)\left[C_n\lambda_n \cosh(\lambda_n y) + D_n\lambda_n \sinh(\lambda_n y)\right]$$

and then setting $y = 0$ and substituting into (3·2·54),

$$\sum_{n=1}^{\infty} \cos(\lambda_n x)C_n\lambda_n = 0$$

The only way this can be satisfied for all x is to take each $C_n = 0$. We could also arrive at this conclusion by multiplying both sides by $\cos(\lambda_m x) dx$ and then integrating from $x = 0$ to $x = L$. Due to orthogonality, only the term for $n = m$ would remain in the sum. The right-hand side is 0. Thus, $C_n = 0$. Therefore (3·2·70) reduces to

$$u(x, y) = \sum_{n=1}^{\infty} \cos(\lambda_n x) D_n \cosh(\lambda_n y) \qquad (3\cdot2\cdot71)$$

Upon setting $y = L$ and substituting (3·2·71) into (3·2·61),

$$\sum_{n=1}^{\infty} \cos(\lambda_n x) D_n \cosh(\lambda_n L) = t_T - t_R$$

To find D_m we will multiply both sides by $\cos(\lambda_m x) dx$ and then integrate from $x = 0$ to $x = L$. This gives

$$\int_{x=0}^{L} \sum_{n=1}^{\infty} \cos(\lambda_n x) D_n \cosh(\lambda_n L) \cos(\lambda_m x) dx = \int_{x=0}^{L} (t_T - t_R) \cos(\lambda_m x) dx$$

From orthogonality, only the term for $n = m$ will remain in the sum. Then, upon also factoring $D_m \cosh(\lambda_m L)$ out of the integral on the left-hand side and $t_T - t_R$ out of the integral on the right-hand side,

$$D_m \cosh(\lambda_m L) \int_{x=0}^{L} \cos^2(\lambda_m x) dx = (t_T - t_R) \int_{x=0}^{L} \cos(\lambda_m x) dx$$

Evaluating these integrals and then solving for D_m gives

$$D_m = 2(t_T - t_R) \frac{(-1)^{m+1}}{\lambda_m L \cosh(\lambda_m L)}$$

These Fourier coefficients may be substituted into (3·2·71) to give $u(x, y)$,

$$u(x, y) = 2(t_T - t_R) \sum_{n=1}^{\infty} \frac{(-1)^{n+1} \cos(\lambda_n x) \cosh(\lambda_n y)}{\lambda_n L \cosh(\lambda_n L)} \qquad (3\cdot2\cdot72)$$

The solution as it now stands is summarized in Table 3•24.

Table 3•24 *Problems for $t(x,y,\theta)$, $u(x,y)$ and $v(x,y,\theta)$.*

Subproblem	Equation	Boundary/Initial Conditions	
$t(x,y,\theta) = t_R + u(x,y) + v(x,y,\theta)$			(3·2·59)
$u(x,y) = 2(t_T - t_R) \sum_{n=1}^{\infty} \dfrac{(-1)^{n+1} \cos(\lambda_n x) \cosh(\lambda_n y)}{\lambda_n L \cosh(\lambda_n L)}$			(3·2·72)
$v(x,y,\theta)$	$k[v_{xx} + v_{yy}] = \rho c v_\theta$		(3·2·43)
		$v_x(0,y,\theta) = 0$	(3·2·47)
		$v(L,y,\theta) = 0$	(3·2·50)
		$v_y(x,0,\theta) = 0$	(3·2·53)
		$v(x,L,\theta) = 0$	(3·2·55)
		$v(x,y,0) = t^{(0)}(x,y) - u(x,y) - t_R$	(3·2·62)

We will now find $v(x,y,\theta)$ using separation of variables. We will first assume that the variables can be separated as

$$v(x, y, \theta) = X(x)Y(y)\Theta(\theta) \tag{3.2.73}$$

Substitution of this assumed solution into (3.2.43) gives

$$k[X''(x)Y(y)\Theta(\theta) + X(x)Y''(y)\Theta(\theta)] = \rho c X(x)Y(y)\Theta'(\theta)$$

Dividing by $kX(x)Y(y)\Theta(\theta)$,

$$\frac{X''(x)}{X(x)} + \frac{Y''(y)}{Y(y)} = \frac{1}{\alpha}\frac{\Theta'(\theta)}{\Theta(\theta)} \tag{3.2.74}$$

The only way that a function of x plus a function of y can always equal a function to θ is if each function is a constant. Therefore we will take

$$\frac{X''(x)}{X(x)} = -\mu^2 \qquad \text{or} \qquad X''(x) + \mu^2 X(x) = 0 \tag{3.2.75}$$

$$\frac{Y''(y)}{Y(y)} = -\gamma^2 \qquad \text{or} \qquad Y''(y) + \gamma^2 Y(y) = 0 \tag{3.2.76}$$

Substituting (3.2.75) and (3.2.76) into (3.2.74) gives

$$-\mu^2 - \gamma^2 = \frac{1}{\alpha}\frac{\Theta'(\theta)}{\Theta(\theta)} \qquad \text{or} \qquad \Theta'(\theta) + (\mu^2 + \gamma^2)\alpha\Theta(\theta) = 0 \tag{3.2.77}$$

Substituting (3.2.73) into the boundary condition (3.2.47) along $x = 0$,

$$X'(0)Y(y)\Theta(\theta) = 0$$

To satisfy this condition we will take

$$X'(0) = 0 \tag{3.2.78}$$

Substituting (3.2.73) into the boundary condition (3.2.50) along $x = L$,

$$X(L)Y(y)\Theta(\theta) = 0$$

To satisfy this condition we will take

$$X(L) = 0 \tag{3.2.79}$$

Similarly, substitution of (3.2.73) into (3.2.53) and (3.2.55) gives

$$Y'(0) = 0 \tag{3.2.80}$$

$$Y(L) = 0 \tag{3.2.81}$$

We now have two eigenproblems to solve. These eigenproblems are summarized in Table 3•25.

Table 3•25 *Eigenproblems for $v(x, y, \theta)$.*

Equation	Boundary Conditions	
$X'' + \mu^2 X = 0$		(3.2.75)
	$X'(0) = 0$	(3.2.78)
	$X(L) = 0$	(3.2.79)
$Y'' + \gamma^2 Y = 0$		(3.2.76)
	$Y'(0) = 0$	(3.2.80)
	$Y(L) = 0$	(3.2.81)

The solution to (3.2.75) that satisfies (3.2.78) and (3.2.79) is given by

$$X_\ell(x) = \cos(\mu_\ell x) \qquad \text{where} \qquad \mu_\ell L = (2\ell - 1)\pi/2 \qquad (3 \cdot 2 \cdot 82)$$

The solution to (3·2·76) that satisfies (3·2·80) and (3·2·81) is given by

$$Y_m(y) = \cos(\gamma_m y) \qquad \text{where} \qquad \gamma_m L = (2m - 1)\pi/2 \qquad (3 \cdot 2 \cdot 83)$$

The solution to (3·2·77) is given by

$$\Theta_{\ell m}(\theta) = \exp\left[-(\mu_\ell^2 + \gamma_m^2)\alpha\theta\right] \qquad (3 \cdot 2 \cdot 84)$$

The separated solutions (3·2·82), (3·2·83) and (3·2·84) may be substituted into (3·2·73) to obtain

$$v_{\ell m}(x, y, \theta) = \cos(\mu_\ell x)\cos(\gamma_m y)\exp\left[-(\mu_\ell^2 + \gamma_m^2)\alpha\theta\right]$$

There is a solution for each combination of ℓ and m. To obtain a more general solution we will multiply by a Fourier coefficient $A_{\ell m}$ and sum over all combinations of ℓ and m. Thus, we will write

$$v(x, y, \theta) = \sum_{\ell=1}^{\infty}\sum_{m=1}^{\infty} A_{\ell m}\cos(\mu_\ell x)\cos(\gamma_m y)\exp\left[-(\mu_\ell^2 + \gamma_m^2)\alpha\theta\right] \qquad (3 \cdot 2 \cdot 85)$$

This solution satisfies the partial differential equation (3·2·43) for $v(x, y, \theta)$ and its four boundary conditions (3·2·47), (3·2·50), (3·2·53) and (3·2·55) shown in Table 3•24. All that remains is to choose the Fourier coefficients $A_{\ell m}$ so that the initial condition (3·2·62) is satisfied. For simplicity we will take $t^{(0)}(x, y) = t_R$. Then, upon setting $\theta = 0$ in (3·2·85) and substituting into (3·2·62), we obtain

$$\sum_{\ell=1}^{\infty}\sum_{m=1}^{\infty} A_{\ell m}\cos(\mu_\ell x)\cos(\gamma_m y) = -u(x, y)$$

To eliminate all terms in the sum over ℓ except ℓ' we will multiply both sides by $\cos(\mu_{\ell'}x)\,dx$ and integrate from $x = 0$ to $x = L$. This gives

$$\sum_{\ell=1}^{\infty}\sum_{m=1}^{\infty} A_{\ell m}\left[\int_{x=0}^{L}\cos(\mu_\ell x)\cos(\mu_{\ell'}x)\,dx\right]\cos(\gamma_m y) = -\int_{x=0}^{L} u(x, y)\cos(\mu_{\ell'}x)\,dx$$

Due to orthogonality, the integral on the left-hand side is 0 except when $\ell = \ell'$. Thus,

$$\sum_{m=1}^{\infty} A_{\ell'm}\left[\int_{x=0}^{L}\cos^2(\mu_{\ell'}x)\,dx\right]\cos(\gamma_m y) = -\int_{x=0}^{L} u(x, y)\cos(\mu_{\ell'}x)\,dx$$

For $\mu_{\ell'}L = (2\ell' - 1)\pi/2$ the integral on the left-hand side is equal to $L/2$. Thus,

$$\frac{L}{2}\sum_{m=1}^{\infty} A_{\ell'm}\cos(\gamma_m y) = -\int_{x=0}^{L} u(x, y)\cos(\mu_{\ell'}x)\,dx$$

To eliminate all terms in the sum except $m = m'$, which will give $A_{\ell'm'}$, we will multiply both sides by $\cos(\gamma_{m'}y)\,dy$ and integrate from $y = 0$ to $y = L$. This gives

$$\frac{L}{2}\sum_{m=1}^{\infty} A_{\ell'm}\int_{y=0}^{L}\cos(\gamma_m y)\cos(\gamma_{m'}y)\,dy$$
$$= -\int_{y=0}^{L}\left[\int_{x=0}^{L} u(x, y)\cos(\mu_{\ell'}x)\,dx\right]\cos(\gamma_{m'}y)\,dy$$

Due to orthogonality, each of the integrals on the left-hand side is 0 except when $m = m'$. Thus,

$$\frac{L}{2}A_{\ell'm'}\int_{y=0}^{L}\cos^2(\gamma_{m'}y)\,dy = -\int_{y=0}^{L}\left[\int_{x=0}^{L}u(x,y)\cos(\mu_{\ell'}x)\,dx\right]\cos(\gamma_{m'}y)\,dy$$

For $\gamma_{m'}L = (2m'-1)\pi/2$ the integral on the left-hand side is equal to $L/2$. Thus,

$$\frac{L^2}{4}A_{\ell'm'} = -\int_{y=0}^{L}\left[\int_{x=0}^{L}u(x,y)\cos(\mu_{\ell'}x)\,dx\right]\cos(\gamma_{m'}y)\,dy$$

Next we will substitute (3·2·72) for $u(x,y)$ to obtain

$$\frac{L^2}{4}A_{\ell'm'} = -\int_{y=0}^{L}\left[\int_{x=0}^{L}2(t_T-t_R)\sum_{n=1}^{\infty}\frac{(-1)^{n+1}\cos(\lambda_n x)\cosh(\lambda_n y)}{\lambda_n L\cosh(\lambda_n L)}\cos(\mu_{\ell'}x)\,dx\right]\cos(\gamma_{m'}y)\,dy$$

This may be rearranged to give

$$\frac{L^2}{4}A_{\ell'm'} = -2(t_T-t_R)\sum_{n=1}^{\infty}\frac{(-1)^{n+1}}{\lambda_n L\cosh(\lambda_n L)}\int_{x=0}^{L}\cos(\lambda_n x)\cos(\mu_{\ell'}x)\,dx\int_{y=0}^{L}\cosh(\lambda_n y)\cos(\gamma_{m'}y)\,dy$$

Since $\lambda_n L = (2n-1)\pi/2$ and $\mu_{\ell'}L = (2\ell'-1)\pi/2$ it can be shown that the integral over x will be 0 except when $n = \ell'$ which is also when $\lambda_n = \mu_{\ell'}$. Thus, there will be only one term in the sum and we can write

$$\frac{L^2}{4}A_{\ell'm'} = -2(t_T-t_R)\frac{(-1)^{\ell'+1}}{\mu_{\ell'}L\cosh(\mu_{\ell'}L)}\int_{x=0}^{L}\cos^2(\mu_{\ell'}x)\,dx\int_{y=0}^{L}\cosh(\mu_{\ell'}y)\cos(\gamma_{m'}y)\,dy$$

For $\mu_{\ell'}L = (2\ell'-1)\pi/2$ the integral over x is equal to $L/2$. Thus,

$$\frac{L}{2}A_{\ell'm'} = -2(t_T-t_R)\frac{(-1)^{\ell'+1}}{\mu_{\ell'}L\cosh(\mu_{\ell'}L)}\int_{y=0}^{L}\cosh(\mu_{\ell'}y)\cos(\gamma_{m'}y)\,dy \quad (3\cdot2\cdot86)$$

From integral tables we can find

$$\int\cosh(ax)\cos(bx)\,dx = \frac{a\sinh(ax)\cos(bx)+b\cosh(ax)\sin(bx)}{a^2+b^2}$$

Thus,

$$\int_{y=0}^{L}\cosh(\mu_{\ell'}y)\cos(\gamma_{m'}y)\,dy$$
$$= \frac{\mu_{\ell'}\sinh(\mu_{\ell'}L)\cos(\gamma_{m'}L)+\gamma_{m'}\cosh(\mu_{\ell'}L)\sin(\gamma_{m'}L)}{\mu_{\ell'}^2+\gamma_{m'}^2}$$

For $\gamma_{m'}L = (2m'-1)\pi/2$, $\cos(\gamma_{m'}L) = 0$ and $\sin(\gamma_{m'}L) = (-1)^{m'+1}$. Thus,

$$\int_{y=0}^{L}\cosh(\beta_{\ell'}y)\cos(\gamma_{m'}y)\,dy = \frac{\gamma_{m'}\cosh(\mu_{\ell'}L)(-1)^{m'+1}}{\mu_{\ell'}^2+\gamma_{m'}^2} \quad (3\cdot2\cdot87)$$

Substituting (3·2·87) into (3·2·86),

$$\frac{L}{2}A_{\ell'm'} = -2(t_T-t_R)\frac{(-1)^{\ell'+1}}{\mu_{\ell'}L\cosh(\mu_{\ell'}L)}\frac{\gamma_{m'}\cosh(\mu_{\ell'}L)(-1)^{m'+1}}{\mu_{\ell'}^2+\gamma_{m'}^2}$$

Upon solving for $A_{\ell'm'}$, rearranging and omitting the primes,

$$A_{\ell m} = -4(t_T-t_R)\frac{(-1)^{\ell+m}}{\mu_\ell L}\frac{\gamma_m L}{(\mu_\ell L)^2+(\gamma_m L)^2} \quad (3\cdot2\cdot88)$$

These are the Fourier coefficients needed in (3·2·85). This completes the solution of this example, shown in Figure 3·34. The analytical results are summarized in Table 3·26. The evaluation of $t(x,y,\theta)$ is certainly a job for a computer.

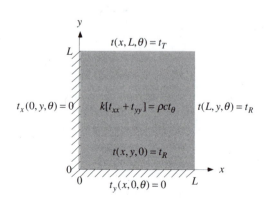

Figure 3·34 *Transient temperature problem whose solution is summarized in Table 3·26.*

Table 3•26 *Solutions for $t(x, y, \theta)$, $u(x, y)$ and $v(x, y, \theta)$.*

$$t(x, y, \theta) = t_R + u(x, y) + v(x, y, \theta) \qquad (3.2.59)$$

$$u(x, y) = 2(t_T - t_R) \sum_{n=1}^{\infty} \frac{(-1)^{n+1} \cos(\lambda_n x) \cosh(\lambda_n y)}{\lambda_n L \cosh(\lambda_n L)} \qquad (3.2.72)$$

$$\lambda_n L = (2n - 1)\pi / 2 \qquad (3.2.68)$$

$$v(x, y, \theta) = \sum_{\ell=1}^{\infty} \sum_{m=1}^{\infty} A_{\ell m} \cos(\mu_\ell x) \cos(\gamma_m y) \exp\left[-(\mu_\ell^2 + \gamma_m^2)\alpha\theta\right] \qquad (3.2.85)$$

$$A_{\ell m} = -4(t_T - t_R) \frac{(-1)^{\ell+m}}{\mu_\ell L} \frac{\gamma_m L}{(\mu_\ell L)^2 + (\gamma_m L)^2} \qquad (3.2.88)$$

$$\mu_\ell L = (2\ell - 1)\pi / 2 \qquad (3.2.82)$$

$$\gamma_m L = (2m - 1)\pi / 2 \qquad (3.2.83)$$

3•2•3 Summary

Sections 3•2•1 and 3•2•2 considered two sample two-dimensional problems. Separation of variables requires a "regular" region (*e.g.*, in Cartesian coordinates, a square or rectangle) and the boundary conditions that can be handled are rather limited (*e.g.*, in the example worked out in Section 3•2•2 the temperature along $x = L$ can at most be quadratic in y).

Even if you are able to obtain an analytical solution to a two-dimensional problem, considerable computation is usually required to obtain numerical results for the temperature distribution. Often a computer is recommended.

Nowadays numerical computer methods are pretty well implemented and many software packages are available. If the region is square or rectangular it might be just as easy to use finite differences (Chapter 8) to solve the problem and not have to worry so much about nonhomogeneous terms, eigenproblems, orthogonality, infinite series and the algebra and calculus that is required. If the region is not square or rectangular the finite-element method (Chapter 9) is a powerful technique for handling the problem. Both of these numerical methods can treat a greater variety of boundary conditions than separation of variables can handle.

Variation of parameters (Section 3•1•7) can be extended to treat time-dependent nonhomogeneous terms, but geometry and boundary conditions are still limited as discussed above.

3•3 SUPPLEMENTARY PROBLEMS

Sections 3•1 and 3•2 contain essentially all the engineering mathematics involved in the method of separation of variables. The purpose of this section is to provide additional examples that will

1. Improve your facility with the separation-of-variables procedure
2. Show how the method can be applied to geometries other than the plane wall
3. Provide a review of Bessel functions

Separation of variables in cylindrical and spherical coordinates follows identically the steps of the preceding sections. Therefore we will not consider lots of boundary conditions but will just a few problems to point out things that may be a bit different than for plane walls.

3•3•1 Solid cylinder, specified-temperature boundary

Consider an infinitely long, solid circular cylinder, Figure 3•35, whose initial temperature is $t^{(0)}$. Suddenly the temperature of its outside surface is changed to t_o. This corresponds to the plane-wall problem considered in Section 3•1•1. The mathematical description of the problem is given in Table 3•27 and shown in Figure 3•36. It is evident from our engineering understanding of the problem that the temperature at $r = 0$ must be finite. Therefore the boundary condition at $r = 0$ is taken to be given by (3.3.2). An equally valid boundary condition at $r = 0$ in a symmetrical problem such as this is to take $t_r(0,\theta) = 0$. However, if the temperature varied with angular position this symmetry condition would not be valid and the finiteness of the temperature at $r = 0$ would have to be used.

Figure 3•35 *Infinitely long, solid circular cylinder, initially at temperature $t^{(0)}$ is suddenly exposed to a step change in surface temperature to t_o.*

Table 3•27 *Mathematical problem description for $t(r,\theta)$.*

Problem	Equation	Boundary/Initial Conditions	
$t(r,\theta)$	$k[t_{rr} + \dfrac{1}{r}t_r] = \rho c t_\theta$		(3.3.1)
		$t(0,\theta) < \infty$	(3.3.2)
		$t(r_o,\theta) = t_o$	(3.3.3)
		$t(r,0) = t^{(0)}$	(3.3.4)

The boundary condition (3.3.2) at $r = 0$ is homogeneous since $Ct(0,\theta)$ is also finite. The boundary condition (3.3.3) at $r = r_o$ is not homogeneous. To handle this nonhomogeneous problem we will assume the solution can be written as

$$t(r,\theta) = a(r) + u(r,\theta) \qquad (3.3.5)$$

We have not considered a $b(\theta)$ term since this is not a run-away problem. Substituting (3.3.5) into (3.3.1),

$$k[a''(r) + u_{rr} + \frac{1}{r}a'(r) + \frac{1}{r}u_r] = \rho c u_\theta$$

This can be satisfied by taking

$$k[u_{rr} + \frac{1}{r}u_r] = \rho c u_\theta \qquad (3.3.6)$$

and

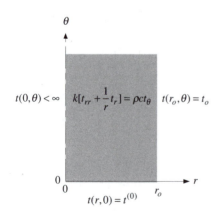

Figure 3•36 *Infinitely long, solid circular cylinder, initially at temperature $t^{(0)}$, is suddenly exposed to a step change in surface temperature to t_o.*

$$a''(r) + \frac{1}{r}a'(r) = 0 \tag{3.3.7}$$

Substituting (3·3·5) into (3·3·2) we obtain

$$a(0) + u(0,\theta) < \infty$$

This can be satisfied by insisting that

$$u(0,\theta) < \infty \tag{3.3.8}$$

and

$$a(0) < \infty \tag{3.3.9}$$

Substitution of (3·3·5) into the boundary condition (3·3·3) at r_o gives

$$a(r_o) + u(r_o,\theta) = t_o$$

Since we need a homogeneous boundary condition for $u(r,\theta)$, we will satisfy this by taking

$$u(r_o,\theta) = 0 \tag{3.3.10}$$

and

$$a(r_o) = t_o \tag{3.3.11}$$

Substitution of the assumed solution (3·3·5) into the initial condition (3·3·4) gives

$$a(r) + u(r,0) = t^{(0)}$$

The initial condition for $u(r,\theta)$ then becomes

$$u(r,0) = t^{(0)} - a(r) \tag{3.3.12}$$

The subproblems for $a(r)$ and $u(r,\theta)$ are summarized in Table 3•28. The homogeneous problem for $u(r,\theta)$ is shown in Figure 3•37.

Table 3•28 *Subproblems for* $t(r,\theta) = a(r) + u(r,\theta)$.

Subproblem	Equation	Boundary/Initial Conditions	
$a(r)$	$a''(r) + \dfrac{1}{r}a'(r) = 0$		(3.3.7)
		$a(0) < \infty$	(3.3.9)
		$a(r_o) = t_o$	(3.3.11)
$u(r,\theta)$	$k[u_{rr} + \dfrac{1}{r}u_r] = \rho c u_\theta$		(3.3.6)
		$u(0,\theta) < \infty$	(3.3.8)
		$u(r_o,\theta) = 0$	(3.3.10)
		$u(r,0) = t^{(0)} - a(r)$	(3.3.12)

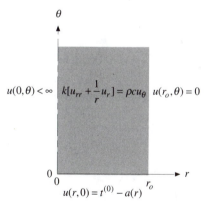

Figure 3•37 *Homogeneous problem for* $u(r,\theta)$ *corresponding to Figure 3•36.*

The solution for $a(r)$ starts by letting $p = da/dr$ in (3·3·7) to obtain

$$\frac{dp}{dr} + \frac{1}{r}p = 0$$

Upon multiplying by $r\,dr$ and recognizing that $r\,dp + p\,dr = d(rp)$, we may integrate to obtain

$$rp = c_1$$

Since $p = da/dr$ we may multiply by dr/r to obtain

$$da = c_1 \frac{dr}{r}$$

This may be integrated to give

$$a(r) = c_1 \ln(r) + c_2$$

To satisfy (3·3·9), we must take $c_1 = 0$. To satisfy (3·3·11) we must have $c_2 = t_o$. Thus,

$$a(r) = t_o \tag{3·3·13}$$

We are now ready to use separation of variables to find $u(r,\theta)$. We will assume that r and θ can be separated and write

$$u(r,\theta) = R(r)\Theta(\theta) \tag{3·3·14}$$

Substitution of this assumed solution into (3·3·6) gives

$$k[R''(r)\Theta(\theta) + \frac{1}{r}R'(r)\Theta(\theta)] = \rho c R(r)\Theta'(\theta)$$

Upon dividing by $kR\Theta$ to separate the variables and letting $k/\rho c = \alpha$,

$$\frac{R''(r)}{R(r)} + \frac{1}{r}\frac{R'(r)}{R(r)} = \frac{1}{\alpha}\frac{\Theta'(\theta)}{\Theta(\theta)}$$

The only way a function of r (left-hand side) can always equal a function of θ (right-hand side) is for both functions to be equal to the same constant. We will take this constant to be $-\lambda^2$ and write

$$R''(r) + \frac{1}{r}R'(r) + \lambda^2 R(r) = 0 \tag{3·3·15}$$

and

$$\Theta'(\theta) + \lambda^2\alpha\Theta(\theta) = 0 \tag{3·3·16}$$

Substituting (3·3·14) into the boundary condition (3·3·8) at $r = 0$,

$$R(0)\Theta(\theta) < \infty$$

This may be satisfied by taking

$$R(0) < \infty \tag{3·3·17}$$

Substituting (3·3·14) into the boundary condition (3·3·10) at $r = r_o$ gives

$$R(r_o)\Theta(\theta) = 0$$

This may be satisfied by taking

$$R(r_o) = 0 \tag{3·3·18}$$

The eigenproblem for $R(r)$ is given by (3·3·15), (3·3·17) and (3·3·18). Equation (3·3·15) is a Bessel equation whose general solution is given by

$$R(r) = AJ_0(\lambda r) + BY_0(\lambda r)$$

To satisfy (3·3·17) we must set $B = 0$ to eliminate $Y_0(\lambda r)$ which approaches $-\infty$ as r approaches 0. To satisfy (3·3·18) we must have

$$R(r_o) = AJ_0(\lambda r_o) = 0$$

Thus we will take

$$J_0(\lambda_n r_o) = 0 \tag{3·3·19}$$

This is the eigencondition for this problem. Only those special values of λr_o (*i.e.*, eigenvalues) that satisfy this condition are allowed. A plot of J_0 is shown in Figure 3•38. Just as with sines and cosines, there are an infinite number of zeros of $J_0(\lambda r_o)$. These zeros are the eigenvalues. The first six values are tabulated in Table D•3 and are also listed in Table 3•29.

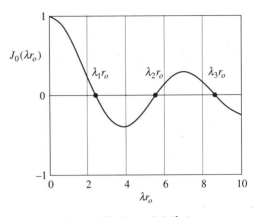

Figure 3•38 *Zeros of $J_0(\lambda r_o)$.*

Table 3•29 *First six roots of $J_0(\lambda_n r_o) = 0$.*

n	$\lambda_n r_o$
1	2.4048
2	5.5201
3	8.6537
4	11.7915
5	14.9309
6	18.0711

The solution to (3•3•15) is given by

$$R_n(r) = A_n J_0(\lambda_n r)$$

The solution to (3•3•16) is given by

$$\Theta_n(\theta) = \exp(-\lambda_n^2 \alpha \theta)$$

The solution for $u(r,\theta)$ will be obtained by substituting these separated solutions into (3•3•14) and then summing over n to obtain

$$u(r,\theta) = \sum_{n=1}^{\infty} A_n J_0(\lambda_n r) \exp(-\lambda_n^2 \alpha \theta) \qquad (3\cdot3\cdot20)$$

where the λ_n are found from the solutions to $J_0(\lambda r_o) = 0$. This solution for $u(r,\theta)$ satisfies the partial differential equation and the boundary conditions. The A_n must still be chosen to satisfy the initial condition (3•3•12) where $a(r)$ is given by (3•3•13). Therefore, at $\theta = 0$, (3•3•8) gives

$$u(r,0) = \sum_{n=1}^{\infty} A_n J_0(\lambda_n r) = t^{(0)} - t_o \qquad (3\cdot3\cdot21)$$

The next step is to make use of orthogonality to evaluate the A_n. For Bessel functions a weighting function is necessary. The simplest way to show what must be done is to go back to the eigenvalue problem as stated by (3•3•15) but multiplied by r. The differential equation and its boundary conditions (3•3•17) and (3•3•18) are given by

$$rR'' + R' + \lambda^2 rR = 0 \qquad R(0) < \infty \qquad R(r_o) = 0$$

If R_m and R_n are the eigenfunctions corresponding to λ_m and λ_n, they must satisfy the differential equation and its boundary conditions. Thus,

$$rR_m'' + R_m' + \lambda_m^2 rR_m = 0 \qquad R_m(0) < \infty \qquad R_m(r_o) = 0$$

$$rR_n'' + R_n' + \lambda_n^2 rR_n = 0 \qquad R_n(0) < \infty \qquad R_n(r_o) = 0$$

If we multiply the first differential equation by R_n, the second differential equation by R_m and then subtract, we obtain

$$r(R_n R_m'' - R_m R_n'') + (R_n R_m' - R_m R_n') + (\lambda_m^2 - \lambda_n^2)rR_m R_n = 0$$

Rewriting,

$$r\frac{d}{dr}(R_n R_m' - R_m R_n') + (R_n R_m' - R_m R_n') = (\lambda_n^2 - \lambda_m^2)rR_m R_n$$

The terms on the left can be combined as the derivative of a product. The equation then becomes

$$\frac{d}{dr}[r(R_n R_m' - R_m R_n')] = (\lambda_n^2 - \lambda_m^2)rR_m R_n$$

Integrating,

$$[r(R_n R_m' - R_m R_n')]_0^{r_o} = (\lambda_n^2 - \lambda_m^2)\int_0^{r_o} rR_m R_n\, dr$$

The left-hand side is zero at the upper limit, since $R_m(r_o) = 0$ and $R_n(r_o) = 0$; it is also zero at the lower limit, since $r = 0$ and R_m, R_n, R_m' and R_n' are all finite. Therefore

$$0 = (\lambda_n^2 - \lambda_m^2)\int_0^{r_o} rR_m R_n\, dr$$

The r appearing in the integrand is called a *weighting function*. For the example being considered here, $R_m = J_0(\lambda_m r)$ and $R_n = J_0(\lambda_n r)$. Thus,

$$\int_0^{r_o} rJ_0(\lambda_m r)J_0(\lambda_n r)\, dr = 0 \qquad \text{for } m \neq n \qquad (3\cdot3\cdot22)$$

In view of (3·3·22), the A_n in (3·3·21) can be evaluated by multiplying (3·3·21) through by $rJ_0(\lambda_m r)$ and integrating between 0 and r_o to give

$$\sum_{n=1}^{\infty} A_n \int_0^{r_o} rJ_0(\lambda_m r)J_0(\lambda_n r)\, dr = (t^{(0)} - t_o)\int_0^{r_o} rJ_0(\lambda_m r)\, dr$$

Since the integral on the left is 0 except when $n = m$, the above reduces to

$$A_m \int_0^{r_o} rJ_0^2(\lambda_m r)\, dr = (t^{(0)} - t_o)\int_0^{r_o} rJ_0(\lambda_m r)\, dr$$

Multiplying each side of the equation by λ_m^2,

$$A_m \int_0^{r_o} (\lambda_m r)J_0^2(\lambda_m r)\, d(\lambda_m r) = (t^{(0)} - t_o)\int_0^{r_o} (\lambda_m r)J_0(\lambda_m r)\, d(\lambda_m r)$$

These integrals may be evaluated from Table B•3 to give

$$A_m \left\{ \frac{(\lambda_m r)^2}{2}\left[J_0^2(\lambda_m r) + J_1^2(\lambda_m r)\right] \right\}_{r=0}^{r_o} = (t^{(0)} - t_o)\left[(\lambda_m r)J_1(\lambda_m r)\right]_{r=0}^{r_o}$$

Inserting the limits, with the lower limits giving zero, yields

$$A_m \frac{(\lambda_m r_o)^2}{2}\left[J_0^2(\lambda_m r_o) + J_1^2(\lambda_m r_o)\right] = (t^{(0)} - t_o)(\lambda_m r_o)J_1(\lambda_m r_o)$$

Upon using the eigencondition (3·3·19) that $J_0(\lambda_n r_o) = 0$,

$$A_m \frac{(\lambda_m r_o)^2}{2}J_1^2(\lambda_m r_o) = (t^{(0)} - t_o)(\lambda_m r_o)J_1(\lambda_m r_o)$$

and finally, upon simplifying and solving for A_m,

$$A_m = \frac{2(t^{(0)} - t_o)}{\lambda_m r_o J_1(\lambda_m r_o)}$$

This may be substituted into (3·3·20) which, in turn, may be substituted into (3·3·5) with $a(r) = t_o$ as given by (3·3·11) to give

$$t(r,\theta) = t_o + 2(t^{(0)} - t_o) \sum_{n=1}^{\infty} \frac{J_0(\lambda_n r) \exp(-\lambda_n^2 \alpha \theta)}{\lambda_n r_o J_1(\lambda_n r_o)} \qquad (3.3.23)$$

where the $(\lambda_n r_o)$ are the roots of $J_0(\lambda_n r_o) = 0$.

It is wise to make sure that the solution checks out dimensionally by normalizing the solution. Upon rearranging (3·3·23),

$$\frac{t(r,\theta) - t_o}{t^{(0)} - t_o} = 2 \sum_{n=1}^{\infty} \frac{J_0(\lambda_n r) \exp[-(\lambda_n r_o)^2 \alpha \theta / r_o^2]}{\lambda_n r_o J_1(\lambda_n r_o)} \qquad (3.3.24)$$

The eigenvalues $\lambda_n r_o$ are dimensionless as is $\alpha \theta / r_o^2$, which is called the *Fourier number*. Thus the solution is completely dimensionless. Also note that, as θ becomes infinite, $t(r,\theta)$ approaches t_o as expected physically.

Observe that even without normalizing before doing the mathematics the final result (3·3·24) is most conveniently expressed in nondimensional terms. The nondimensional time variable is given by $\bar{\theta} = \alpha \theta / r_o^2$.

3•3•2 Solid cylinder, convection boundary

Consider the same problem as in the previous section but with the boundary condition at $r = r_o$ replaced by a convection boundary condition. The mathematical problem is given in Table 3•30 and shown in Figure 3•39.

Table 3•30 *Mathematical problem description for $t(r,\theta)$.*

Problem	Equation	Boundary/Initial Conditions	
$t(r,\theta)$	$k[t_{rr} + \frac{1}{r}t_r] = \rho c t_\theta$		(3.3.25)
		$t(0,\theta) < \infty$	(3.3.26)
		$-kt_r(r_o,\theta) = h[t(r_o,\theta) - t_\infty]$	(3.3.27)
		$t(r,0) = t^{(0)}$	(3.3.28)

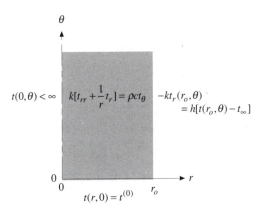

Figure 3•39 *Infinitely long, solid circular cylinder, initially at temperature $t^{(0)}$, is suddenly exposed to a step change in ambient temperature to t_∞.*

Due to the nonhomogeneous boundary condition (3·3·27) we will assume that the solution can be written as

$$t(r,\theta) = a(r) + u(r,\theta) \qquad (3.3.29)$$

Substituting (3·3·29) into (3·3·25) through (3·3·28) sets up problems for $a(r)$ and $u(r,\theta)$. Upon solving for $a(r)$, it turns out that $a(r) = t_\infty$. The problems for $t(r,\theta)$ and $u(r,\theta)$ are then given in Table 3•31. The problem for $u(r,\theta)$ is pictured in Figure 3•40.

Table 3•31 *Mathematical problem description for $t(r,\theta)$ and $u(r,\theta)$.*

Problem	Equation	Boundary/Initial Conditions	
$t(r,\theta) = t_\infty + u(r,\theta)$			
$u(r,\theta)$	$k[u_{rr} + \frac{1}{r}u_r] = \rho c u_\theta$		(3.3.30)
		$u(0,\theta) < \infty$	(3.3.31)
		$-ku_r(r_o,\theta) = hu(r_o,\theta)$	(3.3.32)
		$u(r,0) = t^{(0)} - t_\infty$	(3.3.33)

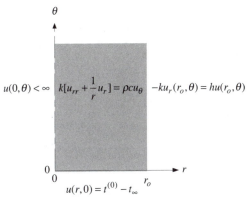

Figure 3•40 *Homogeneous problem for $u(r,\theta)$ corresponding to Figure 3•39.*

We will assume $u(r,\theta) = R(r)\Theta(\theta)$ to separate the variables. The two separated solutions are

$$R(r) = AJ_0(\lambda r) + BY_0(\lambda r) \qquad \text{and} \qquad \Theta(\theta) = \exp(-\lambda^2 \alpha \theta)$$

Again, $B = 0$ since the solution must be finite (3·3·31) at $r = 0$. To satisfy the boundary condition (3·3·32) at $r = r_o$,

$$-kR'(r_o) = hR(r_o)$$

Then, substituting $R(r) = AJ_0(\lambda r)$ and $R'(r) = -A\lambda J_1(\lambda r)$ and then setting $r = r_o$,

$$-kA\lambda J_1(\lambda r) = hAJ_0(\lambda r_o)$$

Upon dividing by kA and multiplying by r_o,

$$(\lambda r_o)J_1(\lambda r_o) = HJ_0(\lambda r_o) \tag{3·3·34}$$

where $H = hr_o / k$. This is the eigencondition for this problem. The first six roots of (3·3·34) can be found in Table D•4. The solution for $u(r,\theta)$ thus far is given as

$$u(r,\theta) = \sum_{n=1}^{\infty} A_n J_0(\lambda_n r)\exp(-\lambda_n^2 \alpha \theta) \tag{3·3·35}$$

To satisfy the initial condition (3·3·33),

$$u(r,0) = \sum_{n=1}^{\infty} A_n J_0(\lambda_n r) = t^{(0)} - t_\infty$$

To evaluate the A_n, multiply through by $rJ_0(\lambda_m r)$ and integrate from $r = 0$ to $r = r_o$. Thus

$$\sum_{n=1}^{\infty} A_n \int_{r=0}^{r_o} rJ_0(\lambda_m r)J_0(\lambda_n r)\,dr = (t^{(0)} - t_\infty)\int_{r=0}^{r_o} rJ_0(\lambda_m r)\,dr$$

It can be shown, using the eigencondition (3·3·34), that the integrals on the left-hand side are all 0 except when $n = m$. Therefore we may write

$$A_m \int_{r=0}^{r_o} rJ_0^2(\lambda_m r)\,dr = (t^{(0)} - t_\infty)\int_{r=0}^{r_o} rJ_0(\lambda_m r)\,dr$$

Multiplying through by λ_m^2, this equation may be written

$$A_m \int_{r=0}^{r_o} (\lambda_m r)J_0^2(\lambda_m r)\,d(\lambda_m r) = (t^{(0)} - t_\infty)\int_{r=0}^{r_o} (\lambda_m r)J_0(\lambda_m r)\,d(\lambda_m r)$$

This may be integrated using Table B•3 to give

$$A_m \left\{ \frac{(\lambda_m r)^2}{2}\left[J_0^2(\lambda_m r) + J_1^2(\lambda_m r)\right] \right\}_{r=0}^{r_o} = (t^{(0)} - t_\infty)\left[(\lambda_m r)J_1(\lambda_m r)\right]_{r=0}^{r_o}$$

Substituting the limits,

$$A_m \frac{(\lambda_m r_o)^2}{2}\left[J_0^2(\lambda_m r_o) + J_1^2(\lambda_m r_o)\right] = (t^{(0)} - t_\infty)(\lambda_m r_o)J_1(\lambda_m r_o)$$

Using the eigencondition (3·3·34) to eliminate $J_1(\lambda_m r_o)$,

$$A_m \frac{(\lambda_m r_o)^2}{2}\left[J_0^2(\lambda_m r_o) + \frac{H^2 J_0^2(\lambda_m r_o)}{(\lambda_m r_o)^2}\right] = (t^{(0)} - t_\infty)HJ_0(\lambda_m r_o)$$

Upon solving for A_m,

$$A_m = \frac{2H(t^{(0)} - t_\infty)}{\left[(\lambda_m r_o)^2 + H^2\right] J_0(\lambda_m r_o)}$$

This can then be substituted into (3·3·35) to obtain $u(r,\theta)$ which can in turn be substituted into (3·3·29) to obtain the complete solution as

$$t(r,\theta) = t_\infty + 2H(t^{(0)} - t_\infty) \sum_{n=1}^{\infty} \frac{J_0(\lambda_n r) \exp(-\lambda_n^2 \alpha \theta)}{\left[(\lambda_n r_o)^2 + H^2\right] J_0(\lambda_n r_o)} \qquad (3\cdot3\cdot36)$$

3·3·3 Hollow cylinder, specified-temperature boundary

In some problems you may not have a solid cylinder but rather one with a hollow center. In this case the Y_0 term cannot be dropped as was done with a solid cylinder. This more-complicated problem is treated in this section.

Consider the hollow cylinder shown in Figure 3•41 having inner radius r_i and outer radius r_o. Initially the cylinder is at temperature $t^{(0)}$. At time zero the inner radius is insulated and the outer surface is stepped to t_o. The mathematical description of the problem is given in Table 3•32 and sketched in Figure 3•42.

Table 3·32 *Mathematical problem description for $t(r,\theta)$.*

Problem	Equation	Boundary/Initial Conditions	
$t(r,\theta)$	$k[t_{rr} + \frac{1}{r}t_r] = \rho c t_\theta$		(3·3·37)
		$t_r(r_i,\theta) = 0$	(3·3·38)
		$t(r_o,\theta) = t_o$	(3·3·39)
		$t(r,0) = t^{(0)}$	(3·3·40)

Due to the nonhomogeneous boundary condition (3·3·39) we will assume that

$$t(r,\theta) = a(r) + u(r,\theta) \qquad (3\cdot3\cdot41)$$

Substituting (3·3·41) into (3·3·37) through (3·3·40) sets up problems for $a(r)$ and $u(r,\theta)$. Upon solving for $a(r)$, it turns out that $a(r) = t_o$. The problems for $t(r,\theta)$ and $u(r,\theta)$ are then given in Table 3•33. The problem for $u(r,\theta)$ is pictured in Figure 3•43.

Table 3·33 *Mathematical problem description for $t(r,\theta)$ and $u(r,\theta)$.*

Problem	Equation	Boundary/Initial Conditions	
$t(r,\theta) = t_o + u(r,\theta)$			(3·3·42)
$u(r,\theta)$	$k[u_{rr} + \frac{1}{r}u_r] = \rho c u_\theta$		(3·3·43)
		$u_r(r_i,\theta) = 0$	(3·3·44)
		$u(r_o,\theta) = 0$	(3·3·45)
		$u(r,0) = t^{(0)} - t_o$	(3·3·46)

We will assume $u(r,\theta) = R(r)\Theta(\theta)$ to separate the variables. The two separated solutions are

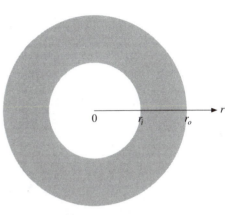

Figure 3·41 *Cross-section of a hollow cylinder.*

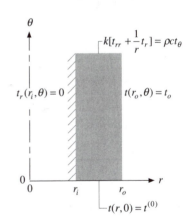

Figure 3·42 *Infinitely long, hollow cylinder, initially at temperature $t^{(0)}$, is suddenly exposed to a step change in surface temperature to t_o.*

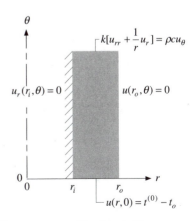

Figure 3·43 *Homogeneous problem for $u(r,\theta)$ corresponding to Figure 3•42.*

$$R(r) = AJ_0(\lambda r) + BY_0(\lambda r) \tag{3.3.47}$$

$$\Theta(\theta) = \exp(-\lambda^2 \alpha \theta) \tag{3.3.48}$$

The boundary conditions that $R(r)$ must satisfy are given by

$$R'(r_i) = 0 \tag{3.3.49}$$

$$R(r_o) = 0 \tag{3.3.50}$$

Differentiating (3.3.47) and substituting it into (3.3.49),

$$R'(r_i) = -A\lambda J_1(\lambda r_i) - B\lambda Y_1(\lambda r_i) = 0 \tag{3.3.51}$$

Substituting (3.3.47) into (3.3.50) gives

$$R(r_o) = AJ_0(\lambda r_o) + BY_0(\lambda r_o) = 0 \tag{3.3.52}$$

The solution of (3.3.51) and (3.3.52) for A and B using determinants gives

$$A = \frac{\begin{vmatrix} 0 & -\lambda Y_1(\lambda r_i) \\ 0 & Y_0(\lambda r_o) \end{vmatrix}}{\begin{vmatrix} -\lambda J_1(\lambda r_i) & -\lambda Y_1(\lambda r_i) \\ J_0(\lambda r_o) & Y_0(\lambda r_o) \end{vmatrix}} \quad \text{and} \quad B = \frac{\begin{vmatrix} -\lambda J_1(\lambda r_i) & 0 \\ J_0(\lambda r_o) & 0 \end{vmatrix}}{\begin{vmatrix} -\lambda J_1(\lambda r_i) & -\lambda Y_1(\lambda r_i) \\ J_0(\lambda r_o) & Y_0(\lambda r_o) \end{vmatrix}}$$

Both A and B are zero unless the determinant in the denominator is zero. Since $A = B = 0$ is a trivial solution to the problem without engineering usefulness, we must set the denominator determinant equal to zero. Thus

$$-\lambda J_1(\lambda r_i) Y_0(\lambda r_o) + J_0(\lambda r_o) \lambda Y_1(\lambda r_i) = 0$$

Dividing by λ and rearranging, the eigencondition may be written as

$$f(\lambda; r_i, r_o) = J_0(\lambda r_o) Y_1(\lambda r_i) - J_1(\lambda r_i) Y_0(\lambda r_o) = 0 \tag{3.3.53}$$

The eigenvalues are determined from (3.3.53). The roots of (3.3.53) are not readily available (if at all). As an example, $f(\lambda; r_i, r_o)$ was calculated as a function of λ for $r_i = 0.5$ and $r_o = 1.0$. The curve is shown in Figure 3•44 and the first four eigenvalues have been found. With appropriate software packages[1] that are now available, this is an easy task. It is not necessary to write a FORTRAN program to find solutions to (3.3.53).

The eigenfunction (3.3.47) may be written as

$$R_n(r) = A_n J_0(\lambda_n r) + B_n Y_0(\lambda_n r)$$

Rather than write the eigenfunction as the sum of two Bessel functions, we will factor out the A_n and write

$$R_n(r) = A_n \left[J_0(\lambda_n r) + \frac{B_n}{A_n} Y_0(\lambda_n r) \right] \tag{3.3.54}$$

It will be convenient to define a *cylinder function* $C_k(\lambda_n r)$ as

$$C_k(\lambda_n r) = J_k(\lambda_n r) + \frac{B_n}{A_n} Y_k(\lambda_n r) \tag{3.3.55}$$

The eigenfunction (3.3.54) may then be written as

$$R_n(r) = A_n C_0(\lambda_n r) \tag{3.3.56}$$

The ratio B_n / A_n in (3.3.55) may be obtained from either (3.3.51) or (3.3.52) as

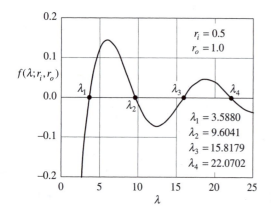

$$r_i = 0.5$$
$$r_o = 1.0$$

$$\lambda_1 = 3.5880$$
$$\lambda_2 = 9.6041$$
$$\lambda_3 = 15.8179$$
$$\lambda_4 = 22.0702$$

Figure 3•44 *Plot of (3.3.53) showing the first four zeros for $r_i = 0.5$ and $r_o = 1.0$.*

---

[1] For example, Klein, S. A. and F. L. Alvarado: *EES – Engineering Equation Solver*, F-Chart Software <http://www.fchart.com/>.

$$\frac{B_n}{A_n} = -\frac{J_1(\lambda_n r_i)}{Y_1(\lambda_n r_i)} = -\frac{J_0(\lambda_n r_o)}{Y_0(\lambda_n r_o)} \qquad (3\cdot3\cdot57)$$

We will need to have the values of the cylinder functions at $r = r_i$ and r_o. Upon dividing (3·3·51) by $-A\lambda$ we may write

$$J_1(\lambda_n r_i) + \frac{B_n}{A_n} Y_1(\lambda_n r_i) = 0$$

By comparison with (3·3·55) we see that this means

$$C_1(\lambda_n r_i) = 0 \qquad (3\cdot3\cdot58)$$

Upon dividing (3·3·52) by A we may write

$$J_0(\lambda_n r_o) + \frac{B_n}{A_n} Y_0(\lambda_n r_o) = 0$$

By comparison with (3·3·55) we see that this means

$$C_0(\lambda_n r_o) = 0 \qquad (3\cdot3\cdot59)$$

The solution for $u(r,\theta) = R(r)\Theta(\theta)$ will then be written in terms of cylinder functions as

$$u(r,\theta) = \sum_{n=1}^{\infty} A_n C_0(\lambda_n r) \exp(-\lambda_n^2 \alpha\theta) \qquad (3\cdot3\cdot60)$$

The values of A_n will be chosen to force the solution (3·3·57) to fit the initial condition (3·3·46). Setting $\theta = 0$ in (3·3·57) and substituting into (3·3·46) gives

$$\sum_{n=1}^{\infty} A_n C_0(\lambda_n r) = t^{(0)} - t_o$$

Cylinder functions are orthogonal with the same weighting function as Bessel functions. To determine the A_n we will multiply both sides by $rC_0(\lambda_m r)\,dr$ and integrate from $r = r_i$ to r_o to obtain

$$\sum_{n=1}^{\infty} A_n \int_{r=r_i}^{r_o} rC_0(\lambda_n r)C_0(\lambda_m r)\,dr = (t^{(0)} - t_o)\int_{r=r_i}^{r_o} rC_0(\lambda_m r)\,dr \qquad (3\cdot3\cdot61)$$

The integral on the left-hand side may be evaluated with the help of the cylinder function integrals in Table B•3. For $n \neq m$ we may write

$$\int rC_0(\lambda_n r)C_0(\lambda_m r)\,dr = \frac{r}{\lambda_n^2 - \lambda_m^2}\left[\lambda_n C_1(\lambda_n r)C_0(\lambda_m r) - \lambda_m C_0(\lambda_n r)C_1(\lambda_m r)\right]$$

At $r = r_o$, the right-hand side will be zero because $C_0(\lambda_n r_o) = 0$ as given by (3·3·59). At $r = r_i$, the right-hand side will be zero because $C_1(\lambda_n r_i) = 0$ as given by (3·3·58). Therefore the integral on the left-hand side of (3·3·61) will be zero for $n \neq m$. Therefore only the term for $n = m$ remains and (3·3·61) simplifies to give

$$A_m \int_{r=r_i}^{r_o} rC_0^2(\lambda_m r)\,dr = (t^{(0)} - t_o)\int_{r=r_i}^{r_o} rC_0(\lambda_m r)\,dr \qquad (3\cdot3\cdot62)$$

The integral on the left-hand side may be evaluated with the help of Table B•3. We may write

$$\int rC_0^2(\lambda_m r)\,dr = -\frac{r^2}{4}\left[C_{-1}(\lambda_m r)C_1(\lambda_m r) - 2C_0^2(\lambda_m r) + C_1(\lambda_m r)C_{-1}(\lambda_m r)\right]$$

At $r = r_o$ we see from (3·3·59) that $C_0(\lambda_m r_o) = 0$. At $r = r_i$ we see from (3·3·58) that $C_1(\lambda_m r_i) = 0$. Therefore we may write

$$\int_{r=r_i}^{r_o} r C_0^2(\lambda_m r)\,dr = -\frac{r_o^2}{4}\Big[C_{-1}(\lambda_m r_o)C_1(\lambda_m r_o) + C_1(\lambda_m r_o)C_{-1}(\lambda_m r_o)\Big]$$
$$+ \frac{r_i^2}{4}\Big[-2C_0^2(\lambda_m r_i)\Big]$$

This may be further simplified by recognizing, from the definition of $C_k(x)$ in Table B•3 and the relations in Table B•1 that $C_{-1}(x) = -C_1(x)$. Therefore we may write

$$\int_{r=r_i}^{r_o} r C_0^2(\lambda_m r)\,dr = -\frac{r_o^2}{4}\Big[-C_1^2(\lambda_m r_o) - C_1^2(\lambda_m r_o)\Big] + \frac{r_i^2}{4}\Big[-2C_0^2(\lambda_m r_i)\Big]$$

This may be further simplified to give

$$\int_{r=r_i}^{r_o} r C_0^2(\lambda_m r)\,dr = \frac{1}{2}\Big[r_o^2 C_1^2(\lambda_m r_o) - r_i^2 C_0^2(\lambda_m r_i)\Big] \tag{3·3·63}$$

The integral on the right-hand side of (3·3·62) may be evaluated with the help of Table B•3. We may write

$$\int_{r=r_i}^{r_o} r C_0(\lambda_m r)\,dr = \frac{1}{\lambda_m^2}\Big[\lambda_m r C_1(\lambda_m r)\Big]_{r=r_i}^{r_o} = \frac{1}{\lambda_m}\Big[r_o C_1(\lambda_m r_o) - r_i C_1(\lambda_m r_i)\Big]$$

Since $C_1(\lambda_m r_i) = 0$ this simplifies to give

$$\int_{r=r_i}^{r_o} r C_0(\lambda_m r)\,dr = \frac{1}{\lambda_m} r_o C_1(\lambda_m r_o) \tag{3·3·64}$$

Substituting (3·3·63) and (3·3·64) into (3·3·62) and solving for A_m gives

$$A_m = (t^{(0)} - t_o)\frac{\dfrac{1}{\lambda_m} r_o C_1(\lambda_m r_o)}{\dfrac{1}{2}\Big[r_o^2 C_1^2(\lambda_m r_o) - r_i^2 C_0^2(\lambda_m r_i)\Big]}$$

This may be rearranged to give

$$A_m = 2(t^{(0)} - t_o)\frac{(\lambda_m r_o)C_1(\lambda_m r_o)}{\Big[(\lambda_m r_o)^2 C_1^2(\lambda_m r_o) - (\lambda_m r_i)^2 C_0^2(\lambda_m r_i)\Big]} \tag{3·3·65}$$

Upon substituting (3·3·65) into (3·3·60) we obtain

$$u(r,\theta) = 2(t^{(0)} - t_o)\sum_{n=1}^{\infty}\frac{(\lambda_n r_o)C_1(\lambda_n r_o)C_0(\lambda_n r)}{\Big[(\lambda_n r_o)^2 C_1^2(\lambda_n r_o) - (\lambda_n r_i)^2 C_0^2(\lambda_n r_i)\Big]}\exp(-\lambda_n^2 \alpha\theta)$$

This may then be substituted into (3·3·42) to obtain the final solution as

$$t(r,\theta) = t_o + 2(t^{(0)} - t_o)\sum_{n=1}^{\infty}\frac{(\lambda_n r_o)C_1(\lambda_n r_o)C_0(\lambda_n r)}{\Big[(\lambda_n r_o)^2 C_1^2(\lambda_n r_o) - (\lambda_n r_i)^2 C_0^2(\lambda_n r_i)\Big]}\exp(-\lambda_n^2 \alpha\theta)$$

where the λ_n are the zeros of (3·3·53).

3•3•4 Solid cylinder, peripheral surface-temperature variation

The steady-state analysis of a solid cylinder, Figure 3•45, whose surface temperature varies with angular position around the circumference of the cylinder is another application of separation of variables to a two-dimensional problem. It also illustrates another boundary condition that we have not discussed. The mathematical description of the problem is given in Table 3•34. For constant conductivity, no temperature gradient in the z-direction, no generation and steady state, (1·1·11) reduces to (3·3·66). Since (3·3·66) is second order in r, two r-boundary conditions are needed. At the center of the cylinder the temperature must be finite (3·3·67). At the surface of the cylinder the temperature will be a specified function angular position $t_o(\phi)$ as given by (3·3·68). Since (3·3·66) is second order in ϕ, two ϕ-boundary conditions are needed. From what we have seen so far, we would expect to specify either t or t_ϕ along two constant values of ϕ. In a circular cylinder there is no surface along constant ϕ where t or t_ϕ is specified as there would be if the cylinder cross-section were a quarter circle or a half circle rather than a full circle. Instead our ϕ-boundary condition will be that, for all r and all ϕ, the temperature at ϕ must be equal to the temperature at $\phi + 2\pi$. This periodic boundary condition is given by (3·3·69). This condition takes the place of two boundary conditions since, if t is periodic, t_ϕ will also be periodic.

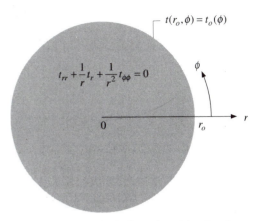

Figure 3•45 *Solid cylinder with peripheral surface-temperature variation.*

Table 3•34 *Mathematical problem description for $t(r,\phi)$.*

Problem	Equation	Boundary Conditions	
$t(r,\phi)$	$t_{rr} + \dfrac{1}{r}t_r + \dfrac{1}{r^2}t_{\phi\phi} = 0$		(3·3·66)
		$t(0,\phi) < \infty$	(3·3·67)
		$t(r_o,\phi) = t_o(\phi)$	(3·3·68)
		$t(r,\phi) = t(r,\phi + 2\pi)$	(3·3·69)

The boundary condition at $r = 0$ is homogeneous, but the boundary condition at $r = r_o$ is nonhomogeneous. However, the ϕ-boundary condition (3·3·69) fits the definition of a homogeneous boundary condition. Therefore the problem is homogeneous enough that we can apply separation of variables and assume

$$t(r,\phi) = R(r)\Phi(\phi) \tag{3·3·70}$$

Substituting (3·3·70) into (3·3·66) gives

$$R''(r)\Phi(\phi) + \frac{1}{r}R'(r)\Phi(\phi) + \frac{1}{r^2}R(r)\Phi''(\phi) = 0$$

Multiplying by r^2, dividing by $R(r)\Phi(\phi)$ and then separating variables,

$$r^2\frac{R''(r)}{R(r)} + r\frac{R'(r)}{R(r)} = -\frac{\Phi''(\phi)}{\Phi(\phi)}$$

The only way that a function of r (left-hand side) can be equal to a function of ϕ (right-hand side) for all r and ϕ is if both sides are equal to the same constant. Since the ϕ-boundary condition is homogeneous, we will choose the constant so that the solution for $\Phi(\phi)$ will be oscillatory. This will give an infinite number of eigenvalues and eigenfunctions that can be used to satisfy the nonhomogeneous boundary condition at $r = r_o$. Thus, we will take the constant to be λ^2 to give

$$r^2 \frac{R''(r)}{R(r)} + r \frac{R'(r)}{R(r)} = \lambda^2 \qquad \text{or} \qquad r^2 R''(r) + r R'(r) - \lambda^2 R(r) = 0 \qquad (3\cdot3\cdot71)$$

and

$$-\frac{\Phi''(\phi)}{\Phi(\phi)} = \lambda^2 \qquad \text{or} \qquad \Phi''(\phi) + \lambda^2 \Phi(\phi) = 0 \qquad (3\cdot3\cdot72)$$

The solution to (3·3·72) is given by

$$\Phi(\phi) = A\sin(\lambda\phi) + B\cos(\lambda\phi) \qquad (3\cdot3\cdot73)$$

In order for (3·3·70) to be periodic to satisfy (3·3·69) we want

$$\Phi(\phi) = \Phi(\phi + 2\pi)$$

Substituting (3·3·73) into this relation,

$$A\sin(\lambda\phi) + B\cos(\lambda\phi) = A\sin[\lambda(\phi + 2\pi)] + B\cos[\lambda(\phi + 2\pi)]$$

This will be true if λ can be chosen so that

$$\sin(\lambda\phi) = \sin[\lambda(\phi + 2\pi)] \qquad (3\cdot3\cdot74)$$

$$\cos(\lambda\phi) = \cos[\lambda(\phi + 2\pi)] \qquad (3\cdot3\cdot75)$$

Using the trigonometric formula for the sine of the sum of two angles, we may write (3·3·74) as

$$\sin(\lambda\phi) = \sin(\lambda\phi)\cos(\lambda 2\pi) + \cos(\lambda\phi)\sin(\lambda 2\pi)$$

For $\lambda = n$ where $n = 0, 1, 2, \cdots$, this reduces to

$$\sin(n\phi) = \sin(n\phi)\cos(n2\pi) + \cos(n\phi)\sin(n2\pi)$$

Since $\cos(n2\pi) = 1$ and $\sin(n2\pi) = 0$ we see that this is an identity. Hence $\lambda = n$ where $n = 0, 1, 2, \cdots$ will satisfy (3·3·74). Similarly, the same choice for λ will also satisfy (3·3·75). Thus the eigenvalues for this problem are given by

$$\lambda_n = n \text{ where } n = 0, 1, 2, \cdots \qquad (3\cdot3\cdot76)$$

Thus the separated solution (3·3·73) for Φ may be written as

$$\Phi_n(\phi) = A_n \sin(n\phi) + B_n \cos(n\phi) \qquad (3\cdot3\cdot77)$$

Next, we must solve (3·3·71) for R. The first thought is that (3·3·71) is a Bessel equation but comparison with (2·4·1) will show it is not. Rather, it is one of a class of equations called *Euler equations*. In general, if you have

$$a_0 r^\upsilon R^{(\upsilon)} + a_1 r^{\upsilon-1} R^{(\upsilon-1)} + \cdots + a_{n-1} r R' + a_n R = 0$$

where $R^{(\upsilon)}$ denotes the υth derivative of R, a change of independent variable from r to z where

$$r = \exp(z) \qquad \text{or} \qquad z = \ln(r) \qquad (3\cdot3\cdot78)$$

will simplify the differential equation. The first derivative of R is given by

$$R' = \frac{dR}{dr} = \frac{dR}{dz}\frac{dz}{dr} = \frac{dR}{dz}\frac{1}{r}$$

The second derivative is given by

$$R'' = \frac{d}{dr}\left[\frac{dR}{dr}\right] = \frac{d}{dr}\left[\frac{dR}{dz}\frac{1}{r}\right] = \frac{d}{dr}\left[\frac{dR}{dz}\right]\frac{1}{r} + \frac{dR}{dz}\frac{d}{dr}\left[\frac{1}{r}\right]$$

$$= \frac{d}{dz}\left[\frac{dR}{dz}\right]\frac{dz}{dr}\frac{1}{r} + \frac{dR}{dz}\frac{-1}{r^2} = \frac{1}{r^2}\frac{d^2R}{dz^2} - \frac{1}{r^2}\frac{dR}{dz}$$

Substituting these results for R' and R'' into (3·3·71) gives

$$r^2\left[\frac{1}{r^2}\frac{d^2R}{dz^2} - \frac{1}{r^2}\frac{dR}{dz}\right] + r\left[\frac{dR}{dz}\frac{1}{r}\right] - \lambda^2 R = 0$$

Removing the brackets and noting that the dR/dz terms cancel, we obtain (after also using the eigencondition that $\lambda = n$)

$$\frac{d^2R}{dz^2} - n^2 R = 0 \tag{3·3·79}$$

For $n = 0$ we may write (3·3·79) as

$$\frac{d^2R_0}{dz^2} = 0$$

This has the solution

$$R_0 = \alpha_0' z + \alpha_0$$

From (3·3·78) we may replace z by $\ln(r)$ to give

$$R_0 = \alpha_0' \ln(r) + \alpha_0 \tag{3·3·80}$$

For $n \neq 0$ the solution to (3·3·79) is given by

$$R_n = \alpha_n \exp(nz) + \alpha_n' \exp(-nz)$$

This may be rewritten as

$$R_n = \alpha_n \exp^n(z) + \alpha_n' \exp^{-n}(z) = \alpha_n[\exp(z)]^n + \alpha_n'[\exp(z)]^{-n}$$

From (3·3·78) we may replace $\exp(z)$ by r to give

$$R_n = \alpha_n r^n + \alpha_n' r^{-n} \tag{3·3·81}$$

We may now write the solution (3·3·70) as

$$t(r,\phi) = R_0(r)\Phi_0(\phi) + \sum_{n=1}^{\infty} R_n(r)\Phi_n(\phi)$$

Substituting (3·3·80) for $R_0(r)$, (3·3·81) for $R_n(r)$ and (3·3·77) for Φ,

$$t(r,\phi) = \left[\alpha_0' \ln(r) + \alpha_0\right]B_0 + \sum_{n=1}^{\infty}\left[\alpha_n r^n + \alpha_n' r^{-n}\right]\left[A_n \sin(n\phi) + B_n \cos(n\phi)\right]$$

We will let $A_0' = \alpha_0' B_0$ and $A_0 = \alpha_0 B_0$ to simplify the first term. We will also write the series as two series. Thus,

$$t(r,\phi) = A_0' \ln(r) + A_0 + \sum_{n=1}^{\infty}\left[\alpha_n r^n + \alpha_n' r^{-n}\right]A_n \sin(n\phi)$$

$$+ \sum_{n=1}^{\infty}\left[\alpha_n r^n + \alpha_n' r^{-n}\right]B_n \cos(n\phi)$$

The coefficients in the two sums may be simplified to give

$$t(r,\phi) = A_0' \ln(r) + A_0 + \sum_{n=1}^{\infty} \left[A_n r^n + A_n' r^{-n} \right] \sin(n\phi)$$

$$+ \sum_{n=1}^{\infty} \left[B_n r^n + B_n' r^{-n} \right] \cos(n\phi)$$

To satisfy the boundary condition (3·3·67) at $r = 0$, we must eliminate each of the terms that go to ∞ at $r = 0$. The solution then reduces to

$$t(r,\phi) = A_0 + \sum_{n=1}^{\infty} A_n r^n \sin(n\phi) + \sum_{n=1}^{\infty} B_n r^n \cos(n\phi) \qquad (3\cdot3\cdot82)$$

So far the solution (3·3·82) satisfies the differential equation (3·3·66) and two boundary conditions (3·3·67) and (3·3·69). We now must satisfy the remaining boundary condition (3·3·68). Setting $r = r_o$ in (3·3·82) and then substituting into (3·3·68) gives

$$t(r_o,\phi) = A_0 + \sum_{n=1}^{\infty} A_n r_o^n \sin(n\phi) + \sum_{n=1}^{\infty} B_n r_o^n \cos(n\phi) = t_o(\phi) \qquad (3\cdot3\cdot83)$$

We next will use orthogonality to evaluate the coefficients. The orthogonality integrals we need are the following:

$$\int_{\phi=0}^{2\pi} \sin(n\phi)\sin(m\phi)\,d\phi = \int_{\phi=0}^{2\pi} \cos(n\phi)\cos(m\phi)\,d\phi = 0 \qquad \text{for } m \neq n$$

$$\int_{\phi=0}^{2\pi} \sin(n\phi)\cos(m\phi)\,d\phi = 0 \qquad \text{for all } m, n$$

$$\int_{\phi=0}^{2\pi} \sin^2(n\phi)\,d\phi = \int_{\phi=0}^{2\pi} \cos^2(n\phi)\,d\phi = \pi \qquad \text{for } n \neq 0$$

$$\int_{\phi=0}^{2\pi} \sin(n\phi)\,d\phi = \int_{\phi=0}^{2\pi} \cos(n\phi)\,d\phi = 0 \qquad \text{for } n \neq 0$$

Multiplying (3·3·83) by $d\phi$, integrating from $\phi = 0$ to 2π and using the appropriate orthogonality integrals,

$$\int_{\phi=0}^{2\pi} A_0\,d\phi = \int_{\phi=0}^{2\pi} t_o(\phi)\,d\phi$$

Since A_0 is a constant it may be factored out of the integral and the integral may be integrated. Upon solving for A_0,

$$A_0 = \frac{1}{2\pi} \int_{\phi=0}^{2\pi} t_o(\phi)\,d\phi \qquad (3\cdot3\cdot84)$$

Note that A_0 is the average temperature on the boundary at $r = r_o$.

To evaluate A_m for $m \geq 1$, multiply (3·3·83) by $\sin(m\phi)\,d\phi$ and integrate from $\phi = 0$ to 2π. Using the appropriate orthogonality integrals, two of the integrals are always zero and the result it given by

$$\sum_{n=1}^{\infty} A_n r_o^n \int_{\phi=0}^{2\pi} \sin(n\phi)\sin(m\phi)\,d\phi = \int_{\phi=0}^{2\pi} t_o(\phi)\sin(m\phi)\,d\phi$$

The integral on the left-hand side is 0 for $n \neq m$ and is equal to π for $n = m$. Therefore only the term for $n = m$ remains in the sum and we may solve for A_m as

$$A_m = \frac{1}{r_o^m \pi} \int_{\phi=0}^{2\pi} t_o(\phi)\sin(m\phi)\,d\phi \qquad (3\cdot3\cdot85)$$

Similarly, to evaluate B_m, multiply (3.3.83) by $\cos(m\phi)d\phi$ and integrate from $\phi = 0$ to 2π. Using the appropriate orthogonality integrals, one finds

$$B_m = \frac{1}{r_o^m \pi} \int_{\phi=0}^{2\pi} t_o(\phi)\cos(m\phi)\,d\phi \tag{3.3.86}$$

For a given $t_o(\phi)$, A_0, A_n and B_n can be found from (3.3.84), (3.3.85) and (3.3.86), respectively. These results can then be substituted into (3.3.82) to obtain the final solution.

3•3•5 Solid sphere, specified surface temperature

Consider a solid sphere initially at a uniform temperature $t^{(0)}$. Then at $\theta = 0$ the surface temperature is stepped to t_o. The mathematical description of the problem is given in Table 3•35 and pictured in Figure 3•46.

Table 3•35 *Mathematical problem description for $t(r,\theta)$.*

Problem	Equation	Boundary/Initial Conditions	
$t(r,\theta)$	$k[t_{rr} + \dfrac{2}{r}t_r] = \rho c t_\theta$		(3.3.87)
		$t(0,\theta) < \infty$	(3.3.88)
		$t(r_o,\theta) = t_o$	(3.3.89)
		$t(r,0) = t^{(0)}$	(3.3.90)

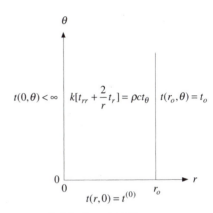

Figure 3•46 *Solid sphere, initially at temperature $t^{(0)}$, is suddenly exposed to a step change in surface temperature to t_o.*

Due to the nonhomogeneous boundary condition (3.3.89) we will take

$$t(r,\theta) = a(r) + u(r,\theta) \tag{3.3.91}$$

Substituting (3.3.91) into (3.3.87) through (3.3.90) sets up problems for $a(r)$ and $u(r,\theta)$. Upon solving for $a(r)$, it turns out that $a(r) = t_o$. The problems for $t(r,\theta)$ and $u(r,\theta)$ are then given in Table 3•36. The homogeneous problem for $u(r,\theta)$ is pictured in Figure 3•47.

Table 3•36 *Mathematical problem description for $t(r,\theta)$ and $u(r,\theta)$.*

Problem	Equation	Boundary/Initial Conditions	
$t(r,\theta) = t_o + u(r,\theta)$			(3.3.92)
$u(r,\theta)$	$k[u_{rr} + \dfrac{2}{r}u_r] = \rho c u_\theta$		(3.3.93)
		$u(0,\theta) < \infty$	(3.3.94)
		$u(r_o,\theta) = 0$	(3.3.95)
		$u(r,0) = t^{(0)} - t_o$	(3.3.96)

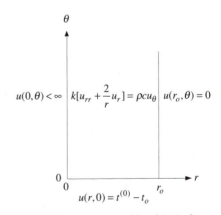

Figure 3•47 *Homogeneous problem for $u(r,\theta)$ corresponding to Figure 3•46.*

One approach to this problem would be to apply separation of variables to (3.3.93) by assuming that

$$u(r,\theta) = R(r)\Theta(\theta) \tag{3.3.97}$$

Substitution into (3.3.93) gives

$$k[R''(r)\Theta(\theta) + \frac{2}{r}R'(r)\Theta(\theta)] = \rho c R(r)\Theta'(\theta)$$

Separating the variables in the usual manner gives

$$\frac{R''(r)}{R(r)} + \frac{2}{r}\frac{R'(r)}{R(r)} = \frac{1}{\alpha}\frac{\Theta'(\theta)}{\Theta(\theta)} = -\lambda^2$$

One separated equation is given by

$$\Theta'(\theta) + \lambda^2\alpha\Theta(\theta) = 0$$

with the solution

$$\Theta(\theta) = \exp(-\lambda^2\alpha\theta) \qquad\qquad (3\cdot3\cdot98)$$

The second separated equation may be written as

$$r^2R''(r) + 2rR'(r) + \lambda^2r^2R(r) = 0 \qquad\qquad (3\cdot3\cdot99)$$

The boundary conditions for (3·3·99) are found by substituting (3·3·97) into (3·3·94) and (3·3·95). This gives

$$R(0) < \infty \qquad\qquad (3\cdot3\cdot100)$$

$$R(r_o) = 0 \qquad\qquad (3\cdot3\cdot101)$$

By comparison with the general Bessel equation (2·4·1), the solution to (3·3·99) is given by

$$R(r) = r^{-1/2}[C_1J_{\frac{1}{2}}(\lambda r) + C_2Y_{\frac{1}{2}}(\lambda r)]$$

or, using the relation that $Y_{\frac{1}{2}}(\lambda r) = -J_{-\frac{1}{2}}(\lambda r)$,

$$R(r) = \frac{1}{\sqrt{r}}[C_1J_{\frac{1}{2}}(\lambda r) - C_2J_{-\frac{1}{2}}(\lambda r)]$$

Recall from (2·2·34) and (2·2·35) that

$$J_{\frac{1}{2}}(x) = \sqrt{\frac{2}{\pi x}}\sin(x) \qquad\text{and}\qquad J_{-\frac{1}{2}}(x) = \sqrt{\frac{2}{\pi x}}\cos(x)$$

Thus

$$R(r) = \frac{1}{\sqrt{r}}\left[C_1\sqrt{\frac{2}{\pi\lambda r}}\sin(\lambda r) - C_2\sqrt{\frac{2}{\pi\lambda r}}\cos(\lambda r)\right]$$

or, upon factoring out the $\sqrt{r}$ in the denominator, we may write

$$R(r) = \frac{1}{r}[C_3\sin(\lambda r) + C_4\cos(\lambda r)]$$

To satisfy (3·3·100) we must take $C_4 = 0$. To satisfy (3·3·101) we must require

$$\sin(\lambda r_o) = 0$$

Thus the eigenvalues are given by

$$\lambda_n r_o = n\pi \qquad\text{for}\qquad n = 1, 2, 3, \ldots \qquad\qquad (3\cdot3\cdot102)$$

and $R(r)$ may be written as

$$R(r) = \frac{1}{r}\sin(\lambda_n r) \qquad\qquad (3\cdot3\cdot103)$$

Substituting (3·3·98) and (3·3·103) into (3·3·97) and summing over all the eigenvalues gives

$$u(r,\theta) = \sum_{n=1}^{\infty}A_n\frac{1}{r}\sin(\lambda_n r)\exp(-\lambda_n^2\alpha\theta) \qquad\qquad (3\cdot3\cdot104)$$

To find A_n we will set $\theta = 0$ and substitute (3·3·104) into the initial condition (3·3·96) to obtain

$$u(r,0) = \sum_{n=1}^{\infty} A_n \frac{1}{r} \sin(\lambda_n r) = t^{(0)} - t_o \tag{3·3·105}$$

Multiplying both sides by $r\sin(\lambda_m r)\,dr$ and integrating from $r = 0$ to r_o,

$$\sum_{n=1}^{\infty} A_n \int_{r=0}^{r_o} \sin(\lambda_n r)\sin(\lambda_m r)\,dr = (t^{(0)} - t_o)\int_{r=0}^{r_o} r\sin(\lambda_m r)\,dr$$

The integral on the left-hand side is zero for $n \neq m$. Therefore only one term in the sum will remain and we can write

$$A_m \int_{r=0}^{r_o} \sin^2(\lambda_m r)\,dr = (t^{(0)} - t_o)\int_{r=0}^{r_o} r\sin(\lambda_m r)\,dr \tag{3·3·106}$$

These integrals may be evaluated as

$$\int_{r=0}^{r_o} \sin^2(\lambda_m r)\,dr = \frac{1}{\lambda_m}\left[\frac{\lambda_m r}{2} - \frac{1}{4}\sin(2\lambda_m r)\right]_{r=0}^{r_o} = \frac{r_o}{2}$$

$$\int_{r=0}^{r_o} r\sin(\lambda_m r)\,dr = \left[\frac{1}{\lambda_m^2}\sin(\lambda_m r) - \frac{r}{\lambda_m}\cos(\lambda_m r)\right]_{r=0}^{r_o} = -\frac{r_o}{\lambda_m}\cos(\lambda_m r_o)$$

Substituting these integrals into (3·3·106) gives

$$A_m \frac{r_o}{2} = -(t^{(0)} - t_o)\frac{r_o}{\lambda_m}\cos(\lambda_m r_o)$$

Solving for A_m,

$$A_m = (t_o - t^{(0)})\frac{2}{\lambda_m}\cos(\lambda_m r_o) \tag{3·3·107}$$

Substituting these coefficients into (3·3·104) gives

$$u(r,\theta) = 2(t_o - t^{(0)})\sum_{n=1}^{\infty} \cos(\lambda_n r_o)\frac{\sin(\lambda_n r)}{\lambda_n r}\exp(-\lambda_n^2 \alpha\theta)$$

This may now be substituted into (3·3·92) to obtain the final solution as

$$t(r,\theta) = t_o + 2(t_o - t^{(0)})\sum_{n=1}^{\infty} \cos(\lambda_n r_o)\frac{\sin(\lambda_n r)}{\lambda_n r}\exp(-\lambda_n^2 \alpha\theta) \tag{3·3·108}$$

Conversion to plane-wall problem   An alternative approach to solving the original problem given in Table 3•35 is to let

$$t(r,\theta) = \frac{1}{r}u(r,\theta) \tag{3·3·109}$$

Then

$$t_r = \frac{1}{r}u_r - \frac{1}{r^2}u \tag{3·3·110}$$

$$t_{rr} = \frac{1}{r}u_{rr} - \frac{1}{r^2}u_r - \frac{1}{r^2}u_r + \frac{2}{r^3}u = \frac{1}{r}u_{rr} - \frac{2}{r^2}u_r + \frac{2}{r^3}u \tag{3·3·111}$$

$$t_\theta = \frac{1}{r}u_\theta \tag{3·3·112}$$

Substituting (3·3·110), (3·3·111) and (3·3·112) into (3·3·87) gives

$$k[\frac{1}{r}u_{rr} - \frac{2}{r^2}u_r + \frac{2}{r^3}u + \frac{2}{r}(\frac{1}{r}u_r - \frac{1}{r^2}u)] = \rho c \frac{1}{r}u_\theta$$

The terms involving u and u_r cancel and after then multiplying by r we obtain

$$ku_{rr} = \rho c u_\theta \tag{3.3.113}$$

Observe that (3.3.113) is just the plane-wall heat equation. Substitution of (3.3.109) into (3.3.88) requires that $u(0,\theta) = 0$. Substitution of (3.3.109) into (3.3.89) requires that $u(r_o,\theta) = r_o t_o$. Substitution of (3.3.109) into the initial condition (3.3.90) requires that $u(r,0) = t^{(0)}r$. The problems for $t(r,\theta)$ and $u(r,\theta)$ are given in Table 3•37. The plane-wall problem for $u(r,\theta)$ is pictured in Figure 3•48.

Table 3•37 *Mathematical problem description for $t(r,\theta)$ and $u(r,\theta)$.*

Problem	Equation	Boundary/Initial Conditions	
$t(r,\theta) = \frac{1}{r}u(r,\theta)$			(3.3.109)
$u(r,\theta)$	$ku_{rr} = \rho c u_\theta$		(3.3.113)
		$u(0,\theta) = 0$	(3.3.114)
		$u(r_o,\theta) = r_o t_o$	(3.3.115)
		$u(r,0) = t^{(0)}r$	(3.3.116)

The nonhomogeneous plane-wall problem for $u(r,\theta)$ can be solved following the steps given in Section 3•1•1. To handle the nonhomogeneous boundary condition ((3.3.115) we will assume

$$u(r,\theta) = a(r) + v(r,\theta) \tag{3.3.117}$$

Substituting (3.3.117) into (3.3.113) through (3.3.116) and constructing a homogeneous problem for $v(r,\theta)$ gives the mathematical problems in Table 3•38. The plane-wall problem for $v(r,\theta)$ is pictured in Figure 3•49.

Table 3•38 *Mathematical problems for $t(r,\theta)$, $a(r)$ and $v(r,\theta)$.*

Problem	Equation	Boundary/Initial Conditions	
$t(r,\theta) = \frac{1}{r}[a(r) + v(r,\theta)]$			(3.3.118)
$a(r)$	$a''(r) = 0$		(3.3.119)
		$a(0) = 0$	(3.3.120)
		$a(r_o) = r_o t_o$	(3.3.121)
$v(r,\theta)$	$kv_{rr} = \rho c v_\theta$		(3.3.122)
		$v(0,\theta) = 0$	(3.3.123)
		$v(r_o,\theta) = 0$	(3.3.124)
		$v(r,0) = t^{(0)}r - a(r)$	(3.3.125)

The solution for $a(r)$ is

$$a(r) = t_o r$$

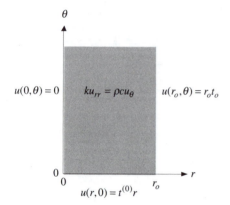

Figure 3•48 *Plane-wall problem for $u(r,\theta)$.*

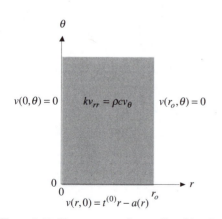

Figure 3•49 *Homogeneous plane-wall problem for $v(r,\theta)$ corresponding to Figure 3•41.*

The updated mathematical problems are now given in Table 3•39. The updated problem for $v(r,\theta)$ is pictured in Figure 3•50.

Table 3•39 *Updated mathematical problems for $t(r,\theta)$ and $v(r,\theta)$.*

Problem	Equation	Boundary/Initial Conditions	
$t(r,\theta) = t_o + \dfrac{1}{r}v(r,\theta)$			(3·3·126)
$v(r,\theta)$	$kv_{rr} = \rho c v_\theta$		(3·3·122)
		$v(0,\theta) = 0$	(3·3·123)
		$v(r_o,\theta) = 0$	(3·3·124)
		$v(r,0) = (t^{(0)} - t_o)r$	(3·3·127)

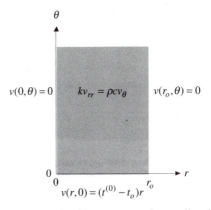

Figure 3•50 *Updated homogeneous plane-wall problem for $v(r,\theta)$ corresponding to Figure 3•41.*

Since the problem for $v(r,\theta)$ is homogeneous we will assume

$$v(r,\theta) = R(r)\Theta(\theta) \tag{3·3·128}$$

Substituting (3·3·128) into (3·3·122), (3·3·123) and (3·3·124) gives

$$R_n(r) = \sin(\lambda_n r)$$

$$\Theta_n(\theta) = \exp(-\lambda_n^2 \alpha\theta)$$

$$\lambda_n r_o = n\pi \qquad \text{where} \qquad n = 1, 2, 3, \cdots$$

We will then write

$$v(r,\theta) = \sum_{n=1}^{\infty} A_n \sin(\lambda_n r)\exp(-\lambda_n^2\alpha\theta) \tag{3·3·129}$$

To satisfy the initial condition (3·3·127) we will set $\theta = 0$ in (3·3·129) and substitute it into (3·3·127) to obtain

$$v(r,0) = \sum_{n=1}^{\infty} A_n \sin(\lambda_n r) = (t^{(0)} - t_o)r$$

This is seen to be the same equation as (3·3·105) whose Fourier coefficients were found to be given by (3·3·107). Thus (3·3·129) becomes

$$v(r,\theta) = 2(t_o - t^{(0)})\sum_{n=1}^{\infty}\frac{1}{\lambda_n}\cos(\lambda_n r_o)\sin(\lambda_n r)\exp(-\lambda_n^2\alpha\theta)$$

Substituting this result into (3·3·126) and rearranging gives the same final solution (3·3·108) as we obtained before.

3•3•6   Reduction to one-dimensional transients

Certain two-dimensional transient problems[1] can be reduced to the solution of two one-dimensional problems. This is not much of a saving in time if neither of the solutions to the one-dimensional problems is available to you. However, many times one or both of the one-dimensional solutions can be found in the literature. Most elementary heat-transfer textbooks [6,7] present several basic solutions in chart form. Clearly, if the one-dimensional solutions are available, a considerable amount of time and effort can be saved.

---

[1] This will also work for three-dimensional transients in rectangular coordinates.

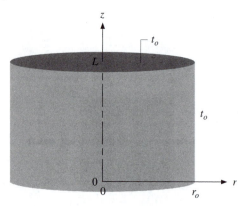

Figure 3•51 *Short circular cylinder, initially at temperature $t^{(0)}$ and suddenly exposed to a step change in surface temperature to t_o.*

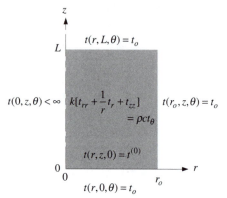

Figure 3•52 *Mathematical model for a short circular cylinder, initially at temperature $t^{(0)}$ and suddenly exposed to a step change in surface temperature to t_o.*

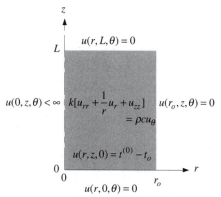

Figure 3•53 *Homogeneous problem for $u(r,z,\theta)$ corresponding to Figure 3•52.*

At the very least, the problem must be linear and homogeneous for this reduction to occur. These are the same restrictions that were necessary for separation of variables to work as discussed in Section 3•1. In fact, the reduction is obtained by separating the variables but in a slightly different way than suggested in Section 3•2•2. The first step in Section 3•2•2 was to separate the time variable from the spatial variables. In this section we will separate the two spatial variables and thereby obtain two subproblems, each involving only one spatial variable and time.

As an example, consider the short cylinder initially at a uniform temperature shown in Figure 3•51. At time zero, the temperature of both ends and the curved surface is suddenly changed to a new value. The mathematical description of the problem is given in Table 3•40 and pictured in Figure 3•52.

Table 3•40 *Mathematical problem description for $t(r,z,\theta)$.*

Problem	Equation	Boundary/Initial Conditions	
$t(r,z,\theta)$	$k[t_{rr} + \dfrac{1}{r}t_r + t_{zz}] = \rho c t_\theta$		(3·3·130)
		$t(0,z,\theta) < \infty$	(3·3·131)
		$t(r_o,z,\theta) = t_o$	(3·3·132)
		$t(r,0,\theta) = t_o$	(3·3·133)
		$t(r,L,\theta) = t_o$	(3·3·134)
		$t(r,z,0) = t^{(0)}$	(3·3·135)

First, it is necessary to obtain a homogeneous problem. This can be done by assuming

$$t(r,z,\theta) = t_o + u(r,z,\theta) \qquad (3·3·136)$$

Substituting (3·3·136) into (3·3·130) through (3·3·135) results in the problem for $u(r,z,\theta)$ given in Table 3•41 and pictured in Figure 3•53.

Table 3•41 *Mathematical problem descriptions for $t(r,z,\theta)$ and $u(r,z,\theta)$.*

Problem	Equation	Boundary/Initial Conditions	
$t(r,z,\theta) = t_o + u(r,z,\theta)$			(3·3·136)
$u(r,z,\theta)$	$k[u_{rr} + \dfrac{1}{r}u_r + u_{zz}] = \rho c u_\theta$		(3·3·137)
		$u(0,z,\theta) < \infty$	(3·3·138)
		$u(r_o,z,\theta) = 0$	(3·3·139)
		$u(r,0,\theta) = 0$	(3·3·140)
		$u(r,L,\theta) = 0$	(3·3·141)
		$u(r,z,0) = t^{(0)} - t_o$	(3·3·142)

The spatial variables can be separated by assuming a product solution of the form

$$u(r,z,\theta) = v(r,\theta)w(z,\theta) \qquad (3·3·143)$$

Substituting this assumed form of the solution into (3·3·137) gives

$$k[v_{rr}w + \frac{1}{r}v_r w + vw_{zz}] = \rho c[v_\theta w + vw_\theta]$$

This may be rearranged to give

$$k[v_{rr} + \frac{1}{r}v_r]w + kw_{zz}v = \rho cv_\theta w + \rho cw_\theta v$$

This equation will be satisfied if the functions $v(r,\theta)$ and $w(z,\theta)$ satisfy the following partial differential equations:

$$k[v_{rr} + \frac{1}{r}v_r] = \rho cv_\theta \qquad\qquad (3\cdot3\cdot144)$$

$$kw_{zz} = \rho cw_\theta \qquad\qquad (3\cdot3\cdot145)$$

These are the differential equations for one-dimensional transients.

Next, (3·3·143) can be substituted into the boundary conditions (3·3·138), (3·3·139), (3·3·140) and (3·3·141) to give

$$u(0,z,\theta) = v(0,\theta)w(z,\theta) < \infty$$

$$u(r_o,z,\theta) = v(r_o,\theta)w(z,\theta) = 0$$

$$u(r,0,\theta) = v(r,\theta)w(0,\theta) = 0$$

$$u(r,L,\theta) = v(r,\theta)w(L,\theta) = 0$$

These relations can be satisfied by putting the following restrictions on $v(r,\theta)$ and $w(z,\theta)$:

$$v(0,\theta) < \infty \qquad\qquad (3\cdot3\cdot146)$$

$$v(r_o,\theta) = 0 \qquad\qquad (3\cdot3\cdot147)$$

$$w(0,\theta) = 0 \qquad\qquad (3\cdot3\cdot148)$$

$$w(L,\theta) = 0 \qquad\qquad (3\cdot3\cdot149)$$

Finally, we must be sure that the initial condition can be satisfied. Substitution of (3·3·143) into (3·3·142) gives

$$u(r,z,0) = v(r,0)w(z,0) = t^{(0)} - t_o$$

One choice regarding the functions $v(r,\theta)$ and $w(z,\theta)$ which will meet this initial condition is to let

$$v(r,0) = t^{(0)} - t_o \qquad\qquad (3\cdot3\cdot150)$$

and

$$w(z,0) = 1 \qquad\qquad (3\cdot3\cdot151)$$

The original two-dimensional transient problem has now been replaced by two one-dimensional transient problems, one for $v(r,\theta)$ and the other for $w(z,\theta)$. These problems are summarized in Table 3•42. The problem for $v(r,\theta)$ describes an infinitely long cylinder of radius r_o, initially at $t^{(0)} - t_o$, whose surface is suddenly stepped to t_o. This problem is pictured in Figure 3•54. The problem for $w(z,\theta)$ describes a plane wall of thickness L, initially at a temperature of 1, whose surfaces are both suddenly stepped to 0. This problem is pictured in Figure 3•55.

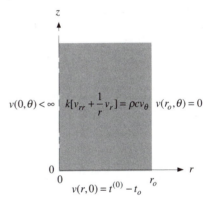

Figure 3•54 *One-dimensional, infinitely long, cylinder problem for $v(r,\theta)$ resulting from Figure 3•52.*

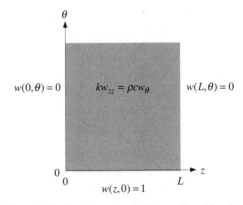

Figure 3•55 *One-dimensional, plane-wall problem for $w(z,\theta)$ resulting from Figure 3•52.*

Table 3•42 *Subproblems for* $t(x,y,\theta)$, $v(r,\theta)$ *and* $w(z,\theta)$.

Subproblem	Equation	Boundary/Initial Conditions	
$t(r,z,\theta) = t_o + u(r,z,\theta)$			(3·3·136)
$u(r,z,\theta) = v(r,\theta)w(z,\theta)$			(3·3·143)
$v(r,\theta)$	$k[v_{rr} + \dfrac{1}{r}v_r] = \rho c v_\theta$		(3·3·144)
		$v(0,\theta) < \infty$	(3·3·146)
		$v(r_o,\theta) = 0$	(3·3·147)
		$v(r,0) = t^{(0)} - t_o$	(3·3·150)
$w(z,\theta)$	$kw_{zz} = \rho c w_\theta$		(3·3·145)
		$w(0,\theta) = 0$	(3·3·148)
		$w(L,\theta) = 0$	(3·3·149)
		$w(z,0) = 1$	(3·3·151)

Both of these problems can be solved by methods discussed earlier in this chapter. The solution for $v(r,\theta)$ was derived in Section 3•3•1 and may be obtained from (3·3·23) as

$$v(r,\theta) = 2(t^{(0)} - t_o)\sum_{n=1}^{\infty} \frac{J_0(\lambda_n r)}{\lambda_n r_o J_1(\lambda_n r_o)}\exp(-\lambda_n^2 \alpha\theta) \qquad (3·3·152)$$

where the $(\lambda_n r_o)$ are the roots of $J_0(\lambda_n r_o) = 0$. The solution for $w(z,\theta)$ works out to be

$$w(z,\theta) = 4\sum_{m=1}^{\infty} \frac{1}{\lambda_m L}\sin(\lambda_m z)\exp(-\lambda_m^2 \alpha\theta) \qquad (3·3·153)$$

where $\lambda_m L = (2m-1)\pi$.

The final solution to the original problem is then obtained by multiplying the solutions for $v(r,\theta)$ and $w(z,\theta)$, (3·3·152) and (3·3·153), to obtain $u(r,z,\theta)$ as given by (3·3·143) which is then substituted into (3·3·136). This will be the same as the solution that can be obtained by separation of variables by assuming $u(r,z,\theta) = R(r)Z(z)\Theta(\theta)$ as suggested in Section 3•2•2. This involves solving two eigenvalue problems and applying orthogonality twice to satisfy the initial condition].[1]

This technique may also be used for transients in rectangular regions by making use of two plane-wall solutions. The convection boundary condition can be handled easily. The extension to a rectangular volume makes use of three plane-wall solutions.

Whenever you want to use this method on a new problem, it is advisable to go through the steps indicated in the above example to be sure that everything works out. Don't simply multiply two solutions together and think that you have the solution to your problem. Check to be sure that the differential equation, the boundary conditions and the initial condition are all satisfied. It is easy to misapply this technique if you are not careful.

---

[1] This is Exercise 3•56.

3•4 SUMMARY REMARKS ON SEPARATION OF VARIABLES

Chapter 3 has provided a fairly comprehensive coverage of the classical method of separation of variables. This technique is probably one of the most useful methods of solving partial differential equations. You have seen it applied to relatively simple problems and then extended to more-difficult problems.

Although you are now able to solve what look to be very complicated problems (to the beginning heat-transfer student anyway), separation of variables has two serious practical limitations. First of all, it is restricted to linear problems. This rules out such interesting heat-transfer phenomena as radiation and temperature-dependent properties.[1] Second, it is restricted to relatively simple geometries in which the boundaries coincide with the coordinates (*e.g.*, plates, cylinders and spheres). This can be a severe engineering restriction since many problems do not have such simple shapes. Often, however, useful results may be obtained by approximating more complicated geometries by one of the simpler ones.

The complicated, messy problems are most often handled on the computer by finite-difference or finite-element methods. These techniques are discussed in chapters 8 and 9.

Before moving on to finite-difference methods, however, three other analytical tools will be discussed: superposition, complex combination and Laplace transforms. These are treated in chapters 4, 5 and 6, respectively. Superposition is used to facilitate the handling of nonhomogeneous terms (provided the problem is linear, of course). Complex combination can be used for a problem with a periodic input when only the sustained portion of the solution is required. Laplace transforms are especially useful for certain geometries (*e.g.*, the semi-infinite solid) which are difficult to handle by separation of variables. These are probably the three most useful analytical tools beyond separation of variables. When a problem cannot be handled analytically by these methods, finite-difference or finite-element methods are usually employed.

---

[1] In certain circumstances these nonlinear problems can be linearized so that separation of variables can be used.

SELECTED REFERENCES

1. Berg, P. W. and J. L. McGregor: *Elementary Partial Differential Equations*, Holden-Day, Inc., San Francisco, 1966.

2. Boyce, W. E. and R. C. DiPrima: *Elementary Differential Equations and Boundary Value Problems*, 4th ed., John Wiley & Sons, Inc., New York, 1986.

3. Hildebrand, F. B.: *Advanced Calculus for Applications*, 2nd ed., Prentice-Hall, Inc., Englewood Cliffs, NJ, 1976.

4. Wylie, C. R. and L. C. Barrett: *Advanced Engineering Mathematics*, 5th ed., McGraw-Hill Book Company, New York, 1982.

5. Jolley, L. B. W.: *Summation of Series*, Dover Publications, Inc., New York, 1961.

6. Carslaw, H. S. and J. C. Jaeger: *Conduction of Heat in Solids*, 2nd ed., Oxford University Press, London, 1959.

7. Holman, J. P.: *Heat Transfer*, 6th ed., McGraw-Hill Book Company, New York, 1986.

8. Kreith, F. and M. S. Bohn: *Principles of Heat Transfer*, 4th ed., Harper & Row, New York, 1986.

9. Van Sant, J. H.: *Conduction Heat Transfer Solutions*, UCRL-52863, National Technical Information Service, Springfield, VA, 1983.

EXERCISES

3•1 Consider the following problem for $t(r,z,\theta)$:

$$k\left[t_{rr} + \frac{1}{r}t_r + t_{zz}\right] + g''' = \rho c t_\theta \qquad t_r(0,z,\theta) = 0$$

$$t(r_o,z,\theta) = 0$$

$$t(r,0,\theta) = 0$$

$$t(r,L,\theta) = 0$$

$$t(r,z,0) = 0$$

Examine both the differential equation and the boundary conditions for homogeneity and linearity.

3•2 Consider the following problem for $t(x,\theta)$:

$$kt_{xx} = \rho c t_\theta \qquad t_x(0,\theta) = 0$$

$$-kt_x(L,\theta) = h[t(L,\theta) - t_\infty]$$

$$t(x,0) = t^{(0)}$$

Examine both the differential equation and the boundary conditions for homogeneity and linearity.

3•3 A plane wall, initially at absolute temperature $T^{(0)}$, is suddenly exposed to *radiation* on each surface from sources at T_∞.
(a) State the differential equation and boundary and initial conditions for this problem.
(b) State whether the method of separation of variables will work. Explain your answer.

3•4 Consider the following problem for $t(x,\theta)$:

$$kt_{xx} + g''' = \rho c t_\theta \qquad t(0,\theta) = t_0$$

$$t(L,\theta) = 0$$

$$t(x,0) = 0$$

Examine both the differential equation and the boundary conditions for homogeneity and linearity.

3•5 Consider the following problem for $t(x,\theta)$:

$$\frac{\partial}{\partial x}\left[(1+at)\frac{\partial t}{\partial x}\right] = \frac{1}{\alpha}\frac{\partial t}{\partial \theta} \qquad t(0,\theta) = t_0$$

$$t(L,\theta) = t_0$$

$$t(x,0) = t^{(0)}$$

Examine both the differential equation and the boundary conditions for homogeneity and linearity.

3•6 For $\lambda_n L = (2n-1)\pi/2$ and $\lambda_m L = (2m-1)\pi/2$, evaluate the following:

(a) $\displaystyle\int_{x=0}^{L} \sin(\lambda_n x)\sin(\lambda_m x)\,dx \qquad$ for $n \neq m$

(b) $\displaystyle\int_{x=0}^{L} \sin(\lambda_n x)\sin(\lambda_m x)\,dx \qquad$ for $n = m$

3•7 Determine the eigenvalues and eigenfunctions for the problem:

$$X'' + \lambda^2 X = 0 \qquad X(0) = 0 \qquad X(L) = 0$$

3•8 Determine the eigenvalues and eigenfunctions for the problem:

$$X'' + \lambda^2 X = 0 \qquad X'(0) = 0 \qquad X'(L) = 0$$

3•9 Show that the two functions

$$P_1(x) = x \qquad \text{and} \qquad P_2(x) = \tfrac{1}{2}(3x^2 - 1)$$

are orthogonal in the interval $-1 \leq x \leq +1$.

3•10 Verify that the Fourier sine series for $1-x$ in the interval $0 < x \leq 1$ is given by

$$1 - x = \frac{2}{\pi}\sum_{n=1}^{\infty} \frac{\sin(n\pi x)}{n}$$

3•11 Expand the function $g(x) = (1-x^2)/2$ in a Fourier series in the interval $0 \le x \le 1$.
(a) Use the eigenfunctions $\sin(n\pi x)$.
(b) Use the eigenfunctions $\cos(n\pi x)$.

3•12 Show that in the interval $0 \le x \le 1$

$$\frac{x^2}{2} - x + \frac{1}{3} = \frac{2}{\pi^2} \sum_{n=1}^{\infty} \frac{\cos(n\pi x)}{n^2}$$

3•13 From the results of Exercise 3•11, show that

$$\sum_{n=1}^{\infty} \frac{(-1)^{n+1}}{n^2} = \frac{\pi^2}{12}$$

3•14 Find the eigencondition and the eigenfunctions for the following problem:

$$kt_{xx} = \rho c t_\theta \qquad\qquad -kt_x(0,\theta) = h[t_\infty - t(0,\theta)]$$
$$-kt_x(L,\theta) = h[t(L,\theta) - t_\infty]$$
$$t(x,0) = t^{(0)}$$

3•15 The solution for the temperature in a plane wall, initially at $t^{(0)}$, whose surface temperatures at $x = 0$ and $x = L$ are each suddenly lowered to t_∞ is given by

$$\frac{t(x,\theta) - t_\infty}{t^{(0)} - t_\infty} = \frac{4}{\pi} \sum_{n=1}^{\infty} \frac{\sin[(2n-1)\pi \frac{x}{L}] \exp[-(2n-1)^2 \pi^2 \frac{\alpha\theta}{L^2}]}{2n-1}$$

(a) Verify that the given solution satisfies the differential equation and boundary conditions by direct substitution.
(b) Verify that the initial condition is also satisfied.
(c) Determine an analytical expression for the total amount of energy (per unit wall area) that has been given off to the surroundings between $\theta = 0$ and $\theta = \tau$.

3•16 A plane wall has an initial temperature distribution given by

$$t(x,0) = t_L + \frac{2}{3}\frac{g'''L^2}{2k} + \frac{g'''L^2}{2k}\frac{4}{\pi^2} \sum_{n=1}^{\infty} \frac{(-1)^{n+1}}{n^2} \cos(n\pi \frac{x}{L})$$

where t_L = outside surface temperature; T
  g''' = uniform generation rate; $E/\Theta\text{-}L^3$
  k = thermal conductivity, $E/\Theta\text{-}L\text{-}T$
  $2L$ = wall thickness; L
  x = distance from center of wall; L

The governing equation for the transient is $kt_{xx} = \rho c t_\theta$. If both surfaces of the plane wall are suddenly insulated at the same time that the generation is suddenly stopped, derive an expression for the wall temperature as a function of position and time.

3•17 Verify that the solution to the plane-wall problem

$$kt_{xx} = \rho c t_\theta \qquad\qquad t(0,\theta) = t_0$$
$$t(L,\theta) = t_L$$
$$t(x,0) = t_L$$

is given by

$$t(x,\theta) = t_0 - (t_0 - t_L)\frac{x}{L}$$
$$- \frac{2}{\pi}(t_0 - t_L)\sum_{n=1}^{\infty} \frac{1}{n}\sin(n\pi \frac{x}{L})\exp(-n^2\pi^2 \frac{\alpha\theta}{L^2})$$

by showing that it satisfies the differential equation and the boundary and initial conditions.

3•18 For the problem of Exercise 3•17, determine the total internal energy (relative to the initial internal energy) in the wall per unit cross-sectional area as a function of time.

3•19 Consider the plane-wall problem given by:

$$kt_{xx} = \rho c t_\theta \qquad\qquad t(0,\theta) = t_0$$
$$t(L,\theta) = t_L$$
$$t(x,0) = t^{(0)}(x)$$

Assuming that $t(x,\theta) = a(x) + b(\theta) + u(x,\theta)$,
(a) Determine the differential equations and the boundary and/or initial conditions that $a(x)$, $b(\theta)$ and $u(x,\theta)$ must satisfy.
(b) Solve for $a(x)$ and $b(\theta)$.
(c) Determine the steady-state solution $u(x,\infty)$.

3•20 A plane wall is initially at a uniform temperature $t^{(0)}$. At $\theta = 0$, the surface temperature at $x = 0$ is changed to t_0 and the surface temperature at $x = L$ is changed to t_L.
(a) State the differential equation and its boundary and initial conditions.
(b) Determine a solution for the transient behavior.

3•21 One side of a plane wall is suddenly exposed to a step change in temperature. The other side is insulated. The mathematical model is:

$$kt_{xx} = \rho c t_\theta \qquad\qquad t(0,\theta) = t_0$$
$$t_x(L,\theta) = 0$$
$$t(x,0) = t^{(0)}$$

Determine the solution for $t(x,\theta)$.

3•22 Consider the plane-wall problem given by:

$$kt_{xx} = \rho c t_\theta \qquad\qquad t(0,\theta) = t^{(0)}$$
$$-kt_x(L,\theta) = h[t(L,\theta) - t_\infty]$$
$$t(x,0) = t^{(0)}$$

Determine the solution for $t(x,\theta)$.

3•23 Consider the plane-wall problem given by:

$$kt_{xx} = \rho c t_\theta$$

$$t(0,\theta) = t_0$$
$$-kt_x(L,\theta) = h[t(L,\theta) - t_\infty]$$
$$t(x,0) = t^{(0)}(x)$$

Assuming that $t(x,\theta) = a(x) + b(\theta) + u(x,\theta)$,

(a) Determine the differential equations and the boundary and/or initial conditions that $a(x)$, $b(\theta)$ and $u(x,\theta)$ should satisfy.
(b) Solve for $a(x)$ and $b(\theta)$.
(c) Determine the steady-state solution $u(x,\infty)$.

3•24 A plane wall, initially at a temperature of $t^{(0)}$, has its surface temperature at $x = 0$ suddenly changed to t_0. The surface at $x = L$ is convectively cooled by a fluid at t_∞. The mathematical model is:

$$kt_{xx} = \rho c t_\theta$$

$$t(0,\theta) = t_0$$
$$-kt_x(L,\theta) = h[t(L,\theta) - t_\infty]$$
$$t(x,0) = t^{(0)}$$

Determine the solution for $t(x,\theta)$.

3•25 A plane wall is initially at a uniform temperature $t^{(0)}$. At time zero uniform energy generation starts. The surface temperatures at $x = 0$ and L are held at their initial values. The mathematical model is:

$$kt_{xx} + g''' = \rho c t_\theta$$

$$t(0,\theta) = t^{(0)}$$
$$t(L,\theta) = t^{(0)}$$
$$t(x,0) = t^{(0)}$$

Determine the solution for $t(x,\theta)$.

3•26 The element in a nuclear reactor is in the form of a flat plate. The governing differential equation and boundary conditions are:

$$kt_{xx} + g''' = \rho c t_\theta \qquad \text{and} \qquad t(0,\theta) = t(L,\theta) = t_0$$

Prior to $\theta = 0$, the energy generation had been going on long enough so that a steady-state temperature distribution had been attained with $t = t_0$ on each face of the plate. At $\theta = 0$ the generation is suddenly stopped and the plate begins to cool. During the cooling process the temperature on each face is held constant at $t = t_0$.

(a) Solve for $t(x,\theta)$ during the cooling process.
(b) Plot nondimensional temperature as a function of position for $\alpha\theta/L^2 = 0.00, 0.01, 0.025, 0.05, 0.10$ and 0.20. Superimpose all curves on one plot.
(c) When will the centerline temperature reach 50 percent response?

3•27 Consider the following problem for $t(x,\theta)$:

$$kt_{xx} + g''' = \rho c t_\theta$$

$$t(0,\theta) = t_0$$
$$t(L,\theta) = t^{(0)}$$
$$t(x,0) = t^{(0)}$$

Determine the solution for $t(x,\theta)$.

3•28 Consider the plane-wall problem given by:

$$kt_{xx} + g''' = \rho c t_\theta$$

$$t(0,\theta) = t_0$$
$$t(L,\theta) = t_L$$
$$t(x,0) = t^{(0)}$$

(a) Explain why it is fine to assume $t(x,\theta) = a(x) + u(x,\theta)$ in this case rather than including $b(\theta)$.
(b) Determine $a(x)$ and $u(x,\theta)$.
(c) Determine $t(x,\theta)$. Express your solution in the form:

$$t(x,\theta) = t^{(0)} + (t_0 - t^{(0)})F(x,\theta) + (t_L - t^{(0)})G(x,\theta) + \frac{g'''L^2}{2k}H(x,\theta)$$

(d) For the special case where $t_0 = t_L = t^{(0)}$, normalize the solution by defining

$$\bar{t} = \frac{t(x,\theta) - t^{(0)}}{g'''L^2/2k} , \qquad \bar{x} = \frac{x}{L} \qquad \text{and} \qquad \bar{\theta} = \frac{\alpha\theta}{L^2}$$

(e) Plot $\bar{t}$ as a function of $\bar{x}$ for $\bar{\theta} = 0.0, 0.01, 0.1$ and 1.0. Use only the first two nonzero terms in the infinite series.

3•29 A plane-wall element in a nuclear reactor has each surface maintained at a constant temperature. At time zero a neutron flux bombards one face and causes thermal generation to begin within the wall. The generation rate decreases exponentially with distance into the wall. The surface temperatures are held constant for all time. The mathematical model is:

$$kt_{xx} + g''' \exp(-ax) = \rho c t_\theta$$

$$t(0,\theta) = t^{(0)}$$
$$t(L,\theta) = t^{(0)}$$
$$t(x,0) = t^{(0)}$$

Determine the solution for $t(x,\theta)$.

3•30 A thin rod is initially at the same temperature as its surroundings. At time zero, the temperature of one end of the rod is stepped to a new value. The rod is sufficiently long so that the heat flux at the other end may be assumed to be zero. The mathematical model is:

$$kAt_{xx} - hp[t - t_\infty] = \rho c A t_\theta$$

$$t(0,\theta) = t_0$$
$$t_x(L,\theta) = 0$$
$$t(x,0) = t^{(0)}$$

Determine the solution for $t(x,\theta)$.

3•31 A plane wall $[k, \rho, c]$ of thickness $2L$ is "contained" between two thin sheets of metal $[\rho_c, c_c]$ at $x = \pm L$. The conductivity of each metal sheet is assumed to be infinite and the thickness of each sheet is δ. The outer surfaces of the metal sheets, at $x = \pm(L+\delta)$, are convectively cooled. The thermal resistance of the metal sheet may be neglected, but its thermal capacitance must be considered. Initially, the wall and the container sheets are at temperature $t^{(0)}$. At time zero the ambient temperature is suddenly changed to t_∞.

(a) Determine the mathematical model for the transient.
(b) Determine the eigencondition. Express your result in terms of λL, $H = hL/k$ and $C = \rho_c c_c \delta / \rho c L$.
(c) Determine the first six eigenvalues for $H = 1$, $C = 1$.

3•32 One side of a plane wall is suddenly exposed to a step change in temperature which decays exponentially back to zero again as time goes on. The other side is insulated. The mathematical model is:

$$kt_{xx} = \rho c t_\theta \qquad \begin{aligned} t(0,\theta) &= t^{(0)} + T\exp(-a\theta) \\ t_x(L,\theta) &= 0 \\ t(x,0) &= t^{(0)} \end{aligned}$$

Determine the solution for $t(x,\theta)$.

3•33 A plane wall is initially at a uniform temperature with no energy generation. At time zero a uniform reaction within the wall begins to take place which increases to a steady value. The surface temperatures are maintained at their initial values. The mathematical model is:

$$kt_{xx} + g'''[1 - \exp(-a\theta)] = \rho c t_\theta \quad \begin{aligned} t(0,\theta) &= t^{(0)} \\ t(L,\theta) &= t^{(0)} \\ t(x,0) &= t^{(0)} \end{aligned}$$

Determine the solution for $t(x,\theta)$.

3•34 Consider the plane-wall problem given by:

$$kt_{xx} = \rho c t_\theta \qquad \begin{aligned} t_x(0,\theta) &= 0 \\ -kt_x(L,\theta) &= q_L''(\theta) \\ t(x,0) &= t^{(0)} \end{aligned}$$

(a) Since the nonhomogeneous term is time dependent, use variation of parameters to determine $t(x,\tau)$. The solution will have two integrals from $\theta = 0$ to τ that involve $q_L''(\theta)$.
(b) Modify your solution so that any series that must be summed converges as $1/(\lambda_n L)^2$. The modified solution will still have two integrals from $\theta = 0$ to τ, but now one integral will involve $q_L''(\theta)$ and the other integral will involve $\dot{q}_L''(\theta)$, the time derivative of $q_L''(\theta)$.
(c) Show that the solution obtained in part (b) satisfies the partial differential equation, boundary conditions and initial condition.

3•35 Consider the plane wall problem given by

$$kt_{xx} = \rho c t_\theta \qquad \begin{aligned} t(0,\theta) &= t^{(0)} + T\cos(\omega\theta) \\ t(L,\theta) &= t^{(0)} \\ t(x,0) &= t^{(0)} \end{aligned}$$

Determine the solution for $t(x,\theta)$.

3•36 Consider a thin rod whose mathematical model is:

$$kAt_{xx} - hp[t - t_\infty] = \rho cAt_\theta \qquad \begin{aligned} t(0,\theta) &= t_0 + T\sin(\omega\theta) \\ t_x(L,\theta) &= 0 \\ t(x,0) &= t^{(0)} \end{aligned}$$

Determine the solution for $t(x,\theta)$.

3•37 Consider the plane-wall problem given by:

$$kt_{xx} = \rho c t_\theta \qquad \begin{aligned} t_x(0,\theta) &= 0 \\ t(L,\theta) &= t_L(\theta) \\ t(x,0) &= t^{(0)}(x) \end{aligned}$$

Note that the nonhomogeneous term is time dependent. In general, the assumption that $t(x,\theta) = a(x) + b(\theta) + u(x,\theta)$ will not work for time-dependent nonhomogeneous terms.

(a) Determine the conditions on $t_L(\theta)$ that are required for the assumption $t(x,\theta) = a(x) + b(\theta) + u(x,\theta)$ to work in this problem.
(b) Determine $a(x)$, $b(\theta)$ and the problem that must be solved for $u(x,\theta)$.

3•38 A two-dimensional, rectangular region has uniform energy generation within. The mathematical model is:

$$k[t_{xx} + t_{yy}] + g''' = 0 \qquad \begin{aligned} t_x(0,y) &= 0 \\ t(L,y) &= t_L \\ t_y(x,0) &= 0 \\ t(x,\ell) &= t_\ell \end{aligned}$$

Determine the solution for $t(x,y)$.

3•39 A two-dimensional, square region has uniform energy generation within. The mathematical model is:

$$k[t_{xx} + t_{yy}] + g''' = 0 \qquad \begin{aligned} t(0,y) &= t_0 \\ t(L,y) &= t_0 \\ t(x,0) &= t_0 \\ t(x,L) &= t_0 \end{aligned}$$

(a) Determine the solution for $t(x,y)$.
(b) Compare the solution for part (a) to the solution for a plane wall with uniform energy generation given by:

$$kt_{xx} + g''' = 0 \qquad t(0) = t(L) = t_0$$

(c) For k, g''' and $L = 1$ and $t_0 = 0$, determine $t(\frac{L}{2},\frac{L}{2})$ for part (a) and $t(\frac{L}{2})$ for part (b). Compare.

3·40 A two-dimensional, square region has specified temperatures on all four sides. The mathematical model is:

$$t_{xx} + t_{yy} = 0$$

$$t(0,y) = t_0$$
$$t(L,y) = t_L$$
$$t(x,0) = t_0$$
$$t(x,L) = t_0$$

Determine the solution for $t(x,\theta)$.

3·41 A two-dimensional, square region has three specified-temperature sides and one adiabatic side. The model is:

$$t_{xx} + t_{yy} = 0$$

$$t_x(0,y) = 0$$
$$t(L,y) = t_L$$
$$t(x,0) = t_0$$
$$t(x,L) = t_0$$

Determine the solution for $t(x,\theta)$.

3·42 A two-dimensional, square region has two specified-temperature sides and two adiabatic sides. The model is:

$$t_{xx} + t_{yy} = 0$$

$$t_x(0,y) = 0$$
$$t(L,y) = t_L$$
$$t_y(x,0) = 0$$
$$t(x,L) = t_0$$

Determine the solution for $t(x,\theta)$.

3·43 Consider the two-dimensional problem given by

$$t_{xx} + t_{yy} = 0$$

$$t_x(0,y) = S(y)$$
$$t(L,y) = P(y)$$
$$t(x,0) = V(x)$$
$$t_y(x,L) = W(x)$$

For an assumed solution given by

$$t(x,y) = a(x)S(y) + b(x)P(y) + u(x,y)$$

(a) Determine the restrictions on $S(y)$ and $P(y)$ for which it is possible to obtain a problem for $u(x,y)$ that is homogeneous in the x direction.
(b) If the restrictions found in part (a) are met, determine the relations that $a(x)$, $b(x)$ and $u(x,y)$ must satisfy.

3·44 A two-dimensional, square region has specified temperature on all four sides. The mathematical model is:

$$t_{xx} + t_{yy} = 0$$

$$t(0,y) = t_0$$
$$t(L,y) = t_0$$
$$t(x,0) = t_0$$
$$t(x,L) = t_0 + T\sin(\pi x/L)$$

Determine the solution for $t(x,y)$.

3·45 A semi-infinite strip extends from $y = 0$ to $y = \infty$. The strip has a finite width L in the x direction. The mathematical model is:

$$t_{xx} + t_{yy} = 0$$

$$t(0,y) = t_0$$
$$t(L,y) = t_0 + T\exp(-ay)$$
$$t(x,0) = t_0$$

Determine the solution for $t(x,y)$.

3·46 Consider the two-dimensional, problem given by:

$$k[t_{xx} + t_{yy}] = \rho c t_\theta$$

$$t_x(0,y,\theta) = 0$$
$$t(L,y,\theta) = t_L$$
$$t_y(x,0,\theta) = 0$$
$$t(x,L,\theta) = t_L$$
$$t(x,y,0) = t^{(0)}(x,y)$$

Assuming that $t(x,y,\theta) = a(x) + b(y) + u(x,y,\theta)$,

(a) Determine $a(x)$ and $b(y)$ that will give a homogeneous problem for $u(x,y,\theta)$. State the differential equation, boundary conditions and initial condition for $u(x,y,\theta)$.
(b) Under what conditions will you be able to solve for $u(x,y,\theta)$ by assuming that $u(x,y,\theta) = v(x,\theta)w(y,\theta)$?
(c) Determine the problems that must be solved for $v(x,\theta)$ and $w(y,\theta)$.

3·47 Consider the two-dimensional problem given by:

$$k[t_{xx} + t_{yy}] = \rho c t_\theta$$

$$t_x(0,y,\theta) = 0$$
$$t_x(L,y,\theta) = 0$$
$$t_y(x,0,\theta) = 0$$
$$t_y(x,L,\theta) = 0$$
$$t(x,y,0) = t^{(0)}(x,y)$$

Since the problem is already homogeneous, start the solution by assuming $t(x,y,\theta) = X(x)Y(y)\Theta(\theta)$. Determine $t(x,y,\theta)$ in-so-far-as-possible.

3·48 Starting with the results of Exercise 3·47,
(a) Complete the solution for the initial condition given by

$$t^{(0)}(x,y) = \begin{cases} T_o \dfrac{L-x}{L} & \text{for } x \geq y \\ T_o \dfrac{L-y}{L} & \text{for } x \leq y \end{cases}$$

(b) If the solution to part (a) were written as

$$t(\bar{x},\bar{y},\bar{\theta}) = \sum_{j=1}^{\infty} a_j(\bar{x},\bar{y})\exp(-\lambda_j\bar{\theta})$$

where $\bar{x} = x/L$, $\bar{y} = y/L$, $\bar{\theta} = \alpha\theta/L^2$ and $\lambda_j < \lambda_{j+1}$, determine λ_j and $a_j(\bar{x},\bar{y})$ for $j = 1$ through 6.

(c) For $T_o = 1$, plot $a_j(\bar{x},0)$ as a functions of $\bar{x}$ for $j = 1$ through 4.

3•49 A square region is initially at a uniform temperature. At time zero one boundary is changed to a new temperature while the other three boundaries are maintained at their initial temperature. The mathematical model is:

$$k[t_{xx} + t_{yy}] = \rho c t_\theta$$

$$t(0, y, \theta) = t_0$$
$$t(L, y, \theta) = t_L$$
$$t(x, 0, \theta) = t_0$$
$$t(x, L, \theta) = t_0$$
$$t(x, y, 0) = t_0$$

Determine the solution for $t(x, y, \theta)$.

3•50 Consider the following problem for $t(x, y, \theta)$:

$$k[t_{xx} + t_{yy}] = \rho c t_\theta$$

$$t_x(0, y, \theta) = 0$$
$$-k t_x(x_o, y, \theta) = h[t(x_o, y, \theta) - t_\infty]$$
$$t_y(x, 0, \theta) = 0$$
$$t(x, y_o, \theta) = t_o$$
$$t(x, y, 0) = t^{(0)}$$

Determine the solution for $t(x, y, \theta)$ by assuming $t(x, y, \theta) = a(x) + b(y) + u(x, y) + v(x, y, \theta)$. *Note:* Part of $v(x, y, \theta)$ may be written as the product of two plane-wall transients whose solutions are related to (3·1·30) and (3·1·66).

3•51 Consider a thick-walled pipe initially at a uniform temperature $t^{(0)}$. Then at time zero the temperatures of both the inside surface and the outside surface are each stepped to a specified value t_o. The mathematical model is:

$$k[t_{rr} + \frac{1}{r} t_r] = \rho c t_\theta$$

$$t(r_i, \theta) = t_o$$
$$t(r_o, \theta) = t_o$$
$$t(r, 0) = t^{(0)}$$

(a) Determine the solution for $t(r, \theta)$.
(b) For $r_i = 0.5$ and $r_o = 1$, determine the first six eigenvalues.
(c) For k, ρ, c and $t^{(0)} = 1$ and $t_o = 0$ along with $r_i = 0.5$ and $r_o = 1$, determine $t(0.75, \theta)$ for $\theta = 0.01$ and 0.1.

3•52 A thin circular fin attached to a pipe is initially at t_∞ in an ambient at t_∞. At $\theta = 0$ the temperature of the fluid flowing through the pipe suddenly changes the temperature of the fin at r_i to t_i. The end of the fin at $r = r_o$ is adiabatic. The mathematical model for the transient is:

$$k\delta[t_{rr} + \frac{1}{r} t_r] = h[t - t_\infty] + \rho c \delta t_\theta$$

$$t(r_i, \theta) = t_i$$
$$t_r(r_o, \theta) = 0$$
$$t(r, 0) = t_\infty$$

(a) Determine $t(r, \theta)$ in-so-far-as-possible. Stop when you reach the point where integrals of modified Bessel functions must be evaluated.
(b) For $r_i = 0.5$ and $r_o = 1$, determine the first six eigenvalues.

3•53 A solid circular rod extends between two walls. The walls are each held at temperature t_0 and the ambient is at t_∞. The thermal conductivity of the rod is k. The heat-transfer coefficient h between the rod and the ambient is uniform.

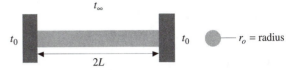

The mathematical problem for the left-hand half of the rod (considering temperature to vary with radius r as well as with distance x along the rod) is:

$$r t_{xx} + r t_{rr} + t_r = 0$$

$$t(0, r) = t_0$$
$$t_x(L, r) = 0$$
$$-k t_r(x, r_o) = h[t(x, r_o) - t_\infty]$$

(a) Determine the solution for $t(x, r)$.
(b) Evaluate the total heat-transfer rate $[E/\Theta]$ between the rod and the ambient.

3•54 In the study of fins, a fin efficiency η is often defined as a performance index.

$$\eta = \frac{\text{Actual heat transfer from fin}}{\text{Heat transfer from fin if entire fin were at wall temperature}}$$

The question might arise in fin design as to whether a "thin rod" approximation is sufficient or whether one must consider temperature gradients normal to the fin length in evaluating the efficiency. A rod of circular cross section is considered in Exercise 3•53. To evaluate the applicability of thin-rod analysis for the results of Exercise 3•53,

(a) Derive an expression for η from thin-rod analysis in terms of mL where $m = \sqrt{hp/kA}$.
(b) Derive an expression for η using the more complete analysis of Exercise 3•53. Express the result in terms of mL and $H = hr_o/k$ (or parameters that only depend on H).
(c) Compare your results by plotting η versus mL for part (a) and for part (b) with $H = 0.1$, 1 and 10.
(d) Discuss your findings.

3•55 Consider the solid-cylinder problem given by:

$$k[t_{rr} + \frac{1}{r} t_r] + g''' = \rho c t_\theta$$

$$t(0, \theta) < \infty$$
$$t(r_o, \theta) = t_o$$
$$t(r, 0) = t^{(0)}$$

(a) Assume that $t(r, \theta) = a(r) + u(r, \theta)$ and determine the problems that $a(r)$ and $u(r, \theta)$ must satisfy.
(b) Determine $a(r)$.
(c) Describe the rate of convergence you expect for the series in $u(r, \theta)$.
(d) Determine $u(r, \theta)$.

3•56 A short cylinder (radius r_o and length L) is initially at a uniform temperature $t^{(0)}$ when its surface temperature is suddenly stepped to a new value t_o. The mathematical model is:

$$k[t_{rr} + \frac{1}{r}t_r + t_{zz}] = \rho c t_\theta$$

$$t(r_o, z, \theta) = t_o$$
$$t(r, 0, \theta) = t_o$$
$$t(r, L, \theta) = t_o$$
$$t(r, z, 0) = t^{(0)}$$

Determine a series solution for $t(r, z, \theta)$ by first constructing a homogeneous problem for $u(r, z, \theta)$ and then assuming $u(r, z, \theta) = R(r)Z(z)\Theta(\theta)$. Compare your answer with the result obtained in Section 3•3•6.

3•57 The mathematical model for the temperature in a solid, circular cylinder of radius r_o and length z_o is:

$$k[t_{rr} + \frac{1}{r}t_r + \frac{1}{r^2}t_{\phi\phi} + t_{zz}] = \rho c t_\theta$$

$$t(r_o, \phi, z, \theta) = t_o$$
$$t_z(r, \phi, 0, \theta) = 0$$
$$t(r, \phi, z_o, \theta) = t_o$$
$$t(r, \phi, z, 0) = t^{(0)}(r, \phi)$$

Determine the solution for $t(r, \phi, z, \theta)$ by first constructing a homogeneous problem for $u(r, \phi, z, \theta)$ and then, since the initial condition is not a function of z, assuming $u(r, \phi, z, \theta) = v(z, \theta)w(r, \phi, \theta)$.

3•58 A rectangular solid initially has a uniform temperature $t^{(0)}$. Its exterior surfaces are then suddenly changed to a temperature t_o. The mathematical model is:

$$k[t_{xx} + t_{yy} + t_{zz}] = \rho c t_\theta$$

$$t_x(0, y, z, \theta) = 0$$
$$t(x_o, y, z, \theta) = t_o$$
$$t_y(x, 0, z, \theta) = 0$$
$$t(x, y_o, z, \theta) = t_o$$
$$t_z(x, y, 0, \theta) = 0$$
$$t(x, y, z_o, \theta) = t_o$$
$$t(x, y, z, 0) = t^{(0)}$$

Determine the solution to this problem by first reducing the problem to three one-dimensional problems by assuming $t(x, y, z, \theta) = t_o + u(x, \theta)v(y, \theta)w(z, \theta)$ and then making use of (3•1•30).

3•59 A two-dimensional, square ($L \times L$) region [k, ρ and c] has adiabatic boundaries along $x = 0$, $y = 0$ and $y = L$. Initially, the temperature within the region is uniformly at $t^{(0)}$. At $\theta = 0$ the temperature along $x = L$ is suddenly changed to vary linearly from t_0 at $y = 0$ to t_L at $y = L$. The generation rate is zero. The three adiabatic boundaries remain adiabatic. The exact solution for $t(x, y, \theta)$ can be expressed as the sum of a steady-state portion $u(x, y)$ and a transient portion $v(x, y, \theta)$ where both $u(x, y)$ and $v(x, y, \theta)$ can be obtained by separation of variables.

(a) Determine the mathematical problems (*i.e.*, differential equation and the boundary and initial conditions) for $u(x, y)$ and for $v(x, y, \theta)$.
(b) Determine the steady-state solution $u(x, y)$.
(c) Determine the transient solution $v(x, y, \theta)$.

For k, ρ, c and $L = 1$ and $t^{(0)} = 0$, $t_0 = 0$ and $t_L = 1$,

(d) Evaluate and plot $u(0, y)$.
(e) Evaluate and plot $v(0, 0, \theta)$ and $v(0, 1, \theta)$ on the same graph.
(f) Evaluate and plot $t(0, 0, \theta)$ and $t(0, 1, \theta)$ on the same graph.

4

SUPERPOSITION

4•0 INTRODUCTION

Since the governing equations in many heat-transfer problems are linear and homogeneous, the solutions to simple problems can be superimposed to find solutions to complicated problems. Superposition has already been introduced in the discussion of separation of variables where an infinite number of partial solutions were added together to fit the initial condition.

In the next two sections the idea of superposition is extended beyond what was employed in Chapter 3. Section 4•1 shows how complicated problems with several nonhomogeneities may be split into several simpler problems that can be added together (superimposed) to give the solution to the original problem. The solution to problems with a time-dependent boundary condition can be generated by using the solution of a problem with a steady boundary condition as shown in Section 4•2.

The problems that can be solved by the methods of sections 4•1 and 4•2 can also be solved by previously discussed methods if one has the time. Frequently, however, an engineer needs an answer in a hurry. These methods permit manipulation of simple solutions that may be found in standard textbooks to obtain answers to a particular problem much more readily than if one were to start from the beginning.

Some words of caution should be mentioned here, however. First, the problem must be *linear*. It would be well to review the comments on linearity which were made in Chapter 3 to be sure that they are understood. Second, the concept of superposition sounds a lot easier to apply than it often turns out to be. One must be very careful to see that the superimposed solutions do indeed add up to satisfy the original problem.

4•1 SIMPLIFICATION OF NONHOMOGENEOUS PROBLEMS

Complicated nonhomogeneous problems can often be solved by splitting the problem into several simpler problems. The solutions to these simpler problems can then be superimposed to obtain the complete solution. As an example of the method, consider a plane wall with energy generation that is a function of both time and position. Time-dependent boundary conditions will also be included.[1] The governing equations are given by

$$kt_{xx} + g'''(x,\theta) = \rho c t_\theta \qquad (4\cdot1\cdot1)$$

$$t(0,\theta) = t_0(\theta) \qquad (4\cdot1\cdot2)$$

---

[1] Time-dependent nonhomogeneous terms were previously discussed in Section 3•1•7.

$$-kt_x(L,\theta) = q_L''(\theta) \tag{4.1.3}$$

$$t(x,0) = t^{(0)}(x) \tag{4.1.4}$$

We want to split this problem up into several simpler subproblems. In general the number of subproblems will be equal to the number of nonhomogeneities in the original problem plus an additional one to take care of the initial condition. Since there are three nonhomogeneous terms and one initial condition in the problem at hand, we will assume $t(x,\theta)$ can be written as the sum of $3 + 1 = 4$ functions and write

$$t(x,\theta) = s(x,\theta) + u(x,\theta) + v(x,\theta) + w(x,\theta) \tag{4.1.5}$$

Upon substituting this assumed form of the solution into (4.1.1), (4.1.2), (4.1.3) and (4.1.4), one obtains

$$k[s_{xx} + u_{xx} + v_{xx} + w_{xx}] + g''' = \rho c[s_\theta + u_\theta + v_\theta + w_\theta]$$

$$s(0,\theta) + u(0,\theta) + v(0,\theta) + w(0,\theta) = t_0(\theta)$$

$$-k[s_x(L,\theta) + u_x(L,\theta) + v_x(L,\theta) + w_x(L,\theta)] = q_L''(\theta)$$

$$s(x,0) + u(x,0) + v(x,0) + w(x,0) = t^{(0)}(x)$$

These four equations can be satisfied if s, u, v and w satisfy the four subproblems shown in Table 4.1.

Table 4.1 *Subproblems for $t(x,\theta) = s(x,\theta) + u(x,\theta) + v(x,\theta) + w(x,\theta)$.*

$s(x,\theta)$	$u(x,\theta)$	$v(x,\theta)$	$w(x,\theta)$
$ks_{xx} + g''' = \rho c s_\theta$	$ku_{xx} = \rho c u_\theta$	$kv_{xx} = \rho c v_\theta$	$kw_{xx} = \rho c w_\theta$
$s(0,\theta) = 0$	$u(0,\theta) = t_0(\theta)$	$v(0,\theta) = 0$	$w(0,\theta) = 0$
$s_x(L,\theta) = 0$	$u_x(L,\theta) = 0$	$-kv_x(L,\theta) = q_L''(\theta)$	$w_x(L,\theta) = 0$
$s(x,0) = 0$	$u(x,0) = 0$	$v(x,0) = 0$	$w(x,0) = t^{(0)}(x)$

Notice that
1. Each subproblem is simpler than the original problem for $t(x,\theta)$. The subproblems for s, u and v have only one nonhomogeneous term.
2. The problems for s, u and v can be solved by variation of parameters.
3. The w-problem is homogeneous and separation of variables will work.
4. It may be possible to find solutions in the literature for some of the subproblems. You are less likely to find a solution for $t(x,\theta)$.

4.2 DUHAMEL'S THEOREM

Thus far, not much has been said regarding problems with time-dependent boundary conditions. These can be of major engineering interest (*e.g.*, internal combustion engine walls, thermal building loads, penetration of daily and annual temperature cycles into the earth). It is therefore evident that special consideration should be given to such boundary conditions.

We have seen in the previous section that it is possible to break up a transient problem with several nonhomogeneous terms and a nonzero initial condition into several problems, each having just one nonhomogeneous term and a zero initial condition, plus a homogeneous problem with a nonzero initial condition. Duhamel's theorem provides a method for solving problems with one time-dependent nonhomogeneous term with a zero initial condition. To use Duhamel's theorem you must be able to find the *fundamental solution* (the same problem but with the time-dependent nonhomogeneous term replaced by a constant nonhomogeneous term). Chapter 3 discussed methods of obtaining solutions to problems with nonhomogeneous terms that were not time dependent.

Rather than trying to be very elegant, Duhamel's theorem will be stated without proof.

Duhamel's Theorem

If $U(x,\theta)$ is the response of a linear system (whose initial condition is 0) to a single nonhomogeneous input given by $G(x)\mathbf{1}(\theta)$, the response $u(x,\theta)$ of the same system to an input $G(x)F(\theta)$ is given by

$$u(x,\theta) = \int_{\tau=0}^{\theta} U(x,\theta-\tau)F'(\tau)\,d\tau + \sum_{i=1}^{I} U(x,\theta-\tau_i)\Delta F_i \qquad (4\cdot2\cdot1)$$

The *unit-step function* $\mathbf{1}(\theta)$ is defined to be 0 for $\theta < 0$ and 1 for $\theta > 0$. From time $\tau = 0$ until time $\tau = \theta$, $F(\tau)$ has a total of I steps. Step i occurs at time τ_i and $\Delta F_i = F(\tau_i^+) - F(\tau_i^-)$ where $F(\tau_i^+)$ is the value of $F(\tau)$ right after step i and $F(\tau_i^-)$ is the value of $F(\tau)$ just prior to step i. In between successive steps $F(\tau)$ is continuous. The integral is the sum of the integrals over the continuous portions of $F(\tau)$. The summation term accounts for the steps ΔF_i in $F(\tau)$ that occur at time τ_i.

To give some credibility to Duhamel's theorem, consider the plane-wall transient given in Table 4•2 and pictured in Figure 4•1. The problem for $u(x,\theta)$ is homogeneous except for a single term $F(\theta)$ in the boundary condition at $x = L$ in this example. The initial condition for $u(x,\theta)$ is 0.

Table 4•2 *Mathematical problem description for $u(x,\theta)$.*

Problem	Equation	Boundary/Initial Conditions	
$u(x,\theta)$	$ku_{xx} = \rho c u_\theta$		$(4\cdot2\cdot2)$
		$u_x(0,\theta) = 0$	$(4\cdot2\cdot3)$
		$u(L,\theta) = F(\theta)$	$(4\cdot2\cdot4)$
		$u(x,0) = 0$	$(4\cdot2\cdot5)$

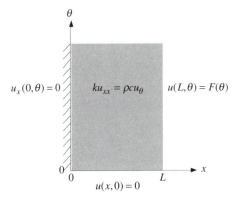

Figure 4•1 *Plane wall, initially at 0, is suddenly exposed to a time-dependent surface temperature at $x = L$.*

The fundamental solution $U(x,\theta)$ needed for Duhamel's theorem is the solution to the problem given in Table 4•3 and pictured in Figure 4•2. It is the same as the problem for $u(x,\theta)$ except that $F(\theta)$ is now 1.

Table 4•3 *Mathematical problem description for $U(x,\theta)$.*

Problem	Equation	Boundary/Initial Conditions	
$U(x,\theta)$	$kU_{xx} = \rho c U_\theta$		$(4\cdot2\cdot6)$
		$U_x(0,\theta) = 0$	$(4\cdot2\cdot7)$
		$U(L,\theta) = 1$	$(4\cdot2\cdot8)$
		$U(x,0) = 0$	$(4\cdot2\cdot9)$

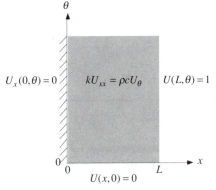

Figure 4•2 *Plane wall, initially at 0, is suddenly exposed to a unit step in surface temperature at $x = L$.*

According to Duhamel's theorem, the solution for $u(x,\theta)$ is given by $(4\cdot2\cdot1)$ where $U(x,\theta)$ is the solution to the problem in Table 4•3. To show that (4·2·1) is the solution, we will show that it satisfies (4·2·2) through (4·2·5). To show that (4·2·2) is satisfied we will differentiate (4·2·1) twice with respect to x and then multiply by k to obtain,

$$ku_{xx}(x,\theta) = \int_{\tau=0}^{\theta} kU_{xx}(x,\theta-\tau)F'(\tau)\,d\tau + \sum_{i=1}^{I} kU_{xx}(x,\theta-\tau_i)\Delta F_i \qquad (4\cdot2\cdot10)$$

To find $u_\theta(x,\theta)$, recall Leibnitz' rule for differentiation of integrals.

Leibnitz' Rule

$$\frac{d}{dx}\int_{a(x)}^{b(x)} F(\xi;x)\,d\xi = \frac{db(x)}{dx}F[b(x);x] + \int_{a(x)}^{b(x)} \frac{\partial}{\partial x}F(\xi;x)\,d\xi - \frac{da(x)}{dx}F[a(x);x]$$

$$(4\cdot2\cdot11)$$

Therefore, differentiating (4·2·1) once with respect to θ gives

$$u_\theta(x,\theta) = U(x,0)F'(\theta) + \int_{\tau=0}^{\theta} U_\theta(x,\theta-\tau)F'(\tau)\,d\tau + \sum_{i=1}^{I} U_\theta(x,\theta-\tau_i)\Delta F_i$$

The initial condition (4·2·9) that $U(x,0)=0$ will eliminate the first term to give (after also multiplying by ρc)

$$\rho c u_\theta(x,\theta) = \int_{\tau=0}^{\theta} \rho c U_\theta(x,\theta-\tau)F'(\tau)\,d\tau + \sum_{i=1}^{I} \rho c U_\theta(x,\theta-\tau_i)\Delta F_i$$

From (4·2·6) we may replace $\rho c U_\theta$ by kU_{xx} to obtain

$$\rho c u_\theta(x,\theta) = \int_{\tau=0}^{\theta} kU_{xx}(x,\theta-\tau)F'(\tau)\,d\tau + \sum_{i=1}^{I} kU_{xx}(x,\theta-\tau_i)\Delta F_i$$

The right-hand side of this result is identical to the right-hand side of (4·2·10). Therefore the left-hand sides are equal and (4·2·2) is satisfied.

To show that the initial condition (4·2·5) is satisfied we will set $\theta=0$ in (4·2·1) to obtain

$$u(x,0) = \int_{\tau=0}^{0} U(x,0)F'(\tau)\,d\tau + U(x,0)\Delta F_0$$

where it has been recognized that, within the interval from $\tau=0$ to $\tau=0$, $\theta-\tau$ can only be equal to 0 and the only step in $F(\tau)$ that could appear in the summation will be at $\tau=0$. The initial condition (4·2·9) for U makes each of the terms on the right-hand side equal to 0. Thus (4·2·5) is satisfied.

To show that the boundary condition (4·2·3) at $x=0$ is satisfied we will differentiate (4·2·1) with respect to x and then set $x=0$ to obtain

$$u_x(0,\theta) = \int_{\tau=0}^{\theta} U_x(0,\theta-\tau)F'(\tau)\,d\tau + \sum_{i=1}^{I} U_x(0,\theta-\tau_i)\Delta F_i$$

From (4·2·7) we see that $U_x(0,\theta)=0$ at all times. Therefore each of the terms on the right-hand side is equal to 0 and (4·2·3) is satisfied.

The most complicated condition to verify is the nonhomogeneous term in the boundary condition at $x=L$. Figure 4•3 depicts an $F(\tau)$ with three steps, at $\tau_1=0$, τ_2 and τ_3. For a step at τ_i, $F(\tau_i)$ has a value of $F^-(\tau_i)$ just prior to the step and a value of $F^+(\tau_i)$ just after the step. Thus, $\Delta F_i = F^+(\tau_i) - F^-(\tau_i)$. Upon setting $x=L$ and $I=3$ in (4·2·1) we obtain

$$u(L,\theta) = \int_{\tau=0}^{\theta} U(L,\theta-\tau)F'(\tau)\,d\tau + \sum_{i=1}^{3} U(L,\theta-\tau_i)\Delta F_i$$

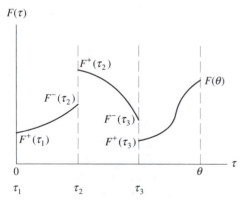

Figure 4•3 *Example forcing function $F(\tau)$ for $0 \le \tau \le \theta$.*

From (4·2·8) we see that $U = 1$ at $x = L$ at any time. Thus we may write

$$u(L,\theta) = \int_{\tau=0}^{\theta} F'(\tau)d\tau + \sum_{i=1}^{3} \Delta F_i$$

The integral may be broken up into three integrals, one for each of the three continuous portions of $F(\tau)$. We may also replace ΔF_i by $F^{+}(\tau_i) - F^{-}(\tau_i)$ for each to the three steps. Thus,

$$u(L,\theta) = \int_{\tau=0}^{\tau_2} F'(\tau)d\tau + \int_{\tau=\tau_2}^{\tau_3} F'(\tau)d\tau + \int_{\tau=\tau_3}^{\theta} F'(\tau)d\tau$$

$$+ [F^{+}(0) - F^{-}(0)] + [F^{+}(\tau_2) - F^{-}(\tau_2)] + [F^{+}(\tau_3) - F^{-}(\tau_3)]$$

Since $F'(\tau)d\tau = dF$ the three integrals may be easily integrated to give

$$u(L,\theta) = [F^{-}(\tau_2) - F^{+}(0)] + [F^{-}(\tau_3) - F^{+}(\tau_2)] + [F(\theta) - F^{+}(\tau_3)]$$

$$+ [F^{+}(0) - F^{-}(0)] + [F^{+}(\tau_2) - F^{-}(\tau_2)] + [F^{+}(\tau_3) - F^{-}(\tau_3)]$$

Inspection of the 12 terms on the right-hand side shows that five pair of terms cancel, $F^{-}(0) = 0$ and only $F(\theta)$ remains. Therefore the boundary condition (4·2·4) at $x = L$ is satisfied. Since we have now shown that (4·2·1) satisfies each of the requirements in Table 4•2, it must be the solution to the problem. Although this "proof" is for a specific problem, similar "proofs" can be given for other problems.

Example A plane wall, initially at $t^{(0)}$, suddenly has its surface temperature at $x = L$ changed to $t_L(\theta)$. The surface at $x = 0$ is adiabatic. The problem is shown in Figure 4•4. This is similar to the problem discussed in Section 3•1•1 except that now the surface temperature at $x = L$ is a specified function of time rather than a constant value. The mathematical description of this problem is given in Table 4•4.

Table 4•4 *Mathematical problem description for $t(x,\theta)$.*

Problem	Equation	Boundary/Initial Conditions	
$t(x,\theta)$	$kt_{xx} = \rho c t_{\theta}$		(4.2.12)
		$t_x(0,\theta) = 0$	(4.2.13)
		$t(L,\theta) = t_L(\theta)$	(4.2.14)
		$t(x,0) = t^{(0)}$	(4.2.15)

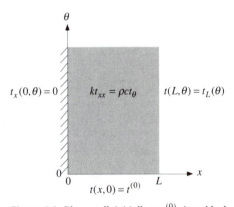

Figure 4•4 *Plane wall, initially at $t^{(0)}$, is suddenly exposed to a time-dependent surface temperature at $x = L$.*

To apply Duhamel's theorem directly to a problem, there can be only one nonhomogeneous term (as in this example) and the initial condition must be 0 (not true in this example). To get around this difficulty we will assume that

$$t(x,\theta) = u(x,\theta) + v(x,\theta) \qquad (4.2.16)$$

where $u(x,\theta)$ will consider the nonhomogeneous term at $x = L$, but have a zero initial condition so that Duhamel's theorem can be used. The function $v(x,\theta)$ will have a nonzero initial condition, but will have homogeneous boundary conditions so that $v(x,\theta)$ can be found by separation of variables.

Upon substituting (4·2·16) into the equations in Table 4•4, the resulting problems for $u(x,\theta)$ and $v(x,\theta)$ are as shown in Figures 4•5 and 4•6. The mathematical descriptions are given in Table 4•5.

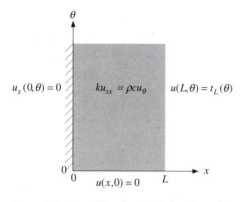

Figure 4·5 *Subproblem for $u(x,\theta)$ for Figure 4•4.*

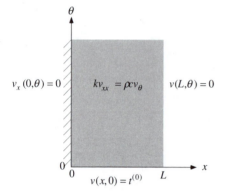

Figure 4·6 *Subproblem for $v(x,\theta)$ for Figure 4•4.*

Table 4·5 *Mathematical problem descriptions for $u(x,\theta)$ and $v(x,\theta)$.*

Problem	Equation	Boundary/Initial Conditions	
$u(x,\theta)$	$ku_{xx} = \rho c u_\theta$		(4·2·17)
		$u_x(0,\theta) = 0$	(4·2·18)
		$u(L,\theta) = t_L(\theta)$	(4·2·19)
		$u(x,0) = 0$	(4·2·20)
$v(x,\theta)$	$kv_{xx} = \rho c v_\theta$		(4·2·21)
		$v_x(0,\theta) = 0$	(4·2·22)
		$v(L,\theta) = 0$	(4·2·23)
		$v(x,0) = t^{(0)}$	(4·2·24)

The fundamental problem we must solve to find the function $U(x,\theta)$ required by Duhamel's theorem is described in Table 4•3. We can obtain $U(x,\theta)$ from the solution found in Section 3•1•1 which may be written as

$$t(x,\theta) = t_L + 2(t^{(0)} - t_L)\sum_{n=1}^{\infty}\frac{\sin(\lambda_n L)}{\lambda_n L}\cos(\lambda_n x)\exp(-\lambda_n^2\alpha\theta)$$

Upon setting $t(x,\theta) = U(x,\theta)$, $t^{(0)} = 0$ and $t_L = 1$ we obtain

$$U(x,\theta) = 1 - 2\sum_{n=1}^{\infty}\frac{\sin(\lambda_n L)}{\lambda_n L}\cos(\lambda_n x)\exp(-\lambda_n^2\alpha\theta)$$

Then, upon replacing θ by $\theta - \tau$,

$$U(x,\theta - \tau) = 1 - 2\sum_{n=1}^{\infty}\frac{\sin(\lambda_n L)}{\lambda_n L}\cos(\lambda_n x)\exp[-\lambda_n^2\alpha(\theta - \tau)]$$

Substitution into Duhamel's theorem gives (for only one step at $\theta = 0$),

$$u(x,\theta) = \int_{\tau=0}^{\theta}\left\{1 - 2\sum_{n=1}^{\infty}\frac{\sin(\lambda_n L)}{\lambda_n L}\cos(\lambda_n x)\exp[-\lambda_n^2\alpha(\theta - \tau)]\right\}t_L'(\tau)\,d\tau$$

$$+\left\{1 - 2\sum_{n=1}^{\infty}\frac{\sin(\lambda_n L)}{\lambda_n L}\cos(\lambda_n x)\exp(-\lambda_n^2\alpha\theta)\right\}t_L(0)$$

Upon integrating as much as possible without having a specific $t_L(\tau)$,

$$u(x,\theta) = t_L(\theta) - t_L(0)$$

$$-2\sum_{n=1}^{\infty}\frac{\sin(\lambda_n L)}{\lambda_n L}\cos(\lambda_n x)\exp(-\lambda_n^2\alpha\theta)\int_{\tau=0}^{\theta}\exp(\lambda_n^2\alpha\tau)t_L'(\tau)\,d\tau$$

$$+t_L(0) - 2t_L(0)\sum_{n=1}^{\infty}\frac{\sin(\lambda_n L)}{\lambda_n L}\cos(\lambda_n x)\exp(-\lambda_n^2\alpha\theta)$$

Upon canceling $t_L(0)$,

$$u(x,\theta) = t_L(\theta) - 2\sum_{n=1}^{\infty}\frac{\sin(\lambda_n L)}{\lambda_n L}\cos(\lambda_n x)\exp(-\lambda_n^2\alpha\theta)\int_{\tau=0}^{\theta}\exp(\lambda_n^2\alpha\tau)t_L'(\tau)\,d\tau$$

$$-2t_L(0)\sum_{n=1}^{\infty}\frac{\sin(\lambda_n L)}{\lambda_n L}\cos(\lambda_n x)\exp(-\lambda_n^2\alpha\theta)$$

SELECTED REFERENCES

1. Carslaw, H. S. and J. C. Jaeger: *Conduction of Heat in Solids*, 2nd ed., Oxford University Press, London, 1959.

2. Wylie, C. R. and L. C. Barrett: *Advanced Engineering Mathematics*, 5th ed., McGraw-Hill Book Company, New York, 1982.

EXERCISES

4·1 Consider the following problem for $t(x,\theta)$:

$$kt_{xx} + g''' = \rho ct_\theta \qquad \begin{aligned} t(0,\theta) &= t_0 \\ t(L,\theta) &= t_L \\ t(x,0) &= t^{(0)} \end{aligned}$$

Determine $t(x,\theta)$ by breaking the problem up into two subproblems. The solution for each subproblem should be obtained from the answers to exercises 3·20 and 3·25 by suitable substitutions.

4·2 Consider the following problem for $t(x,y)$:

$$t_{xx} + t_{yy} = 0 \qquad \begin{aligned} t(0,y) &= T_1 \\ t(L,y) &= T_2 \\ t(x,0) &= T_3 \\ t(x,L) &= T_4 \end{aligned}$$

Determine $t(x,y)$ by breaking the problem up into four subproblems, each having only one nonhomogeneous term. The solution for each subproblem should be obtained from the answer to Exercise 3·40 by suitable substitutions.

4·3 Consider the following problem for $t(x,y)$:

$$t_{xx} + t_{yy} = 0 \qquad \begin{aligned} t(0,y) &= T_1 \\ t(L,y) &= T_2 \\ t(x,0) &= T_1 \\ t(x,L) &= T_2 \end{aligned}$$

Determine $t(x,y)$ by breaking the problem up into two subproblems. The solution for each subproblem should be obtained from the answer to Exercise 3·40 by suitable substitutions.

4·4 Consider the following problem for $t(x,\theta)$:

$$kt_{xx} + g''' = \rho ct_\theta \qquad \begin{aligned} t(0,\theta) &= t_0 \\ -kt_x(L,\theta) &= h[t(L,\theta) - t_\infty] \\ t(x,0) &= t^{(0)} \end{aligned}$$

Determine the basic subproblems (each containing at most one nonhomogeneity) you could use to solve this problem.

4·5 A plane wall is initially at $t^{(0)}$. Beginning at time 0 its surface temperature at $x = L$ increases linearly until time τ_1 when is suddenly stepped back to $t^{(0)}$ for the remainder of the transient. The governing equations are given by

$$kt_{xx} = \rho ct_\theta \qquad \begin{aligned} t_x(0,\theta) &= 0 \\ t(L,\theta) &= \begin{cases} t^{(0)} + a\theta & \text{for } 0 \le \theta < \theta_1 \\ t^{(0)} & \text{for } \theta_1 < \theta \end{cases} \\ t(x,0) &= t^{(0)} \end{aligned}$$

(a) Determine the solution for $0 \le \theta \le \theta_1$ and for $\theta_1 \le \theta$. Verify that your two solutions are identical at $\theta = \theta_1$.

(b) If your solution in part (a) contains a series that is only a function of x, determine the algebraic function of x the series represents. An algebraic function is helpful because it is easier to evaluate than an infinite series. *Hint:* Replace the infinite series by $f(x)$ and substitute the solution into the partial differential equation and its boundary conditions to obtain an ordinary differential equation and boundary conditions. Solve for $f(x)$.

4·6 The functions $U(x,\theta)$, $V(x,\theta)$, $W(x,\theta)$ and $S(x,\theta)$ are known functions that solve the following problems:

$$U(x,\theta): \quad kU_{xx} + 1 = \rho cU_\theta \qquad V(x,\theta): \quad kV_{xx} = \rho cV_\theta$$
$$U(0,\theta) = 0 \qquad\qquad\qquad V(0,\theta) = 1$$
$$U_x(L,\theta) = 0 \qquad\qquad\qquad V(L,\theta) = 0$$
$$U(x,0) = 0 \qquad\qquad\qquad V(x,0) = 0$$

$$W(x,\theta): \quad kW_{xx} = \rho cW_\theta \qquad S(x,\theta): \quad kS_{xx} + 1 = \rho cS_\theta$$
$$W(0,\theta) = 1 \qquad\qquad\qquad S(0,\theta) = 0$$
$$W_x(L,\theta) = 0 \qquad\qquad\qquad S(L,\theta) = 0$$
$$W(x,0) = 0 \qquad\qquad\qquad S(x,0) = 0$$

By using superposition and Duhamel's theorem, determine the solution to the following problem for $t(x,\theta)$ in terms of $U(x,\theta)$, $V(x,\theta)$, $W(x,\theta)$ and $S(x,\theta)$:

$$kt_{xx} + G\theta = \rho ct_\theta \qquad \begin{aligned} t(0,\theta) &= t_1 + t_0 \exp(-a\theta) \\ t_x(L,\theta) &= 0 \\ t(x,0) &= t^{(0)} \end{aligned}$$

4·7 Consider the plane wall problem given by

$$kt_{xx} = \rho ct_\theta \qquad \begin{aligned} t(0,\theta) &= t^{(0)} + T\cos(\omega\theta) \\ t(L,\theta) &= t^{(0)} \\ t(x,0) &= t^{(0)} \end{aligned}$$

Determine $t(x,\theta)$ using Duhamel's theorem. Write the solution as the sum of a sustained part plus a transient part which decays to zero as time goes on.

4•8 One side of a plane wall is suddenly exposed to a step change in temperature which decays exponentially back to zero again as time goes on. The other side is insulated. The normalized problem might be as shown.

$$kt_{xx} = \rho c t_\theta \qquad t(0,\theta) = t^{(0)} + T\exp(-a\theta)$$
$$t_x(L,\theta) = 0$$
$$t(x,0) = t^{(0)}$$

(a) Using Duhamel's theorem, determine $t(x,\theta)$.
(b) In a practical case you might be concerned about the insulation melting if the initially applied temperature at $x = 0$ were greater than the temperature the insulation at $x = L$ could stand and decayed slowly. For $aL^2/\alpha = 1.0$, plot normalized temperature $\bar{t}(L,\bar{\theta})$ as a function of normalized time $\bar{\theta}$. Determine the maximum value of $\bar{t}(L,\bar{\theta})$ and the value of $\bar{\theta}$ at maximum $\bar{t}(L,\bar{\theta})$.

4•9 If $U(x,\theta)$ satisfies

$$U_{xx}(x,\theta) + G(x)\mathbf{1}(\theta) = U_\theta(x,\theta) \qquad U(0,\theta) = 0$$
$$U(1,\theta) = 0$$
$$U(x,0) = 0$$

Show by direct substitution that

$$u(x,\theta) = \int_{\tau=0}^{\theta} F(\tau)\frac{\partial U(x,\theta-\tau)}{\partial\theta}\,d\tau$$

is a solution to the following problem:

$$u_{xx}(x,\theta) + G(x)F(\theta) = u_\theta(x,\theta) \qquad u(0,\theta) = 0$$
$$u(1,\theta) = 0$$
$$u(x,0) = 0$$

4•10 A plane wall, insulated on one surface, has a time-dependent heat-flux input imposed on the other surface. The mathematical model is given by

$$kt_{xx} = \rho c t_\theta \qquad -kt_x(0,\theta) = q_0'' f(\theta)$$
$$t_x(L,\theta) = 0$$
$$t(x,0) = t^{(0)}$$

The maximum heat flux input q_0'' is a constant. The function $f(\theta)$ is described by the sketch.

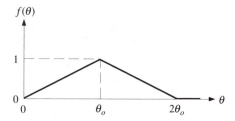

Using Duhamel's theorem, determine $t(x,\theta)$ for $\theta > 2\theta_o$.

4•11 Solve Exercise 3•33 using Duhamel's theorem and the solution to Exercise 3•25.

4•12 Consider a thin rod whose mathematical model is given by

$$kAt_{xx} - hp[t - t_\infty] = \rho cAt_\theta \qquad t(0,\theta) = t_0 + T\sin(\omega\theta)$$
$$t_x(L,\theta) = 0$$
$$t(x,0) = t^{(0)}$$

Using Duhamel's theorem and the solution to Exercise 3•30, determine the solution for $t(x,\theta)$.

4•13 The solution to the following problem

$$u_{xx} = u_\theta \qquad u(0,\theta) = 1$$
$$u(1,\theta) = 0$$
$$u(x,0) = 0$$

is given by

$$u(x,\theta) = 1 - x - \frac{2}{\pi}\sum_{n=1}^{\infty}\frac{\sin(n\pi x)}{n}\exp(-n^2\pi^2\theta)$$

If, instead of a step in temperature, the surface at $x = 0$ is made to undergo a series of pulses as given by,

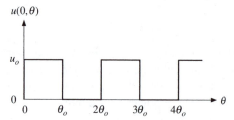

Determine an analytical expression for the normalized temperature $u(x,\theta)$ for $3\theta_o < \theta < 4\theta_o$.

4•14 Initially, a plane-wall nuclear-reactor fuel element [k, ρ, c] is operating steadily at a uniform generation rate g_i'''. The surface at $x = 0$ is adiabatic and the surface at $x = L$ is convectively cooled by the surroundings [h, t_∞]. Suddenly at $\theta = 0$ the generation rate is reduced to g_o''' and then begins to decay exponentially to 0. The sketch shows the generation rate as a function of time.

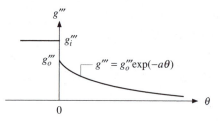

Determine $t(x,\theta)$ for this transient by assuming $t(x,\theta) = t_\infty + u(x,\theta) + w(x,\theta)$ where $u(x,\theta)$ can be obtained using Duhamel's theorem, $U(x,\theta)$ and $F(\tau)$. Sketch $F(\tau)$ for the Duhamel solution for both $\tau < 0$ and $\tau > 0$. Determine $U(x,\theta)$ and $w(x,\theta)$ from solutions in Carslaw and Jaeger [1]. State the pages and equation numbers in [1] that contain the two most appropriate solutions and "translate" these solutions into the notation used in this exercise and text.

CHAPTER

5

COMPLEX COMBINATION

5•0 INTRODUCTION

There are engineering problems in which periodic boundary conditions must be considered. Examples might occur in the study of the oscillatory behavior of internal combustion engine cylinder walls or in the penetration of the daily and annual temperature cycles into the earth. Solutions to problems with periodic boundary conditions can be obtained by variation of parameters as discussed in Section 3•3•7 or by Duhamel's theorem as discussed in Section 4•2. In each of these approaches a complete solution is obtained which can be written as the sum of a periodic, sustained solution plus a transient solution that decays to zero as time increases. Often there is a lot of work in obtaining the complete solution. In many applications, however, the transient decays so rapidly that the engineer is primarily interested in the sustained solution. In these cases *complex combination* may be used, with a great saving in time and effort,[1] to obtain only the sustained, periodic response of the system to a periodic disturbance.

If $u(x,\theta)$ is the sustained, periodic solution to the problem of interest (as yet $u(x,\theta)$ is unknown and is the sought-after function) which has a periodic disturbance such as $A\sin(\omega\theta)$ or $A\cos(\omega\theta)$, the steps in arriving at the solution $u(x,\theta)$ are listed below:

1. Construct a problem for $v(x,\theta)$ corresponding identically to the u problem but with the periodic disturbance replaced by $A\cos(\omega\theta)$ or $A\sin(\omega\theta)$. That is, the disturbance in the v problem is 90° out of phase with the u problem.

2. Construct a problem for $w(x,\theta) = u(x,\theta) + iv(x,\theta)$ by multiplying the v problem by $i = \sqrt{-1}$ and adding it to the u problem.[2]

3. Assume a periodic, product solution for w of the form

   $$w(x,\theta) = X(x)\exp(\pm i\omega\theta)$$

   to eliminate θ from the w problem.

4. Solve the resulting X problem for $X(x)$.

5. Write $w(x,\theta) = X(x)\exp(\pm i\omega\theta)$ as the sum of real and imaginary parts,

   $$w(x,\theta) = u(x,\theta) + iv(x,\theta)$$

   The real part of $w(x,\theta)$ is the sought-after solution $u(x,\theta)$. The imaginary part is the solution for $v(x,\theta)$.

   To use this method the problem must be completely homogeneous except for the periodic disturbance. Sections 5•1 and 5•2 discuss specific examples to illustrate the steps outlined above.

---

[1] This method may not be any faster than using Duhamel's theorem if the fundamental solution is already available.

[2] The w problem is the "complex combination" of the u problem and the v problem.

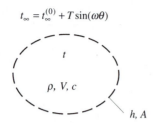

$$t_\infty = t_\infty^{(0)} + T\sin(\omega\theta)$$

t

ρ, V, c

h, A

Figure 5•1 *Lumped system in an ambient with a sinusoidal variation in temperature.*

5•1 LUMPED SYSTEMS

In Section 1•2•5 the complete solution was obtained for the temperature-time curve of a lumped system that was suddenly immersed in an oscillating temperature ambient. In this section, the same problem will be discussed, but only the sustained portion of the solution will be obtained.

Suppose a solid, initially at $t = t^{(0)}$, is suddenly placed in an ambient whose temperature is fluctuating sinusoidally with time as indicated in Figure 5•1. For a sufficiently small Biot number, a lumped analysis may be made. The governing differential equation was given by (1•2•24) as

$$\frac{dt}{d\theta} + \frac{hA}{\rho Vc}t = \frac{hA}{\rho Vc}t_\infty \qquad (5\cdot1\cdot1)$$

Upon substituting the expression for t_∞, the differential equation (5•1•1) can be written as

$$\frac{dt}{d\theta} + \frac{hA}{\rho Vc}t = \frac{hA}{\rho Vc}\left[t_\infty^{(0)} + T\sin(\omega\theta)\right] \qquad (5\cdot1\cdot2)$$

The initial condition is that $t(0) = t^{(0)}$.

To make the problem homogeneous, except for the $\sin(\omega\theta)$ term, we will let

$$t(\theta) = t_\infty^{(0)} + u(\theta) \qquad (5\cdot1\cdot3)$$

The differential equation (5•1•2) then becomes

$$\frac{du}{d\theta} + \frac{hA}{\rho Vc}\left[t_\infty^{(0)} + u\right] = \frac{hA}{\rho Vc}\left[t_\infty^{(0)} + T\sin(\omega\theta)\right]$$

or, upon canceling $t_\infty^{(0)}$,

$$\frac{du}{d\theta} + \frac{hA}{\rho Vc}u = \frac{hAT}{\rho Vc}\sin(\omega\theta) \qquad (5\cdot1\cdot4)$$

The initial condition for u is given by

$$u(0) = t^{(0)} - t_\infty^{(0)} \qquad (5\cdot1\cdot5)$$

In this problem, the periodic disturbance appears in the differential equation itself—not as a boundary condition as will be discussed in Section 5•2.

The v problem will be identical to the u problem except that the equation is given by

$$\frac{dv}{d\theta} + \frac{hA}{\rho Vc}v = \frac{hAT}{\rho Vc}\cos(\omega\theta) \qquad (5\cdot1\cdot6)$$

rather than (5•1•4). The initial condition is the same as (5•1•5),

$$v(0) = t^{(0)} - t_\infty^{(0)} \qquad (5\cdot1\cdot7)$$

Next, the complex combination of the u problem and the v problem will be formed by defining

$$w = u + iv \qquad (5\cdot1\cdot8)$$

The differential equation for w is found by multiplying (5•1•6) by $i = \sqrt{-1}$ and adding it to (5•1•4) to give

$$\frac{dw}{d\theta} + \frac{hA}{\rho Vc}w = \frac{hAT}{\rho Vc}[\sin(\omega\theta) + i\cos(\omega\theta)] \qquad (5\cdot1\cdot9)$$

The initial condition becomes

$$w(0) = (t^{(0)} - t_\infty^{(0)})(1 + i) \tag{5.1.10}$$

It is now essential to recall some complex-variable theory. A complex quantity can be written as

$$\exp(i\phi) = \cos(\phi) + i\sin(\phi) \tag{5.1.11}$$

Thus,

$$\exp(-i\phi) = \cos(-\phi) + i\sin(-\phi)$$

or, using trigonometric identities for sine and cosine of negative quantities,

$$\exp(-i\phi) = \cos(\phi) - i\sin(\phi) \tag{5.1.12}$$

Multiplying (5.1.12) by i gives

$$i\exp(-i\phi) = i\cos(\phi) + \sin(\phi)$$

or

$$i\exp(-i\phi) = \sin(\phi) + i\cos(\phi) \tag{5.1.13}$$

Therefore (5.1.9) can be rewritten as

$$\frac{dw}{d\theta} + \frac{hA}{\rho Vc}w = \frac{hAT}{\rho Vc}i\exp(-i\omega\theta) \tag{5.1.14}$$

The initial condition may be restated as

$$w(0) = (t^{(0)} - t_\infty^{(0)})(1 + i) \tag{5.1.15}$$

This conversion of $\sin(\omega\theta)$ and $\cos(\omega\theta)$ into exponential form is essential in this method.

The next step is to assume a complex exponential solution for w. Thus we will assume

$$w = W\exp(-i\omega\theta) \tag{5.1.16}$$

where W is a constant amplitude to be determined so that (5.1.16) satisfies the differential equation (5.1.14).

If (5.1.16) is now substituted into (5.1.14) we obtain

$$-i\omega W\exp(-i\omega\theta) + \frac{hA}{\rho Vc}W\exp(-i\omega\theta) = \frac{hAT}{\rho Vc}i\exp(-i\omega\theta)$$

Observe that (5.1.16) is constructed so that the time dependence will cancel out of the problem when (5.1.16) and (5.1.14) are combined. Thus, upon dividing by $\exp(-i\omega\theta)$ and solving for W,

$$W = \frac{\dfrac{hAT}{\rho Vc}i}{\dfrac{hA}{\rho Vc} - i\omega}$$

Multiplying numerator and denominator by the *complex conjugate* of the denominator allows W to be rewritten without complex numbers in the denominator as

$$W = \frac{\dfrac{hAT}{\rho Vc}i}{\dfrac{hA}{\rho Vc} - i\omega}\frac{\dfrac{hA}{\rho Vc} + i\omega}{\dfrac{hA}{\rho Vc} + i\omega} = \frac{hAT}{\rho Vc}\frac{i\dfrac{hA}{\rho Vc} - \omega}{\left(\dfrac{hA}{\rho Vc}\right)^2 + \omega^2}$$

Therefore (5.1.16) becomes

$$w = \frac{hAT}{\rho Vc} \frac{i \dfrac{hA}{\rho Vc} - \omega}{\left(\dfrac{hA}{\rho Vc}\right)^2 + \omega^2} \exp(-i\omega\theta)$$

Notice that in obtaining the above expression for w no attempt was made to satisfy the initial condition as stated by (5·1·15). As observed in Section 1•2•5, the initial condition was contained only in the transient term and had no influence upon the sustained solution. Therefore, since the method of complex combination gives only the sustained solution, the initial condition is not expected to enter the solution.

The solution for w must now be written as the sum of a real and an imaginary part. First, the solution may be rewritten as

$$w = \frac{\dfrac{hAT}{\rho Vc}}{\left(\dfrac{hA}{\rho Vc}\right)^2 + \omega^2} \left[\frac{hA}{\rho Vc} i \exp(-i\omega\theta) - \omega \exp(-i\omega\theta) \right]$$

Substituting (5·1·12) and (5·1·13) into this expression for w gives

$$w = \frac{\dfrac{hAT}{\rho Vc}}{\left(\dfrac{hA}{\rho Vc}\right)^2 + \omega^2} \left[\frac{hA}{\rho Vc} \{\sin(\omega\theta) + i\cos(\omega\theta)\} - \omega\{\cos(\omega\theta) - i\sin(\omega\theta)\} \right]$$

or, upon combining real and imaginary terms,

$$w = \frac{\dfrac{hAT}{\rho Vc} \left[\left\{ \dfrac{hA}{\rho Vc}\sin(\omega\theta) - \omega\cos(\omega\theta) \right\} + i\left\{ \dfrac{hA}{\rho Vc}\cos(\omega\theta) + \omega\sin(\omega\theta) \right\} \right]}{\left(\dfrac{hA}{\rho Vc}\right)^2 + \omega^2}$$

$$(5\cdot1\cdot17)$$

Since $w = u + iv$, the solution for u can be observed from (5·1·17) to be

$$u = \frac{\dfrac{hAT}{\rho Vc}}{\left(\dfrac{hA}{\rho Vc}\right)^2 + \omega^2} \left[\frac{hA}{\rho Vc}\sin(\omega\theta) - \omega\cos(\omega\theta) \right]$$

This may be rearranged to give

$$u = \frac{\dfrac{hAT}{\rho Vc}}{\sqrt{\left(\dfrac{hA}{\rho Vc}\right)^2 + \omega^2}} \left[\frac{\dfrac{hA}{\rho Vc}}{\sqrt{\left(\dfrac{hA}{\rho Vc}\right)^2 + \omega^2}}\sin(\omega\theta) - \frac{\omega}{\sqrt{\left(\dfrac{hA}{\rho Vc}\right)^2 + \omega^2}}\cos(\omega\theta) \right]$$

As was done in Section 1•2•5, a phase angle beta can be defined as shown in Figure 5•2. The solution can be rearranged to obtain

$$u = \frac{(hA/\rho Vc)T}{\sqrt{(hA/\rho Vc)^2 + \omega^2}} \left[\cos(\beta)\sin(\omega\theta) - \sin(\beta)\cos(\omega\theta) \right]$$

Upon making use of the trigonometric identity for $\sin(x - y)$, the solution can be rewritten as

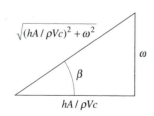

Figure 5•2 *Definition of phase angle.*

$$u = \frac{(hA/\rho Vc)T}{\sqrt{(hA/\rho Vc)^2 + \omega^2}} \sin(\omega\theta - \beta)$$

Substituting this result into (5·1·3) gives

$$t(\theta) = t_\infty^{(0)} + \frac{(hA/\rho Vc)T}{\sqrt{(hA/\rho Vc)^2 + \omega^2}} \sin(\omega\theta - \beta) \quad \text{where } \beta = \tan^{-1}\left[\frac{\omega}{hA/\rho Vc}\right]$$

This is identical to the sustained solution as given by (1·2·43). The initial condition does not enter the solution because it dies out with the transient term. The sustained solution is independent of the initial condition.

5•2 THE SEMI-INFINITE SOLID

The semi-infinite solid geometry is simply a plane wall whose face at $x = L$ has been moved to $x \to \infty$ as shown in Figure 5•3. At first glance it might appear that this geometry could have no practical engineering application. Actually, however, it is often quite useful.

One engineering application of the semi-infinite solid configuration is in finite geometries where, for short times, the heating or cooling effects at the surface have not yet been felt very far into the material. An example might be a step change in the surface temperature of a plane wall at small enough times so that the center temperature is still at its initial value. A step change in the surface temperature of a semi-infinite solid will be considered in Chapter 6 since the Laplace transformation is the efficient way to handle the problem.

Another problem of engineering interest is the penetration of the daily and annual temperature cycles into the earth's surface. The earth can be considered as a semi-infinite solid since its radius is so much larger than the depth to which the thermal fluctuations penetrate. This problem is most easily handled in this section using complex combination.

The governing partial differential equation is given by

$$kt_{xx} = \rho ct_\theta \tag{5.2.1}$$

with the boundary conditions that

$$t(0,\theta) = t_\infty + T\cos(\omega\theta) \tag{5.2.2}$$

and

$$\lim_{x\to\infty} t(x,\theta) = t_\infty \tag{5.2.3}$$

The initial condition is again unimportant to the sustained solution. The problem may be pictured as shown in Figure 5•4.

First, it is essential to make the problem homogeneous except for the $\cos(\omega\theta)$ term. This may be accomplished by defining

$$t(x,\theta) = t_\infty + u(x,\theta) \tag{5.2.4}$$

Upon substituting (5·2·4) into (5·2·1), the problem for u then becomes

$$ku_{xx} = \rho cu_\theta \tag{5.2.5}$$

Substituting (5·2·4) into (5·2·2) gives

$$u(0,\theta) = T\cos(\omega\theta) \tag{5.2.6}$$

and substituting (5·2·4) into (5·2·3) gives

$$\lim_{x\to\infty} u(x,\theta) = 0 \tag{5.2.7}$$

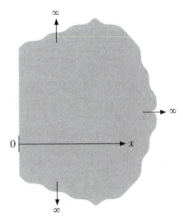

Figure 5•3 *The semi-infinite solid.*

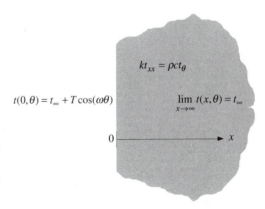

Figure 5•4 *The semi-infinite solid with a cosine variation in surface temperature.*

The corresponding v problem is then defined by

$$kv_{xx} = \rho c v_\theta \tag{5.2.8}$$

with

$$v(0,\theta) = T\sin(\omega\theta) \tag{5.2.9}$$

and

$$\lim_{x\to\infty} v(x,\theta) = 0 \tag{5.2.10}$$

The complex combination $w = u + iv$ is constructed by multiplying (5·2·8) through (5·2·10) by i and adding them to equations (5·2·5) through (5·2·7), respectively. The resulting problem for w is then given by

$$kw_{xx} = \rho c w_\theta \tag{5.2.11}$$

with

$$w(0,\theta) = u(0,\theta) + iv(0,\theta) = T\cos(\omega\theta) + iT\sin(\omega\theta)$$

which becomes, upon using (5·1·11),

$$w(0,\theta) = T\exp(i\omega\theta) \tag{5.2.12}$$

The boundary condition as x becomes infinite is

$$\lim_{x\to\infty} w(x,\theta) = 0 \tag{5.2.13}$$

Next, a complex exponential solution is assumed for $w(x,\theta)$ of the form

$$w(x,\theta) = X(x)\exp(i\omega\theta) \tag{5.2.14}$$

The function $X(x)$ will be determined so that the differential equation (5·2·11) and its boundary conditions (5·2·12) and (5·2·13) will be satisfied.

Substituting (5·2·14) into (5·2·11) gives

$$kX''\exp(i\omega\theta) = \rho c X i\omega \exp(i\omega\theta)$$

Again, the assumed solution is constructed so that the time dependence will cancel out of the above equation. Upon canceling $\exp(i\omega\theta)$, rearranging the equation and introducing $\alpha = k/\rho c$,

$$X'' - \frac{i\omega}{\alpha}X = 0 \tag{5.2.15}$$

Since this equation is second order, it requires two boundary conditions. At $x = 0$, (5·2·14) can be combined with (5·2·12) to give

$$w(0,\theta) = X(0)\exp(i\omega\theta) = T\exp(i\omega\theta)$$

or

$$X(0) = T \tag{5.2.16}$$

Similarly, (5·2·13) and (5·2·14) yield

$$\lim_{x\to\infty} X(x) = 0 \tag{5.2.17}$$

The solution of (5·2·15) is now obtained by the usual methods applied for ordinary differential equations with constant coefficients.[1] Thus the solution to (5·2·15) may be written as

---

[1] Following Appendix A·2·3, the general solution to $y''(x) + \kappa y(x) = 0$ can be shown to be given by $y(x) = A\exp(ix\sqrt{\kappa}) + B\exp(-ix\sqrt{\kappa})$.

$$X(x) = A \exp\left(ix\sqrt{\frac{-i\omega}{\alpha}} \right) + B \exp\left(-ix\sqrt{\frac{-i\omega}{\alpha}} \right)$$

Since $i\sqrt{-i} = i^2\sqrt{i} = -\sqrt{i}$, this may be rewritten as

$$X(x) = A \exp\left(-x\sqrt{\frac{i\omega}{\alpha}} \right) + B \exp\left(x\sqrt{\frac{i\omega}{\alpha}} \right) \qquad (5\cdot2\cdot18)$$

Before applying the boundary conditions, it is helpful to review some more complex-variable theory. Recall that a complex number z may be expressed as

$$z = r \exp(i\phi) \qquad (5\cdot2\cdot19)$$

This is shown graphically in Figure 5•5. Then z^α is found as

$$z^\alpha = r^\alpha \exp(i\alpha\phi) \qquad (5\cdot2\cdot20)$$

From (5·2·19),

$$i = \exp(i\pi/2) \qquad (5\cdot2\cdot21)$$

and from (5·2·20),

$$\sqrt{i} = \exp(i\pi/4) \qquad (5\cdot2\cdot22)$$

This may be represented pictorially as in Figure 5•6. In terms of its real and imaginary components, $\sqrt{i}$ may thus be represented as

$$\sqrt{i} = \frac{1}{\sqrt{2}} + i\frac{1}{\sqrt{2}} \qquad (5\cdot2\cdot23)$$

Therefore

$$\exp\left(x\sqrt{\frac{i\omega}{\alpha}} \right) = \exp\left[x\sqrt{\frac{\omega}{\alpha}}\left(\frac{1}{\sqrt{2}} + i\frac{1}{\sqrt{2}} \right) \right] = \exp\left(x\sqrt{\frac{\omega}{2\alpha}} \right) \exp\left(ix\sqrt{\frac{\omega}{2\alpha}} \right)$$

or

$$\exp\left(x\sqrt{\frac{i\omega}{\alpha}} \right) = \exp\left(x\sqrt{\frac{\omega}{2\alpha}} \right)\left[\cos\left(x\sqrt{\frac{\omega}{2\alpha}} \right) + i\sin\left(x\sqrt{\frac{\omega}{2\alpha}} \right) \right] \qquad (5\cdot2\cdot24)$$

This is seen to become infinite as x becomes infinite. Consequently, for (5·2·17) to be satisfied, B must be taken to be zero in (5·2·18). The solution (5·2·18) then becomes

$$X(x) = A \exp\left(-x\sqrt{\frac{i\omega}{\alpha}} \right) \qquad (5\cdot2\cdot25)$$

When (5·2·25) is substituted into the other boundary condition (5·2·16),

$$X(0) = T = A$$

Thus, with the help of (5·2·23), the final solution for X is

$$X(x) = T \exp\left[-x\sqrt{\frac{\omega}{\alpha}}\left(\frac{1}{\sqrt{2}} + i\frac{1}{\sqrt{2}} \right) \right] = T \exp\left(-x\sqrt{\frac{\omega}{2\alpha}} \right) \exp\left(-ix\sqrt{\frac{\omega}{2\alpha}} \right) \qquad (5\cdot2\cdot26)$$

Substituting (5·2·26) into (5·2·14) gives the solution for $w(x,\theta)$ as

$$w(x,\theta) = T \exp\left(-x\sqrt{\frac{\omega}{2\alpha}} \right) \exp\left(-ix\sqrt{\frac{\omega}{2\alpha}} \right) \exp(i\omega\theta) \qquad (5\cdot2\cdot27)$$

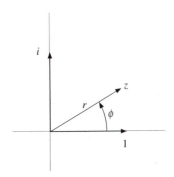

Figure 5•5 *Polar representation of a complex number* $z = r \exp(i\phi)$.

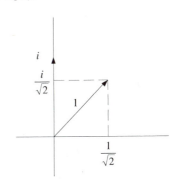

Figure 5•6 *Polar representation of* $\sqrt{i}$.

This expression for $w(x,\theta)$ must now be written in terms of a real and an imaginary part. Thus, rewriting (5·2·27),

$$w(x,\theta) = T \exp\left(-x\sqrt{\frac{\omega}{2\alpha}}\right)\exp\left[i\left(\omega\theta - x\sqrt{\frac{\omega}{2\alpha}}\right)\right]$$

Or, upon using (5·1·11),

$$w(x,\theta) = T \exp\left(-x\sqrt{\frac{\omega}{2\alpha}}\right)\left[\cos\left(\omega\theta - x\sqrt{\frac{\omega}{2\alpha}}\right) + i\sin\left(\omega\theta - x\sqrt{\frac{\omega}{2\alpha}}\right)\right] \quad (5·2·28)$$

Since $w = u + iv$, it is observed from (5·2·28) that $u(x,\theta)$ must be given by

$$u(x,\theta) = T \exp\left(-x\sqrt{\frac{\omega}{2\alpha}}\right)\cos\left(\omega\theta - x\sqrt{\frac{\omega}{2\alpha}}\right) \quad (5·2·29)$$

The solution for $t(x,\theta)$ is then found by substituting (5·2·29) into (5·2·4) to give

$$t(x,\theta) = t_\infty + T \exp\left(-x\sqrt{\frac{\omega}{2\alpha}}\right)\cos\left(\omega\theta - x\sqrt{\frac{\omega}{2\alpha}}\right) \quad (5·2·30)$$

Several interesting observations can be made from this solution. First, the phase shift is a function of depth x. Thus there are positions under the earth's surface that are increasing in temperature even though it is winter at the surface. Second, the amplitude of the oscillations decreases with depth and larger frequencies damp out faster than smaller frequencies. Thus the daily temperature cycle does not penetrate as deeply into the ground as does the annual temperature cycle. It can also be observed that the oscillations damp out nearer to the surface in materials with small thermal diffusivities (*e.g.*, nickel) than they do in large-diffusivity materials (*e.g.*, copper).

It should be noted that this problem was solved without normalization. Normalization could have been used, however, to provide some significant insights into the problem without ever solving it. Application of normalization in this case is discussed in Section 7•2•3, which should be read at this point to increase your understanding of the use of normalization as an analytical tool.

5•3 SUMMARY REMARKS ON COMPLEX COMBINATION

The method of complex combination has yielded the sustained, periodic solutions to two representative examples. These are not the complete solutions to either problem but were easier to obtain. Often, significant conclusions can be drawn from the sustained solutions alone which may answer the problem of interest to the engineer.

The examples discussed in Sections 5•1 and 5•2 should be sufficient to indicate how the method is applied. The only new area of mathematics involved is some simple complex-variable theory.

SELECTED REFERENCES

1. Arpaci, V. S.: *Conduction Heat Transfer*, Addison-Wesley Publishing Company, Inc., Reading, MA, 1966.

2. Churchill, R. V. and J. W. Brown: *Complex Variables and Applications*, 4th ed., McGraw-Hill Book Company, New York, 1984.

3. Schneider, P. J.: *Conduction Heat Transfer*, Addison-Wesley Publishing Company, Inc., Reading, MA, 1955.

EXERCISES

5•1 Using the method of complex combination, determine the sustained solution to the problem:

$$\frac{dt}{d\theta} + \frac{hA}{\rho Vc}t = \frac{hA}{\rho Vc}\left[t_\infty^{(0)} + T\cos(\omega\theta)\right] \qquad \text{with} \qquad t(0) = t^{(0)}$$

5•2 Using the method of complex combination, determine the sustained solution for the problem described by

$$kt_{xx} = \rho ct_\theta \qquad\qquad t(0,\theta) = t_\infty + T\sin(\omega\theta)$$
$$\lim_{x\to\infty} t(x,\theta) = t_\infty$$
$$t(x,0) = t_\infty$$

5•3 From the sustained semi-infinite-solid solution found in Section 5•2, calculate the ratio of the penetration depth (into the earth) of the annual temperature cycle to that of the daily temperature cycle.

5•4 Using the method of complex combination, determine the sustained solution for the semi-infinite-solid problem described by

$$kt_{xx} + g'''\cos(\omega\theta) = \rho ct_\theta \qquad t(0,\theta) = t_0$$
$$t(\infty,\theta) < \infty$$

5•5 A semi-infinite solid is subjected to a periodic heat flux at its surface. The mathematical model is given by

$$kt_{xx} = \rho ct_\theta \qquad\qquad -kt_x(0,\theta) = Q_0''\cos(\omega\theta)$$
$$\lim_{x\to\infty} t(x,\theta) = t^{(0)}$$

Using the method of complex combination, determine the sustained solution.

5•6 A semi-infinite solid is subjected to a periodic ambient temperature given by

$$t_\infty(\theta) = t_\infty^{(0)} + T\cos(\omega\theta)$$

For a finite surface heat-transfer coefficient, the differential equation and boundary conditions are

$$kt_{xx} = \rho ct_\theta \qquad\qquad kt_x(0,\theta) = h[t(0,\theta) - t_\infty(\theta)]$$
$$\lim_{x\to\infty} t(x,\theta) = t_\infty^{(0)}$$

(a) Using the method of complex combination, determine the sustained solution.
(b) Determine the solution for $h = 0$.
(c) Determine the solution for $h = \infty$. Compare to (5·2·30).
(d) Plot $(t_\infty(\theta) - t_\infty^{(0)})/T$ and $(t(0,\theta) - t_\infty^{(0)})/T$ as functions of $\omega\theta$ for one cycle for $(h/k)\sqrt{2\alpha/\omega} = 0.1$, 1.0 and 10.

5•7 Consider an infinitely long, semi-infinite hollow cylinder. The differential equation and boundary conditions are

$$k\left[t_{rr} + \frac{1}{r}t_r\right] = \rho ct_\theta \qquad\qquad t(r_i,\theta) = t^{(0)} + T\cos(\omega\theta)$$
$$\lim_{r\to\infty} t(r,\theta) = t^{(0)}$$

Using the method of complex combination, determine the sustained solution. *Hint:* Ordinary Bessel functions of a complex argument are related to modified Bessel functions. Modified Bessel functions are related to Kelvin functions.

CHAPTER

6

LAPLACE TRANSFORMS

6•0 INTRODUCTION

The Laplace transformation is a well-known technique for solving linear differential equations. Many of the problems in the earlier chapters could just as well have been solved by this method. There are, however, significant problems in which previously studied methods are difficult to apply. Semi-infinite solid transients and heat-exchanger transients are examples which lend themselves more readily to solution by Laplace transforms than by other methods.

The Laplace transform method lends itself particularly well to the solution of time-dependent problems. A transformation of the time variable essentially removes time from the problem. The transformed equation is thereby easier to solve than the original equation. As was the case in studying previous analytical methods, a thorough treatment will not be given. Rather, an introduction will be presented that will enable the engineer to solve many of the more elementary problems and give a familiarity with the method. A more comprehensive presentation is reserved for an advanced mathematics course.

In Section 6•1, the transformation is defined and applied to functions with engineering interest. The inversion of transformed functions is discussed in Section 6•2. Then in Section 6•3, the application of the Laplace transformation to the solution of ordinary differential equations is presented. The method is extended to the solution of partial differential equations in Section 6•4.

6•1 THE LAPLACE TRANSFORMATION

This section defines the Laplace transformation of a function. Particular attention will be paid to the transformation of derivatives because we will be using this method to solve differential equations. We must also consider the transformation of functions of two variables because we will apply this method to partial differential equations.

6•1•1 Definition

Consider a function $f(t)$. This function can be multiplied by $\exp(-st)$ and integrated with respect to t from zero to infinity. The new function, which will be a function of the parameter s, will be denoted by $\hat{f}(s)$ or by $\pounds[f(t)]$. That is,

Laplace transformation

$$\hat{f}(s) = \pounds[f(t)] = \int_{t=0}^{\infty} \exp(-st)f(t)\,dt \qquad (6\cdot1\cdot1)$$

The function $\hat{f}(s)$ is called the *transform of* $f(t)$. The functions $f(t)$ and $\hat{f}(s)$ are called a *transform pair*. Equation (6·1·1) is the definition of the Laplace transformation. A few examples will help illustrate this definition.

Example 1 Consider the function $f(t) = 1$. From (6·1·1),

$$\hat{f}(s) = \int_{t=0}^{\infty} \exp(-st) 1 \, dt$$

Integrating and substituting the limits of integration,

$$\hat{f}(s) = -\left[\frac{1}{s}\exp(-st)\right]_{t=0}^{\infty} = \frac{1}{s}$$

Thus the Laplace transform of $f(t) = 1$ is $\hat{f}(s) = 1/s$, or

$$£[1] = \frac{1}{s} \tag{6.1.2}$$

Example 2 Consider the function $f(t) = \exp(at)$. Then from (6·1·1),

$$\hat{f}(s) = \int_{t=0}^{\infty} \exp(-st)\exp(at) \, dt$$

Integrating,

$$\hat{f}(s) = \left[\frac{1}{a-s}\exp\{(a-s)t\}\right]_{t=0}^{\infty}$$

Since we are at liberty to choose the parameter s, it will be chosen to be algebraically greater than a. This will ensure that the upper limit does not cause $\hat{f}(s)$ to be infinite. Substituting the limits of integration gives

$$\hat{f}(s) = \frac{1}{s-a} \qquad\qquad s > a$$

This, then, is the transform of $\exp(at)$.

Example 3 Consider the sum of two functions, $h(t) = f(t) + g(t)$. Substitution into (6·1·1) gives

$$\hat{h}(s) = \int_{t=0}^{\infty} \exp(-st)[f(t) + g(t)] \, dt$$

$$= \int_{t=0}^{\infty} \exp(-st)f(t) \, dt + \int_{t=0}^{\infty} \exp(-st)g(t) \, dt$$

From (6·1·1) it is recognized that the above can be written as

$$\hat{h}(s) = \hat{f}(s) + \hat{g}(s)$$

Thus the Laplace transform of the sum of two functions is the sum of the Laplace transforms of each individual function. It also follows that the transform of $cf(t)$ is simply $c\hat{f}(s)$.

Simply stated, this example has shown that

$$£[f + g] = £[f] + £[g] \tag{6.1.3}$$

and

$$£[cf] = c£[f] \tag{6.1.4}$$

There are many relations such as these which will be useful to the engineering analyst. These are tabulated in Appendix E. This table is by no means complete, but rather it contains the transform pairs commonly found in the analysis of the equations discussed in conduction heat transfer. Usually the engineer will make use of such a table and will not rederive a relation each time it is required. Additional tables may be found in [1 through 7].

6•1•2 Transformation of ordinary derivatives

The transformation of derivatives is naturally quite important in the solution of differential equations. The procedure is just the same as transforming any function, *i.e.*, substitute into (6·1·1). Thus

$$\hat{f}'(s) = \pounds[f'(t)] = \int_{t=0}^{\infty} \exp(-st)f'(t)\,dt$$

Integrating by parts,

$$\hat{f}'(s) = \left[f(t)\exp(-st)\right]_{t=0}^{\infty} + s\int_{t=0}^{\infty} \exp(-st)f(t)\,dt$$

As long as $f(\infty)$ is finite,[1] the upper limit of the integrated term is zero and only the term from the lower limit remains. The remaining integral is recognized as the transform of $f(t)$. Therefore the above reduces to

$$\hat{f}'(s) = -f(0) + s\hat{f}(s) \tag{6·1·5}$$

Similarly, the transform of the second derivative is given by

$$\hat{f}''(s) = \pounds[f''(t)] = \int_{t=0}^{\infty} \exp(-st)f''(t)\,dt$$

Integrating by parts,

$$\hat{f}''(s) = \left[f'(t)\exp(-st)\right]_{t=0}^{\infty} + s\int_{t=0}^{\infty} \exp(-st)f'(t)\,dt$$

As long as $f'(\infty)$ is finite, the upper limit again is zero for the integrated terms. The integral is recognized as the transform of the first derivative which has just been found. Thus

$$\hat{f}''(s) = -f'(0) + s[-f(0) + s\hat{f}(s)] = -f'(0) - sf(0) + s^2\hat{f}(s) \tag{6·1·6}$$

Higher-order derivatives can be transformed in the same manner.

6•1•3 Transformation of functions of two variables

The transformation of functions of more than one variable is essential in the solution of partial differential equations. The transform of $f(x,y)$ can be taken with respect to either variable. Transforming the x variable gives

$$\pounds_x[f(x,y)] = \int_{x=0}^{\infty} \exp(-sx)f(x,y)\,dx = \hat{f}(s;y) \tag{6·1·7}$$

In this transformation, the variable x is replaced by the parameter s.

The y transform of $f(x,y)$ would correspondingly be given by

$$\pounds_y[f(x,y)] = \int_{y=0}^{\infty} \exp(-sy)f(x,y)\,dy = \hat{f}(x;s) \tag{6·1·8}$$

In this transformation, the variable y is replaced by the parameter s.

The transform of partial derivatives must also be considered. The x transform of $f_x(x,y)$ is found by substituting $f_x(x,y)$ into (6·1·7) to obtain

$$\pounds_x[f_x(x,y)] = \int_{x=0}^{\infty} \exp(-sx)\frac{\partial f(x,y)}{\partial x}\,dx \tag{6·1·9}$$

A parts integration, holding y constant, yields

$$\pounds_x[f_x(x,y)] = \left[f(x,y)\exp(-sx)\right]_{x=0}^{\infty} + s\int_{x=0}^{\infty} \exp(-sx)f(x,y)\,dx$$

Again, for $f(\infty,y)$ being finite, this reduces to

---

[1] Actually, it is only necessary that $\lim_{t\to\infty}[f(t)\exp(-st)] = 0$.

$$\pounds_x[f_x(x,y)] = -f(0,y) + s\hat{f}(s;y) \tag{6·1·10}$$

Here, $\hat{f}(s;y)$ is considered to be a function of y with s treated as a constant parameter. The y variable has just been carried through the transformation as if it were a constant.

The x transform of $f_y(x,y)$ is obtained by substituting $f_y(x,y)$ into (6·1·7),

$$\pounds_x[f_y(x,y)] = \int_{x=0}^{\infty} \exp(-sx) \frac{\partial f(x,y)}{\partial y} dx$$

Next, recalling Leibnitz' rule for the differentiation of integrals given by (4·2·11), the above may be written as

$$\pounds_x[f_y(x,y)] = \frac{d}{dy} \int_{x=0}^{\infty} \exp(-sx) f(x,y) dx$$

The integral may be recognized as the x transform of $f(x,y)$. Thus

$$\pounds_x[f_y(x,y)] = \frac{d\hat{f}(s;y)}{dy} \tag{6·1·11}$$

Second derivatives may be found in a similar manner. Thus

$$\pounds_x[f_{xx}(x,y)] = -f_x(0,y) - sf(0,y) + s^2\hat{f}(s;y) \tag{6·1·12}$$

Equation (6·1·12) is analogous to (6·1·6) for ordinary derivatives. In addition,

$$\pounds_y[f_{xx}(x,y)] = \frac{d^2\hat{f}(x;s)}{dx^2} \tag{6·1·13}$$

follows directly from (6·1·11). Mixed derivatives could also be transformed, but they are not usually of much interest in heat transfer.

6•2 INVERSION OF TRANSFORMED FUNCTIONS

As will be seen when we discuss the use of the Laplace transformation in solving differential equations, it will be necessary to be able to determine what the original function is if you are given its transform. This is called *inverting the transformed function* or *finding the inverse of the function*. The easiest and most expedient way is to look through a table of transform pairs to see if it contains the transform you are trying to invert. If you are lucky, you will find it. More typically, the transform you want will not be in the table. In these cases, the ideas discussed in sections 6•2•1 and 6•2•2 will be of value in obtaining an inversion.

6•2•1 Partial fractions

Partial fractions is a useful tool to use in conjunction with tables of transform pairs to obtain inverse transforms. This technique is used to split complicated fractions not found in the tables into simpler fractions which can be found in tables.

We are all familiar with adding several fractions together to obtain a combined fraction as shown below.

$$\frac{1}{x} + \frac{1}{x+1} - \frac{1}{x-2} = \frac{x^2 - 4x - 2}{x^3 - x^2 - 2x}$$

The method of partial fractions tells us how to go from a "complicated" fraction, such as on the right-hand side above, to the sum of several "simple" fractions, as on the left-hand side.

In each of the three cases discussed below, the numerator is of a lower degree than the denominator. If this is not the case, you can simply divide the numerator by the denominator. As indicated below, the remainder will be a fraction whose numerator has a lower degree than the denominator.

$$\frac{x^3 + x^2 - 1}{x^2 - 1} = x + 1 + \frac{x}{x^2 - 1}$$

The method partial fractions can then be used to simplify the remainder.

Case 1: Distinct linear factors in the denominator As an example, let us consider the following fraction:

$$\frac{x^2 - 4x - 2}{x(x+1)(x-2)}$$

This fraction can be written as the sum of several fractions whose least common denominator is the denominator in the above fraction. Thus we write

$$\frac{x^2 - 4x - 2}{x(x+1)(x-2)} = \frac{A}{x} + \frac{B}{x+1} + \frac{C}{x-2} \qquad (6\cdot2\cdot1)$$

where A, B and C are to be determined.

First we will clear fractions by multiplying by $x(x+1)(x-2)$ to get

$$x^2 - 4x - 2 = A(x+1)(x-2) + Bx(x-2) + Cx(x+1) \qquad (6\cdot2\cdot2)$$

Since this relation must be true for any x, the quick way to solve for A, B and C is to make a judicious choice for x that will allow one of these constants to be determined immediately. Then another choice for x will be made to evaluate another constant. For example, setting $x = 0$ in $(6\cdot2\cdot2)$ gives

$$-2 = -2A \qquad \text{or} \qquad A = 1$$

setting $x = -1$ in $(6\cdot2\cdot2)$ gives

$$1 + 4 - 2 = -1(-3)B \qquad \text{or} \qquad B = 1$$

and setting $x = 2$ in $(6\cdot2\cdot2)$ gives

$$4 - 8 - 2 = 2(3)C \qquad \text{or} \qquad C = -1$$

Therefore $(6\cdot2\cdot1)$ becomes

$$\frac{x^2 - 4x - 2}{x(x+1)(x-2)} = \frac{1}{x} + \frac{1}{x+1} - \frac{1}{x-2} \qquad (6\cdot2\cdot3)$$

Another way to evaluate A, B and C from $(6\cdot2\cdot2)$ would have been to equate coefficients of like powers of x on either side of the equation. Rearranging $(6\cdot2\cdot2)$,

$$x^2 - 4x - 2 = A(x^2 - x - 2) + B(x^2 - 2x) + C(x^2 + x)$$
$$= (A + B + C)x^2 + (-A - 2B + C)x + (-2A)$$

Equating the coefficients of like powers of x,

$$A + B + C = 1$$
$$-A - 2B + C = -4$$
$$-2A = -2$$

The solution to this set of equations is $A = 1$, $B = 1$ and $C = -1$, as before.

Case 2: Linear factors in the denominator, some repeated When some of the linear factors in the denominator are repeated, the procedure is almost the same as in Case 1 except that a term for each power of the repeated factor must be included. As an example,

$$\frac{3x+2}{x(x+1)^2} = \frac{A}{x} + \frac{B}{(x+1)^2} + \frac{C}{x+1} \qquad (6\cdot2\cdot4)$$

Clearing fractions,

$$3x+2 = A(x+1)^2 + Bx + Cx(x+1)$$

Rearranging,

$$3x+2 = A(x^2+2x+1) + Bx + C(x^2+x)$$
$$= (A+C)x^2 + (2A+B+C)x + A$$

Equating coefficients of like powers of x,

$$A + C = 0$$
$$2A + B + C = 3$$
$$A = 2$$

These give $A = 2$, $B = 1$ and $C = -2$. Thus

$$\frac{3x+2}{x(x+1)^2} = \frac{2}{x} + \frac{1}{(x+1)^2} - \frac{2}{x+1} \qquad (6\cdot2\cdot5)$$

Case 3: Quadratic factors in the denominator With quadratic factors the general scheme is the same with the exception that in the numerator of the partial fraction having the quadratic factor you must use the form $Ax + B$. For example,

$$\frac{3x^2+x+13}{(x-1)(x^2+16)} = \frac{A}{x-1} + \frac{Bx+C}{x^2+16} \qquad (6\cdot2\cdot6)$$

Clearing fractions,

$$3x^2 + x + 13 = A(x^2+16) + (Bx+C)(x-1)$$

Rearranging,

$$3x^2 + x + 13 = Ax^2 + 16A + Bx^2 + Cx - Bx - C$$
$$= (A+B)x^2 + (C-B)x + (16A-C)$$

Equating coefficients of like powers of x,

$$A + B = 3$$
$$C - B = 1$$
$$16A - C = 13$$

The solution to these equations is $A = 1$, $B = 2$ and $C = 3$. Thus

$$\frac{3x^2+x+13}{(x-1)(x^2+16)} = \frac{1}{x-1} + \frac{2x+3}{x^2+16} \qquad (6\cdot2\cdot7)$$

To demonstrate the utility of partial fractions in obtaining the inverse of a transformed function, let us consider the following example of the transform of $h(t)$:

$$\hat{h}(s) = \frac{a}{(s-a)(s^2+a^2)}$$

This may be expanded by partial fractions as follows:

$$\hat{h}(s) = \frac{a}{(s-a)(s^2+a^2)} = \frac{A}{s-a} + \frac{Bs+C}{s^2+a^2} = \frac{A}{s-a} + B\frac{s}{s^2+a^2} + C\frac{1}{s^2+a^2}$$

Multiplying by $(s-a)(s^2+a^2)$ to clear of fractions,

$$a = A(s^2+a^2) + Bs(s-a) + C(s-a)$$

Rearranging to combine like powers of s,

$$a = (A+B)s^2 + (C-Ba)s + (Aa^2 - Ca)$$

Equating coefficients of like powers of s gives

$$A + B = 0$$

$$C - Ba = 0$$

$$Aa^2 - Ca = a$$

Solving these three algebraic equations simultaneously for A, B and C,

$$A = \frac{1}{2a} \qquad\qquad B = -\frac{1}{2a} \qquad\qquad C = -\frac{1}{2}$$

Thus

$$\hat{h}(s) = \frac{1}{2a}\frac{1}{s-a} - \frac{1}{2a}\frac{s}{s^2+a^2} - \frac{1}{2}\frac{1}{s^2+a^2} = \frac{1}{2a}\left[\frac{1}{s-a} - \frac{s}{s^2+a^2} - \frac{a}{s^2+a^2}\right]$$

The inverses of each of the three separate terms within the brackets can be found directly in Appendix E as pairs 3, 5 and 4. Using these three pairs along with pair 9, the result is

$$h(t) = \frac{1}{2a}[\exp(at) - \cos(at) - \sin(at)]$$

6•2•2 Real convolution theorem

Many theorems could be discussed regarding the Laplace transformation. The most useful of these in the solution of the heat-transfer problems in this book is the real convolution theorem. It states that

Real Convolution Theorem

If $\hat{f}(s)$ and $\hat{g}(s)$ are the transforms of $f(t)$ and $g(t)$, respectively, then the function $h(t)$, whose transform is $\hat{h}(s) = \hat{f}(s)\hat{g}(s)$, is given by

$$h(t) = \int_{\sigma=0}^{t} f(\sigma)g(t-\sigma)d\sigma \qquad\qquad (6\cdot2\cdot8)$$

The function $h(t)$ is called the *real convolution of the functions* $f(t)$ *and* $g(t)$.

This theorem is useful because one is always faced, during the solution of differential equations, with the task of finding $h(t)$ when $\hat{h}(s)$ is given. Often this may simply be accomplished by using a table of transform pairs. In many cases, however, the particular relation that is required is not in the table. If in these cases $\hat{h}(s)$ can be written as the product of two functions $\hat{f}(s)$ and $\hat{g}(s)$, whose inverses can each be found in a table, the real convolution theorem can be employed to obtain $h(t)$.

The proof of this theorem is presented in the remainder of this section, not because of the importance of the proof itself, but because the technique used in interchanging the order of integration is one of which the engineer should be aware.

If

$$\hat{h}(s) = \hat{f}(s)\hat{g}(s)$$

then, by definition of $\hat{f}(s)$ and $\hat{g}(s)$, the above may be written as

$$\hat{h}(s) = \int_{\sigma=0}^{\infty} \exp(-s\sigma)f(\sigma)d\sigma \int_{\tau=0}^{\infty} \exp(-s\tau)g(\tau)d\tau$$

$$= \int_{\sigma=0}^{\infty}\int_{\tau=0}^{\infty} \exp[-s(\sigma+\tau)]f(\sigma)g(\tau)d\tau\,d\sigma$$

Next, consider σ to be fixed and change variables by letting $t = \sigma + \tau$ and $dt = d\tau$ to eliminate τ from the integral. Then

$$\hat{h}(s) = \int_{\sigma=0}^{\infty}\int_{t=\sigma}^{\infty} \exp(-st)f(\sigma)g(t-\sigma)dt\,d\sigma$$

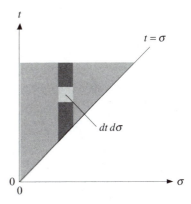

The next step is to change the order of integration to integrate first over σ rather than over t. Since the limits of the t integration contain σ, extra care must be taken. The above double integral may be pictured as an integral over an area as shown in Figure 6•1.

In the integration over t between σ and ∞, the differential area covers the narrow cross-hatched strip in the $t\sigma$ plane. In the second integration (over σ) between 0 and ∞, the narrow strip moves over the entire shaded region above the $t = \sigma$ line. It is important that the integration be over the same area in the $t\sigma$ plane after interchanging the order of integration. Thus the picture of the interchanged integration would appear as shown in Figure 6•2. The integration over σ covers the narrow cross-hatched strip between 0 and $\sigma = t$. The integration over t between 0 and ∞ then covers the same area as the original double integral. The integral thus can be written as

$$\hat{h}(s) = \int_{t=0}^{\infty}\int_{\sigma=0}^{t} \exp(-st)f(\sigma)g(t-\sigma)d\sigma\,dt$$

$$= \int_{t=0}^{\infty} \exp(-st)\left[\int_{\sigma=0}^{t} f(\sigma)g(t-\sigma)d\sigma\right]dt$$

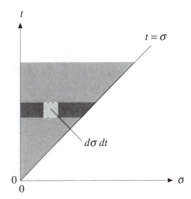

Figure 6•2 *Double integral with integration first over σ and then over t.*

The integral over t can be recognized as the Laplace transform of the function in brackets. Also, $\hat{h}(s)$ is the Laplace transform of $h(t)$. Thus the above may be written as

$$\pounds[h(t)] = \pounds\left[\int_{\sigma=0}^{t} f(\sigma)g(t-\sigma)d\sigma\right]$$

Therefore

$$h(t) = \int_{\sigma=0}^{t} f(\sigma)g(t-\sigma)d\sigma$$

must be the inverse transform of

$$\hat{h}(s) = \hat{f}(s)\hat{g}(s)$$

This completes the proof of the real convolution theorem.

Example Consider the same example as was discussed at the end of Section 6•2•1 where the inverse of the following expression was desired:

$$\hat{h}(s) = \frac{a}{(s-a)(s^2+a^2)}$$

This can be treated as a product of

$$\hat{f}(s) = \frac{a}{s^2 + a^2} \qquad \text{and} \qquad \hat{g}(s) = \frac{1}{s - a}$$

The inverses of these are given in Appendix E as pairs 4 and 3. Thus,

$$f(t) = \sin(at) \qquad \text{and} \qquad g(t) = \exp(at)$$

Thus the real convolution theorem gives

$$h(t) = \int_{\sigma=0}^{t} \sin(a\sigma)\exp[a(t - \sigma)]d\sigma = \exp(at)\int_{\sigma=0}^{t} \sin(a\sigma)\exp(-a\sigma)d\sigma$$

The integral may be evaluated using integral tables to give

$$h(t) = \frac{\exp(at) - \sin(at) - \cos(at)}{2a}$$

This is the same result as obtained previously by partial fractions.

6•3 SOLUTION TO ORDINARY DIFFERENTIAL EQUATIONS

The ordinary differential equations that arise in heat transfer can be solved by previously discussed methods with very little difficulty. The Laplace transformation will not offer major simplification. It is advisable to consider ordinary differential equations first, however, to demonstrate the procedure before considering partial differential equations. Briefly, here is the diagrammed procedure:

1. Ordinary differential equation and initial condition for $t(\theta)$

    ↓ Laplace transformation

2. Algebraic equation for $\hat{t}(s)$

    ↓ algebra

3. Solution for $\hat{t}(s)$

    ↓ tables, theorems

4. Solution for $t(\theta)$

Thus the ordinary differential equation for $t(\theta)$ and its initial condition is converted to an algebraic equation that is then solved for $\hat{t}(s)$. The most difficult step is the final one in which $\hat{t}(s)$ is inverted to obtain $t(\theta)$. If the necessary transform pair can be found in the tables, the task is easy. When it is not in the tables, considerably more ingenuity is demanded.

Example The governing differential equation for the lumped-parameter-transient problem of Section 1•2•5 may be obtained from (1·2·24) as

$$\frac{dt}{d\theta} + \frac{hA}{\rho Vc}t = \frac{hA}{\rho Vc}t_\infty \tag{6·3·1}$$

with the initial condition that

$$t(0) = t^{(0)} \tag{6·3·2}$$

The first step is to transform the differential equation by multiplying through by $\exp(-s\theta)$ and integrating from $\theta = 0$ to $\theta = \infty$. Thus

$$\int_{\theta=0}^{\infty} \exp(-s\theta)\frac{dt}{d\theta}d\theta + \frac{hA}{\rho Vc}\int_{\theta=0}^{\infty} \exp(-s\theta)t\,d\theta = \frac{hA}{\rho Vc}t_\infty\int_{\theta=0}^{\infty} \exp(-s\theta)\,d\theta$$

where t_∞ has been assumed to be constant.

The first integral can be integrated by parts and the right-hand side may be integrated to give

$$\left[\exp(-s\theta)t(\theta)\right]_{\theta=0}^{\infty} + s\int_{\theta=0}^{\infty}\exp(-s\theta)t\,d\theta + \frac{hA}{\rho Vc}\int_{\theta=0}^{\infty}\exp(-s\theta)t\,d\theta$$

$$= \frac{hA}{\rho Vc}t_{\infty}\left[\frac{\exp(-s\theta)}{-s}\right]_{\theta=0}^{\infty}$$

Since, based upon our physical knowledge of this problem, $t(\theta)$ is never infinite, the integrated term on the left-hand side is zero at the upper limit. The two remaining integrals in the above are recognized to be simply $\hat{t}(s)$. Thus the above reduces to the algebraic equation

$$-t(0) + \left(s + \frac{hA}{\rho Vc}\right)\hat{t}(s) = \frac{hA}{\rho Vc}t_{\infty}\frac{1}{s}$$

This may be solved for $\hat{t}(s)$ to give

$$\hat{t}(s) = \frac{t^{(0)} + \dfrac{hA}{\rho Vc}t_{\infty}\dfrac{1}{s}}{s + \dfrac{hA}{\rho Vc}} = t^{(0)}\frac{1}{s + \dfrac{hA}{\rho Vc}} + \frac{hA}{\rho Vc}t_{\infty}\frac{1}{s + \dfrac{hA}{\rho Vc}}\frac{1}{s} \qquad (6\cdot3\cdot3)$$

where the initial condition (6·3·2) has been used to replace $t(0)$.

Next (6·3·3) must be inverted. The first term may be inverted by using Appendix E directly (pair 3). The real convolution theorem can be used to invert the second term since the inverse of each of the terms in the product are in Appendix E (pairs 1 and 3). Thus we may write

$$t(\theta) = t^{(0)}\exp\left(-\frac{hA}{\rho Vc}\theta\right) + \frac{hA}{\rho Vc}t_{\infty}\int_{\sigma=0}^{\theta}\exp\left(-\frac{hA}{\rho Vc}\sigma\right)1\,d\sigma$$

Upon integrating and substituting limits,

$$t(\theta) = t^{(0)}\exp\left(-\frac{hA}{\rho Vc}\theta\right) - t_{\infty}\left[\exp\left(-\frac{hA}{\rho Vc}\theta\right) - 1\right]$$

Upon rearranging,

$$t(\theta) = t_{\infty} + (t^{(0)} - t_{\infty})\exp\left(-\frac{hA}{\rho Vc}\theta\right) \qquad (6\cdot3\cdot4)$$

This is identical to the solution (1·2·31) obtained in Section 1•2•5.

Another way to invert (6·3·3) is to first use partial fractions on the second term on the right-hand side. Then (6·3·3) becomes[1]

$$\hat{t}(s) = t^{(0)}\frac{1}{s + \dfrac{hA}{\rho Vc}} + t_{\infty}\left[\frac{1}{s} - \frac{1}{s + \dfrac{hA}{\rho Vc}}\right]$$

Upon combining the first and the last terms,

$$\hat{t}(s) = t_{\infty}\frac{1}{s} + (t^{(0)} - t_{\infty})\frac{1}{s + \dfrac{hA}{\rho Vc}}$$

These two terms may be inverted using pairs 1 and 3 from Appendix E to again arrive at (6·3·4).

---

[1] Using partial fractions $\dfrac{a}{s(s+a)} = \dfrac{A}{s} + \dfrac{B}{s+a} = \dfrac{1}{s} - \dfrac{1}{s+a}$.

6•4 SOLUTION TO PARTIAL DIFFERENTIAL EQUATIONS

The use of the Laplace transformation in the solution of partial differential equations is extremely convenient and efficient in many cases. The procedure is much the same as for the solution of ordinary differential equations. The steps[1] are shown below:

1. Partial differential equation, boundary and initial conditions for $t(x,\theta)$

  $\downarrow$ Laplace transformation with respect to θ

2. Ordinary differential equation for $\hat{t}(x;s)$

  $\downarrow$ ordinary differential equation theory

3. Solution for $\hat{t}(x;s)$

  $\downarrow$ tables, theorems

4. Solution for $t(x,\theta)$

These steps will be demonstrated in the next two sections. Section 6•4•1 considers a previously studied problem regarding a plane-wall transient. Section 6•4•2 considers a semi-infinite solid problem that has not previously been discussed in this book because the Laplace transformation is so much more efficient than other methods of solution.

6•4•1 The plane wall

Consider the plane-wall problem discussed in Section 3•1•1 and shown in Figure 6•3. The governing differential equation and its boundary and initial conditions are restated here as

$$kt_{xx} = \rho ct_\theta \qquad (6\cdot4\cdot1)$$

The boundary conditions are

$$t_x(0,\theta) = 0 \qquad (6\cdot4\cdot2)$$

$$t(L,\theta) = t_L \qquad (6\cdot4\cdot3)$$

and the initial condition is that

$$t(x,0) = t^{(0)}(x) \qquad (6\cdot4\cdot4)$$

Although this is a nonhomogeneous problem because of (6•4•3), the Laplace-transform procedure does not require any special methods.

The first step in the solution is to transform (6•4•1) with respect to θ by multiplying through by $\exp(-s\theta)$ and integrating from $\theta = 0$ to $\theta = \infty$. Thus

$$k\int_{\theta=0}^{\infty}\exp(-s\theta)t_{xx}\,d\theta = \rho c\int_{\theta=0}^{\infty}\exp(-s\theta)t_\theta\,d\theta \qquad (6\cdot4\cdot5)$$

Since the limits of the left-hand integral are not functions of x, the x derivatives may simply be taken out of the integral according to Leibnitz' rule (4•2•11). The right-hand integral may be integrated by parts, holding x fixed. We may then write (6•4•5) as

$$k\frac{d^2}{dx^2}\int_{\theta=0}^{\infty}\exp(-s\theta)t\,d\theta = \rho c\left\{\left[\exp(-s\theta)t\right]_{\theta=0}^{\infty} + s\int_{\theta=0}^{\infty}\exp(-s\theta)t\,d\theta\right\}$$

The integrals in this equation are both $\hat{t}(x;s)$, the θ transform of $t(x,\theta)$. Thus, after also substituting the limits on the integrated term, we obtain

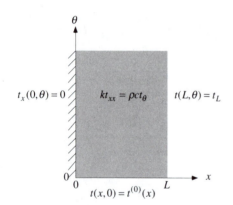

$$t_x(0,\theta) = 0 \qquad kt_{xx} = \rho ct_\theta \qquad t(L,\theta) = t_L$$

$$t(x,0) = t^{(0)}(x)$$

Figure 6•3 *Plane wall, initially at temperature $t^{(0)}(x)$, is suddenly exposed to a step change in surface temperature at $x = L$.*

---

[1] The Laplace transformation is intended to be applied to the time variable, not the position variable.

$$k\frac{d^2\hat{t}(x;s)}{dx^2} = \rho c\left\{-t(x,0) + s\hat{t}(x;s)\right\}$$

Dividing by k, introducing $\alpha = k/\rho c$, using (6·4·4) and rearranging,

$$\frac{d^2\hat{t}(x;s)}{dx^2} - \frac{s}{\alpha}\hat{t}(x;s) = -\frac{1}{\alpha}t^{(0)}(x) \tag{6·4·6}$$

This is an ordinary differential equation for $\hat{t}(x;s)$ which is considered to be a function of x with the constant parameter s carried along. The initial condition has already been incorporated into (6·4·6). The two boundary conditions (6·4·2) and (6·4·3) of the original partial differential equation will now be transformed to provide the boundary conditions for (6·4·6).

Transforming (6·4·2) in the usual manner by multiplying through by $\exp(-s\theta)$ and integrating gives

$$\int_{\theta=0}^{\infty}\exp(-s\theta)\left[\frac{\partial t(x,\theta)}{\partial x}\right]_{x=0}d\theta = 0$$

Using Leibnitz' rule,

$$\left[\frac{d}{dx}\int_{\theta=0}^{\infty}\exp(-s\theta)t(x,\theta)\,d\theta\right]_{x=0} = 0$$

The integral is seen to be $\hat{t}(x;s)$ the transform of $t(x,\theta)$. Thus,

$$\left[\frac{d\hat{t}(x;s)}{dx}\right]_{x=0} = 0 \tag{6·4·7}$$

Similarly, (6·4·3) may be transformed to give

$$\hat{t}(L;s) = \int_{\theta=0}^{\infty}\exp(-s\theta)t_L\,d\theta$$

Using pair 1 from Appendix E to evaluate the transform of t_L,

$$\hat{t}(L;s) = \frac{t_L}{s} \tag{6·4·8}$$

The problem has now been reduced to the solution of (6·4·6) with boundary conditions given by (6·4·7) and (6·4·8).

For $t^{(0)}(x) = t^{(0)}$, a constant, the general solution to (6·4·6) may be written as

$$\hat{t}(x;s) = t^{(0)}\frac{1}{s} + A\sinh(x\sqrt{s/\alpha}) + B\cosh(x\sqrt{s/\alpha}) \tag{6·4·9}$$

To satisfy (6·4·7) we will differentiate with respect to x

$$\frac{d\hat{t}(x;s)}{dx} = A\sqrt{s/\alpha}\cosh(x\sqrt{s/\alpha}) + B\sqrt{s/\alpha}\sinh(x\sqrt{s/\alpha})$$

Setting $x = 0$ and then substituting into (6·4·7), we obtain $A = 0$. Next we will set $x = L$ and $A = 0$ in (6·4·9) and substitute into (6·4·8) to obtain

$$\hat{t}(L;s) = t^{(0)}\frac{1}{s} + B\cosh(L\sqrt{s/\alpha}) = \frac{t_L}{s}$$

Solving for B,

$$B = \frac{t_L - t^{(0)}}{s}\frac{1}{\cosh(L\sqrt{s/\alpha})}$$

Substituting this expression for B and $A = 0$ into (6·4·9) gives

$$\hat{t}(x;s) = t^{(0)}\frac{1}{s} + (t_L - t^{(0)})\frac{\cosh(x\sqrt{s/\alpha})}{s\cosh(L\sqrt{s/\alpha})}$$

The final step is to invert $\hat{t}(x;s)$ to find the desired solution for $t(x,\theta)$. The first term may be inverted using pair 1 from Appendix E. The second term may be inverted by using pair 32 with x replaced by $x/\sqrt{\alpha}$, a replaced by $L/\sqrt{\alpha}$ and λ_n replaced by $(2n-1)\pi\sqrt{\alpha}/2L$. Thus,

$$t(x,\theta) = t^{(0)} + (t_L - t^{(0)})\left\{1 + \frac{4}{\pi}\sum_{n=1}^{\infty}\frac{(-1)^n}{(2n-1)}\cos[(2n-1)\frac{\pi}{2}\frac{x}{L}]\exp[-(2n-1)^2\frac{\pi^2}{4}\frac{\alpha\theta}{L^2}]\right\}$$

Removing the brackets,

$$t(x,\theta) = t_L + (t_L - t^{(0)})\frac{4}{\pi}\sum_{n=1}^{\infty}\frac{(-1)^n}{(2n-1)}\cos[(2n-1)\frac{\pi}{2}\frac{x}{L}]\exp[-(2n-1)^2\frac{\pi^2}{4}\frac{\alpha\theta}{L^2}]$$

Rearranging,

$$\frac{t(x,\theta) - t_L}{t^{(0)} - t_L} = \frac{4}{\pi}\sum_{n=1}^{\infty}\frac{(-1)^{n+1}}{(2n-1)}\cos[(2n-1)\frac{\pi}{2}\frac{x}{L}]\exp[-(2n-1)^2\frac{\pi^2}{4}\frac{\alpha\theta}{L^2}]$$

This result agrees with the solution (3·1·31) found in Section 3•1•1.

6•4•2 The semi-infinite solid

This physical configuration has already been considered in Section 5•2 when discussing periodic boundary conditions. In the current section a step change in surface temperature will be considered. This problem is not conveniently handled by separation of variables since the second boundary condition is at infinity rather than at some finite distance. The Laplace transform technique is the best method of handling this problem.

The governing partial differential equation, boundary conditions and initial condition are shown in Figure 6•4 and given by

$$kt_{xx} = \rho c t_\theta \tag{6.4.10}$$

$$t(0,\theta) = t_0 \tag{6.4.11}$$

$$\lim_{x\to\infty} t(x,\theta) = t^{(0)} \tag{6.4.12}$$

$$t(x,0) = t^{(0)} \tag{6.4.13}$$

The first step in the solution is to transform (6·4·10) with respect to θ by multiplying through by $\exp(-s\theta)$ and integrating from $\theta = 0$ to $\theta = \infty$. Thus

$$k\int_{\theta=0}^{\infty}\exp(-s\theta)t_{xx}\,d\theta = \rho c\int_{\theta=0}^{\infty}\exp(-s\theta)t_\theta\,d\theta \tag{6.4.14}$$

Since the limits of the left-hand integral are not functions of x, the x derivatives may simply be taken out of the integral according to Leibnitz' rule (4·2·11). The right-hand integral may be integrated by parts, holding x fixed. We may then write (6·4·14) as

$$k\frac{d^2}{dx^2}\int_{\theta=0}^{\infty}\exp(-s\theta)t\,d\theta = \rho c\left\{[\exp(-s\theta)t]_{\theta=0}^{\infty} + s\int_{\theta=0}^{\infty}\exp(-s\theta)t\,d\theta\right\}$$

The integrals in this equation are both $\hat{t}(x;s)$, the θ transform of $t(x,\theta)$. Thus, after also substituting the limits on the integrated term, we obtain

$$k\frac{d^2\hat{t}(x;s)}{dx^2} = \rho c\left\{-t(x,0) + s\hat{t}(x;s)\right\}$$

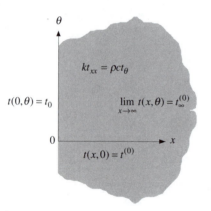

Figure 6•4 *The semi-infinite solid with a step change in surface temperature.*

θ

$kt_{xx} = \rho c t_\theta$

$t(0,\theta) = t_0$

$\lim_{x\to\infty} t(x,\theta) = t_\infty^{(0)}$

0

x

$t(x,0) = t^{(0)}$

Dividing by k, introducing $\alpha = k/\rho c$ and using (6·4·13),

$$\frac{d^2 \hat{t}(x;s)}{dx^2} - \frac{s}{\alpha}\hat{t}(x;s) = -\frac{1}{\alpha}t^{(0)} \tag{6·4·15}$$

Transforming the boundary condition (6·4·11) at $x = 0$ by multiplying through by $\exp(-s\theta)$ and integrating gives

$$\left[\int_{\theta=0}^{\infty}\exp(-s\theta)t(x,\theta)\,d\theta\right]_{x=0} = t_0\int_{\theta=0}^{\infty}\exp(-s\theta)\,d\theta$$

The left-hand integral is seen to be $\hat{t}(x;s)$ the transform of $t(x,\theta)$. The integral on the right-hand side may be evaluated using Appendix E, pair 1. Thus,

$$\hat{t}(0;s) = t_0\,\frac{1}{s} \tag{6·4·16}$$

Similarly, (6·4·12) may be transformed to give

$$\lim_{x\to\infty}\int_{\theta=0}^{\infty}\exp(-s\theta)t(x,\theta)\,d\theta = t^{(0)}\frac{1}{s}$$

The integral is again $\hat{t}(x;s)$. Thus,

$$\lim_{x\to\infty}\hat{t}(x;s) = t^{(0)}\frac{1}{s} \tag{6·4·17}$$

The problem has now been transformed to finding the solution of (6·4·15) with boundary conditions given by (6·4·16) and (6·4·17).

The general solution to (6·4·15) may be written as

$$\hat{t}(x;s) = t^{(0)}\frac{1}{s} + A\exp(-x\sqrt{s/\alpha}) + B\exp(+x\sqrt{s/\alpha}) \tag{6·4·18}$$

As x becomes infinite, $\hat{t}(x;s)$ will become infinite unless we take $B = 0$. Taking $B = 0$ and letting x become infinite in (6·4·18), we see that (6·4·17) is also satisfied. To satisfy the boundary condition (6·4·16) at $x = 0$ we will let $x = 0$ in (6·4·18) and substitute into (6·4·16) to obtain

$$t^{(0)}\frac{1}{s} + A = t_0\frac{1}{s}$$

Solving for A,

$$A = (t_0 - t^{(0)})\frac{1}{s}$$

The transformed solution (6·4·18) then becomes

$$\hat{t}(x;s) = t^{(0)}\frac{1}{s} + (t_0 - t^{(0)})\frac{1}{s}\exp(-x\sqrt{s/\alpha})$$

The first term can be inverted using pair 1 in Appendix E. The second term can be inverted using pair 20, taking $a = x/\sqrt{\alpha}$. The solution is then given in terms of the *complimentary error function* as

$$t(x,\theta) = t^{(0)} + (t_0 - t^{(0)})\mathrm{erfc}\!\left(\frac{x}{2\sqrt{\alpha\theta}}\right) \tag{6·4·19}$$

The complimentary error function $\mathrm{erfc}(x)$ is defined by the following integral:

$$\mathrm{erfc}(x) = \frac{2}{\sqrt{\pi}}\int_{\sigma=x}^{\infty}\exp(-\sigma^2)\,d\sigma \tag{6·4·20}$$

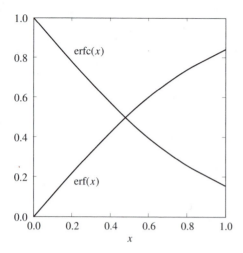

Figure 6•5 *The error functions.*

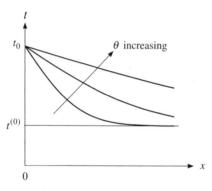

Figure 6•6 *Temperature distributions in a semi-infinite solid.*

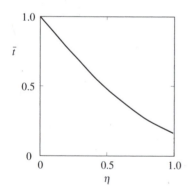

Figure 6•7 *Normalized temperature distribution in a semi-infinite solid.*

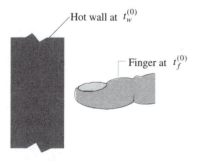

Figure 6•8 *Finger about to touch a hot wall.*

It is related to the ordinary error function $\mathrm{erf}(x)$ which is defined as

$$\mathrm{erf}(x) = \frac{2}{\sqrt{\pi}} \int_{\sigma=0}^{x} \exp(-\sigma^2)\,d\sigma \qquad (6\cdot4\cdot21)$$

Therefore

$$\mathrm{erf}(x) + \mathrm{erfc}(x) = \frac{2}{\sqrt{\pi}} \int_{\sigma=0}^{\infty} \exp(-\sigma^2)\,d\sigma$$

From a set of integral tables, it can be shown that

$$\int_{\sigma=0}^{\infty} \exp(-\sigma^2)\,d\sigma = \frac{\sqrt{\pi}}{2} \qquad (6\cdot4\cdot22)$$

Thus the error functions are related by

$$\mathrm{erf}(x) + \mathrm{erfc}(x) = 1 \qquad (6\cdot4\cdot23)$$

Error functions are common in the analysis of regions of infinite extent and therefore are worthy of attention. Numerical values for these functions can be found in [7] or almost any other set of tables. Plots of the error functions are shown in Figure 6•5.

The solution to the semi-infinite solid problem for a step change in the surface temperature may then be evaluated. The sketch in Figure 6•6 shows the temperature distribution for three different values of time.

Figure 6•6 is not the most compact presentation of the solution that can be made. The solution (6·4·19) can be rearranged to give

$$\frac{t(x,\theta) - t^{(0)}}{t_0 - t^{(0)}} = \mathrm{erfc}\left(\frac{x}{2\sqrt{\alpha\theta}}\right)$$

The right-hand side is a function of the combined variable $x/2\sqrt{\alpha\theta}$. Upon letting $\eta = x/2\sqrt{\alpha\theta}$ and writing the nondimensional temperature ratio on the left-hand side as $\bar{t}(\eta)$ gives

$$\bar{t}(\eta) = \mathrm{erfc}(\eta)$$

The nondimensional temperature is plotted as a function of $\eta = x/2\sqrt{\alpha\theta}$ in Figure 6•7. Each curve in Figure 6•6 falls on the same curve in Figure 6•7.

The reduction of the solution to a function of one variable η is an important simplification. Had we known this was possible ahead of time, the governing partial differential equation for $t(x,\theta)$ could have been transformed into an ordinary differential equation for $t(\eta)$. Since ordinary differential equations are much easier to solve, this would be a considerable simplification. One method of telling ahead of time whether variables can be combined is based on normalization. This is discussed in Section 7•3.

Contact temperature One of the interesting applications of the semi-infinite-solid solution is the estimation of *contact temperature*. The contact temperature indicates how "hot" or how "cold" an object will feel upon touching it. For example, consider a hot wall with a finger about to touch it as shown in Figure 6•8. For very short times after finger contact, both the wall and the finger will act as semi-infinite solids since the region affected by the contact will be quite small compared to the entire wall and finger. Indeed, it is just such "short-time" applications as this where the semi-infinite-solid model is most useful since it is rare that a solid is actually semi-infinite in practice.

The analytical model of this problem will be two semi-infinite solids, initially at different temperatures $t_w^{(0)}$ and $t_f^{(0)}$, that are suddenly brought into contact at zero time (see Figure 6•9). It will be assumed that there is no contact resistance between the finger and the wall, and therefore each solid will instantaneously come to some intermediate contact temperature t_c at the interface; consequently (6·4·19) can be directly applied to each solid.

Observe that the higher the contact temperature the warmer the wall will appear to the finger. Also, the contact temperature must be such that the heat flow leaving the wall at the interface is the same as that which enters the finger.

The solution to the semi-infinite-solid problem is given by (6·4·19) as

$$t(x,\theta) = t^{(0)} + (t_c - t^{(0)})\,\mathrm{erfc}(\frac{x}{2\sqrt{\alpha\theta}}) \tag{6·4·24}$$

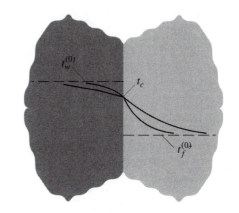

Figure 6•9 *Two semi-infinite solids, initially at different temperatures, suddenly brought into contact.*

where t_c = contact temperature has been substituted for t_0. The heat flow per unit surface area *out* of the solid is given by

$$q_{out}'' = k\frac{\partial t}{\partial x}\Big|_{x=0}$$

Substituting (6·4·24) with the complimentary error function replaced by its integral definition (6·4·20),

$$q_{out}'' = \left[k\frac{\partial}{\partial x}\left\{t^{(0)} + (t_c - t^{(0)})\frac{2}{\sqrt{\pi}}\int_{\sigma=\frac{x}{2\sqrt{\alpha\theta}}}^{\infty} \exp(-\sigma^2)\,d\sigma\right\}\right]_{x=0}$$

This may be differentiated, recalling Leibnitz' rule (4·2·11), to give

$$q_{out}'' = \frac{2}{\sqrt{\pi}}\left[k(t_c - t^{(0)})\left\{-\frac{\partial}{\partial x}\left(\frac{x}{2\sqrt{\alpha\theta}}\right)\exp(-\frac{x^2}{4\alpha\theta})\right\}\right]_{x=0}$$

Upon differentiating the term in parentheses and then setting $x = 0$,

$$q_{out}'' = \frac{-2}{\sqrt{\pi}}\left[k(t_c - t^{(0)})\frac{1}{2\sqrt{\alpha\theta}}\right] = \frac{k(t^{(0)} - t_c)}{\sqrt{\pi\alpha\theta}}$$

It is also convenient to substitute $\alpha = k/\rho c$ and simplify to give

$$q_{out}'' = \frac{\sqrt{k\rho c}\,(t^{(0)} - t_c)}{\sqrt{\pi\theta}} \tag{6·4·25}$$

The heat flow out of the wall is then given by

$$q_{out}''(wall) = \frac{\sqrt{(k\rho c)_w}\,(t_w^{(0)} - t_c)}{\sqrt{\pi\theta}}$$

The heat flow into the finger is given by

$$q_{in}''(finger) = \frac{\sqrt{(k\rho c)_f}\,(t_c - t_f^{(0)})}{\sqrt{\pi\theta}}$$

Since these two heat fluxes must be equal at all times,

$$\frac{\sqrt{(k\rho c)_w}\,(t_w^{(0)} - t_c)}{\sqrt{\pi\theta}} = \frac{\sqrt{(k\rho c)_f}\,(t_c - t_f^{(0)})}{\sqrt{\pi\theta}}$$

The denominators cancel and the relation may be solved to give the contact temperature as

$$t_c = \frac{t_w^{(0)}\sqrt{(k\rho c)_w} + t_f^{(0)}\sqrt{(k\rho c)_f}}{\sqrt{(k\rho c)_w} + \sqrt{(k\rho c)_f}} \qquad (6\cdot4\cdot26)$$

The contact temperature depends upon the $k\rho c$ products of the two solids. The larger the $k\rho c$ product of the wall is relative to $(k\rho c)_f$, the closer the contact temperature will be to the initial wall temperature. Some representative values of the $k\rho c$ product are presented in Table 6•1.

Table 6•1 *The $k\rho c$ product for several materials.*

Material	$k\rho c$ $Btu^2 / ft^4\text{-}F^2\text{-}hr$	Feel
Copper	12,000	Hot
Aluminum	4,670	
Steel, mild, 1 %	1,400	
Teflon	65.6	
Brick, fireclay	16.7	
Wood	2.3	
Asbestos	0.6	Cool
Skin	15–60	

Example Consider a hot object at 300 F and your finger at 100 F. Assuming the skin $k\rho c$ product is 36 $Btu^2 / ft^4\text{-}F^2\text{-}hr$, Table 6•2 gives the contact temperatures for three materials. These data should agree with your physical experience that copper feels hotter to touch than asbestos at the same temperature.

Table 6•2 *Contact temperatures for hot object at 300 F and finger at 100 F.*

Material	t_c, F
Copper	290
Brick	181
Asbestos	123

SELECTED REFERENCES

1. Abramowitz, M. and I. A. Stegun (eds.): *Handbook of Mathematical Functions*, Applied Mathematics Series 55, National Bureau of Standards, 1964. [Dover, 1965].

2. Campbell, G. A. and R. M. Foster: *Fourier Integrals for Practical Applications*, D. Van Nostrand Company, Inc., New York, 1948.

3. Carslaw, H. S. and J. C. Jaeger: *Conduction of Heat in Solids*, 2nd ed., Oxford University Press, London, 1959.

4. Churchill, R. V.: *Operational Mathematics*, McGraw-Hill Book Company, New York, 1958.

5. Doetsch, G.: *Guide to the Applications of Laplace Transforms*, D. Van Nostrand Company, Inc., New York, 1963.

6. Erdélyi, A.: *Tables of Integral Transforms*, vol. 1, McGraw-Hill Book Company, New York, 1954.

7. Spiegel, M. R.: *Schaum's Outline of Theory and Problems of Laplace Transforms*, McGraw-Hill Book Company, New York, 1965.

8. Flügge, W.: *Four-Place Tables of Transcendental Functions*, McGraw-Hill Book Company, New York, 1954.

EXERCISES

6•1 Show that

$$\pounds[\exp(-at)] = \frac{1}{s+a}$$

6•2 Show that

$$\pounds[\cos(bt)] = \frac{s}{s^2 + b^2}$$

6•3 Verify that

$$\pounds[\sinh(bt)] = \frac{b}{s^2 - b^2}$$

6•4 Verify that

$$\pounds_x[f_{xx}(x,y)] = -f_x(0,y) - sf(0,y) + s^2 \hat{f}(s;y)$$

6•5 Show that

$$\pounds_y[f_{xx}(x,y)] = \frac{d^2 \hat{f}(x;s)}{dx^2}$$

6•6 Find $f(t)$ if

$$\hat{f}(s) = \frac{s+1}{s^2 + s - 6}$$

Hint: $s^2 + s - 6$ may be factored into $(s-2)(s+3)$.

6•7 Using the real convolution theorem, find $h(t)$ if

$$\hat{h}(s) = \frac{1}{s^2 - 3s + 2}$$

6•8 Energy generation suddenly begins in a small piece of material. As the material heats up, it loses energy to the ambient. The governing differential equation is

$$\frac{dt}{d\theta} + \frac{hA}{\rho V c}(t - t_\infty) = \frac{g'''}{\rho c}$$

The initial condition is that $t(0) = t_\infty$. Using Laplace transforms, determine $t(\theta)$.

6•9 The differential equation for a lumped solid suddenly immersed in an ambient whose temperature is fluctuating is given by

$$\frac{dt}{d\theta} + \frac{hA}{\rho V c}(t - t_\infty) = 0$$

where $t_\infty = T_0 + T\cos(\omega\theta)$. The initial condition is given by $t(0) = t^{(0)}$. Using Laplace transforms, determine $t(\theta)$.

6•10 A plane wall suddenly subjected to a constant, nonzero heat-flux is described by

$$kt_{xx} = \rho c t_\theta \qquad\qquad t_x(0,\theta) = 0$$
$$-kt_x(L,\theta) = q_L''$$
$$t(x,0) = t^{(0)}$$

Using Laplace transforms, determine $t(x,\theta)$. Compare your solution to the one obtained in Section 3•1•5.

6•11 Consider the plane-wall problem given by

$$kt_{xx} = \rho c t_\theta \qquad\qquad t(0,\theta) = t_0 \exp(-a\theta)$$
$$t(L,\theta) = t_L$$
$$t(x,0) = t^{(0)}$$

Using Laplace transforms, determine $t(x,\theta)$.

6•12 A plane wall subjected to a step change in the temperature of one surface is described by

$$kt_{xx} = \rho c t_\theta \qquad\qquad t(0,\theta) = t_0$$

$$t(L,\theta) = t^{(0)}$$

$$t(x,0) = t^{(0)}$$

(a) Using Laplace transforms, determine $t(x,\theta)$.

For very short times the effect of the step change in the surface temperature has not penetrated very deeply, and the plane wall might be modeled as a semi-infinite solid. The solution to the semi-infinite model is given by (6·4·19).

(b) For $\bar{x} = x/L = 0.1$, compare the solution to part (a) with the semi-infinite-solid solution by plotting $\bar{t} = (t - t^{(0)})/(t_0 - t^{(0)})$ as a function of $\bar{\theta} = \alpha\theta/L^2$ for both solutions out to $\bar{\theta} = 1$. For how large a value of $\bar{\theta}$ is the semi-infinite-solid solution acceptable at this $\bar{x}$? For how small a value of $\bar{\theta}$ is a single term in the exact series solution acceptable?

6•13 A semi-infinite solid is suddenly exposed to a step change in the ambient temperature. The governing equations are given by

$$kt_{xx} = \rho c t_\theta \qquad\qquad -kt_x(0,\theta) = h[t_\infty - t(0,\theta)]$$

$$\lim_{x\to\infty} t(x,\theta) = t^{(0)}$$

$$t(x,0) = t^{(0)}$$

(a) Determine $t(x,\theta)$.
(b) Take the limit of the solution in part (a) to determine $t(x,\theta)$ as the heat-transfer coefficient h becomes infinite.

6•14 A plane wall, initially conducting heat at steady state, is suddenly insulated at $x = 0$. The governing equations are

$$kt_{xx} = \rho c t_\theta \qquad\qquad t_x(0,\theta) = 0$$

$$t(L,\theta) = t_L$$

$$t(x,0) = t_0 + (t_L - t_0)\frac{x}{L}$$

Using Laplace transforms, determine a solution for $t(x,\theta)$ that is valid for short times.

6•15 A semi-infinite solid is suddenly subjected to a constant heat-flux input q_0'' on its exposed face. Initially the solid is at temperature $t^{(0)}$.
(a) Determine $t(x,\theta)$.
(b) Determine the limiting solution as time increases.

6•16 Show that the contact temperature is independent of time when two semi-infinite solids at different temperatures are brought into contact.
Hint: Let t_1 be the temperature in body 1 and t_2 be the temperature in body 2 and then find the simultaneous solution to the pair of partial differential equations

$$k_1\frac{\partial^2 t_1}{\partial x^2} = \rho_1 c_1 \frac{\partial t_1}{\partial \theta} \qquad\qquad k_2\frac{\partial^2 t_2}{\partial x^2} = \rho_2 c_2 \frac{\partial t_2}{\partial \theta}$$

with the boundary conditions that

$$t_1(0,\theta) = t_2(0,\theta)$$

$$\left[k_1\frac{\partial t_1(x,\theta)}{\partial x}\right]_{x=0} = \left[k_2\frac{\partial t_2(x,\theta)}{\partial x}\right]_{x=0}$$

$$\lim_{x\to-\infty} t_1(x,\theta) = t_1^{(0)} \qquad\qquad \lim_{x\to+\infty} t_2(x,\theta) = t_2^{(0)}$$

and with the initial conditions that

$$t_1(x,0) = t_1^{(0)} \qquad\qquad t_2(x,0) = t_2^{(0)}$$

Note: Be sure to validate the boundary conditions used. Apply the Laplace transform to obtain the solution.

6•17 The surface temperature of a semi-infinite solid is raised linearly with time. The governing equations are

$$k\frac{\partial^2 t}{\partial x^2} = \rho c \frac{\partial t}{\partial \theta} \qquad\qquad t(0,\theta) = t^{(0)} + b\theta$$

$$\lim_{x\to\infty} t(x,\theta) = t^{(0)}$$

$$t(x,0) = t^{(0)}$$

Determine the solution for $t(x,\theta)$.

NORMALIZATION

7•0 INTRODUCTION

Although you may have been exposed to normalization, or at least to the use of nondimensional groups (*e.g.*, Biot number, Reynolds number, Prandtl number, Nusselt number), it is not anticipated that you have thought of normalization as an analytical tool. Consequently you should pay particular attention to the comments made in this chapter.

The material in this chapter is designed to be read along with the material in the first six chapters, since certain topics which arise naturally in the earlier chapters are discussed in more detail here. This discussion of normalization is also designed to be self-contained so that the reader can find all the comments on this subject in one chapter.

Three important uses of normalization are discussed: parameter simplification in Section 7•1, engineering approximations in Section 7•2 and combination of variables in Section 7•3. The material in Section 7•1 is most advantageously read after reviewing Chapter 1. Section 7•2 can be read most effectively along with the material in chapters 3 and 5. Section 7•3 should be read while studying the semi-infinite solid in Chapter 6.

7•1 PARAMETER SIMPLIFICATION

In many engineering problems the number of dimensional variables and parameters is quite large and therefore unwieldy. Consequently the mathematics and the presentation of results is also cumbersome. Normalization can be used to arrange these quantities into nondimensional groups. The number of nondimensional groups will be smaller in number than the original number of variables and parameters. This simplifies the mathematics and presentation of results.

This section discusses the idea of normalization and presents examples as illustrations. It does not present any foolproof way to go about normalizing a problem. Indeed this would be impossible because there are so many approaches leading to acceptable results. Several examples are given to illustrate some of the more important ideas. Experience is the best teacher of how best to normalize a problem.

7•1•1 Simplification of results

In many cases the engineer will find it convenient to go through the analysis of a problem in dimensional form to obtain the final result. In practically all these cases, however, there will be too many parameters and variables to make any sort of a generally useful plot of the results unless the parameters can somehow be grouped into a smaller number. This is the first application of normalization.

Consider the example of energy generation in a plane wall discussed in Section 1•2•3. The final result for the temperature distribution was given by (1·2·12) as

$$t - t_L = \frac{g'''}{2k}(L^2 - x^2) \qquad (7·1·1)$$

To make a general plot of temperature as a function of position, one must consider that temperature t is a function of t_L, g''', k, L and x. This is far too complicated for simple presentation in dimensional form. The best that could be done is to plot temperature as a function of position for various values of t_L, holding g''', k and L fixed. Additional plots would be required to show the effects of varying g''', k and L.

A great simplification can be obtained if (7·1·1) is rearranged in nondimensional form. Upon factoring the L^2 out of the parentheses, the following is obtained:

$$t - t_L = \frac{g'''L^2}{2k}\left[1 - \left(\frac{x}{L}\right)^2\right]$$

Since the term in brackets is dimensionless, the term $g'''L^2/2k$ must have the dimensions of temperature to make the equation dimensionally correct. Thus the equation can be written in nondimensional form as

$$\frac{t - t_L}{g'''L^2/2k} = 1 - \left(\frac{x}{L}\right)^2 \qquad (7·1·2)$$

The solution can be written in shorthand notation by defining the dimensionless variables

$$\bar{t} = \frac{t - t_L}{g'''L^2/2k} \qquad \text{and} \qquad \bar{x} = \frac{x}{L}$$

With these definitions the solution becomes

$$\bar{t} = 1 - \bar{x}^2 \qquad (7·1·3)$$

Notice that in (7·1·1) the temperature was a function of position and four other parameters (t_L, g''', k and L). In (7·1·3) the normalized temperature is a function only of the normalized position. Normalization has brought about a reduction of four parameters. This is a considerable saving.

In this form, a general plot can easily be made as shown in Figure 7•1. This plot contains all values of the parameters which appear in (7·1·1). It is more difficult to use for a specific case than a special plot for that particular case, but its value is in its generality. This single curve can be used for all values of t_L, g''', k and L.

It should be observed that the group $g'''L^2/2k$ which magically appeared in the problem (and had the dimensions of temperature) has physical significance. It is not simply a meaningless group of parameters. It is the maximum temperature difference that occurs within the wall. As such it may have engineering importance if melting or thermal stresses are a concern.

Another illustrative example is the lumped-system transient discussed in Section 1•2•5. The final result (1·2·31) was normalized by defining

$$\bar{t} = \frac{t(\theta) - t_\infty}{t^{(0)} - t_\infty} \qquad \text{and} \qquad \bar{\theta} = \frac{hA}{\rho Vc}\theta = \frac{\theta}{\rho Vc/hA} \qquad (7·1·4)$$

The normalized solution could then be written as

$$\bar{t}(\bar{\theta}) = \exp(-\bar{\theta}) \qquad (7·1·5)$$

This result was presented as a single curve in Figure 1•34.

These illustrations show how normalization can be put to good use after the differential equation has already been solved. Effective presentation of results is almost always done in normalized form.

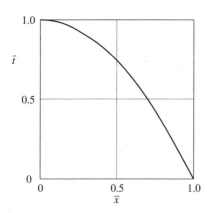

Figure 7•1 *Normalized temperature distribution in a plane wall with uniform energy generation.*

7•1•2　Simplification of differential equations

It will often be of value to the engineer to put a differential equation into its simplest form prior to doing any of the mathematics necessary to solve it. This is especially useful when there are a large number of parameters involved and when there is apt to be quite a bit of work to obtain the final solution. Normalization will reduce the number of parameters and also save on the amount of writing required to solve the problem.

Example 1　Consider the uniform generation of energy in a plane wall. The differential equation for $t(x)$ was given in Section 1•2•3, (1·2·8), as

$$\frac{d^2 t}{dx^2} + \frac{g'''}{k} = 0 \tag{7·1·6}$$

The boundary conditions are that

$$\frac{dt}{dx} = 0 \qquad \text{at } x = 0$$

$$t = t_L \qquad \text{at } x = L$$

The procedure is to normalize the variables first by defining the dimensionless variables

$$\bar{t} = \frac{t - t_b}{t_n} \qquad \text{and} \qquad \bar{x} = \frac{x}{x_n}$$

where t_n and x_n are "normalizing constants" to chosen later. The selection of $t - t_b$, where t_b is a "base" temperature, in the numerator is based on the experience that temperature difference is what is important in conduction heat transfer rather than temperature level itself. Now

$$\frac{dt}{dx} = \frac{d}{dx}(t - t_b) = \frac{d}{dx}(t_n \bar{t}) = t_n \frac{d\bar{t}}{dx}$$

and from the chain rule of calculus,

$$\frac{d\bar{t}}{dx} = \frac{d\bar{t}}{d\bar{x}} \frac{d\bar{x}}{dx} = \frac{d\bar{t}}{d\bar{x}} \frac{1}{x_n}$$

Then upon combining the above relations,

$$\frac{dt}{dx} = \frac{t_n}{x_n} \frac{d\bar{t}}{d\bar{x}}$$

The careful analyst will observe that the dimensions of both sides of the expression for the first derivative are T/L. Similarly,

$$\frac{d^2 t}{dx^2} = \frac{d}{dx}\left(\frac{dt}{dx}\right) = \frac{d}{d\bar{x}}\left(\frac{dt}{dx}\right)\frac{d\bar{x}}{dx} = \frac{d}{d\bar{x}}\left(\frac{t_n}{x_n}\frac{d\bar{t}}{d\bar{x}}\right)\frac{1}{x_n} = \frac{t_n}{x_n^2}\frac{d^2\bar{t}}{d\bar{x}^2}$$

Again observe that the dimensions on both sides are identical $[T/L^2]$. The differential equation (7·1·6) then becomes

$$\frac{t_n}{x_n^2}\frac{d^2\bar{t}}{d\bar{x}^2} + \frac{g'''}{k} = 0$$

Next the differential equation is made dimensionless by multiplying by x_n^2/t_n to give

$$\frac{d^2\bar{t}}{d\bar{x}^2} + \frac{g''' x_n^2}{k t_n} = 0$$

This must be dimensionless because both $\bar{t}$ and $\bar{x}$ and consequently the second derivative are dimensionless. Thus, if one term in the equation is dimensionless, all terms must be dimensionless.

The selection of x_n and t_n is next. The choice should be based on either of two reasons (*and* in the order given below).

1. *Physical significance* The engineer should not lose sight of the physics involved in the problem. All of the physically significant constants will appear either in the differential equation or in its boundary conditions (or initial condition). A boundary condition might be given at a special value of x (*e.g.*, $x = L$). Then $x_n = L$ would have physical significance. There might be a specified step change in temperature (*e.g.*, $t_n - t_\infty$) that could be used for t_n.

2. *Mathematical simplicity* There will often be several coefficients which appear in the differential equation and its boundary and/or initial conditions. These coefficients will consist of the normalizing constants and physical properties. The engineer might as well make the mathematics as painless as possible. This usually means that as many as possible of these coefficients are set equal to unity by the choice of the remaining normalizing constants.

For the problem at hand then, we first check to see if there is any physical basis for the selection of x_n, t_n or t_b. In this problem L is a physically significant length. Observe that one of the boundary conditions is given at $x = L$. Therefore choose $x_n = L$. There is no physically significant temperature to choose for t_n but the t_L that appears in the boundary condition at $x = L$. is a good choice for t_b. Since there is no physically significant temperature to choose for t_n, take $t_n = g'''L^2/k$ for reasons of mathematical simplicity. This choice makes the differential equation quite "clean" and "neat" because the constant term will be unity. A dimensional check will reveal that $g'''L^2/k$ has the required dimensions of temperature. The value $g'''L^2/k$ will also have physical significance; it will be related to the maximum temperature difference involved in the problem. Until you actually solve the problem, you cannot be sure of the exact relation. This in itself is quite a lot of information considering that the differential equation hasn't even been solved yet. It turns out that $g'''L^2/k$ is twice the maximum temperature difference.

With these choices of x_n and t_n, the differential equation reduces to

$$\frac{d^2\bar{t}}{d\bar{x}^2} + 1 = 0$$

In normalized form the boundary conditions become

$$\frac{d\bar{t}}{d\bar{x}} = 0 \qquad \text{at } \bar{x} = 0$$

$$\bar{t} = \frac{t_L - t_b}{g'''L^2/k} \qquad \text{at } \bar{x} = 1$$

As a result of the normalization procedure the parameters t_L, g''', k and L have all been hidden. The problem has been reduced to finding $\bar{t}$ as a function only of $\bar{x}$ with no other parameters present. The solution to the problem can be obtained by the same methods used to solve the dimensional form of the problem. The result is given by

$$\bar{t} = \tfrac{1}{2}(1 - \bar{x}^2) \tag{7·1·8}$$

Observe that (7·1·8) is in a slightly different form from (7·1·3). Either one is adequate. By normalizing the differential equation before solving it, the solution in arriving at (7·1·8) is a little neater than in arriving at (7·1·3) because t_L, g''', k and L did not have to be written as much. In all probability the analyst will use a little of each approach. Make what seem to be logical choices of x_n and t_n before solving the problem and then, having obtained the solution, check to see if any additional significance can be given to the choices for x_n and t_n. You may even want to redefine your variables. In this example, since $t_n = g'''L^2/k$ turns out to be twice the maximum temperature difference, you might redefine $t_n = g'''L^2/2k$.

Example 2 As a second example, consider the lumped-system transient discussed in Section 1•2•5. The governing dimensional differential equation was given by (1·2·24) as

$$\frac{dt}{d\theta} + \frac{hA}{\rho Vc}t = \frac{hA}{\rho Vc}t_\infty \tag{7·1·9}$$

where t_∞ was taken to be a constant value. The initial condition was that $t = t^{(0)}$ at $\theta = 0$.

Again, the procedure is to define nondimensional variables as follows:

$$\bar{t} = \frac{t - t_b}{t_n} \qquad \text{and} \qquad \bar{\theta} = \frac{\theta}{\theta_n}$$

Substituting these definitions into the differential equation,

$$\frac{t_n}{\theta_n}\frac{d\bar{t}}{d\bar{\theta}} + \frac{hA}{\rho Vc}(t_b + t_n\bar{t}) = \frac{hA}{\rho Vc}t_\infty$$

A physically significant choice for the base temperature is to take $t_b = t_\infty$. This choice is also mathematically nice since the terms containing t_b and t_∞ in the differential equation will cancel. The differential equation can be made dimensionless by multiplying by θ_n/t_n to give

$$\frac{d\bar{t}}{d\bar{\theta}} + \frac{hA\theta_n}{\rho Vc}\bar{t} = 0$$

Observe that t_n has dropped out of the differential equation. Thus it need not yet be specified. Since there is no physically significant time to choose for θ_n, take $\theta_n = \rho Vc/hA$ for mathematical simplicity. Whenever a value is chosen for one of the normalizing constants based on mathematical simplicity, it can also be expected to have physical importance. This is will be discussed in Section 7•2•2.

The normalized differential equation is then given by

$$\frac{d\bar{t}}{d\bar{\theta}} + \bar{t} = 0$$

The initial condition becomes

$$\bar{t}(0) = \frac{t^{(0)} - t_\infty}{t_n}$$

The choice of t_n is relatively straightforward. Based both on physical grounds and on mathematical simplicity, take $t_n = t^{(0)} - t_\infty$, the maximum temperature difference in the problem. The initial condition then becomes

$$\bar{t}(0) = 1$$

The solution is obtained by the same method used in Section 1•2•5. The result is identical to (7·1·5).

Example 3 Normalization may also be used with partial differential equations. Consider a plane wall, initially at a uniform temperature $t^{(0)}$, whose surface temperatures are each stepped to a new value t_L at time zero. The governing dimensional partial differential equation is given by

$$k \frac{\partial^2 t}{\partial x^2} = \rho c \frac{\partial t}{\partial \theta} \tag{7.1.11}$$

The boundary and initial conditions are that

$$t = t_L \qquad \text{at} \qquad x = \pm L \tag{7.1.12}$$

and

$$t = t^{(0)} \qquad \text{at} \qquad \theta = 0 \tag{7.1.13}$$

The first step in normalization is to define nondimensional variables to use in place of t, x and θ. Thus

$$\bar{t} = \frac{t - t_b}{t_n} \quad , \qquad \bar{x} = \frac{x}{x_n} \quad \text{and} \quad \bar{\theta} = \frac{\theta}{\theta_n}$$

The differential equation (7.1.11) then becomes

$$k \frac{t_n}{x_n^2} \frac{\partial^2 \bar{t}}{\partial \bar{x}^2} = \rho c \frac{t_n}{\theta_n} \frac{\partial \bar{t}}{\partial \bar{\theta}}$$

At this point it is wise to check the units to be sure they agree with (7.1.11). Observe that t_n / x_n^2 has the same units as $\partial^2 t / \partial x^2$ and that $\partial^2 \bar{t} / \partial \bar{x}^2$ is dimensionless. Thus the units check on the left-hand side. A similar verification can be made on the right-hand side. Multiplying by $x_n^2 / k t_n$ to nondimensionalize the equation and replacing $k / \rho c$ by α gives

$$\frac{\partial^2 \bar{t}}{\partial \bar{x}^2} = \frac{x_n^2}{\alpha \theta_n} \frac{\partial \bar{t}}{\partial \bar{\theta}}$$

In this case, there is a physically significant length L to choose for x_n. Inspection of the differential equation, the boundary conditions and the initial condition does not reveal any physically significant time to select for θ_n. In the interest of mathematical simplicity θ_n is then taken to be L^2 / α. As will be discussed in Section 7•2•2, the choice of $\theta_n = L^2 / \alpha$ can also be shown to have physical implications in this problem. With these choices for x_n and θ_n the differential equation reduces to

$$\frac{\partial^2 \bar{t}}{\partial \bar{x}^2} = \frac{\partial \bar{t}}{\partial \bar{\theta}} \tag{7.1.14}$$

To completely normalize the problem, the boundary and initial conditions must also be normalized. At $x = \pm L$ (now, at $\bar{x} = \pm 1$) we have

$$\bar{t}(\pm 1, \bar{\theta}) = \frac{t_L - t_b}{t_n} \tag{7.1.15}$$

A physically significant and mathematically nice base temperature would be to take $t_b = t_L$. The initial condition (7.1.13) becomes

$$\bar{t}(\bar{x}, 0) = \frac{t^{(0)} - t_b}{t_n} = \frac{t^{(0)} - t_L}{t_n}$$

A logical choice would be to take $t_n = t^{(0)} - t_L$, the maximum temperature difference. Thus the initial condition becomes

$$\bar{t} = 1 \tag{7.1.16}$$

Observe that in dimensional form $t = t(x,\theta;k,\rho,c,L,t_L,t^{(0)})$, whereas in nondimensional form $\bar{t} = \bar{t}(\bar{x},\bar{\theta})$. The parameters k, ρ, c, L, t_L and $t^{(0)}$ have all been absorbed by the normalization. This is the most compact form for the problem.

It should be pointed out here that, although normalization may be a mathematical convenience, normalization is never required to carry out the mathematical solution to a problem. The engineering analyst should weigh the advantages of normalization against the time it takes to convert from a dimensional model to a nondimensional one and then back to dimensional form afterward if it is required.

7•2 ENGINEERING APPROXIMATIONS

Normalization can be quite helpful in making engineering approximations or estimates. Often an engineer can find out all that it is necessary to know about a problem simply by normalizing it. Normalization can also suggest approximations which can be made to simplify further analysis. The next three sections discuss ways in which normalization alone can lead to new insights into a problem.

7•2•1 Biot number significance

The plane wall is a commonly proposed model in many engineering analyses because of its simplicity. The analytical treatment necessary to study the transient behavior of this simple model can be markedly different depending on the numerical values of the parameters involved. A thin sheet of aluminum suddenly immersed in a cool air stream would be analyzed much differently than a thick sheet of plastic suddenly immersed in a cool water stream. Normalization of the governing equations can suggest simplifications in the analysis.

Consider a plane wall, initially at temperature $t^{(0)}$, which is suddenly immersed in an ambient whose temperature is t_∞. A sketch of the problem is shown in Figure 7•2. The internal temperature distribution will change from its initial uniform condition to a final steady-state value equal to the ambient temperature. The differential equation describing this problem is

$$k\frac{\partial^2 t}{\partial x^2} = \rho c \frac{\partial t}{\partial \theta} \qquad (7 \cdot 2 \cdot 1)$$

The boundary condition at $x = 0$ is observed from symmetry to be

$$\left.\frac{\partial t}{\partial x}\right|_{x=0} = 0 \qquad (7 \cdot 2 \cdot 2)$$

The convection boundary condition at $x = L$ is given by

$$-k\left.\frac{\partial t}{\partial x}\right|_{x=L} = h[t(L,\theta) - t_\infty] \qquad (7 \cdot 2 \cdot 3)$$

The initial condition is that at $\theta = 0$

$$t(x,0) = t^{(0)} \qquad (7 \cdot 2 \cdot 4)$$

These four equations give the mathematical formulation in dimensional quantities for this problem. Their solution will provide the temperature as a function of x and θ. The mathematical solution could be carried out at this point using the methods discussed in Chapter 3 (see Exercise 3•17); normalization is not necessary. In this section, however, we will normalize to see the new insights which can be obtained from normalization alone.

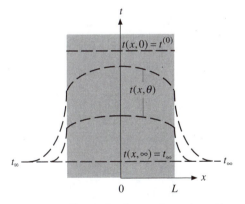

Figure 7•2 *Sketch of a plane-wall transient with surface convection.*

Following the steps outlined in Section 7•1•2 for normalizing the governing equations, we will first define the nondimensional variables

$$\bar{t} = \frac{t - t_\infty}{t_n} \quad , \quad \bar{x} = \frac{x}{x_n} \quad \text{and} \quad \bar{\theta} = \frac{\theta}{\theta_n}$$

where t_n, x_n and θ_n are as yet undetermined normalizing constants. A choice for the base temperature of $t_b = t_\infty$ has already been made in the definition of $\bar{t}$.

When these dimensionless variables are introduced into (7·2·1), the differential equation becomes

$$k \frac{t_n}{x_n^2} \frac{\partial^2 \bar{t}}{\partial \bar{x}^2} = \rho c \frac{t_n}{\theta_n} \frac{\partial \bar{t}}{\partial \bar{\theta}}$$

Multiplying by $x_n^2 / k t_n$ to make the equation dimensionless and replacing $k / \rho c$ by α gives

$$\frac{\partial^2 \bar{t}}{\partial \bar{x}^2} = \frac{x_n^2}{\alpha \theta_n} \frac{\partial \bar{t}}{\partial \bar{\theta}} \tag{7·2·5}$$

Introduction of the dimensionless variables into (7·2·2) and dividing by t_n / x_n gives

$$\left. \frac{\partial \bar{t}}{\partial \bar{x}} \right|_{x=0} = 0 \tag{7·2·6}$$

The boundary condition (7·2·3) at $x = L$ becomes

$$-k \frac{t_n}{x_n} \left. \frac{\partial \bar{t}}{\partial \bar{x}} \right|_{x=L} = h t_n \bar{t}(L, \theta)$$

Multiplication by $x_n / k t_n$ will make this boundary condition dimensionless. Thus,

$$-\left. \frac{\partial \bar{t}}{\partial \bar{x}} \right|_{x=L} = \frac{h x_n}{k} \bar{t}(L, \theta) \tag{7·2·7}$$

Introduction of normalized variables into the initial condition (7·2·4) gives

$$\bar{t}(x, 0) = \frac{t^{(0)} - t_\infty}{t_n} \tag{7·2·8}$$

Now we must select values for t_n, x_n and θ_n. First, are there any physically significant quantities that can be used? These quantities would appear somewhere in the statement of the problem. That is, they would appear in the differential equation or its boundary or initial conditions. In this case the length L is significant in that it is related to the thickness of the wall; it is the geometrical "size" of the problem. It is also the position at which one of the boundary conditions is given. Therefore we will take $x_n = L$. There is no significant time appearing in the statement of the problem. Consequently θ_n will be chosen as L^2 / α to simplify the differential equation (7·2·5). The choice of $t_n = t^{(0)} - t_\infty$ is now clear from both physical and mathematical arguments. Physically, it is the greatest temperature difference that ever appears in the problem. Mathematically, it reduces the initial condition to a very concise statement. With these choices for t_n, x_n and θ_n the normalized equations reduce to

$$\frac{\partial^2 \bar{t}}{\partial \bar{x}^2} = \frac{\partial \bar{t}}{\partial \bar{\theta}} \tag{7·2·9}$$

$$\left.\frac{\partial \bar{t}}{\partial \bar{x}}\right|_{\bar{x}=0} = 0 \qquad\qquad\qquad (7\cdot2\cdot10)$$

$$-\left.\frac{\partial \bar{t}}{\partial \bar{x}}\right|_{\bar{x}=1} = \frac{hL}{k}\bar{t}(1,\bar{\theta}) = H\bar{t}(1,\bar{\theta}) \qquad\qquad (7\cdot2\cdot11)$$

$$\bar{t}(x,0) = 1 \qquad\qquad\qquad (7\cdot2\cdot12)$$

Observe that a dimensionless group $H = hL/k$ has appeared. This group is called the *Biot number*. Groups such as this that appear in the course of normalizing a problem are quite important. The significance of the Biot number will be discussed as soon as the normalizing procedure has been completed.

Equations (7·2·9) through (7·2·12) are the normalized equations that describe the transient. Observe the reduced number of parameters. Now $\bar{t} = \bar{t}(\bar{x},\bar{\theta};H)$, whereas in dimensional form $t = t(x,\theta;L,k,\rho,c,h,t_\infty,t^{(0)})$. Seven parameters have been reduced to one.

Without even solving the equations, a lot can be learned by simply considering the dimensionless parameter H which has appeared during the normalization. Simplifications can often be made for either very small values or very large values of the dimensionless parameters.

A very small value of H, approaching zero, would reduce the boundary condition at the surface (7·2·11) to

$$-\left.\frac{\partial \bar{t}}{\partial \bar{x}}\right|_{\bar{x}=1} = 0$$

This could be interpreted as being an insulated condition. However, this would not be a very useful interpretation since it would imply that there could be no heat transfer between the solid and the ambient. Consequently no change in temperature would occur. A much more valuable interpretation would be to say that, since the gradient at the surface is so small (and the gradient is the largest at this point in the solid), the internal temperatures must be practically uniform. In this case the internal variation of temperature with x could be overlooked. This means that temperature is only a function of time, and its behavior can be described by an ordinary differential equation rather than by a partial differential equation. This is the lumped-system formulation in Section 1•1•3.

A very large value of H, approaching infinity, would reduce the surface boundary condition (7·2·11) to

$$\bar{t}(1,\bar{\theta}) = -\frac{1}{H}\left.\frac{\partial \bar{t}}{\partial \bar{x}}\right|_{\bar{x}=1} = 0$$

The interpretation of this is that the surface temperature is immediately changed to the ambient value. While this does not lead to a drastic simplification (as was the case for small values of H), the mathematical solution is made a little easier.[1] The final result is a little simpler in that one less parameter, H, is involved.

---

[1] The solution for finite values of H involves eigenvalues satisfying the condition $\lambda \sin(\lambda) = H\cos(\lambda)$. These are tabulated in Table D•1 only up to $H = 100$. For larger values of H, a specified-temperature boundary condition could be used in place of the convective one. This leads to the eigencondition that $\cos(\lambda) = 0$ or $\lambda = (2n+1)\pi/2$ where $n = 0, 1, 2, \cdots$. Observe how closely the first eigenvalue in Table D•1 approaches $\lambda = \pi/2 = 1.5708$ as H increases.

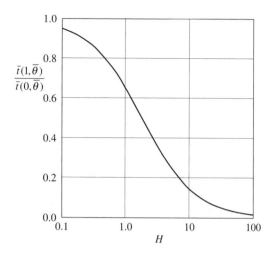

Figure 7•3 *Influence of Biot number on internal temperature differences.*

Figure 7•3 shows the effect of H on the ratio of normalized surface temperature to normalized centerplane temperature.[1] Observe that for $H \leq 0.1$ the ratio is 0.95 or greater. This is usually considered uniform enough to justify a lumped-system engineering analysis. As H increases, the temperature variation within the wall increases as shown by the decreasing temperature ratio. At $H = 100$ the surface temperature is practically zero before the centerplane temperature has changed. This is the basis for assuming a step change in surface temperature for large values of H.

Additional insight into the significance of the Biot number can be seen from rearranging it slightly as follows:

$$H = hA\frac{L}{kA} = \frac{L/kA}{1/hA}$$

where the conduction area has been introduced. Observe that the numerator is the steady-state conduction resistance of a plane wall that is L units thick. The denominator is the convection resistance at the surface. A small value of H then is interpreted as a small internal thermal resistance in comparison to the external resistance. If the internal resistance is small enough, it can be neglected (the lumped-system analysis). A large value of H simply means that the internal resistance is large in comparison to the external resistance. In this case the internal resistance cannot be neglected, and internal temperature distributions must be considered.

7•2•2 Response-time estimates

Occasionally an engineer may only need an order-of-magnitude estimate of a transient response time. This can often be learned through normalization.

Normalizing constants (*e.g.*, x_n, θ_n and t_n) always contain some physically important information. This information may be immediately evident, as in the problem considered in Section 7•2•1 where $x_n = L$ pertained to the wall thickness and $t_n = t^{(0)} - t_\infty$ was the maximum temperature difference that occurred in the problem. Or, it may be less evident, as in the selection of $\theta_n = L^2/\alpha$ which was made on the basis of mathematical simplicity in that it "cleaned up" the differential equation. However θ_n has physical content. It is representative of the time scale involved in the problem. It tells whether the response time will be in terms of seconds, minutes, hours or days.

A plot of normalized centerplane temperature $\bar{t}(0,\theta)$ as a function of unnormalized time θ is shown in Figure 7•4 for the problem of Section 7•2•1. Three different values of the Biot number H are shown. One should observe from this graph that the time θ_n is a physically representative time. Certainly if you were asked to give an order-of-magnitude estimate of the length of time it would take for this plane wall to respond to a change in ambient temperature, you would select θ_n rather than $\theta_n/10$ or $10\theta_n$. At a value of $\theta_n/10$ the transient has barely begun. And at a value of $10\theta_n$ the transient is practically over except for small Biot numbers. Note that θ_n does not represent a particular amount of response because of the effect of the Biot number. Thus θ_n is not really a time constant, but rather it is a "significant time" (or "time scale") of the problem.

In summary, even though θ_n was originally chosen for purely mathematical reasons, it still is representative of the time scale involved in the problem. Thus it has physical significance.

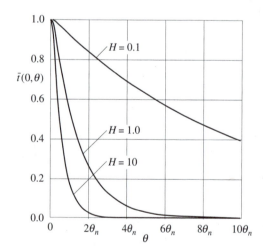

Figure 7•4 *Centerplane response of a plane wall (distributed analysis with $\theta_n = L^2/\alpha$).*

---

[1] The analytical methods used to obtain this curve were discussed in Section 3•1•4. In the numerical evaluation it has been assumed that $\bar{\theta} \quad 0.2$ and only the first term in the series is used.

Normalization of the lumped-system equation also yields information regarding the time scale of a problem in which a lumped analysis is adequate. Example 2 in Section 7•1•2 considered the normalization procedure. During the normalization, a value of $\theta_n = \rho Vc/hA$ was selected on the basis of mathematical beauty to simplify the differential equation (7·1·9). Since in this particular problem there is no other parameter, one single curve is enough to present the entire solution. Consequently, a time of θ_n always represents the same point in the response (rather than different values as in Figure 7•4). As shown in Figure 7•5, θ_n is the time for the response to reach $1/e$. It is called the *time constant* of the system. Thus in the lumped-system analysis the physical significance of θ_n is much more specific and definite than it was in the distributed analysis.

Further insight into θ_n for both the distributed and the lumped analyses can be obtained by writing them as the product of a thermal capacitance and a thermal resistance. For the distributed case,

$$\theta_n = \frac{L^2}{\alpha} = \frac{\rho c L^2}{k} = \rho c L A \frac{L}{kA} = \rho Vc \frac{L}{kA}$$

represents the product of the thermal capacitance time the internal thermal resistance. Recall that, as the Biot number increased, the internal resistance became the dominant resistance, and consequently it has turned up in θ_n. For the lumped case,

$$\theta_n = \frac{\rho Vc}{hA} = \rho Vc \frac{1}{hA}$$

represents the product of the capacitance time the external resistance. For small Biot numbers the internal resistance is neglected in comparison to the external resistance. Consequently the dominant external resistance shows up in θ_n.

Thus thermal capacitance and thermal resistance are important concepts in transient analyses. The larger they are, the larger the value of θ_n, and the longer the response time.

In summary, the physical size of θ_n can have important significance to the engineer even though its precise meaning is not known. For example, in selecting instrumentation to measure such a response, a value of $\theta_n = 1$ sec would suggest much different recording equipment than a value of $\theta_n = 1$ min or 1 hr. The ease with which this information was found should also be noted.

7•2•3 Penetration-depth estimate

In Section 5•2 a solution was obtained for the temperature fluctuation in a semi-infinite solid due to a periodic surface-temperature fluctuation. It was found, after considerable analysis, that these fluctuations damped out as the distance into the solid increased. The analytical solution could be used to determine the depth of penetration of these fluctuations. A much quicker estimate of this penetration depth can be obtained through the use of normalization.

The governing differential equation and its boundary conditions were given by (5·2·1), (5·2·2) and (5·2·3). Following the usual normalization procedure, dimensionless variables can be defined as

$$\bar{t} = \frac{t - t_\infty}{t_n} \quad , \quad \bar{x} = \frac{x}{x_n} \quad \text{and} \quad \bar{\theta} = \frac{\theta}{\theta_n}$$

Substituting these into the differential equation (5·2·1),

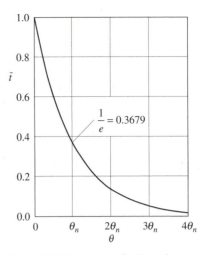

Figure 7•5 *Time constant for lumped-system transient ($\theta_n = \rho Vc/hA$).*

$$k \frac{t_n}{x_n^2} \frac{\partial^2 \bar{t}}{\partial \bar{x}^2} = \rho c \frac{t_n}{\theta_n} \frac{\partial \bar{t}}{\partial \bar{\theta}}$$

Multiplying by $x_n^2 / k t_n$ to make the equation dimensionless and replacing $k / \rho c$ by α gives

$$\frac{\partial^2 \bar{t}}{\partial \bar{x}^2} = \frac{x_n^2}{\alpha \theta_n} \frac{\partial \bar{t}}{\partial \bar{\theta}}$$

The boundary condition (5·2·2) at the surface $x = 0$ becomes

$$\bar{t}(0, \bar{\theta}) = \frac{T}{t_n} \cos(\omega \theta_n \bar{\theta})$$

The boundary condition as x becomes infinite becomes

$$\lim_{\bar{x} \to \infty} \bar{t}(\bar{x}, \bar{\theta}) = 0 \tag{7·2·13}$$

Now we must look at the differential equation and its boundary conditions for physically significant quantities to use for t_n, x_n and θ_n. The choice for t_n, based on physical reasons, is to use the amplitude of the oscillations at the surface. Thus we will take $t_n = T$. The period of the imposed fluctuations, $2\pi / \omega$, has physical meaning. Therefore we will take $\theta_n = 2\pi / \omega$. Since there is no significant length in this problem, we will take $x_n^2 = \alpha \theta_n = 2\pi \alpha / \omega$ for mathematical simplicity to clean up the differential equation. The differential equation and the boundary condition at $x = 0$ become

$$\frac{\partial^2 \bar{t}}{\partial \bar{x}^2} = \frac{\partial \bar{t}}{\partial \bar{\theta}} \tag{7·2·14}$$

$$\bar{t}(0, \bar{\theta}) = \cos(2\pi \bar{\theta}) \tag{7·2·15}$$

The problem is now completely normalized, as stated by (7·2·13), (7·2·14) and (7·2·15).

The only characteristic constant chosen on the basis of mathematical simplicity was $x_n = \sqrt{2\pi \alpha / \omega}$. As with all characteristic constants, this should have a physical interpretation. In this case x_n is representative of the depth of penetration of the oscillations. It is the length scale associated with this problem in much the same way as θ_n was associated with the time scale in Section 7•2•2.

A plot of the amplitude of the oscillations as a function of distance is shown in Figure 7•6. At a depth of x_n the amplitude is about 17 percent of the surface value. Certainly $x_n / 10$ or $10 x_n$ would not be as representative of the penetration depth as x_n. Thus one concludes that x_n gives a good order-of-magnitude estimate of the penetration depth.

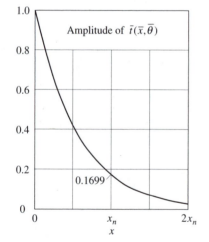

Figure 7•6 *Oscillation amplitude as a function of distance into a semi-infinite solid ($x_n = \sqrt{2\pi \alpha / \omega}$).*

7•2•4   Numerical calculations

One extremely practical use of normalization arises when making numerical calculations. As a result of the physical significance that is built into the normalization procedure and the physical significance that the mathematically defined terms possess, the most important values of the variables will usually be between 0 and 1. This is valuable information when deciding what values of the independent variables should be used in making numerical calculations of the solution.

As an example, consider the problem discussed in Section 7•2•1. The dimensional solution to that problem can be found as

$$t(x,\theta) = t_\infty + 2(t^{(0)} - t_\infty)\sum_{n=1}^{\infty}\frac{h\sin(\lambda_n L)\cos(\lambda_n x)}{\lambda_n[hL + k\sin^2(\lambda_n L)]}\exp(-\lambda_n^2\alpha\theta)$$

In this form one has really no idea of what values of time (in hours) should be used to make numerical calculations. Values of x, of course, would be between 0 and L. But, without normalization, you are in the dark as to significant values of θ. In normalized form the solution becomes

$$\bar{t}(\bar{x},\bar{\theta}) = 2H\sum_{n=1}^{\infty}\frac{\sin(\lambda_n L)\cos(\lambda_n L\bar{x})}{\lambda_n L[H + \sin^2(\lambda_n L)]}\exp[-(\lambda_n L)^2\bar{\theta}]$$

and values of $\bar{\theta}$ on the order of 1.0 will most likely be of interest. When evaluating this solution as a function of time, you might try $\bar{\theta} = 1.0$ to see where you stand. Then you might fill in points by taking $\bar{\theta} = 0.5$, 2.0 and 4.0. Additional points could be added where desired after looking at the results of these first calculations.

In summary, Section 7•2 has shown how normalization can be used as a tool to increase one's understanding of a complicated problem. It can be used to

1. Discover the important parameters (*e.g.*, H).
2. Suggest simplifications (*e.g.*, $H \to 0$ or ∞).
3. Estimate the most important range of variables that are unbounded (*e.g.*, $\bar{\theta}$ runs between 0 and ∞ but, practically speaking, values of $\bar{\theta}$ between 0.1 and 10 will usually be enough).

Having discovered the significant variables and parameters, Section 7•2•4 shows how helpful this information can be when making numerical calculations.

7•3 COMBINATION OF VARIABLES

Normalization can be used to determine whether or not the independent variables of a partial differential equation may be combined in such a way as to reduce the number of variables. This would be a great simplification. A partial differential equation with two independent variables would then be transformed into an ordinary differential equation with only one independent variable.

As an example of how this works, let us consider the semi-infinite solid which is subjected to a step change in its surface temperature. This problem was discussed in Section 6•4•2. The dimensional governing equations are given as

$$k\frac{\partial^2 t}{\partial x^2} = \rho c\frac{\partial t}{\partial\theta} \tag{7.3.1}$$

with

$$t(0,\theta) = t_0 \tag{7.3.2}$$

and

$$\lim_{x\to\infty} t(x,\theta) = t^{(0)} \tag{7.3.3}$$

where the initial condition is that

$$t(x,0) = t^{(0)} \tag{7.3.4}$$

Following the usual procedure for normalization, define the following normalized variables:

$$\bar{t} = \frac{t - t^{(0)}}{t_n} \quad , \quad \bar{x} = \frac{x}{x_n} \quad \text{and} \quad \bar{\theta} = \frac{\theta}{\theta_n} \tag{7.3.5}$$

Upon substituting these into (7·3·1), multiplying by $x_n^2 / k t_n$ and replacing $k / \rho c$ by α,

$$\frac{\partial^2 \bar{t}}{\partial \bar{x}^2} = \frac{x_n^2}{\alpha \theta_n} \frac{\partial \bar{t}}{\partial \bar{\theta}} \tag{7.3.6}$$

Introduction of the dimensionless variables into (7·3·2) and dividing by t_n / x_n gives

$$\bar{t}(0, \bar{\theta}) = \frac{t_0 - t^{(0)}}{t_n} \tag{7.3.7}$$

The boundary condition (7·3·3) for large x becomes

$$\lim_{\bar{x} \to \infty} t(\bar{x}, \bar{\theta}) = 0 \tag{7.3.8}$$

The initial condition (7·3·4) becomes

$$\bar{t}(\bar{x}, 0) = 0 \tag{7.3.9}$$

The next step is to see if there are any physically significant quantities to choose for t_n, x_n or θ_n. Upon searching (7·3·6) through (7·3·9) for these quantities, the choice of $t_n = t_0 - t^{(0)}$ seems obvious from both physical and mathematical considerations. No distance shows up as in the plane wall, and no time shows up as in the semi-infinite solid with a periodic surface temperature. There is nothing at all which has physical significance to choose for x_n or θ_n.

Based on mathematical simplicity, we will then choose x_n and θ_n to be related by $x_n^2 / \alpha \theta_n = 1$ to simplify the differential equation. Thus the normalized equations become

$$\frac{\partial^2 \bar{t}}{\partial \bar{x}^2} = \frac{\partial \bar{t}}{\partial \bar{\theta}} \tag{7.3.10}$$

$$\bar{t}(0, \bar{\theta}) = 1 \tag{7.3.11}$$

$$\lim_{\bar{x} \to \infty} t(\bar{x}, \bar{\theta}) = 0 \tag{7.3.12}$$

$$\bar{t}(\bar{x}, 0) = 0 \tag{7.3.13}$$

Since no physical significance can be given to either x_n or θ_n and, by choosing $x_n^2 / \alpha \theta_n = 1$ for mathematical simplicity, both x_n and θ_n have been eliminated from the problem, *combined variable* ξ can be found. One possible choice would be to define ξ as

$$\xi = \frac{\bar{x}^2}{\bar{\theta}} = \frac{(x / x_n)^2}{\theta / \theta_n} = \frac{x^2}{\theta} \frac{\theta_n}{x_n^2} = \frac{x^2}{\alpha \theta}$$

By using $\xi = x^2 / \alpha \theta$ as a combined variable, the physically meaningless quantities x_n and θ_n will be eliminated from the problem. A more convenient (*i.e.,* conventional) choice,[1] it turns out, is to define the combined variable η as

---

[1] This means that someone has already worked the problem and determined a more suitable choice. One could certainly work the problem by using ξ instead of η (see Exercise 7•10).

$$\eta = \frac{1}{2}\sqrt{\xi} = \frac{\bar{x}}{2\sqrt{\bar{\theta}}} = \frac{x}{2\sqrt{\alpha\theta}} \tag{7·3·14}$$

The combined variable η will be used in place of $\bar{x}$ and $\bar{\theta}$ in (7·3·10).

From the chain rule of calculus we can write

$$\frac{\partial \bar{t}}{\partial \bar{x}} = \frac{\partial \bar{t}}{\partial \eta}\frac{\partial \eta}{\partial \bar{x}}$$

From (7·3·14) we can see that

$$\frac{\partial \eta}{\partial \bar{x}} = \frac{1}{2\sqrt{\bar{\theta}}}$$

Thus the expression for the first derivative becomes

$$\frac{\partial \bar{t}}{\partial \bar{x}} = \frac{\partial \bar{t}}{\partial \eta}\frac{1}{2\sqrt{\bar{\theta}}}$$

The second derivative can similarly be written as

$$\frac{\partial^2 \bar{t}}{\partial \bar{x}^2} = \frac{\partial}{\partial \bar{x}}\left(\frac{\partial \bar{t}}{\partial \bar{x}}\right) = \frac{\partial}{\partial \bar{x}}\left(\frac{\partial \bar{t}}{\partial \eta}\frac{1}{2\sqrt{\bar{\theta}}}\right) = \frac{1}{2\sqrt{\bar{\theta}}}\frac{\partial}{\partial \bar{x}}\left(\frac{\partial \bar{t}}{\partial \eta}\right)$$

Using the chain rule once again,

$$\frac{\partial^2 \bar{t}}{\partial \bar{x}^2} = \frac{1}{2\sqrt{\bar{\theta}}}\frac{\partial}{\partial \eta}\left(\frac{\partial \bar{t}}{\partial \eta}\right)\frac{\partial \eta}{\partial \bar{x}} = \frac{1}{2\sqrt{\bar{\theta}}}\frac{\partial^2 \bar{t}}{\partial \eta^2}\frac{1}{2\sqrt{\bar{\theta}}}$$

Therefore

$$\frac{\partial^2 \bar{t}}{\partial \bar{x}^2} = \frac{1}{4\bar{\theta}}\frac{\partial^2 \bar{t}}{\partial \eta^2} \tag{7·3·15}$$

The time derivative can be written as

$$\frac{\partial \bar{t}}{\partial \bar{\theta}} = \frac{\partial \bar{t}}{\partial \eta}\frac{\partial \eta}{\partial \bar{\theta}}$$

and, from (7·3·14), we can write

$$\frac{\partial \eta}{\partial \bar{\theta}} = \frac{\bar{x}}{2}\left(-\frac{1}{2}\right)\bar{\theta}^{-3/2} = -\frac{\bar{x}}{4\bar{\theta}^{3/2}}$$

Thus

$$\frac{\partial \bar{t}}{\partial \bar{\theta}} = \frac{\partial \bar{t}}{\partial \eta}\left(-\frac{\bar{x}}{4\bar{\theta}^{3/2}}\right) \tag{7·3·16}$$

Now (7·3·15) and (7·3·16) can be substituted into (7·3·10) to give

$$\frac{1}{4\bar{\theta}}\frac{\partial^2 \bar{t}}{\partial \eta^2} = \frac{\partial \bar{t}}{\partial \eta}\left(-\frac{\bar{x}}{4\bar{\theta}^{3/2}}\right)$$

or

$$\frac{\partial^2 \bar{t}}{\partial \eta^2} = \frac{\partial \bar{t}}{\partial \eta}\left(-\frac{\bar{x}}{\sqrt{\bar{\theta}}}\right)$$

Now, from (7·3·14),

$$\frac{\bar{x}}{\sqrt{\bar{\theta}}} = 2\eta$$

Thus the differential equation becomes

$$\frac{\partial^2 \bar{t}}{\partial \eta^2} + 2\eta \frac{\partial \bar{t}}{\partial \eta} = 0$$

Since $\bar{t}$ will be a function only of η, this is really an ordinary differential equation. As such the partial derivatives can be replaced by ordinary derivatives. Thus

$$\frac{d^2 \bar{t}}{d\eta^2} + 2\eta \frac{d\bar{t}}{d\eta} = 0 \qquad (7\cdot3\cdot17)$$

Since (7·3·17) is second order in η, it requires only two boundary conditions. The original problem for $\bar{t}(\bar{x}, \bar{\theta})$ has three side conditions, (7·3·11), (7·3·12) and (7·3·13). These original three conditions for $\bar{t}(\bar{x}, \bar{\theta})$ must therefore collapse into only two conditions for $\bar{t}(\eta)$ if combination of variables is to work. The first condition (7·3·11) has $\bar{x} = 0$ and from (7·3·14) we see that this gives $\eta = 0$ for any $\bar{\theta}$. Thus (7·3·11) becomes

$$\bar{t}(0) = 1 \qquad (7\cdot3\cdot18)$$

The second boundary condition for (7·3·17) will encompass both (7·3·12) and (7·3·13). Observe, from (7·3·14), that η is infinite for $\bar{x}$ being infinite and also for $\bar{\theta} = 0$. Thus (7·3·12) and (7·3·13) become identical and are given by

$$\bar{t}(\infty) = 0 \qquad (7\cdot3\cdot19)$$

The problem has now been completely normalized and transformed into an ordinary differential equation (7·3·17) with two boundary conditions (7·3·18) and (7·3·19). This is a considerable saving since ordinary differential equations are much easier to solve than partial differential equations.

Since only derivatives of $\bar{t}$ appear in (7·3·17), the substitution of $p = d\bar{t}/d\eta$ will reduce the order of the equation. Thus

$$\frac{dp}{d\eta} + 2\eta p = 0$$

The variables may be separated as follows:

$$\frac{dp}{p} = -2\eta \, d\eta$$

and then one integration may be carried out:

$$\ln(p) = -\eta^2 + \ln(c_1)$$

or, upon taking antilogarithms,

$$\frac{d\bar{t}}{d\eta} = p = c_1 \exp(-\eta^2)$$

This may again be integrated to give[1]

$$\bar{t}(\eta) - \bar{t}(0) = c_1 \int_{\sigma=0}^{\eta} \exp(-\sigma^2) \, d\sigma$$

Since $\bar{t}(0) = 1$ from (7·3·18), this becomes

---

[1] Note the dummy variable of integration that is being used. The validity of this result can be checked by differentiating it to see if the correct expression is obtained for $d\bar{t}/d\eta$ (recall Leibnitz' rule).

$$\bar{t}(\eta) = 1 + c_1 \int_{\sigma=0}^{\eta} \exp(-\sigma^2) d\sigma$$

From (7·3·19), c_1 may be evaluated by setting $\eta = \infty$ in the expression for $\bar{t}(\eta)$. Thus

$$0 = 1 + c_1 \int_{\sigma=0}^{\infty} \exp(-\sigma^2) d\sigma$$

and so

$$c_1 = -\frac{1}{\int_{\sigma=0}^{\infty} \exp(-\sigma^2) d\sigma}$$

Thus the solution becomes

$$\bar{t}(\eta) = 1 - \frac{\int_{\sigma=0}^{\eta} \exp(-\sigma^2) d\sigma}{\int_{\sigma=0}^{\infty} \exp(-\sigma^2) d\sigma} \qquad (7\cdot3\cdot20)$$

These integrals can be evaluated by numerical methods if necessary. In this particular case, however, they can be recognized as being related to the error function defined by (6·4·21). Using (6·4·21) and (6·4·22), the solution (7·3·20) may be written as

$$\bar{t}(\eta) = 1 - \text{erf}(\eta) \qquad (7\cdot3\cdot21)$$

Since $\bar{t} = (t - t^{(0)})/(t_0 - t^{(0)})$ and $\eta = x/2\sqrt{\alpha\theta}$, this can be recognized as the same solution that was obtained in Section 6•4•2.

In summary, we have seen the use of normalization in determining whether it was possible to obtain a solution to a partial differential equation by combination of variables. In this particular case, the original equation was reduced to an ordinary differential equation that was solved by routine methods. This was a great simplification.

The use of combination of variables to obtain a *similarity solution* has many engineering applications. Similarity is discussed in much more detail in [1].

SELECTED REFERENCE

1. Hansen, A. G.: *Similarity Analyses of Boundary Value Problems in Engineering*, Prentice-Hall, Inc., Englewood Cliffs, NJ, 1964.

EXERCISES

7•1 Normalize the following differential equation and its boundary conditions:

$$kA\frac{d^2t}{dx^2} - hp(t - t_\infty) = 0 \qquad t(0) = t_0$$

$$\frac{dt}{dx}\bigg|_{x=L} = 0$$

7•2 Normalize the differential equation and boundary conditions of Exercise 1•15.

7•3 The governing ordinary differential equation for the temperature distribution along a thin rod during extrusion is given by

$$kA\frac{d^2t}{dx^2} - \dot{m}_f c\frac{dt}{dx} - hp(t - t_\infty) = 0$$

where k = thermal conductivity of rod
   A = cross-sectional area of rod
   $\dot{m}_f$ = mass rate of flow of rod out of extrusion die
   c = specific heat of rod
   h = convective heat-transfer coefficient around rod
   p = perimeter of rod
   t_∞ = ambient temperature
   t = rod temperature
   x = axial position

At $x = 0$, the position where the rod leaves the extrusion die, the rod temperature is given as t_0. At a large distance from $x = 0$ the rod has essentially cooled to t_∞. In deriving the above, axial conduction, axial convection and convective heat loss were considered to be the only important energy flows. Radial temperature variations were neglected.

(a) Normalize the equation and its boundary conditions to establish a suitable set of nondimensional variables and parameters.
(b) State how you would solve the equation.
(c) Make a normalized sketch of what you think the solution would look like. Clearly label the coordinates and any parameters that appear. This sketch should be a qualitative picture showing the general behavior of temperature with distance and the influence of the various parameters you have obtained in part *a*.
(d) Obtain the analytical solution for this problem.

7•4 Normalize the following differential equation and its boundary conditions:

$$\frac{d^2t}{dr^2} + \frac{1}{r}\frac{dt}{dr} = 0 \qquad t(r_i) = t_i$$

$$t(r_o) = t_o$$

7•5 Normalize the following problem to determine a suitable set of nondimensional variables and parameters:

$$k\frac{\partial^2 t}{\partial x^2} = \rho c\frac{\partial t}{\partial \theta} \qquad -k\frac{\partial t}{\partial x}\bigg|_{x=0} = q_0''$$

$$\frac{\partial t}{\partial x}\bigg|_{x=L} = 0$$

$$t(x,0) = t^{(0)}$$

7•6 Normalize the following differential equation and its boundary and initial conditions to simplify the variables and parameters:

$$k\frac{\partial^2 t}{\partial x^2} = \rho c\frac{\partial t}{\partial \theta} \qquad t(0,\theta) = t^{(0)} + T\exp(-A\theta)$$

$$\frac{\partial t}{\partial x}\bigg|_{x=L} = 0$$

$$t(x,0) = t^{(0)}$$

7•7 The differential equation describing a thermocouple that is suddenly inserted into an ambient air stream with an oscillating temperature is given by

$$\rho Vc\frac{dt}{d\theta} + hA(t - t_\infty) = 0$$

where $t_\infty = t^{(0)} + T\sin(\omega\theta)$. The initial condition is that $t(0) = t^{(0)}$. Normalize the problem to establish a suitable set of nondimensional variables and parameters. Without solving the equation, what simplifications can be obtained for large and small values of the parameters? What can you say regarding the design of a thermocouple for measuring oscillatory temperatures?

7•8   Consider a plane wall in a fluid whose temperature is oscillating. The governing equations are

$$k\frac{\partial^2 t}{\partial x^2} = \rho c \frac{\partial t}{\partial \theta} \qquad \frac{\partial t}{\partial x}\bigg|_{x=0} = 0$$

$$-k\frac{\partial t}{\partial x}\bigg|_{x=L} = h\left\{t(L,\theta) - \left[t^{(0)} + T\cos(\omega\theta)\right]\right\}$$

$$t(x,0) = t^{(0)}$$

Normalize these equations to find a set of nondimensional variables and parameters. Discuss any simplifications that can be obtained for large or small values of the parameters.

7•9   By using normalization alone (*i.e.*, without ever solving the differential equation), estimate the ratio of the penetration depth (into the earth) of the annual temperature cycle to that of the daily temperature cycle. Compare your result to the exact solution (Exercise 5•3).

7•10 Carry out the solution to the problem discussed in Section 7•3 using the combined variable $\xi = x^2/\alpha\theta$ rather than η as defined by (7•3•14). Verify that the same final solution is obtained.

7•11 A semi-infinite solid (with a finite surface heat-transfer coefficient) is subjected to a step change in ambient temperature. The governing equations are given by

$$k\frac{\partial^2 t}{\partial x^2} = \rho c \frac{\partial t}{\partial \theta} \qquad -k\frac{\partial t}{\partial x}\bigg|_{x=0} = h\left[t_\infty - t(0,\theta)\right]$$

$$t(\infty,\theta) = t^{(0)}$$

$$t(x,0) = t^{(0)}$$

(a) Normalize the problem to obtain the nondimensional variables and parameters you would use to present the solution to this problem.

(b) Determine whether combination of variables will work. If so, obtain the ordinary differential equation and its boundary conditions.

7•12 The surface temperature of a semi-infinite solid is raised linearly with time. The governing equations are

$$k\frac{\partial^2 t}{\partial x^2} = \rho c \frac{\partial t}{\partial \theta} \qquad t(0,\theta) = t^{(0)} + b\theta$$

$$t(\infty,\theta) = t^{(0)}$$

$$t(x,0) = t^{(0)}$$

(a) Normalize the problem to obtain the nondimensional variables and parameters you would use to present the solution to this problem.

(b) Determine whether combination of variables will work. If so, obtain the ordinary differential equation and its boundary conditions.

7•13 The boundary-layer equations for laminar flow of an incompressible fluid flowing along a flat plate are given as

$$\frac{\partial u}{\partial x} + \frac{\partial v}{\partial y} = 0$$

$$u\frac{\partial u}{\partial x} + v\frac{\partial u}{\partial y} = \nu\frac{\partial^2 u}{\partial y^2}$$

where u and v are the velocities in the x and y directions, respectively, and ν is the kinematic viscosity. The boundary conditions are the following:

$$u(0,y) = u_\infty$$

$$u(x,0) = 0 \qquad\qquad v(x,0) = 0$$

$$u(x,\infty) = u_\infty$$

Simultaneous solution of the two partial differential equations would provide $u = u(x,y)$ and $v = v(x,y)$.

(a) Normalize the problem to obtain the nondimensional variables and parameters you would use to present the solution to this problem. *Hint:* Take the undetermined normalizing constant to be x_n.

(b) Determine whether combination of variables will work. If so, obtain the two ordinary differential equations and their boundary conditions.

8

FINITE DIFFERENCES

8•0 INTRODUCTION

Widespread use of the digital computer has made finite-difference methods extremely valuable for solving problems that are not susceptible to the analytical methods discussed in earlier chapters. As we have noticed, analytical methods are usually restricted to very simple geometries and boundary conditions. For the more complex problems, finite differences is a feasible method of attack.

It should be mentioned here, however, that, although the majority of practical problems may require finite-difference methods to obtain detailed answers, analytical methods are still important. In setting up a complex problem using finite differences, limiting cases are often considered as a check on the computations. These limiting cases often have analytical solutions that can be used for comparison to the finite-difference results. Asymptotic, analytical solutions can often be used in conjunction with the computer to provide better solutions.

After some of the mathematical concepts necessary to understand the material in this chapter are given in Section 8•1, the discretization of the spatial problem that leads to the finite-difference formulation will be examined in Section 8•2. Steady-state and transient solutions to these problems will then be examined in sections 8•3 and 8•4 respectively. Section 8•5 provides some analysis of transient problems to provide more insight into what is happening in finite-difference solutions. Finally, an introduction to some more advanced problems will be given in Section 8•6.

8•1 FUNDAMENTAL CONCEPTS

The analytical formulation of a two-dimensional conduction problem involved the selection of a system, an energy balance, the rate equations, and the passing to the limit as the system shrank to zero size. The result of this process was a partial differential equation as shown in Chapter 1. In the finite-difference formulation of the same problem the system is not allowed to shrink to zero size and the result is a system of simultaneous ordinary differential equations rather than a partial differential equation.

To effectively handle systems of simultaneous ordinary differential equations we must understand some things about matrices and matrix algebra. Matrix algebra is discussed in Section 8•1•1. In order not to shrink the system to zero we will make approximations to the rate equations. The common approximations are discussed in Section 8•1•2.

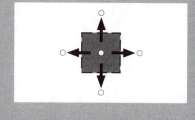

8•1•1 Matrix algebra

Systems of linear equations are a common occurrence in finite-difference methods. It is thus essential that we have a convenient way to represent such systems of equations. The use of matrix notation will be extremely helpful to us as we study finite-difference methods. This section is intended to provide you with enough of an introduction to matrices so that you will be able to understand what is to come later in the chapter. Additional matrix manipulations will be discussed later when they are required.

For simplicity, let us consider the following system of two algebraic equations for the two unknowns u_1 and u_2:

$$a_{11}u_1 + a_{12}u_2 = r_1$$
$$a_{21}u_1 + a_{22}u_2 = r_2$$

(8·1·1)

The coefficients a_{11}, a_{12}, a_{21}, and a_{22} and the right-hand sides r_1 and r_2 are constants. The matrix representation for these equations is

$$\begin{bmatrix} a_{11} & a_{12} \\ a_{21} & a_{22} \end{bmatrix} \begin{bmatrix} u_1 \\ u_2 \end{bmatrix} = \begin{bmatrix} r_1 \\ r_2 \end{bmatrix}$$

(8·1·2)

The coefficients a_{ij} in the algebraic equations are collected together as the *entries* in the *coefficient matrix*. The unknowns u_i are written in the *column matrix* or *vector* alongside the coefficient matrix. The right-hand sides r_i are put in another column matrix or vector on the other side of the equal sign. This matrix equation has the same meaning as the system of (8·1·1) and avoids the writing of the unknowns as many times. An even shorter notation is to let single symbols stand for each of the three matrices. Thus we might write (8·1·2) as

$$\mathbf{Au = r}$$

(8·1·3)

where $\mathbf{A}$ stands for the square coefficient matrix, and $\mathbf{u}$ and $\mathbf{r}$ stand for the column matrices (vectors) containing the u_i and r_i.

We will find it useful to be familiar with a few of the mathematical manipulations that can be done with matrices. This will enable us to handle systems of equations in a more efficient manner. Some of these operations are discussed in the remainder of this section.

The left-hand sides of (8·1·2) and (8·1·3) can be thought of as matrix multiplication. The entries in the product matrix, the r_i in this case, are given by the following rule:

$$r_i = \sum_{j=1}^{2} a_{ij}u_j$$

(8·1·4)

As a numerical illustration of matrix multiplication, consider the following example:

$$\mathbf{A} = \begin{bmatrix} 1 & 2 \\ 3 & 4 \end{bmatrix} \qquad \mathbf{u} = \begin{bmatrix} 2 \\ 3 \end{bmatrix}$$

$$\mathbf{Au} = \begin{bmatrix} 1 & 2 \\ 3 & 4 \end{bmatrix} \begin{bmatrix} 2 \\ 3 \end{bmatrix} = \begin{bmatrix} 1 \cdot 2 + 2 \cdot 3 \\ 3 \cdot 2 + 4 \cdot 3 \end{bmatrix} = \begin{bmatrix} 8 \\ 18 \end{bmatrix}$$

Square matrices (same number of rows as columns) may also be multiplied together ($\mathbf{P = AB}$) by the following rule:

$$p_{ik} = \sum_{j} a_{ij}b_{jk}$$

(8·1·5)

As a numerical illustration, consider the matrices

$$A = \begin{bmatrix} 1 & 2 \\ 3 & 4 \end{bmatrix} \quad B = \begin{bmatrix} -5 & 3 \\ 4 & 2 \end{bmatrix}$$

Then

$$P = AB = \begin{bmatrix} 1 & 2 \\ 3 & 4 \end{bmatrix}\begin{bmatrix} -5 & 3 \\ 4 & 2 \end{bmatrix} = \begin{bmatrix} 1(-5)+2(4) & 1(3)+2(2) \\ 3(-5)+4(4) & 3(3)+4(2) \end{bmatrix} = \begin{bmatrix} 3 & 7 \\ 1 & 17 \end{bmatrix}$$

It is also important to know that, in general, $AB \neq BA$. For example, if the above multiplication is carried out in the reverse order,

$$BA = \begin{bmatrix} -5 & 3 \\ 4 & 2 \end{bmatrix}\begin{bmatrix} 1 & 2 \\ 3 & 4 \end{bmatrix} = \begin{bmatrix} -5(1)+3(3) & -5(2)+3(4) \\ 4(1)+2(3) & 4(2)+2(4) \end{bmatrix} = \begin{bmatrix} 4 & 2 \\ 10 & 16 \end{bmatrix}$$

Notice that $BA \neq AB$. This means that we must be careful to keep matrix multiplications in the correct order.

A matrix with m rows and n columns is called an $m \times n$ *matrix*. Two matrices can be multiplied together only if the number of columns in the first matrix is equal to the number of rows in the second matrix. The resulting product is a matrix with the same number of rows as the first matrix and the same number of columns as the second matrix. That is

$$A_{(m \times n)} \cdot B_{(n \times p)} = C_{(m \times p)} \tag{8·1·6}$$

This rule can be extended to the product of three or more matrices as follows:

$$A_{(m \times n)} \cdot B_{(n \times p)} \cdot C_{(p \times q)} = D_{(m \times q)} \tag{8·1·7}$$

This triple product may be computed as either $(AB)C$ or $A(BC)$.

If A and B are each $m \times n$ matrices, we say A and B have equal size. Matrices of equal size may be added together ($S = A + B$) simply by adding the corresponding entries according to the following rule:

$$s_{ij} = a_{ij} + b_{ij} \tag{8·1·8}$$

Thus,

$$S = A + B = \begin{bmatrix} 1 & 2 \\ 3 & 4 \end{bmatrix} + \begin{bmatrix} -5 & 3 \\ 4 & 2 \end{bmatrix} = \begin{bmatrix} 1-5 & 2+3 \\ 3+4 & 4+2 \end{bmatrix} = \begin{bmatrix} -4 & 5 \\ 7 & 6 \end{bmatrix}$$

It can be shown that $AB + AC = A(B+C)$ and that $AC + BC = (A+B)C$.

Matrices may be multiplied by a constant (often called a scalar) by simply multiplying each of the entries by the constant. That is,

$$3A = 3\begin{bmatrix} 1 & 2 \\ 3 & 4 \end{bmatrix} = \begin{bmatrix} 3 & 6 \\ 9 & 12 \end{bmatrix}$$

A *diagonal matrix* is a matrix that has nonzero entries only along its main diagonal. An example of a diagonal matrix would be

$$\begin{bmatrix} 1 & 0 & 0 \\ 0 & 4 & 0 \\ 0 & 0 & 2 \end{bmatrix}$$

A special diagonal matrix I is the *identity matrix*. Its diagonal entries are all equal to 1. That is, for a 3×3 case,

$$I = \begin{bmatrix} 1 & 0 & 0 \\ 0 & 1 & 0 \\ 0 & 0 & 1 \end{bmatrix}$$

The *transpose* $\mathbf{A}^T$ of matrix $\mathbf{A}$ is found by interchanging rows and columns in $\mathbf{A}$. That is, the rows in a matrix $\mathbf{A}$ are the columns in $\mathbf{A}^T$. Thus, if

$$\mathbf{A} = \begin{bmatrix} a_{11} & a_{12} \\ a_{21} & a_{22} \end{bmatrix} \qquad \text{then} \qquad \mathbf{A}^T = \begin{bmatrix} a_{11} & a_{21} \\ a_{12} & a_{22} \end{bmatrix}$$

The transpose of a matrix product is related to the transposes of the two matrices by

$$(\mathbf{AB})^T = \mathbf{B}^T \mathbf{A}^T \tag{8·1·9}$$

This can be demonstrated by considering

$$\mathbf{AB} = \begin{bmatrix} a_{11} & a_{12} \\ a_{21} & a_{22} \end{bmatrix}\begin{bmatrix} b_{11} & b_{12} \\ b_{21} & b_{22} \end{bmatrix} = \begin{bmatrix} a_{11}b_{11} + a_{12}b_{21} & a_{11}b_{12} + a_{12}b_{22} \\ a_{21}b_{11} + a_{22}b_{21} & a_{21}b_{12} + a_{22}b_{22} \end{bmatrix}$$

$$\mathbf{B}^T\mathbf{A}^T = \begin{bmatrix} b_{11} & b_{21} \\ b_{12} & b_{22} \end{bmatrix}\begin{bmatrix} a_{11} & a_{21} \\ a_{12} & a_{22} \end{bmatrix} = \begin{bmatrix} b_{11}a_{11} + b_{21}a_{12} & b_{11}a_{21} + b_{21}a_{22} \\ b_{12}a_{11} + b_{22}a_{12} & b_{12}a_{21} + b_{22}a_{22} \end{bmatrix}$$

Observe that the expression for $\mathbf{B}^T\mathbf{A}^T$ is indeed the transpose of $\mathbf{AB}$.

A matrix is said to be *symmetric* if it is equal to its own transpose. This means that entry a_{ij} must equal entry a_{ji}. An example of a symmetric matrix would be

$$\begin{bmatrix} 1 & 3 & 0 \\ 3 & 4 & 5 \\ 0 & 5 & 2 \end{bmatrix}$$

The *inverse* $\mathbf{A}^{-1}$ of a matrix $\mathbf{A}$ is another matrix such that, when $\mathbf{A}$ and its inverse $\mathbf{A}^{-1}$ are multiplied together, their product is the identity matrix $\mathbf{I}$. That is,

$$\mathbf{AA}^{-1} = \mathbf{I} \qquad \text{or} \qquad \mathbf{A}^{-1}\mathbf{A} = \mathbf{I}$$

For example, the inverse of the matrix $\mathbf{A}$ that we have been using is

$$\mathbf{A}^{-1} = \begin{bmatrix} -2.0 & 1.0 \\ 1.5 & -0.5 \end{bmatrix}$$

This may be verified by multiplying $\mathbf{A}$ and $\mathbf{A}^{-1}$,

$$\mathbf{AA}^{-1} = \begin{bmatrix} 1 & 2 \\ 3 & 4 \end{bmatrix}\begin{bmatrix} -2.0 & 1.0 \\ 1.5 & -0.5 \end{bmatrix} = \begin{bmatrix} 1 & 0 \\ 0 & 1 \end{bmatrix}$$

The *determinant* of a matrix $\mathbf{A}$ is the determinant of the entries in $\mathbf{A}$. For example, expanding a 3×3 matrix $\mathbf{A}$ in terms of its first column,

$$\det(\mathbf{A}) = \begin{vmatrix} a_{11} & a_{12} & a_{13} \\ a_{21} & a_{22} & a_{23} \\ a_{31} & a_{32} & a_{33} \end{vmatrix} = a_{11}\begin{vmatrix} a_{22} & a_{23} \\ a_{32} & a_{33} \end{vmatrix} - a_{21}\begin{vmatrix} a_{12} & a_{13} \\ a_{32} & a_{33} \end{vmatrix} + a_{31}\begin{vmatrix} a_{12} & a_{13} \\ a_{22} & a_{23} \end{vmatrix}$$

$$= a_{11}(a_{22}a_{33} - a_{32}a_{23}) - a_{21}(a_{12}a_{33} - a_{32}a_{13}) + a_{31}(a_{12}a_{23} - a_{22}a_{13})$$

$$\tag{8·1·10}$$

Although the above concepts of matrix multiplication and addition, the identity matrix, the inverse, the transpose and the determinant have all been illustrated with small examples, these ideas are all valid for $n \times n$ matrices. The computations are more involved, however. The remainder of this chapter will use matrix representation to simplify the presentation of systems of equations.

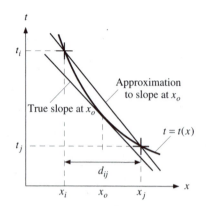

Figure 8•1 *Approximation of $t_x\big|_{x_o}$.*

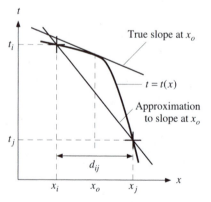

Figure 8•2 *Approximation of $t_x\big|_{x_o}$ when $t = t(x)$ has large curvature between x_i and x_j.*

8•1•2 Approximation of conduction rate equations

Heat transfer in a solid depends upon the rate equations for conduction, energy generation and energy storage. The forms for these rate equations suitable for the analytical formulation of the problem were discussed in Section 1•1•2. Now we want to approximate these equations in a way that is suitable for the finite-difference formulation of the problem.

Conduction The rate of conduction through a solid depends upon the temperature gradient as given by (1·1·2). Figure 8•1 shows a curve of temperature as a function of position. The straight line tangent to the curve is the slope at position x_o. This is the slope you would use in (1·1·2) to calculate the heat flow at x_o. The slope at x_o may be approximated in terms of the temperatures t_i and t_j at locations x_i and x_j on either side of x_o as shown in Figure 8•1. We will write

$$\frac{\partial t}{\partial x}\bigg|_{x_o} \cong \frac{t_j - t_i}{x_j - x_i}$$

The rate of conduction in the x direction at x_o will be approximated as

$$q\big|_{x_o} \cong -k_{ij}A_{ij}\frac{t_j - t_i}{x_j - x_i} = k_{ij}A_{ij}\frac{t_i - t_j}{x_j - x_i}$$

where

k_{ij} = conductivity of material between locations x_i and x_j

A_{ij} = conduction area half-way between locations x_i and x_j

We will then let q_{ij} be the conduction heat-transfer rate from position x_i toward position x_j and write

$$q_{ij} = \frac{k_{ij}A_{ij}}{d_{ij}}(t_i - t_j) \qquad (8\cdot1\cdot11)$$

where

$d_{ij} \equiv |x_j - x_i|$ is the distance between nodes i and j and

$k_{ij}A_{ij}/d_{ij}$ = conduction conductance between position x_i and position x_j.

To obtain the heat transfer from x_j to x_i, we can switch the subscripts in (8·1·11) and write

$$q_{ji} = \frac{k_{ji}A_{ji}}{d_{ji}}(t_j - t_i) \qquad (8\cdot1\cdot12)$$

Since $k_{ji} = k_{ij}$, $A_{ij} = A_{ij}$ and $d_{ji} = d_{ij}$, the conductance in (8·1·12) is equal to the conductance in (8·1·11), and $q_{ji} = -q_{ij}$.

When there is a large curvature in $t(x)$ between x_i and x_j, as shown in Figure 8•2, the approximation to the slope at x_o is quite different than the true slope. The approximation given by (8·1·11) is no longer very adequate. In such cases one would want to reduce the distance between x_i and x_j, to obtain a better approximation to the slope at $x_o = \frac{1}{2}(x_i + x_j)$.

Conduction is a little more complicated to evaluate when temperature is a function of x and y. Figure 8•3 shows a portion of the x-y plane with nine regularly spaced nodal points indicated. The dashed line is the boundary of the system around node i for which we will later want to write an energy balance. The system boundaries are the perpendicular bisectors to the lines from node i to each of the surrounding nodes j, k, ℓ, and m. We are interested in approximating the heat-transfer rate from node i toward node j. Assuming unit depth normal to the paper, we may write

$$q_{ij} = -\int_{y=(y_i+y_m)/2}^{(y_k+y_i)/2} k\left(\frac{x_i+x_j}{2},y\right)t_x\left(\frac{x_i+x_j}{2},y\right)dy$$

We will approximate the temperature gradient at the boundary between nodes i and j by writing

$$-t_x\left(\frac{x_i+x_j}{2},y\right) = \frac{t_i-t_j}{d_{ij}}$$

This approximation will be assumed to be valid along the entire boundary at $x = (x_i + x_j)/2$. We can then approximate the heat-transfer rate as

$$q_{ij} = \frac{t_i-t_j}{d_{ij}}\int_{y=(y_i+y_m)/2}^{(y_k+y_i)/2} k\left(\frac{x_i+x_j}{2},y\right)dy$$

We must next evaluate k along $x = (x_i + x_j)/2$. As we will see later, the line connecting nodes i and j could be a boundary between two different materials. Suppose material a is below this line and material b is above the line. We may then write

$$q_{ij} = \frac{t_i-t_j}{d_{ij}}\left[k_a\int_{y=(y_i+y_m)/2}^{y_i} dy + k_b\int_{y=y_i}^{(y_k+y_i)/2} dy\right]$$

Upon integrating and substituting limits,

$$q_{ij} = \frac{t_i-t_j}{d_{ij}}\left[k_a\left\{y_i-\frac{y_i+y_m}{2}\right\}+k_b\left\{\frac{y_k+y_i}{2}-y_i\right\}\right]$$

Simplifying the terms within the braces,

$$q_{ij} = \frac{t_i-t_j}{d_{ij}}\left[k_a\left\{\frac{y_i-y_m}{2}\right\}+k_b\left\{\frac{y_k-y_i}{2}\right\}\right]$$

For uniform spacing $d_{im} = d_{ik} = d$ of the nodes in the y direction, each term in braces is $d/2$ and for unit depth $A_{ij} = d$. Thus, we may write

$$q_{ij} = \frac{t_i-t_j}{d_{ij}}\left[k_a\left\{\frac{A_{ij}}{2}\right\}+k_b\left\{\frac{A_{ij}}{2}\right\}\right]=\left(\frac{k_a+k_b}{2}\right)\frac{A_{ij}}{d_{ij}}(t_i-t_j)$$

This approximation replaces (8·1·11) when conductivity varies in the y direction as it might in two-dimensional problems.

Generation The energy generation rate in a volume surrounding position i will be approximated as

$$g_i = g_i''' V_i = (g''' V)_i \qquad (8·1·13)$$

where g_i''' is the generation rate per unit volume at location i and V_i is the volume of the region.

Storage The rate of energy storage in a volume surrounding position i will be approximated as

$$\frac{dE_i}{d\theta} = \frac{dU_i}{d\theta} = \frac{d(\rho_i V_i u_i)}{d\theta} = \rho_i V_i \frac{du_i}{d\theta} = \rho_i V_i c_i \frac{dt_i}{d\theta} = (\rho c V)_i \frac{dt_i}{d\theta} \qquad (8·1·14)$$

This relation assumes the temperature t_i at position i is a suitable average temperature for the volume V_i surrounding position i.

In Section 8•2 we will see how these approximate rate equations [(8·1·11), (8·1·13), and (8·1·14)] can be used to obtain a finite-difference formulation for two-dimensional conduction problems.

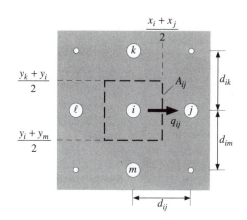

Figure 8•3 *Approximation of q_{ij} for two-dimensional problems.*

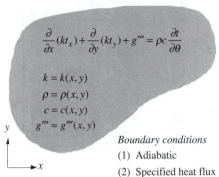

$$\frac{\partial}{\partial x}(kt_x) + \frac{\partial}{\partial y}(kt_y) + g''' = \rho c \frac{\partial t}{\partial \theta}$$

$$k = k(x, y)$$
$$\rho = \rho(x, y)$$
$$c = c(x, y)$$
$$g''' = g'''(x, y)$$

Boundary conditions
(1) Adiabatic
(2) Specified heat flux
Initial condition   (3) Convection
$t(x, y, 0) = t^{(0)}(x, y)$ (4) Specified temperature

Figure 8•4 *A general two-dimensional, transient conduction problem.*

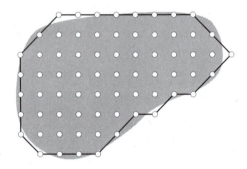

Figure 8•5 *Placement of nodal points in a square pattern showing approximation of region boundaries.*

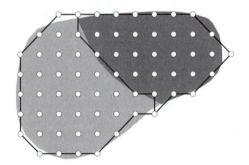

Figure 8•6 *Region with two materials will also have internal boundaries to approximate.*

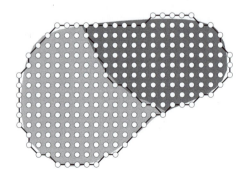

Figure 8•7 *A finer pattern of nodal points improves the boundary approximation.*

8•2 DISCRETIZING THE SPATIAL PROBLEM

One of the main reasons that analytical methods fail in practice is that they can only be easily applied when the geometry is very simple. An irregular shape such as the one shown in Figure 8•4 would be hard to treat. Rectangles and circles would be two examples where analytical methods might be expected to have some chance of success since the boundaries coincide with the coordinate system. On the other hand, a turbine blade (airfoil cross-section) geometry would be very cumbersome to handle analytically if it could be done at all.

Shapes that are made of more than one material (composite materials) are also difficult to handle analytically because the properties depend on position. Although Figure 8•4 might imply continuously varying k, ρ, c and g''', the position dependence could be due to the shape having a different material in one portion than in another portion. An example would be a nuclear reactor which contains fuel rods with one set of thermal properties and moderator with a second set of thermal properties. It should also be observed that the geometry is also apt to be complicated in a nuclear reactor.

In this section we will consider the general class of problems pictured in Figure 8•4. Observe that the geometry is not very simple and that the material properties vary with position. In general, an exact solution of the partial differential equation would not be possible for this problem.

The first step in the finite-difference formulation is to *discretize* the spatial problem. This is most easily done by choosing a set of regularly spaced nodal points as shown in Figure 8•5. The spacing in the x-direction is uniform and equal to the spacing in the y-direction which gives a square pattern for the nodal points. A rectangular pattern could almost as easily have been chosen. An irregular pattern, however, would complicate the problem considerably. The actual boundary of the region is approximated as closely as possible by connecting the nodal points by straight line segments. The approximate region, whose boundary is made up of straight-line segments is the region with which we will be working. You may think of the approximate region as a "rough machining" of the actual region.

If a region is made of two materials, as indicated by the shading in Figure 8•6, there will be an internal boundary between the two materials. This internal boundary will also be approximated by connecting the nodal points by straight line segments as shown in Figure 8•6.

Thus, there will be *interior nodes* that do not lie on any boundary and *boundary nodes* that lie on either an internal material boundary or on the external boundary of the region. Section 8•2•1 will discuss interior nodes and Section 8•2•2 will consider boundary nodes.

A better approximation to the boundaries (*i.e.*, "finer machining") can be obtained by halving the spacing between nodal points as shown in Figure 8•7. You should note, however, that halving the spacing between nodal points increases the number of nodal points by about a factor of four (from 69 to 257 by actual count for this problem). This will entail a lot more work but will give better answers as we will see later. Alternatively one could depart from uniform nodal point spacing near the boundary to better approximate the boundary. This special treatment at the boundary is rather cumbersome however. The finite-element method discussed in Chapter 9 is far better suited to handle irregular two-dimensional problems than the finite-difference method we are discussing in this chapter. Thus we will continue to assume a regular nodal-point pattern in this chapter and wait until Chapter 9 to worry about irregular nodal-point locations.

8•2•1 Interior nodes

An interior node is a node that does not lie on any boundary. A typical interior nodal point i is shown in Figure 8•8. It is surrounded by nodal points j, k, ℓ and m. Except for a brief discussion later in Section 8•6•1, we will only consider uniformly spaced nodes in this chapter. The horizontal distance between nodes, as well as the vertical distance between nodes, will be denoted by d. The conduction areas for q_{ij}, q_{ik}, $q_{i\ell}$ and q_{im} will each be A. The volume of the system around node i is V. For unit depth into the page, $A = d \cdot 1$ and $V = d^2 \cdot 1$.

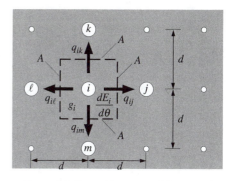

Figure 8•8 *Typical interior nodal point and surrounding nodes.*

An ordinary differential equation relating the temperatures t_i, t_j, t_k, t_ℓ and t_m can be obtained by making an energy balance on the shaded system surrounding node i. The sides of the system bisect the distances between the nodal points as shown in Figure 8•8. An energy balance on system i is given by

$$g_i = \frac{dE_i}{d\theta} + q_{im} + q_{i\ell} + q_{ij} + q_{ik}$$

The conduction terms will be approximated following (8·1·11). Since node i is an interior node, the conductivities in all four conduction rate equations will be the same. That is, $k_{ij} = k_{ik} = k_{i\ell} = k_{im} = k$. The generation and storage terms follow (8·1·13) and (8·1·14), respectively. Upon substituting the rate equations into the energy balance we obtain

$$g'''V = \rho c V \frac{dt_i}{d\theta} + \frac{kA}{d}(t_i - t_m) + \frac{kA}{d}(t_i - t_\ell) + \frac{kA}{d}(t_i - t_j) + \frac{kA}{d}(t_i - t_k)$$

For convenience we will rearrange this equation to give:

$$g'''V = \rho c V \frac{dt_i}{d\theta} + \left[-\frac{kA}{d}t_m - \frac{kA}{d}t_\ell + 4\frac{kA}{d}t_i - \frac{kA}{d}t_j - \frac{kA}{d}t_k \right] \qquad (8.2.1)$$

Next we will simplify (8·2·1) and prepare for matrix notation by defining

$$g_i = g'''V \ , \quad c_{ii} = \rho c V$$

$$\kappa_{im} = -\frac{kA}{d} \ , \quad \kappa_{i\ell} = -\frac{kA}{d} \ , \quad \kappa_{ii} = 4\frac{kA}{d} \ , \quad \kappa_{ij} = -\frac{kA}{d} \ , \quad \kappa_{ik} = -\frac{kA}{d} \qquad (8.2.2)$$

The coefficient κ_{ii} of temperature t_i is the sum of the conductances from node i to the four adjacent nodes surrounding node i. Notice also that the sum $\kappa_{im} + \kappa_{i\ell} + \kappa_{ii} + \kappa_{ij} + \kappa_{ik} = 0$. The value g_i is the rate of energy generation within nodal system i and c_{ii} is the thermal capacitance of nodal system i. Upon introducing this notation into (8·2·1) and switching left- and right-hand sides we obtain

$$c_{ii}\frac{dt_i}{d\theta} + \left[\kappa_{im}t_m + \kappa_{i\ell}t_\ell + \kappa_{ii}t_i + \kappa_{ij}t_j + \kappa_{ik}t_k \right] = g_i \qquad (8.2.3)$$

This result is typical of any interior node. You may think of (8·2·3) as an ordinary differential equation for temperature t_i at node i as a function of time θ. The equation is complicated by the fact that it also contains the temperatures of the four surrounding nodes. We will have a similar differential equation for each nodal energy balance. Thus, we are deriving a coupled set of differential equations that must be solved simultaneously.

In order to talk more efficiently about the set of differential equations we are deriving, it is convenient to introduce matrix notation and the idea of a matrix differential equation. First, we will introduce **t** as the column matrix or vector of the N nodal temperatures. We will define **t** and its time derivative $d\mathbf{t}/d\theta$ by

$$
\mathbf{t} = \begin{bmatrix} t_1 \\ t_m \\ t_\ell \\ t_i \\ t_j \\ t_k \\ t_N \end{bmatrix}
\qquad
\frac{d\mathbf{t}}{d\theta} = \dot{\mathbf{t}} = \begin{bmatrix} \dot{t}_1 \\ \dot{t}_m \\ \dot{t}_\ell \\ \dot{t}_i \\ \dot{t}_j \\ \dot{t}_k \\ \dot{t}_N \end{bmatrix}
\tag{8·2·4}
$$

The storage term on the left-hand side of (8·2·3) can then be written as

$$
\begin{bmatrix} & & & \\ & & c_{ii} & \\ & & & \end{bmatrix}
\begin{bmatrix} \dot{t}_1 \\ \dot{t}_m \\ \dot{t}_\ell \\ \dot{t}_i \\ \dot{t}_j \\ \dot{t}_k \\ \dot{t}_N \end{bmatrix} = \mathbf{C}^{(i)}\dot{\mathbf{t}}
\tag{8·2·5}
$$

Every entry in $\mathbf{C}^{(i)}$ is 0 except for c_{ii} on the main diagonal. Upon carrying out the matrix multiplication you see that the only nonzero term in the vector $\mathbf{C}^{(i)}\dot{\mathbf{t}}$ is $c_{ii}\dot{t}_i$ that appears in row i. We will call $\mathbf{C}^{(i)}$ the *nodal capacitance matrix* for node i. The remaining terms on the left-hand side of (8·2·3) can be represented in matrix form as

$$
\begin{bmatrix} & & & & & \\ \kappa_{im} & \kappa_{i\ell}\ \kappa_{ii}\ \kappa_{ij} & \kappa_{ik} & & \\ & & & & \end{bmatrix}
\begin{bmatrix} t_1 \\ t_m \\ t_\ell \\ t_i \\ t_j \\ t_k \\ t_N \end{bmatrix} = \mathbf{K}^{(i)}\mathbf{t}
\tag{8·2·6}
$$

Again all of the entries in $\mathbf{K}^{(i)}$ are 0 except for the ones shown. For node i these nonzero entries are all in row i. Multiplication of the matrices shows that the entry in row i of the vector $\mathbf{K}^{(i)}\mathbf{t}$ will be

$$(\mathbf{K}^{(i)}\mathbf{t})_i = \kappa_{im}t_m + \kappa_{i\ell}t_\ell + \kappa_{ii}t_i + \kappa_{ij}t_j + \kappa_{ik}t_k$$

The matrix $\mathbf{K}^{(i)}$ will be called the *nodal conduction matrix* for node i. The right-hand side of (8·2·3) can be written as a vector $\mathbf{g}^{(i)}$ whose entries are all 0 except for g_i in row i. We will call $\mathbf{g}^{(i)}$ the *nodal generation vector* for node i. Thus (8·2·3), representing an energy balance on system i, can be written using matrix notation as

$$\mathbf{C}^{(i)}\dot{\mathbf{t}} + \mathbf{K}^{(i)}\mathbf{t} = \mathbf{g}^{(i)} \tag{8.2.7}$$

The matrices $\mathbf{C}^{(i)}$, $\mathbf{K}^{(i)}$ and $\mathbf{g}^{(i)}$ contain information about the material properties near node i and the size of the system around i. The nodal capacitance matrix $\mathbf{C}^{(i)}$ contains the thermal capacitance $(\rho cV)_i$ of nodal system i. The nodal conduction matrix $\mathbf{K}^{(i)}$ contains the conduction conductances between node i and the four surrounding nodes. The nodal generation vector $\mathbf{g}^{(i)}$ contains the energy generation rate in system i.

Next we will move on and write an energy balance on system j next door to system i as shown in Figure 8•9. Assuming that node j is also an interior node that does not lie on a material boundary, we would find that

$$g'''V = \rho cV\frac{dt_j}{d\theta} + \frac{kA}{d}(t_j - t_n) + \frac{kA}{d}(t_j - t_i) + \frac{kA}{d}(t_j - t_p) + \frac{kA}{d}(t_j - t_q)$$

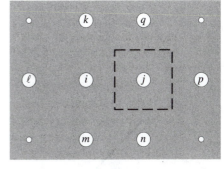

Figure 8•9 *Interior nodal point j next to node i.*

Upon rearranging and defining notation similar to (8·2·2), we obtain

$$c_{jj}\frac{dt_j}{d\theta} + \left[\kappa_{jn}t_n + \kappa_{ji}t_i + \kappa_{jj}t_j + \kappa_{jp}t_p + \kappa_{jq}t_q\right] = g_j$$

or, in matrix form,

$$\mathbf{C}^{(j)}\dot{\mathbf{t}} + \mathbf{K}^{(j)}\mathbf{t} = \mathbf{g}^{(j)} \tag{8.2.8}$$

We can then add (8·2·8) to (8·2·7) to get

$$[\mathbf{C}^{(i)} + \mathbf{C}^{(j)}]\dot{\mathbf{t}} + [\mathbf{K}^{(i)} + \mathbf{K}^{(j)}]\mathbf{t} = \mathbf{g}^{(i)} + \mathbf{g}^{(j)} \tag{8.2.9}$$

In order to see how things are developing let us write out (8·2·9) with the capacitance, conduction, and generation matrices in expanded form as

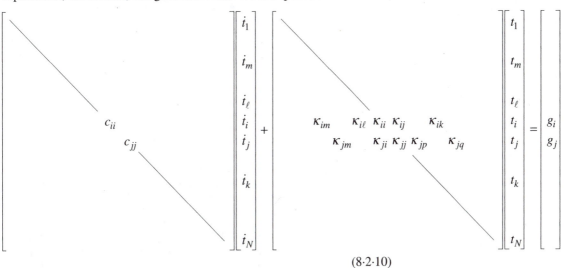

$$(8.2.10)$$

In practice, one does not store each $\mathbf{C}^{(i)}$, $\mathbf{K}^{(i)}$ and $\mathbf{g}^{(i)}$ separately in the computer and then add them together to find the *global* $\mathbf{C}$, $\mathbf{K}$ and $\mathbf{g}$. This would require too much storage and computation. Rather, as each nonzero entry in $\mathbf{C}^{(i)}$, $\mathbf{K}^{(i)}$ and $\mathbf{g}^{(i)}$ is found, it is simply stored in the appropriate location of the global $\mathbf{C}$, $\mathbf{K}$ and $\mathbf{g}$.

All nonzero entries from node j are in row j. The global capacitance matrix **C** that is developing has nonzero entries only on the main diagonal. The global conduction matrix **K** will be a *banded* matrix since **K** will have nonzero entries along the main diagonal and in a band to the left and to the right of the main diagonal, but will have only zeroes outside of the band. The number of diagonals (main diagonal plus off-diagonals to the right of the main diagonal) needed to include all of the nonzero entries on and to the right of the main diagonal is called the *upper bandwidth* of the matrix. If node 1 were adjacent to node N you would have a nonzero κ_{1N} entry in the upper right-hand corner of **K** and a nonzero κ_{N1} entry in the lower left-hand corner and the upper bandwidth of **K** would be N. Smart node numbering can avoid this. Nodes should be numbered to make the upper bandwidth small to reduce computer storage and computation.

Each interior node that does not lie on a material boundary will make a contribution to the global matrices similar to what we have just seen for nodes i and j. The global matrix differential equation

$$\mathbf{C\dot{t} + Kt = g} \tag{8.2.11}$$

will then contain entries in every row corresponding to these interior nodes. This matrix differential equation (8.2.11) is shown below in expanded form to show its features.

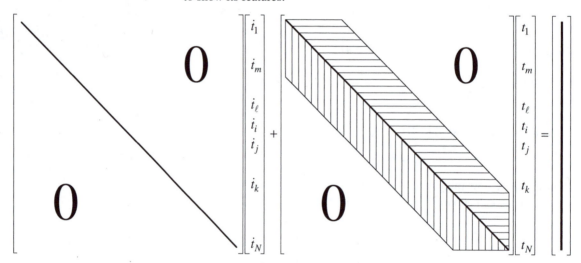

It is also important to observe that **K** will be symmetric since

$$\kappa_{ij} = -\frac{k_{ij}A_{ij}}{d_{ij}} = \frac{kA}{d} = -\frac{k_{ji}A_{ji}}{d_{ji}} = \kappa_{ji}$$

Since **K** is symmetric, the entries in **K** in the vertical columns below the main diagonal will be equal to the entries in **K** in the corresponding rows to the right of the main diagonal. It is not necessary to store the entries below the main diagonal. For a large problem, this is a significant saving in computer storage.

In summary, the finite-difference discretization of the spatial problem produces a system of coupled, first-order, ordinary differential equations that must later be solved simultaneously. Rather than writing down each separate nodal equation, we have used matrix notation to represent the entire system of equations by a single, global matrix differential equation (8.2.11). The symmetry and bandedness of **K** are important features that will be exploited to obtain more efficient computer solutions.

Section 8•2•2 will complete the spatial discretization by considering nodes along material boundaries.

8·2·2 Boundary nodes

Nodal systems on material boundaries are somewhat more complicated. A node on an internal boundary between two materials will involve two sets of thermal properties. For a node on an external boundary, the shape of the nodal system will not be square as it was for an interior nodal system. Nodal systems on an external boundary must incorporate the boundary conditions for the problem. It is also possible to have an internal node involving more than two materials or a node on an external boundary that involves more than one material.

Nodes on internal boundaries For an interior node that lies on a boundary between two materials, there are six possible ways that the material boundary can pass through the nodal system as shown in Figure 8•10. Each case must be treated separately. The horizontal and vertical distances between nodes is still d. The volume of each of these six nodal systems is still V and for unit depth $V = d^2 \cdot 1$. The conduction area in a single material is either A or $A/2$ and for unit depth $A = d \cdot 1$.

As an example of an interior boundary node let us consider the situation shown in Figure 8•10(e). An enlarged view is shown in Figure 8•11. An energy balance on this system is given by

$$g_i = \frac{dE_i}{d\theta} + q_{im} + q_{i\ell} + q_{ij} + q_{ik}$$

For this system, $5/8$ of the volume is material 1 and $3/8$ of the volume is material 2. Therefore the generation rate equation is given by

$$g_i = g_1''' \frac{5}{8}V + g_2''' \frac{3}{8}V = \left(\frac{5}{8}g_1''' + \frac{3}{8}g_2'''\right)V$$

and the rate equation for energy storage is given by

$$\frac{dE_i}{d\theta} = \left(\rho_1 \frac{5}{8}Vc_1 + \rho_2 \frac{3}{8}Vc_2\right)\frac{dt_i}{d\theta} = \left(\frac{5}{8}\rho_1 c_1 + \frac{3}{8}\rho_2 c_2\right)V\frac{dt_i}{d\theta}$$

The rate equations for q_{im} and $q_{i\ell}$ involve only material 1 and the rate equation for q_{ij} involves only material 2 and are given by

$$q_{im} = \frac{k_1 A}{d}(t_i - t_m) , \quad q_{i\ell} = \frac{k_1 A}{d}(t_i - t_\ell) , \quad \text{and} \quad q_{ij} = \frac{k_2 A}{d}(t_i - t_j)$$

The rate equation for q_{ik} involves both materials since there is a parallel heat-flow path from node i to node k. One path is through material 1; the second path is through material 2. The rate equation for q_{ik} is given by

$$q_{ik} = k_1 \frac{A}{2}\frac{(t_i - t_k)}{d} + k_2 \frac{A}{2}\frac{(t_i - t_k)}{d} = \left(\frac{k_1 + k_2}{2}\right)\frac{A}{d}(t_i - t_k)$$

Upon substituting all of the rate equations into the energy balance,

$$\left(\frac{5}{8}g_1''' + \frac{3}{8}g_2'''\right)V = \left(\frac{5}{8}\rho_1 c_1 + \frac{3}{8}\rho_2 c_2\right)V\frac{dt_i}{d\theta} + \frac{k_1 A}{d}(t_i - t_m) + \frac{k_1 A}{d}(t_i - t_\ell)$$
$$+ \frac{k_2 A}{d}(t_i - t_j) + \left(\frac{k_1 + k_2}{2}\right)\frac{A}{d}(t_i - t_k)$$

For convenience we will rearrange this equation to give:

$$\left(\frac{5}{8}g_1''' + \frac{3}{8}g_2'''\right)V = \left(\frac{5}{8}\rho_1 c_1 + \frac{3}{8}\rho_2 c_2\right)V\frac{dt_i}{d\theta} + \left[-\frac{k_1 A}{d}t_m - \frac{k_1 A}{d}t_\ell + \left(\frac{5}{2}k_1 + \frac{3}{2}k_2\right)\frac{A}{d}t_i - \frac{k_2 A}{d}t_j - \left(\frac{k_1 + k_2}{2}\right)\frac{A}{d}t_k\right]$$

$$(8 \cdot 2 \cdot 12)$$

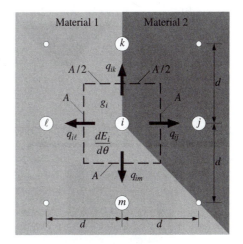

Figure 8•10 *Node i on an internal boundary between two materials.*

Figure 8•11 *Node i on an internal boundary between two materials as in Figure 8•10(e).*

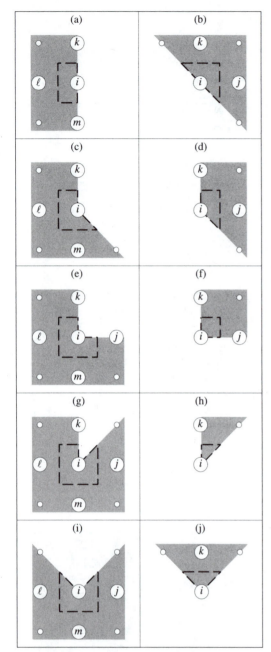

Figure 8·12 *Node i on an external boundary.*

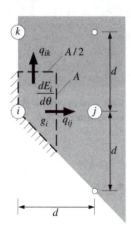

Figure 8·13 *Node i on an adiabatic external boundary.*

Next we will simplify (8·2·12) and prepare for matrix notation by defining

$$g_i = \left(\frac{5}{8}g_1''' + \frac{3}{8}g_2'''\right)V \;, \quad c_{ii} = \left(\frac{5}{8}\rho_1 c_1 + \frac{3}{8}\rho_2 c_2\right)V$$

$$\kappa_{im} = -\frac{k_1 A}{d}\;, \quad \kappa_{i\ell} = -\frac{k_1 A}{d}\;, \quad \kappa_{ii} = \left(\frac{5}{2}k_1 + \frac{3}{2}k_2\right)\frac{A}{d}\;,$$

$$\kappa_{ij} = -\frac{k_2 A}{d}\;, \quad \kappa_{ik} = -\left(\frac{k_1 + k_2}{2}\right)\frac{A}{d} \tag{8·2·13}$$

Upon introducing this notation into (8·2·12) and switching left- and right-hand sides we obtain

$$c_{ii}\frac{dt_i}{d\theta} + \left[\kappa_{im}t_m + \kappa_{i\ell}t_\ell + \kappa_{ii}t_i + \kappa_{ij}t_j + \kappa_{ik}t_k\right] = g_i \tag{8·2·14}$$

This differential equation has the same form as the previous differential equations for nodes that were not on a material boundary. The character of the system of differential equations has not changed. The capacitance matrix **C** is still diagonal. The conduction matrix **K** is still banded and symmetric. Each row in **K** still sum to zero.

Nodes on external boundaries The ten possible shapes that a nodal system on an external boundary system could have are shown in Figure 8·12. Each case must be treated separately. The horizontal and vertical distances between nodes is still d. The volume of these 10 nodal systems is either $\frac{1}{8}V$, $\frac{2}{8}V$, $\frac{3}{8}V$, $\frac{4}{8}V$, $\frac{3}{8}V$, $\frac{6}{8}V$ or $\frac{7}{8}V$ and for unit depth $V = d^2 \cdot 1$. The conduction area is either A or $A/2$ and for unit depth $A = d \cdot 1$.

Adiabatic As an example of an adiabatic boundary let us consider the system shown in Figure 8·12(d). An enlarged view is shown in Figure 8·13. The volume of this system is 3/8 of V where, for unit depth, $V = d^2 \cdot 1$. One conduction area is A and the other is $A/2$ where, for unit depth, $A = d \cdot 1$. An energy balance on this system is given by

$$g_i = \frac{dE_i}{d\theta} + q_{ij} + q_{ik}$$

No energy transfer across the adiabatic external boundary is included in the balance. Introducing the rate equations into the energy balance gives

$$g'''\frac{3}{8}V = \rho c \frac{3}{8}V\frac{dt_i}{d\theta} + \frac{kA}{d}(t_i - t_j) + \frac{kA/2}{d}(t_i - t_k)$$

Upon rearranging we obtain,

$$g'''\frac{3}{8}V = \rho c \frac{3}{8}V\frac{dt_i}{d\theta} + \left[\left(\frac{kA}{d} + \frac{kA}{2d}\right)t_i - \frac{kA}{d}t_j - \frac{kA}{2d}t_k\right]$$

Simplifying the notation as we did in (8·2·12), (8·2·13) and (8·2·14) gives

$$c_{ii}\frac{dt_i}{d\theta} + \left[\kappa_{ii}t_i + \kappa_{ij}t_j + \kappa_{ik}t_k\right] = g_i \tag{8·2·15}$$

This result is very similar to (8·2·14) with the exception that $\kappa_{im}t_m$ and $\kappa_{i\ell}t_\ell$ do not appear. Also, c_{ii}, κ_{ii}, κ_{ik} and g_i have different numerical values than in (8·2·14). However, in matrix notation, this equation has the same form as (8·2·7).

This nodal differential equation (8·2·15) for boundary node i can be added into the global differential equation (8·2·10) in the same manner as the interior nodal equations were. It will appear in row i of the global matrix equation (8·2·11).

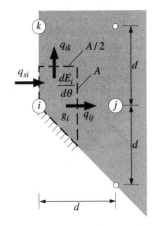

Specified heat flux Figure 8•14 is an enlargement of Figure 8•12(d) that assumes there is a uniform specified heat flux input along part of the boundary. The area S_i of the nodal system that is exposed to the specified heat flux input is given by $S_i = A/2$ where, for unit depth, $A = d \cdot 1$. The energy balance is given by

$$g_i + q_{si} = \frac{dE_i}{d\theta} + q_{ij} + q_{ik}$$

The rate equation for the specified heat flow will be written as the product of a specified heat flux q''_{si} and the surface area S_i exposed to the heat flux. That is,

$$q_{si} = q''_{si}S_i = q''_{si}\frac{A}{2}$$

Substitution of all of the rate equations into the energy balance gives

$$g'''\frac{3}{8}V + q''_{si}\frac{A}{2} = \rho c\frac{3}{8}V\frac{dt_i}{d\theta} + \frac{kA}{d}(t_i - t_j) + \frac{kA/2}{d}(t_i - t_k)$$

This may be rearranged and simplified to give

$$c_{ii}\frac{dt_i}{d\theta} + \left[\kappa_{ii}t_i + \kappa_{ij}t_j + \kappa_{ik}t_k\right] = g_i + q_i \qquad (8\cdot2\cdot16)$$

where the new term q_i that did not appear in (8·2·15) is given by

$$q_i = q_{si} = q''_{si}S_i = q''_{si}\frac{A}{2}$$

In expanded matrix form, (8·2·16) may be written as

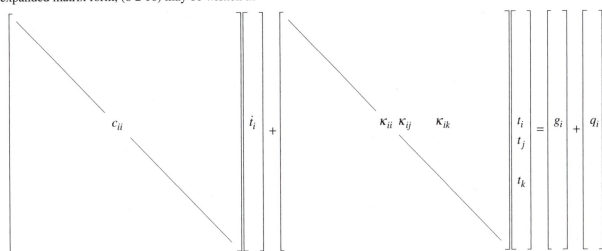

In matrix notation we would write (8·2·16) as

$$\mathbf{C}^{(i)}\dot{\mathbf{t}} + \mathbf{K}^{(i)}\mathbf{t} = \mathbf{g}^{(i)} + \mathbf{q}^{(i)} \qquad (8\cdot2\cdot17)$$

We see that the effect of a specified heat flux at the surface shows up in a *nodal heat-flow vector* $\mathbf{q}^{(i)}$ being added to $\mathbf{g}^{(i)}$. The global matrix equation (8·2·11) will now include a *global heat-flow vector* $\mathbf{q}$ and be written as

$$\mathbf{C}\dot{\mathbf{t}} + \mathbf{K}\mathbf{t} = \mathbf{g} + \mathbf{q} \qquad (8\cdot2\cdot18)$$

This equation now includes the interior nodes and the boundary nodes along both adiabatic and specified nonzero-heat-flux boundaries. An adiabatic boundary could be considered to be a special case of a specified heat flux boundary with $q''_{si} = 0$.

Figure 8•14 *A typical node on an external boundary where the heat flux is specified.*

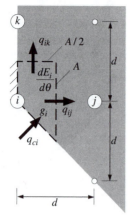

Figure 8•15 *A typical node on a convective external boundary.*

Convection The convection boundary node shown in Figure 8•15 is an enlargement of Figure 8•12(d). The convection heat-transfer rate q_{ci} is taken to be positive when flowing into the solid. The convection heat transfer is assumed to be entering only along the diagonal portion of the boundary. The area S_i of the nodal system that is exposed to the convection input is given by $S_i = A\sqrt{2}/2$ where, for unit depth, $A = d \cdot 1$. The energy balance is then given by

$$g_i + q_{ci} = \frac{dE_i}{d\theta} + q_{ij} + q_{ik}$$

Upon substituting the rate equations we get

$$g'''\frac{3}{8}V + h_{ci}\frac{A\sqrt{2}}{2}(t_{\infty i} - t_i) = \rho c\frac{3}{8}V\frac{dt_i}{d\theta} + \frac{kA}{d}(t_i - t_j) + \frac{kA/2}{d}(t_i - t_k)$$

where the ambient temperature outside of node i is $t_{\infty i}$ and is assumed to be given. If we now rearrange and introduce notation similar to (8·2·13), this equation may be written as

$$c_{ii}\frac{dt_i}{d\theta} + \left[(\kappa_{ii} + h_{ii})t_i + \kappa_{ij}t_j + \kappa_{ik}t_k\right] = g_i + h_i \qquad (8\cdot2\cdot19)$$

where the convection conductance h_{ii} is given by

$$h_{ii} = h_{ci}S_i = h_{ci}A\sqrt{2}/2$$

and h_i is given by

$$h_i = h_{ci}S_i t_{\infty i} = h_{ci}(A\sqrt{2}/2)t_{\infty i}$$

This differential equation (8·2·19) can be viewed in matrix notation as

$$\mathbf{C}^{(i)}\dot{\mathbf{t}} + [\mathbf{K}^{(i)} + \mathbf{H}^{(i)}]\mathbf{t} = \mathbf{g}^{(i)} + \mathbf{h}^{(i)} \qquad (8\cdot2\cdot20)$$

where the new matrix $\mathbf{H}^{(i)}$ and the new vector $\mathbf{h}^{(i)}$ are given by

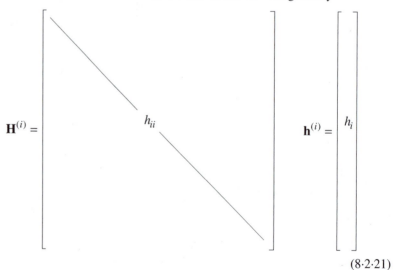

$$(8\cdot2\cdot21)$$

In practice, one would not actually store $\mathbf{H}^{(i)}$ and $\mathbf{h}^{(i)}$ in the computer, but simply add h_{ii} to κ_{ii} in $\mathbf{K}$ and add h_i to g_i in $\mathbf{g}$.

Each boundary node with convection would contribute not only to $\mathbf{C}$, $\mathbf{K}$ and $\mathbf{g}$, but also to $\mathbf{H}$ (a diagonal convection matrix) and $\mathbf{h}$ (a convection vector). Thus the global equation (8·2·18) can now be written as

$$\mathbf{C}\dot{\mathbf{t}} + (\mathbf{K} + \mathbf{H})\mathbf{t} = \mathbf{g} + \mathbf{q} + \mathbf{h} \qquad (8\cdot2\cdot22)$$

Convection modifies the diagonal of $\mathbf{K}$ and $\mathbf{g}$.

Specified temperature When the temperature at node i is specified to be T_i (upper case T will indicate a specified constant value of temperature) we really do not have to include node i in the system of equations. Instead of considering N equations we could work with $N-1$ equations since we do not have to find its temperature. One drawback with this approach is that node i will often be in the middle of the set of nodes and if you decide later to change the boundary condition you would have a major job to renumber the nodes to include node i if the specified-temperature condition were removed. A second drawback is that you often want to find the heat flow at a specified-temperature node and it is convenient if the node is included in the set of nodes. Generally, the number of specified-temperature nodes is small relative to the total number of nodes and the savings obtained from having a few less equations to solve is not worth the special treatment it would require.

Since the temperature of node i is known, we don't need to determine a differential equation for that node. The algebraic equation

$$t_i = T_i$$

could be used. Row i in the system of equations could then be given by

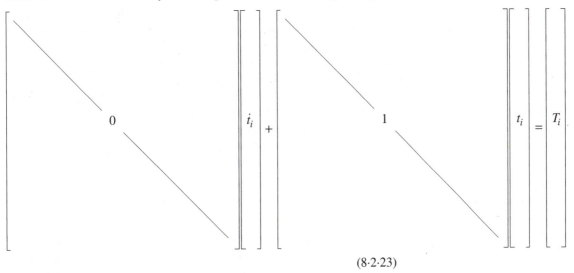

$$(8\cdot2\cdot23)$$

Observe that row i simply gives $t_i = T_i$.

There are several computational considerations that make (8·2·23) an unwise choice:

(1) $(\mathbf{K}+\mathbf{H})$ is no longer a symmetric matrix since there will be nonzero off-diagonal entries in column i that are not symmetric to the off-diagonal entries in row i which are all zero. This increases computer storage requirements and computational effort.

(2)   An algebraic equation is now embedded in the midst of a system of ordinary differential equations. This will require special handling.

(3)   Setting the diagonal term in $(\mathbf{K}+\mathbf{H})$ equal to 1 may not be in balance with the rest of the entries in $(\mathbf{K}+\mathbf{H})$ and hence could cause numerical problems.

(4)   Setting $c_{ii} = 0$ might also be awkward.

To avoid these difficulties we will take another approach. Although heat transfer with the outside is usually needed to maintain a specified temperature T_i for node i, we will never-the-less ignore this heat transfer and derive the governing equation for node i just as we have previously done for other nodes. The governing equations for all nodal systems will be derived before making any adjustments for specified-temperature nodes.

After deriving the equations for all N nodal systems, we may write the system of equations $\mathbf{C\dot{t}} + (\mathbf{K+H})\mathbf{t} = (\mathbf{g+h+q})$ as

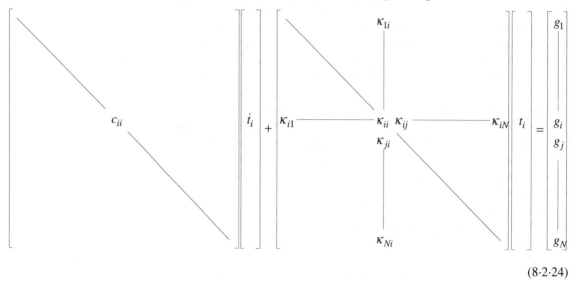

$$(8\cdot2\cdot24)$$

where, to shorten notation, κ denotes an entry in $(\mathbf{K+H})$ now instead of just $\mathbf{K}$ and g denotes an entry in $(\mathbf{g+h+q})$ instead of just $\mathbf{g}$. The entries in row i are indicated. The entries in column i are also indicated since they are the coefficients that multiply t_i in the other rows.

If the temperature of node i in (8·2·24) has a specified value T_i, t_i is not really an unknown. We will transfer the terms involving $t_i = T_i$ in all rows except row i from the left-hand side of (8·2·24) to the right-hand side. The coefficients of these terms are the off-diagonal entries in column i in $(\mathbf{K+H})$. The off-diagonal entries in column i must then be set equal to zero. These modifications to (8·2·24) may be shown as

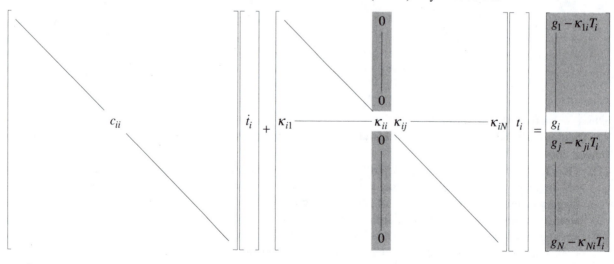

$$(8\cdot2\cdot25)$$

The shading in (8·2·25) indicates where changes have been made. There is no change on the right-hand side in row i at this stage. The modified $(\mathbf{K+H})$ in (8·2·25) is not symmetric since all of the off-diagonal entries in column i are now zero whereas the off-diagonal entries in row i are not all zero. Similar modifications should be made for all specified-temperature nodes before going on to the next step.

To restore the symmetry of the modified $(\mathbf{K+H})$ and to ensure that row i will give $t_i = T_i$, we will now set each of the off-diagonal entries in row i equal to zero and change the entry in row i of the right-hand side to $\kappa_{ii}T_i$ to give

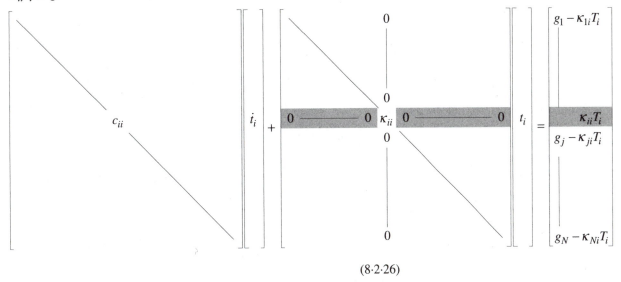

$$(8\cdot2\cdot26)$$

The shading in (8·2·26) shows where changes were made. The equation in row i is now uncoupled from the other equations and is given by

$$c_{ii}\frac{dt_i}{d\theta} + \kappa_{ii}t_i = \kappa_{ii}T_i \qquad (8\cdot2\cdot27)$$

When the initial value of t_i is T_i the solution to (8·2·27) will be $t_i = T_i$ for all θ. Thus, instead of an algebraic equation (8·2·23) for this node, we will write a differential equation and take the initial condition for node i to be $t_i(0) = T_i$, the specified value of temperature at node i. Similar changes must be made in all rows corresponding to specified-temperature nodes.

One disadvantage of this method of handling specified-temperature nodes is that in large problems there may be many transfers that have to be made in order to restore symmetry. An alternative method that does not require any off-diagonal modifications is to multiply c_{ii} and κ_{ii} in (8·2·24) by a very large number, say 10^8, and replace g_i in the right-hand-side vector by $10^8\kappa_{ii}T_i$. The differential equation (8·2·24) for node i then becomes (after slight rearrangement)

$$10^8 c_{ii}\frac{dt_i}{d\theta} + 10^8\kappa_{ii}t_i + \sum_{\substack{j=1\\j\neq i}}^{N}\kappa_{ij}t_j = 10^8\kappa_{ii}T_i$$

This equation is still symmetric with the other equations in (8·2·24) because the off-diagonal entries have not been modified. The effect of the off-diagonal entries has been greatly reduced however in comparison to the diagonal terms and to the right-hand side. Its solution, for $t_i(0) = T_i$, would be almost $t_i = T_i$, but not precisely, since there would be some effect of the off-diagonal terms. If you can be happy with a solution that is not precisely correct (and certainly engineers are often willing to accept approximate solutions) then this method may work out quite well for you. A possible disadvantage of this alternative method is that the diagonal entries, $10^8 c_{ii}$ and $10^8\kappa_{ii}$, are considerably out of scale with the other entries in $\mathbf{C}$ and $(\mathbf{K} + \mathbf{H})$. This might cause some difficulties in the numerical computations required to solve the system of equations.

8•2•3 Initial condition

The initial condition is approximated simply by taking each of the nodal temperatures to equal the given initial temperature at that location. Thus, if the exact initial temperature distribution is given by

$$t(x,y,0) = t^{(0)}(x,y) \tag{8.2.28}$$

we will then write the initial condition at node i as

$$t(x_i,y_i,0) = t^{(0)}(x_i,y_i) \qquad \text{or} \qquad t_i(0) = t_i^{(0)} \tag{8.2.29}$$

Thus, if there are no specified temperatures, the initial value of the nodal temperature vector is given by

$$\mathbf{t}(0) = \mathbf{t}^{(0)} = \begin{bmatrix} t_1^{(0)} \\ \vdots \\ t_{i-1}^{(0)} \\ t_i^{(0)} \\ t_{i+1}^{(0)} \\ \vdots \\ t_N^{(0)} \end{bmatrix} \tag{8.2.30}$$

If there is a sudden step at $\theta = 0$ at node i from $t_i^{(0)}$ to T_i, the initial temperature vector would then have to be modified to

$$\mathbf{t}(0) = \mathbf{t}^{(0)} = \begin{bmatrix} t_1^{(0)} \\ \vdots \\ t_{i-1}^{(0)} \\ T_i \\ t_{i+1}^{(0)} \\ \vdots \\ t_N^{(0)} \end{bmatrix} \tag{8.2.31}$$

Since the value of the temperature at node i has been specified, the value of the initial temperature at node i has been changed from whatever it was just before the step at $\theta = 0$ to its new, specified value T_i after the step. As mentioned in Section 8•2•2 regarding (8.2.24), this change in the initial condition for a specified-temperature node i is needed so that the solution to the differential equation (8.2.24) for node i will be T_i for all θ. Similar changes to $\mathbf{t}^{(0)}$ must be made for each node whose temperature is specified.

8•2•4   Summary

The discretization process has used N nodal points to approximate the region of interest. Then N energy balances were written, one for each of the N nodal systems. This gave a system of N simultaneous, first-order, ordinary differential equations instead of a single partial differential equation and its boundary conditions. Although not the way we derived the set of equations in the previous two sections, the process of obtaining the system of ordinary differential equations can be considered in three steps:

1. Assuming the entire boundary is adiabatic, write an energy balance for each of the N nodal systems. The resulting system of N ordinary differential equations may be written using matrix notation as

$$\mathbf{C}\dot{\mathbf{t}} + \mathbf{K}\mathbf{t} = \mathbf{g}$$

2. To incorporate specified-heat-flux and/or convective boundaries, modify the system of equations by adding $\mathbf{q}$, $\mathbf{Ht}$ and $\mathbf{h}$ to give

$$\mathbf{C}\dot{\mathbf{t}} + (\mathbf{K} + \mathbf{H})\mathbf{t} = (\mathbf{g} + \mathbf{q} + \mathbf{h}) \tag{8·2·32}$$

Thus far, no modifications for specified temperatures have been made. Upon rearranging this equation,

$$\mathbf{g} + \mathbf{q} + (\mathbf{h} - \mathbf{Ht}) = \mathbf{C}\dot{\mathbf{t}} + \mathbf{K}\mathbf{t} \tag{8·2·33}$$

Vector $\mathbf{g}$ contains the generation rates within each of the N systems. The vector $\mathbf{q}$ contains the specified heat flows into each nodal system from outside the solid. The vector $(\mathbf{h} - \mathbf{Ht})$ contains the rate of convection into each nodal system. The entries in $\mathbf{q}$ and $(\mathbf{h} - \mathbf{Ht})$ are 0 except for rows corresponding to nodes along a boundary with a specified nonzero heat flux or convection. The vector $\mathbf{C}\dot{\mathbf{t}}$ contains the rates of energy storage within each of the N nodal systems. And finally, the vector $\mathbf{K}\mathbf{t}$ contains the net conduction rate out of each of the N systems and into the surrounding nodal systems. Thus, (8·2·33) simply states that the sum of the energy generation plus the specified-heat-flux input and convective inflow to each nodal system must be either stored within the system or conducted out of the system.

3. Modify the system of equations to account for nodes having a specified temperature. For such a node, this was done by disregarding the energy balance for the nodal system and replacing it by a differential equation whose solution is the desired specified temperature when the specified temperature is its initial condition. To maintain symmetry it was then necessary to move some of the terms in other rows of (8·2·32) from the left-hand side of the equation to the right-hand side, but this did not actually change these rows. After these modifications, (8·2·32) and its initial condition may be written as

$$\mathbf{Ç}\dot{\mathbf{t}} + \mathbf{S}\mathbf{t} = \mathbf{r} \qquad\qquad \mathbf{t}(0) = \mathbf{t}^{(0)} \tag{8·2·34}$$

Matrix $\mathbf{Ç} = \mathbf{C}$ unless changes to $\mathbf{C}$ were made for specified temperatures. Matrix $\mathbf{S}$ is $(\mathbf{K} + \mathbf{H})$ modified for specified temperatures. Vector $\mathbf{r}$ is $(\mathbf{g} + \mathbf{q} + \mathbf{h})$ modified for specified temperatures.

This completes the discretization of the spatial problem to replace the governing partial differential equation and its boundary conditions by a system of ordinary differential equations. The solution of (8·2·34) will give $\mathbf{t}(\theta)$, the nodal-point temperatures as functions of time. The steady-state solution of (8·2·34) will be discussed in Section 8•3 and the transient solution will be discussed in sections 8•4 and 8•5.

8•2•5 Heat-flow computation

The computation of heat flows is often of interest to the engineer. This can be done once the nodal temperatures **t** have been found. For interior nodal points (*e.g.*, see Figure 8•7) we can compute conduction heat flow from nodal system i to nodal system ℓ by using the approximate rate equation to give

$$q_{i\ell} = \frac{kA}{d}(t_i - t_\ell) \tag{8.2.35}$$

Once t_i and t_ℓ have been found, $q_{i\ell}$ can be computed without difficulty.

We are more often interested in computing the heat flow into the region from the outside. For an adiabatic boundary this heat flow is obviously 0. For the specified-heat-flux boundary shown in Figure 8•11,

$$q_{si} = q''_{si} S_i \tag{8.2.36}$$

where q''_{si} is the specified heat flux input and S_i is the boundary area of nodal system i that is exposed to the specified heat flux.

For the convective boundary shown in Figure 8•12 we can write the convection rate equation to give

$$q_{ci} = h_{ci} S_i (t_{\infty i} - t_i) \tag{8.2.37}$$

where h_{ci} is the heat-transfer coefficient, S_i is the convection area, and $t_{\infty i}$ is the ambient temperature for nodal system i. Once t_i has been found, q_{ci} can be computed.

For a specified-temperature node we can write an energy balance to find the heat flow into the region. For the specified-temperature node shown in Figure 8•16, the energy balance may be written as

$$g_i + q_{oi} = \frac{dE_i}{d\theta} + q_{ij} + q_{ik}$$

where q_{oi} is the heat flow into the nodal system that we are trying to find. Upon substituting the rate equations we get

$$g''' \frac{3}{8} V + q_{oi} = \rho c \frac{3}{8} V \frac{dt_i}{d\theta} + \frac{kA}{d}(t_i - t_j) + \frac{kA/2}{d}(t_i - t_k)$$

Since $t_i = T_i$ is a constant value, the derivative of t_i with respect to θ is 0. We may then rearrange to solve for q_{oi} to obtain

$$q_{oi} = \frac{kA}{d}(t_i - t_j) + \frac{kA}{2d}(t_i - t_k) - g''' \frac{3}{8} V \tag{8.2.38}$$

Once the temperatures have been obtained the heat flow can be computed.

As discussed above, there are three ways to determine the heat flow into a boundary node depending upon the boundary condition (specified heat flux, convection, or specified temperature). The computation of nodal heat inputs may be unified by using the matrix differential equation (8.2.33) we have derived. As summarized in Section 8•2•4, (8.2.33) resulted from an energy balance on each of the N nodal points except that in this equation the specified-temperature nodes have been assumed to be adiabatic. If we also now let **q** contain the heat flows into specified-temperature nodes that are required to maintain the specified temperature, (8.2.33) can also be used for specified-temperature nodes. Upon rearranging (8.2.33),

$$\mathbf{q} + (\mathbf{h} - \mathbf{Ht}) = \mathbf{C\dot{t}} + \mathbf{Kt} - \mathbf{g}$$

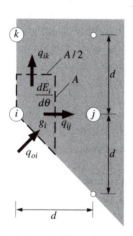

Figure 8•16 *A typical boundary node on an external specified-temperature boundary.*

The left-hand side gives the heat inflow into each of the N nodal systems. The vector $\mathbf{q}$ now contains the specified heat flows (known) plus the heat flows (unknown) required to maintain the specified temperatures. The vector $(\mathbf{h} - \mathbf{Ht})$ contains the convective heat inflows. If we define $\mathbf{q}_o$ to be a vector containing the heat flows into each nodal system from outside the region we may then write

$$\mathbf{q}_o = \mathbf{q} + (\mathbf{h} - \mathbf{Ht}) \tag{8.2.39}$$

or, from the energy balance,

$$\mathbf{q}_o = \mathbf{C\dot{t}} + \mathbf{Kt} - \mathbf{g} \tag{8.2.40}$$

If there are no specified temperatures, all boundary nodes will have either a specified heat flux or convection and $\mathbf{q}_o$ may be found from (8.2.39) once the temperatures have been found since all of the entries in $\mathbf{q}$, $\mathbf{h}$ and $\mathbf{H}$ are known. For an internal node or a node along an adiabatic boundary, the entry in $\mathbf{q}_o$ will be 0. For a node i along a nonadiabatic boundary the entry in $\mathbf{q}_o$ will be $q_i + (h_i - h_{ii}t_i)$ where q_i is from (8.2.16) and h_i and h_{ii} are from (8.2.19).

When there are specified temperatures, (8.2.39) cannot be used since there are unknown heat flows in $\mathbf{q}$ (those required to maintain the specified temperatures). In this case we will use (8.2.40) to find the nodal heat inflows. Equation (8.2.40) can also be used when there are no specified temperatures. All of the entries in $\mathbf{C}$, $\mathbf{K}$ and $\mathbf{g}$ are known and once we have found the temperatures we will know $\mathbf{t}$ and can find $\mathbf{\dot{t}}$ by rearranging (8.2.34) to give

$$\mathbf{\c{C}\dot{t}} = \mathbf{r} - \mathbf{St} \tag{8.2.41}$$

This may be solved for $\mathbf{\dot{t}}$ and substituted into (8.2.40) along with $\mathbf{C}$, $\mathbf{K}$, $\mathbf{t}$ and $\mathbf{g}$ to find $\mathbf{q}_o$. If no changes to $\mathbf{C}$ were made in treating the specified-temperature nodes then $\mathbf{\c{C}} = \mathbf{C}$ and (8.2.41) could be solved for the product $\mathbf{\c{C}\dot{t}}$, rather than $\mathbf{\dot{t}}$, which could then be substituted into (8.2.40) to find $\mathbf{q}_o$. It would not be necessary to actually find $\mathbf{\dot{t}}$.

For steady-state problems the time derivatives are all 0. Thus $\mathbf{\dot{t}} = \mathbf{0}$ and (8.2.40) reduces to

$$\mathbf{q}_o = \mathbf{Kt} - \mathbf{g} \tag{8.2.42}$$

8•3 STEADY-STATE PROBLEMS

In steady-state problems the time derivative in (8.2.34) is $\mathbf{0}$ and the governing system of equations (8.2.34) reduces to

$$\mathbf{St} = \mathbf{r} \tag{8.3.1}$$

This is no longer a system of differential equations, but is now a system of algebraic equations. No attention needs to be paid to the capacitance matrix $\mathbf{C}$ or to the initial condition $\mathbf{t}^{(0)}$ since neither of these plays a role in the steady-state solution.

In Section 8•3•1 we will look at a specific example to see how the governing system of algebraic equations is obtained. Sections 8•3•2 and 8•3•3 will discuss two ways in which these equations may be effectively solved on the computer. Finally, in Section 8•3•4 we will consider a computer program to solve a steady-state problem and look at the numerical results of the example in Section 8•3•1.

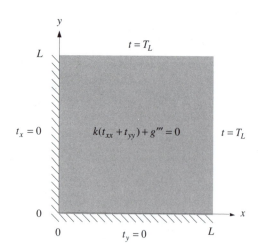

Figure 8•17 *Steady-state energy generation in a square region.*

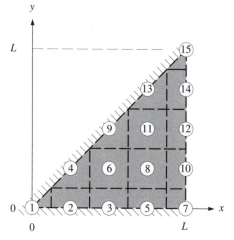

Figure 8•18 *Discretization of region shown in Figure 8•17.*

8•3•1 Discretization example

Let us consider the steady-state generation of energy in the square region shown in Figure 8•17. The exact solution of the governing partial differential equation for this problem can be obtained using the methods of Chapter 3.

In order to reduce the amount of work it is important to take advantage of symmetry whenever possible. The example shown in Figure 8•17 is symmetric about the diagonal line passing through the points (0, 0) and (L, L). Hence this line will be adiabatic and we need only discretize one-half of the region as shown in Figure 8•18. Observe that 15 nodal points (node numbers are encircled) have been used to cover half of the square region whereas 25 nodal points would have been required to cover the entire square using the same nodal-point spacing.

You should also observe that the nodal-point numbers have not been assigned in order in the x-direction. Instead they have been numbered in order following diagonal lines across the region (*e.g.*, see nodes 7, 8 and 9). These diagonal lines traverse the region in the narrowest direction. The reason for numbering the nodes in this way is to reduce the bandwidth of the **K** matrix that will be derived and hence reduce computer storage requirements and the amount of computation required. The maximum difference between adjacent nodal-point numbers is 3 which will mean that **K** will have three off-diagonal rows above the main diagonal and three off-diagonals below the main diagonal. If the points had been numbered horizontally (*e.g.*, 1, 2, 3, 4 and 5 along the x-axis) the maximum difference between adjacent nodal-point numbers would have been 4. This would require an additional off-diagonal row above and below the main diagonal in **K**.

The next step is to construct the finite-difference system of algebraic equations, $\mathbf{Kt} = \mathbf{g}$, for this steady-state example. Since there are 15 nodes, the conduction matrix **K** will be a 15×15, banded, symmetric matrix. The band will consist of the main diagonal and three off-diagonals above the main diagonal and three below the main diagonal. To start, let us zero all of the entries in the band within which we will be interested. We can depict this by writing

$$
\begin{bmatrix}
0 & 0 & 0 & 0 & & & & & & & & & & & \\
0 & 0 & 0 & 0 & 0 & & & & & & & & & & \\
0 & 0 & 0 & 0 & 0 & 0 & & & & & & & & & \\
0 & 0 & 0 & 0 & 0 & 0 & 0 & & & & & & & & \\
 & 0 & 0 & 0 & 0 & 0 & 0 & 0 & & & & & & & \\
 & & 0 & 0 & 0 & 0 & 0 & 0 & 0 & & & & & & \\
 & & & 0 & 0 & 0 & 0 & 0 & 0 & 0 & & & & & \\
 & & & & 0 & 0 & 0 & 0 & 0 & 0 & 0 & & & & \\
 & & & & & 0 & 0 & 0 & 0 & 0 & 0 & 0 & & & \\
 & & & & & & 0 & 0 & 0 & 0 & 0 & 0 & 0 & & \\
 & & & & & & & 0 & 0 & 0 & 0 & 0 & 0 & 0 & \\
 & & & & & & & & 0 & 0 & 0 & 0 & 0 & 0 & 0 \\
 & & & & & & & & & 0 & 0 & 0 & 0 & 0 & 0 \\
 & & & & & & & & & & 0 & 0 & 0 & 0 & 0 \\
 & & & & & & & & & & & 0 & 0 & 0 & 0 \\
\end{bmatrix}
\begin{bmatrix}
t_1 \\ t_2 \\ t_3 \\ t_4 \\ t_5 \\ t_6 \\ t_7 \\ t_8 \\ t_9 \\ t_{10} \\ t_{11} \\ t_{12} \\ t_{13} \\ t_{14} \\ t_{15}
\end{bmatrix}
=
\begin{bmatrix}
0 \\ 0 \\ 0 \\ 0 \\ 0 \\ 0 \\ 0 \\ 0 \\ 0 \\ 0 \\ 0 \\ 0 \\ 0 \\ 0 \\ 0
\end{bmatrix}
$$

(8·3·2)

All of the entries in **K** outside of the band will also be 0, but there will be no need to consider them. Hence only the band is shown to be filled with zeros at this point.

The next step is to determine the entries in row 1 of (8·3·2) by writing an energy balance on the system surrounding node 1. For this system the conduction area between nodes 1 and 2 is $\frac{L}{8}$ (for unit depth in the z-direction), the distance between the nodes is $\frac{L}{4}$ and the volume of the system is $\frac{1}{2}\cdot\frac{L}{8}\cdot\frac{L}{8} = \frac{1}{128}L^2$ (for unit depth in the z-direction). The energy balance (using approximate rate equations) is then given by

$$g'''\frac{L^2}{128} = \frac{k(L/8)}{L/4}(t_1 - t_2)$$

or, after some cancellation and rearrangement,

$$\frac{k}{2}[t_1 - t_2] = \frac{g'''L^2}{128}$$

It is important to note that when you are manipulating finite-difference equations, you should not multiply a single equation by anything because the symmetry of the system of equations will be destroyed. When this equation for node 1 is inserted into row 1 of (8·3·2) we get

$$\frac{k}{2}\begin{bmatrix} 1 & -1 & 0 & 0 & & & & & & & & & & & \\ 0 & 0 & 0 & 0 & 0 & & & & & & & & & & \\ 0 & 0 & 0 & 0 & 0 & 0 & & & & & & & & & \\ 0 & 0 & 0 & 0 & 0 & 0 & 0 & & & & & & & & \\ & 0 & 0 & 0 & 0 & 0 & 0 & 0 & & & & & & & \\ & & 0 & 0 & 0 & 0 & 0 & 0 & 0 & & & & & & \\ & & & 0 & 0 & 0 & 0 & 0 & 0 & 0 & & & & & \\ & & & & 0 & 0 & 0 & 0 & 0 & 0 & 0 & & & & \\ & & & & & 0 & 0 & 0 & 0 & 0 & 0 & 0 & & & \\ & & & & & & 0 & 0 & 0 & 0 & 0 & 0 & 0 & & \\ & & & & & & & 0 & 0 & 0 & 0 & 0 & 0 & 0 & \\ & & & & & & & & 0 & 0 & 0 & 0 & 0 & 0 & 0 \\ & & & & & & & & & 0 & 0 & 0 & 0 & 0 & 0 \\ & & & & & & & & & & 0 & 0 & 0 & 0 & 0 \\ & & & & & & & & & & & 0 & 0 & 0 & 0 \end{bmatrix}\begin{bmatrix} t_1 \\ t_2 \\ t_3 \\ t_4 \\ t_5 \\ t_6 \\ t_7 \\ t_8 \\ t_9 \\ t_{10} \\ t_{11} \\ t_{12} \\ t_{13} \\ t_{14} \\ t_{15} \end{bmatrix} = \frac{g'''L^2}{128}\begin{bmatrix} 1 \\ 0 \\ 0 \\ 0 \\ 0 \\ 0 \\ 0 \\ 0 \\ 0 \\ 0 \\ 0 \\ 0 \\ 0 \\ 0 \\ 0 \end{bmatrix}$$

$$(8\cdot3\cdot3)$$

where $\frac{k}{2}$ and $\frac{1}{128}g'''L^2$ have been factored out for convenience in writing the matrix equation.

The second row is obtained by writing an energy balance on node 2. Nodal-point spacing is still $\frac{L}{4}$. The conduction area between nodes 1 and 2 and between 2 and 3 is $\frac{L}{8}$ (for unit depth) and between nodes 2 and 4 it is $\frac{L}{4}$. The volume of system 2 is $\frac{L}{4}\cdot\frac{L}{8} = \frac{1}{32}L^2$. The energy balance is then given by

$$g'''\frac{L^2}{32} = \frac{k(L/8)}{L/4}(t_2 - t_1) + \frac{k(L/8)}{L/4}(t_2 - t_3) + \frac{k(L/4)}{L/4}(t_2 - t_4)$$

or, after some cancellation and rearrangement,

$$\frac{k}{2}[-t_1 + 4t_2 - t_3 - 2t_4] = \frac{g'''L^2}{128}[4]$$

When this equation for node 2 is inserted into row 2 of (8·3·3) we obtain

$$\frac{k}{2}\begin{bmatrix} 1 & -1 & 0 & 0 & & & & & & & & & & & \\ -1 & 4 & -1 & -2 & 0 & & & & & & & & & & \\ 0 & 0 & 0 & 0 & 0 & 0 & & & & & & & & & \\ 0 & 0 & 0 & 0 & 0 & 0 & 0 & & & & & & & & \\ & 0 & 0 & 0 & 0 & 0 & 0 & 0 & & & & & & & \\ & & 0 & 0 & 0 & 0 & 0 & 0 & 0 & & & & & & \\ & & & 0 & 0 & 0 & 0 & 0 & 0 & 0 & & & & & \\ & & & & 0 & 0 & 0 & 0 & 0 & 0 & 0 & & & & \\ & & & & & 0 & 0 & 0 & 0 & 0 & 0 & 0 & & & \\ & & & & & & 0 & 0 & 0 & 0 & 0 & 0 & 0 & & \\ & & & & & & & 0 & 0 & 0 & 0 & 0 & 0 & 0 & \\ & & & & & & & & 0 & 0 & 0 & 0 & 0 & 0 & 0 \\ & & & & & & & & & 0 & 0 & 0 & 0 & 0 & 0 \\ & & & & & & & & & & 0 & 0 & 0 & 0 & 0 \\ & & & & & & & & & & & 0 & 0 & 0 & 0 \end{bmatrix}\begin{bmatrix} t_1 \\ t_2 \\ t_3 \\ t_4 \\ t_5 \\ t_6 \\ t_7 \\ t_8 \\ t_9 \\ t_{10} \\ t_{11} \\ t_{12} \\ t_{13} \\ t_{14} \\ t_{15} \end{bmatrix} = \frac{g'''L^2}{128}\begin{bmatrix} 1 \\ 4 \\ 0 \\ 0 \\ 0 \\ 0 \\ 0 \\ 0 \\ 0 \\ 0 \\ 0 \\ 0 \\ 0 \\ 0 \\ 0 \end{bmatrix}$$

(8·3·4)

The third row is derived from an energy balance on node 3 which is quite similar to the balance for node 2. The equation is given by

$$\frac{k}{2}[-t_2 + 4t_3 - t_5 - 2t_6] = \frac{g'''L^2}{128}[4]$$

Upon insertion of this equation for node 3 into row 3 of (8·3·4) we get

$$\frac{k}{2}\begin{bmatrix} 1 & -1 & 0 & 0 & & & & & & & & & & & \\ -1 & 4 & -1 & -2 & 0 & & & & & & & & & & \\ 0 & -1 & 4 & 0 & -1 & -2 & & & & & & & & & \\ 0 & 0 & 0 & 0 & 0 & 0 & 0 & & & & & & & & \\ & 0 & 0 & 0 & 0 & 0 & 0 & 0 & & & & & & & \\ & & 0 & 0 & 0 & 0 & 0 & 0 & 0 & & & & & & \\ & & & 0 & 0 & 0 & 0 & 0 & 0 & 0 & & & & & \\ & & & & 0 & 0 & 0 & 0 & 0 & 0 & 0 & & & & \\ & & & & & 0 & 0 & 0 & 0 & 0 & 0 & 0 & & & \\ & & & & & & 0 & 0 & 0 & 0 & 0 & 0 & 0 & & \\ & & & & & & & 0 & 0 & 0 & 0 & 0 & 0 & 0 & \\ & & & & & & & & 0 & 0 & 0 & 0 & 0 & 0 & 0 \\ & & & & & & & & & 0 & 0 & 0 & 0 & 0 & 0 \\ & & & & & & & & & & 0 & 0 & 0 & 0 & 0 \\ & & & & & & & & & & & 0 & 0 & 0 & 0 \end{bmatrix}\begin{bmatrix} t_1 \\ t_2 \\ t_3 \\ t_4 \\ t_5 \\ t_6 \\ t_7 \\ t_8 \\ t_9 \\ t_{10} \\ t_{11} \\ t_{12} \\ t_{13} \\ t_{14} \\ t_{15} \end{bmatrix} = \frac{g'''L^2}{128}\begin{bmatrix} 1 \\ 4 \\ 4 \\ 0 \\ 0 \\ 0 \\ 0 \\ 0 \\ 0 \\ 0 \\ 0 \\ 0 \\ 0 \\ 0 \\ 0 \end{bmatrix}$$

(8·3·5)

Continuing on, we next go to node 4. Nodal-point spacing is still $\frac{L}{4}$. The conduction area between nodes 2 and 4 and between 6 and 4 is $\frac{L}{4}$. The volume of system 4 is $\frac{1}{2} \cdot \frac{L}{4} \cdot \frac{L}{4} = \frac{1}{32}L^2$. The energy balance is given by

$$\frac{g'''L^2}{32} = \frac{k(L/4)}{L/4}(t_4 - t_2) + \frac{k(L/4)}{L/4}(t_4 - t_6)$$

or,

$$\frac{k}{2}[-2t_2 + 4t_4 - 2t_6] = \frac{g'''L^2}{128}[4]$$

Inserting this equation for node 4 into row 4 of (8·3·5) gives

$$
\frac{k}{2}
\begin{bmatrix}
1 & -1 & 0 & 0 & & & & & & & & & & & \\
-1 & 4 & -1 & -2 & 0 & & & & & & & & & & \\
0 & -1 & 4 & 0 & -1 & -2 & & & & & & & & & \\
0 & -2 & 0 & 4 & 0 & -2 & 0 & & & & & & & & \\
& 0 & 0 & 0 & 0 & 0 & 0 & 0 & & & & & & & \\
& & 0 & 0 & 0 & 0 & 0 & 0 & 0 & & & & & & \\
& & & 0 & 0 & 0 & 0 & 0 & 0 & 0 & & & & & \\
& & & & 0 & 0 & 0 & 0 & 0 & 0 & 0 & & & & \\
& & & & & 0 & 0 & 0 & 0 & 0 & 0 & 0 & & & \\
& & & & & & 0 & 0 & 0 & 0 & 0 & 0 & 0 & & \\
& & & & & & & 0 & 0 & 0 & 0 & 0 & 0 & 0 & \\
& & & & & & & & 0 & 0 & 0 & 0 & 0 & 0 & 0 \\
& & & & & & & & & 0 & 0 & 0 & 0 & 0 & 0 \\
& & & & & & & & & & 0 & 0 & 0 & 0 & 0 \\
& & & & & & & & & & & 0 & 0 & 0 & 0 \\
\end{bmatrix}
\begin{bmatrix}
t_1 \\ t_2 \\ t_3 \\ t_4 \\ t_5 \\ t_6 \\ t_7 \\ t_8 \\ t_9 \\ t_{10} \\ t_{11} \\ t_{12} \\ t_{13} \\ t_{14} \\ t_{15}
\end{bmatrix}
= \frac{g'''L^2}{128}
\begin{bmatrix}
1 \\ 4 \\ 4 \\ 4 \\ 0 \\ 0 \\ 0 \\ 0 \\ 0 \\ 0 \\ 0 \\ 0 \\ 0 \\ 0 \\ 0
\end{bmatrix}
\tag{8·3·6}
$$

This process of writing energy balances to find the entries in **K** and **g** is continued until each node has been incorporated. The resulting matrix equation is given by

$$
\frac{k}{2}
\begin{bmatrix}
1 & -1 & 0 & 0 & & & & & & & & & & & \\
-1 & 4 & -1 & -2 & 0 & & & & & & & & & & \\
0 & -1 & 4 & 0 & -1 & -2 & & & & & & & & & \\
0 & -2 & 0 & 4 & 0 & -2 & 0 & & & & & & & & \\
& 0 & -1 & 0 & 4 & 0 & -1 & -2 & & & & & & & \\
& & -2 & -2 & 0 & 8 & 0 & -2 & -2 & & & & & & \\
& & & 0 & -1 & 0 & 2 & 0 & 0 & -1 & & & & & \\
& & & & -2 & -2 & 0 & 8 & 0 & -2 & -2 & & & & \\
& & & & & -2 & 0 & 0 & 4 & 0 & -2 & 0 & & & \\
& & & & & & -1 & -2 & 0 & 4 & 0 & -1 & 0 & & \\
& & & & & & & -2 & -2 & 0 & 8 & -2 & -2 & 0 & \\
& & & & & & & & 0 & -1 & -2 & 4 & 0 & -1 & 0 \\
& & & & & & & & & 0 & -2 & 0 & 4 & -2 & 0 \\
& & & & & & & & & & 0 & -1 & -2 & 4 & -1 \\
& & & & & & & & & & & 0 & 0 & -1 & 1 \\
\end{bmatrix}
\begin{bmatrix}
t_1 \\ t_2 \\ t_3 \\ t_4 \\ t_5 \\ t_6 \\ t_7 \\ t_8 \\ t_9 \\ t_{10} \\ t_{11} \\ t_{12} \\ t_{13} \\ t_{14} \\ t_{15}
\end{bmatrix}
= \frac{g'''L^2}{128}
\begin{bmatrix}
1 \\ 4 \\ 4 \\ 4 \\ 4 \\ 8 \\ 2 \\ 8 \\ 4 \\ 4 \\ 8 \\ 4 \\ 4 \\ 4 \\ 1
\end{bmatrix}
\tag{8·3·7}
$$

The conduction matrix **K** is on the left-hand side and the generation vector **g** appears on the right-hand side. The matrix **K** is symmetric as promised. Also, observe that the sum of the entries in each row of **K** is 0 and, due to the symmetry of **K**, the sum of the entries in each column of **K** is also 0. The sum of the entries in **g** is $\frac{1}{2}g'''L^2$ which is the total generation rate in the region shown in Figure 8•18.

Since the derivation of (8·3·7) assumed that the boundary along $x = L$ was adiabatic, rather than specified temperature, we must modify (8·3·7).

Since nodes 7, 10, 12, 14 and 15 have specified temperatures, we will first transfer all of the off-diagonal entries in columns 7, 10, 12, 14 and 15 to the right-hand side of the matrix equation to obtain

$$
\frac{k}{2}
\begin{bmatrix}
1 & -1 & 0 & 0 & & & & & & & & & & & \\
-1 & 4 & -1 & -2 & 0 & & & & & & & & & & \\
0 & -1 & 4 & 0 & -1 & -2 & & & & & & & & & \\
0 & -2 & 0 & 4 & 0 & -2 & 0 & & & & & & & & \\
 & 0 & -1 & 0 & 4 & 0 & -1 & -2 & & & & & & & \\
 & & -2 & -2 & 0 & 8 & 0 & -2 & -2 & & & & & & \\
 & & & 0 & -1 & 0 & 2 & 0 & 0 & -1 & & & & & \\
 & & & & -2 & -2 & 0 & 8 & 0 & -2 & -2 & & & & \\
 & & & & & -2 & 0 & 0 & 4 & 0 & -2 & 0 & & & \\
 & & & & & & -1 & -2 & 0 & 4 & 0 & -1 & 0 & & \\
 & & & & & & & -2 & -2 & 0 & 8 & -2 & -2 & 0 & \\
 & & & & & & & & 0 & -1 & -2 & 4 & 0 & -1 & 0 \\
 & & & & & & & & & 0 & -2 & 0 & 4 & -2 & 0 \\
 & & & & & & & & & & 0 & -1 & -2 & 4 & -1 \\
 & & & & & & & & & & & 0 & 0 & -1 & 1
\end{bmatrix}
\begin{bmatrix}
t_1 \\ t_2 \\ t_3 \\ t_4 \\ t_5 \\ t_6 \\ t_7 \\ t_8 \\ t_9 \\ t_{10} \\ t_{11} \\ t_{12} \\ t_{13} \\ t_{14} \\ t_{15}
\end{bmatrix}
=
\frac{g'''L^2}{128}
\begin{bmatrix}
1 \\ 4 \\ 4 \\ 4 \\ 4 \\ 8 \\ 2 \\ 8 \\ 4 \\ 4 \\ 8 \\ 4 \\ 4 \\ 4 \\ 1
\end{bmatrix}
+
\frac{k}{2}
\begin{bmatrix}
 \\ \\ \\
0T_7 \\
1T_7 \\
0T_7 \\
1T_{10} \\
0T_7 + 2T_{10} \\
0T_7 + 0T_{10} + 0T_{12} \\
1T_7\qquad\; + 1T_{12} \\
0T_{10} + 2T_{12} + 0T_{14} \\
1T_{10}\qquad\; + 1T_{14} + 0T_{15} \\
0T_{10} + 0T_{12} + 2T_{14} + 0T_{15} \\
1T_{12}\qquad\; + 1T_{15} \\
0T_{12} + 1T_{14}
\end{bmatrix}
\tag{8·3·8}
$$

The shading on the left-hand side of (8·3·8) indicates the terms that are being transferred. The shading on the right-hand side of (8·3·8) indicates the entries where changes have been made. The modifications to the right-hand side are shown in a separate vector but, in practice, the computer would make the numerical changes directly in **g**. Symbols T_7, T_{10}, T_{12}, T_{14} and T_{15} have been used to indicate the changes needed if the specified values of temperature were different at each node.

To complete the transfer, the shaded entries on the left-hand side of (8·3·8) will next be set equal to zero to give

$$
\frac{k}{2}
\begin{bmatrix}
1 & -1 & 0 & 0 & & & & & & & & & & & \\
-1 & 4 & -1 & -2 & 0 & & & & & & & & & & \\
0 & -1 & 4 & 0 & -1 & -2 & & & & & & & & & \\
0 & -2 & 0 & 4 & 0 & -2 & 0 & & & & & & & & \\
 & 0 & -1 & 0 & 4 & 0 & 0 & -2 & & & & & & & \\
 & & -2 & -2 & 0 & 8 & 0 & -2 & -2 & & & & & & \\
 & & & 0 & -1 & 0 & 2 & 0 & 0 & 0 & & & & & \\
 & & & & -2 & -2 & 0 & 8 & 0 & 0 & -2 & & & & \\
 & & & & & -2 & 0 & 0 & 4 & 0 & -2 & 0 & & & \\
 & & & & & & 0 & -2 & 0 & 4 & 0 & 0 & 0 & & \\
 & & & & & & & -2 & -2 & 0 & 8 & 0 & -2 & 0 & \\
 & & & & & & & & 0 & 0 & -2 & 4 & 0 & 0 & 0 \\
 & & & & & & & & & 0 & -2 & 0 & 4 & 0 & 0 \\
 & & & & & & & & & & 0 & 0 & -2 & 4 & 0 \\
 & & & & & & & & & & & 0 & 0 & 0 & 1
\end{bmatrix}
\begin{bmatrix}
t_1 \\ t_2 \\ t_3 \\ t_4 \\ t_5 \\ t_6 \\ t_7 \\ t_8 \\ t_9 \\ t_{10} \\ t_{11} \\ t_{12} \\ t_{13} \\ t_{14} \\ t_{15}
\end{bmatrix}
=
\frac{g'''L^2}{128}
\begin{bmatrix}
1 \\ 4 \\ 4 \\ 4 \\ 4 \\ 8 \\ 2 \\ 8 \\ 4 \\ 4 \\ 8 \\ 4 \\ 4 \\ 4 \\ 1
\end{bmatrix}
+
\frac{k}{2}
\begin{bmatrix}
 \\ \\ \\
0T_7 \\
1T_7 \\
0T_7 \\
1T_{10} \\
0T_7 + 2T_{10} \\
0T_7 + 0T_{10} + 0T_{12} \\
1T_7\qquad\; + 1T_{12} \\
0T_{10} + 2T_{12} + 0T_{14} \\
1T_{10}\qquad\; + 1T_{14} + 0T_{15} \\
0T_{10} + 0T_{12} + 2T_{14} + 0T_{15} \\
1T_{12}\qquad\; + 1T_{15} \\
0T_{12} + 1T_{14}
\end{bmatrix}
$$

where the shading indicates the entries that have been set to zero.

Setting the off-diagonal entries in columns 7, 10, 12, 14 and 15 in **K** equal to zero destroys the symmetry of **K**. To restore the symmetry of **K** we will now set each of the off-diagonal entries in rows 7, 10, 12, 14 and 15 equal to zero to obtain

$$
\frac{k}{2}
\begin{bmatrix}
1 & -1 & 0 & 0 & & & & & & & & & & & \\
-1 & 4 & -1 & -2 & 0 & & & & & & & & & & \\
0 & -1 & 4 & 0 & -1 & -2 & & & & & & & & & \\
0 & -2 & 0 & 4 & 0 & -2 & 0 & & & & & & & & \\
& 0 & -1 & 0 & 4 & 0 & 0 & -2 & & & & & & & \\
& & -2 & -2 & 0 & 8 & 0 & -2 & -2 & & & & & & \\
& & & 0 & 0 & 0 & 2 & 0 & 0 & 0 & & & & & \\
& & & & -2 & -2 & 0 & 8 & 0 & 0 & -2 & & & & \\
& & & & & -2 & 0 & 0 & 4 & 0 & -2 & 0 & & & \\
& & & & & & 0 & 0 & 0 & 4 & 0 & 0 & 0 & & \\
& & & & & & & -2 & -2 & 0 & 8 & 0 & -2 & 0 & \\
& & & & & & & & 0 & 0 & 0 & 4 & 0 & 0 & 0 \\
& & & & & & & & & 0 & -2 & 0 & 4 & 0 & 0 \\
& & & & & & & & & & 0 & 0 & 0 & 4 & 0 \\
& & & & & & & & & & & 0 & 0 & 0 & 1 \\
\end{bmatrix}
\begin{bmatrix} t_1 \\ t_2 \\ t_3 \\ t_4 \\ t_5 \\ t_6 \\ t_7 \\ t_8 \\ t_9 \\ t_{10} \\ t_{11} \\ t_{12} \\ t_{13} \\ t_{14} \\ t_{15} \end{bmatrix}
= \frac{g'''L^2}{128}
\begin{bmatrix} 1 \\ 4 \\ 4 \\ 4 \\ 4 \\ 8 \\ 2 \\ 8 \\ 4 \\ 4 \\ 8 \\ 4 \\ 4 \\ 4 \\ 1 \end{bmatrix}
+ \frac{k}{2}
\begin{bmatrix}
\\
\\
\\
0T_7 \\
1T_7 \\
0T_7 \\
1T_{10} \\
0T_7 + 2T_{10} \\
0T_7 + 0T_{10} + 0T_{12} \\
1T_7 \qquad + 1T_{12} \\
0T_{10} + 2T_{12} + 0T_{14} \\
1T_{10} \qquad + 1T_{14} + 0T_{15} \\
0T_{10} + 0T_{12} + 2T_{14} + 0T_{15} \\
1T_{12} \qquad + 1T_{15} \\
0T_{12} + 1T_{14}
\end{bmatrix}
$$

Finally, modifications will have to be made to the right-hand side in rows 7, 10, 12, 14 and 15 to ensure that the solution for these nodes will be the desired specified temperatures. In each of these specified-temperature rows the energy generation and any other modifications on the right-hand side will be disregarded. The right-hand side of each specified-temperature row i will be set equal to $\kappa_{ii}T_i$ to give

$$
\frac{k}{2}
\begin{bmatrix}
1 & -1 & 0 & 0 & & & & & & & & & & & \\
-1 & 4 & -1 & -2 & 0 & & & & & & & & & & \\
0 & -1 & 4 & 0 & -1 & -2 & & & & & & & & & \\
0 & -2 & 0 & 4 & 0 & -2 & 0 & & & & & & & & \\
& 0 & -1 & 0 & 4 & 0 & 0 & -2 & & & & & & & \\
& & -2 & -2 & 0 & 8 & 0 & -2 & -2 & & & & & & \\
& & & 0 & 0 & 0 & 2 & 0 & 0 & 0 & & & & & \\
& & & & -2 & -2 & 0 & 8 & 0 & 0 & -2 & & & & \\
& & & & & -2 & 0 & 0 & 4 & 0 & -2 & 0 & & & \\
& & & & & & 0 & 0 & 0 & 4 & 0 & 0 & 0 & & \\
& & & & & & & -2 & -2 & 0 & 8 & 0 & -2 & 0 & \\
& & & & & & & & 0 & 0 & 0 & 4 & 0 & 0 & 0 \\
& & & & & & & & & 0 & -2 & 0 & 4 & 0 & 0 \\
& & & & & & & & & & 0 & 0 & 0 & 4 & 0 \\
& & & & & & & & & & & 0 & 0 & 0 & 1 \\
\end{bmatrix}
\begin{bmatrix} t_1 \\ t_2 \\ t_3 \\ t_4 \\ t_5 \\ t_6 \\ t_7 \\ t_8 \\ t_9 \\ t_{10} \\ t_{11} \\ t_{12} \\ t_{13} \\ t_{14} \\ t_{15} \end{bmatrix}
= \frac{g'''L^2}{128}
\begin{bmatrix} 1 \\ 4 \\ 4 \\ 4 \\ 4 \\ 8 \\ 0 \\ 8 \\ 4 \\ 0 \\ 8 \\ 0 \\ 4 \\ 0 \\ 0 \end{bmatrix}
+ \frac{k}{2}
\begin{bmatrix}
\\
\\
\\
0T_7 \\
1T_7 \\
0T_7 \\
2T_7 \\
0T_7 + 2T_{10} \\
0T_7 + 0T_{10} + 0T_{12} \\
4T_{10} \\
0T_{10} + 2T_{12} + 0T_{14} \\
4T_{12} \\
0T_{10} + 0T_{12} + 2T_{14} + 0T_{15} \\
4T_{14} \\
1T_{15}
\end{bmatrix}
\tag{8·3·9}
$$

where the shading indicates the entries that have been changed.

We have now completed the discretization process and have arrived at a system of algebraic equations to solve. The final modified versions of **K** and **g** in (8·3·9) will be called **S** and **r** respectively. Sections 8·3·2 and 8·3·3 discuss computer-oriented techniques for solving such a system of equations. Section 8·3·4 shows how you can obtain numerical results and compares these results to the exact solution of the governing partial differential equation for this problem.

8•3•2 Gaussian elimination

The system of equations (8·3·1) may be written out as

$$
\begin{bmatrix}
s_{11} & s_{12} & s_{13} & s_{14} & \cdots & & s_{1N} \\
s_{21} & s_{22} & s_{23} & s_{24} & & & \\
s_{31} & s_{32} & s_{33} & s_{34} & & & \\
s_{41} & s_{42} & s_{43} & s_{44} & & & \\
\vdots & & & & \ddots & & \vdots \\
s_{N1} & & & \cdots & & & s_{NN}
\end{bmatrix}
\begin{bmatrix}
t_1 \\ t_2 \\ t_3 \\ t_4 \\ \vdots \\ t_N
\end{bmatrix}
=
\begin{bmatrix}
r_1 \\ r_2 \\ r_3 \\ r_4 \\ \vdots \\ r_N
\end{bmatrix}
\tag{8.3.10}
$$

One of the standard procedures for solving such a linear system of algebraic equations is called *gaussian elimination*. In this technique the first equation is used to eliminate all of the entries in the first column below s_{11}. The first step for example would be to multiply the first row in (8·3·10) by s_{21}/s_{11} and then subtract it from the second row thereby eliminating the s_{21} entry in row 2. The right-hand side would also be modified by this process. The first row could then be multiplied by s_{31}/s_{11} and subtracted from the third row to eliminate s_{31} from row 3. This could be continued until all entries in the first column below the main diagonal have been eliminated. Observe that, for a banded system of equations, all of the entries below the band in the first column are already 0 and nothing need be done to them. For example only three entries in the first column of (8·3·9) need be operated on. The modified second row (0 in column 1) can then be used to eliminate all the entries below the main diagonal in the second column in the third through the last row. This elimination procedure is continued until you arrive at a system of equations that looks as follows in matrix form:

$$
\begin{bmatrix}
s_{11} & s_{12} & s_{13} & s_{14} & \cdots & & s_{1N} \\
 & s'_{22} & s'_{23} & s'_{24} & & & \\
 & & s'_{33} & s'_{34} & & & \\
 & & & s'_{44} & & & \\
 & & & & \ddots & & \vdots \\
 & & \mathbf{0} & & & & s'_{NN}
\end{bmatrix}
\begin{bmatrix}
t_1 \\ t_2 \\ t_3 \\ t_4 \\ \vdots \\ t_N
\end{bmatrix}
=
\begin{bmatrix}
r_1 \\ r'_2 \\ r'_3 \\ r'_4 \\ \vdots \\ r'_N
\end{bmatrix}
\tag{8.3.11}
$$

where the primed entries designate entries that have been changed in value by the elimination process. The system of equations is now called *upper triangular*. It can readily be solved by *back substitution*. The last equation can be solved for t_N to give

$$t_N = r'_N / s'_{NN}$$

The expression for t_N can be substituted into the next-to-last equation to find t_{N-1} as

$$t_{N-1} = \frac{(r'_{N-1} - s'_{N-1,N} t_N)}{s'_{N-1,N-1}}$$

This back substitution process can be continued until all of the temperatures back to t_1 have been computed. Gaussian elimination is easy to program on the computer. As an engineer you will undoubtedly be able to find such a computer subroutine already written for you. All that you will have to do is incorporate it into the program you are writing.

Since it is often helpful to study a specific numerical example, let us consider the solution to

$$\begin{bmatrix} 4 & -2 \\ -2 & 10 \end{bmatrix} \begin{bmatrix} t_1 \\ t_2 \end{bmatrix} = \begin{bmatrix} 18 \\ 36 \end{bmatrix}$$

Multiplying the first equation by $\frac{1}{2}$ and adding it to the second equation gives

$$\begin{bmatrix} 4 & -2 \\ 0 & 9 \end{bmatrix} \begin{bmatrix} t_1 \\ t_2 \end{bmatrix} = \begin{bmatrix} 18 \\ 45 \end{bmatrix}$$

The coefficient matrix has now become upper triangular. Next we will divide the last equation by 9 to get

$$\begin{bmatrix} 4 & -2 \\ 0 & 1 \end{bmatrix} \begin{bmatrix} t_1 \\ t_2 \end{bmatrix} = \begin{bmatrix} 18 \\ 5 \end{bmatrix}$$

Multiplying the last equation by 2 and adding it to the first equation gives

$$\begin{bmatrix} 4 & 0 \\ 0 & 1 \end{bmatrix} \begin{bmatrix} t_1 \\ t_2 \end{bmatrix} = \begin{bmatrix} 28 \\ 5 \end{bmatrix}$$

Dividing the first equation by 4 gives

$$\begin{bmatrix} 1 & 0 \\ 0 & 1 \end{bmatrix} \begin{bmatrix} t_1 \\ t_2 \end{bmatrix} = \begin{bmatrix} 7 \\ 5 \end{bmatrix} \tag{8.3.12}$$

The coefficient matrix is now an identity matrix and the vector on the right-hand side now contains the solution.

These operations are equivalent to writing S as

$$S = LU \tag{8.3.13}$$

where L is a *lower-triangular matrix* with ones along the main diagonal and U is an upper-triangular matrix. In this example,

$$S = \begin{bmatrix} 4 & -2 \\ -2 & 10 \end{bmatrix} = \begin{bmatrix} 1.0 & 0.0 \\ -0.5 & 1.0 \end{bmatrix} \begin{bmatrix} 4.0 & -2.0 \\ 0.0 & 9.0 \end{bmatrix} = LU \tag{8.3.14}$$

Observe that the row multiplications or divisions by –0.5, 9.0, 2.0 and 4.0 that were necessary to arrive at (8.3.12) appear in L and U. The expression given by (8.3.13) is called the LU *decomposition* of S. If one knows the decomposition of S, (8.3.1) can be rewritten as

$$LUt = r$$

If we then define $Ut = y$, the solution can be broken down into two parts:

$$Ly = r \tag{8.3.15}$$

which can be solved for y and then

$$Ut = y \tag{8.3.16}$$

can then be solved for **t** In the numerical example we have been considering, (8·3·15) would be

$$\begin{bmatrix} 1.0 & 0.0 \\ -0.5 & 1.0 \end{bmatrix} \begin{bmatrix} y_1 \\ y_2 \end{bmatrix} = \begin{bmatrix} 18 \\ 36 \end{bmatrix}$$

This can be solved by *forward substitution* to give

$$\mathbf{y} = \begin{bmatrix} y_1 \\ y_2 \end{bmatrix} = \begin{bmatrix} 18 \\ 45 \end{bmatrix}$$

Then (8·3·16) becomes

$$\begin{bmatrix} 4.0 & -2.0 \\ 0.0 & 9.0 \end{bmatrix} \begin{bmatrix} t_1 \\ t_2 \end{bmatrix} = \begin{bmatrix} 18 \\ 45 \end{bmatrix}$$

which can be solved by back substitution to give

$$\mathbf{t} = \begin{bmatrix} t_1 \\ t_2 \end{bmatrix} = \begin{bmatrix} 7 \\ 5 \end{bmatrix}$$

The gaussian elimination process has been carried out in two parts. First **S** was decomposed to find **L** and **U** in (8·3·13). Then two triangular systems of equations were solved, (8·3·15) by forward substitution and then (8·3·16) by backward substitution.

It is important to observe that the matrix **S** that arises in finite-difference computations will be symmetric. Hence only the diagonal and the upper triangle entries in **S** need be stored in the computer. In our little example, three storage locations would be required to store the 4, 10 and −2. The **LU** decomposition requires four storage locations because −0.5, 4.0, 9.0 and −2.0 must all be stored. This is as many storage locations as would be required to store the complete original matrix without making use of symmetry. In a large problem this essentially doubles the computer storage requirement.

There is a modification that can be made to the general decomposition of **S** (8·3·13) for symmetric **S** that will save on computer storage space. For symmetric **S** it is possible to write

$$\mathbf{S} = \mathbf{LDL}^T \tag{8·3·17}$$

where **L** is the same **L** as in (8·3·13) and **D** is a diagonal matrix. In our little example,

$$\mathbf{S} = \begin{bmatrix} 4 & -2 \\ -2 & 10 \end{bmatrix} = \begin{bmatrix} 1.0 & 0.0 \\ -0.5 & 1.0 \end{bmatrix} \begin{bmatrix} 4.0 & 0.0 \\ 0.0 & 9.0 \end{bmatrix} \begin{bmatrix} 1.0 & -0.5 \\ 0.0 & 1.0 \end{bmatrix} = \mathbf{LDL}^T$$

Since **L** always will have ones on the main diagonal, the only numbers in this decomposition that must be stored are −0.5, 4.0 and 9.0. They can be stored as

$$\begin{bmatrix} 4.0 & -0.5 \\ 0.0 & 9.0 \end{bmatrix}$$

This is the same computer storage requirement as for the storage of a symmetric **S** (the diagonal and upper triangle). The decomposition given by (8·3·17) can be used to rewrite (8·3·1) as

$$\mathbf{LDL}^T \mathbf{t} = \mathbf{r}$$

By defining $\mathbf{L}^T \mathbf{t} = \mathbf{y}$ and $\mathbf{Dy} = \mathbf{z}$ this problem is equivalent to solving the three simpler problems:

$$\mathbf{L}\mathbf{z} = \mathbf{r} \tag{8.3.18}$$

which can then be solved for $\mathbf{z}$.

$$\mathbf{D}\mathbf{y} = \mathbf{z} \tag{8.3.19}$$

which can in turn be solved for $\mathbf{y}$ and finally

$$\mathbf{L}^T\mathbf{t} = \mathbf{y} \tag{8.3.20}$$

which can be solved for $\mathbf{t}$. Thus the Gaussian elimination process is equivalent to decomposing $\mathbf{S}$ to find $\mathbf{L}$ and $\mathbf{D}$ in (8·3·17) and then solving a set of three backward or forward substitution processes, (8·3·18), (8·3·19) and (8·3·20), to find the final solution.

In still another modification it is possible to rewrite (8·3·17) as

$$\mathbf{S} = \mathbf{L}\mathbf{D}^{1/2}\mathbf{D}^{1/2}\mathbf{L}^T$$

This would run into trouble when $\mathbf{D}$ contained negative numbers which would have to have their square root taken. This should not happen in finite differences. In our little numerical example,

$$\mathbf{S} = \begin{bmatrix} 4 & -2 \\ -2 & 10 \end{bmatrix} = \begin{bmatrix} 1.0 & 0.0 \\ -0.5 & 1.0 \end{bmatrix} \begin{bmatrix} 2 & 0 \\ 0 & 3 \end{bmatrix} \begin{bmatrix} 2 & 0 \\ 0 & 3 \end{bmatrix} \begin{bmatrix} 1.0 & -0.5 \\ 0.0 & 1.0 \end{bmatrix} = (\mathbf{L}\mathbf{D}^{1/2})(\mathbf{D}^{1/2}\mathbf{L}^T)$$

Upon carrying out the first pair of matrix multiplications and the second pair of matrix multiplications as indicated by the parentheses this becomes

$$\mathbf{S} = \begin{bmatrix} 4 & -2 \\ -2 & 10 \end{bmatrix} = \begin{bmatrix} 2 & 0 \\ -1 & 3 \end{bmatrix} \begin{bmatrix} 2 & -1 \\ 0 & 3 \end{bmatrix}$$

Observe that the two matrices on the right-hand side are transposes of each other. Thus we have seen that it is possible to write

$$\mathbf{S} = \mathbf{U}^T\mathbf{U}$$

where $\mathbf{U}$ is an upper-triangular matrix. It is not the same $\mathbf{U}$ that appears in (8·3·13). The direct decomposition of $\mathbf{S}$ into $\mathbf{U}^T\mathbf{U}$ is discussed in the next section.

8•3•3 Cholesky decomposition

The matrix $\mathbf{S}$ that arises in finite-difference computations will always be able to be decomposed as

$$\mathbf{S} = \mathbf{U}^T\mathbf{U} \tag{8·3·21}$$

The method for doing this is called *Cholesky square-root decomposition*.

Let us look at a 3×3 example to get the idea of how $\mathbf{U}$ may be found. We want to write

$$\begin{bmatrix} s_{11} & s_{12} & s_{13} \\ s_{21} & s_{22} & s_{23} \\ s_{31} & s_{32} & s_{33} \end{bmatrix} = \begin{bmatrix} u_{11} & & \\ u_{12} & u_{22} & \\ u_{13} & u_{23} & u_{33} \end{bmatrix}\begin{bmatrix} u_{11} & u_{12} & u_{13} \\ & u_{22} & u_{23} \\ & & u_{33} \end{bmatrix}$$

$$= \begin{bmatrix} u_{11}^2 & u_{11}u_{12} & u_{11}u_{13} \\ u_{11}u_{12} & u_{12}^2 + u_{22}^2 & u_{12}u_{13} + u_{22}u_{23} \\ u_{11}u_{13} & u_{12}u_{13} + u_{22}u_{23} & u_{13}^2 + u_{23}^2 + u_{33}^2 \end{bmatrix}$$

Note that $\mathbf{U}^T\mathbf{U}$ is symmetric. Therefore $\mathbf{S}$ must be symmetric to be able to write $\mathbf{S} = \mathbf{U}^T\mathbf{U}$. By comparing entries in the first row of $\mathbf{S}$ with those in $\mathbf{U}^T\mathbf{U}$ we find that

$$u_{11}^2 = s_{11} \quad \text{so that} \quad u_{11} = \sqrt{s_{11}}$$

$$u_{11}u_{12} = s_{12} \quad \text{so that} \quad u_{12} = s_{12}/u_{11}$$

$$u_{11}u_{13} = s_{13} \quad \text{so that} \quad u_{13} = s_{13}/u_{11}$$

A comparison of the second row (in the upper triangle) in $\mathbf{U}^T\mathbf{U}$ with the second row in $\mathbf{S}$ gives

$$u_{12}^2 + u_{22}^2 = s_{22} \quad \text{so that} \quad u_{22} = \sqrt{s_{22} - u_{12}^2}$$

$$u_{12}u_{13} + u_{22}u_{23} = s_{23} \quad \text{so that} \quad u_{23} = \frac{s_{23} - u_{12}u_{13}}{u_{22}}$$

And finally, from the third row we find

$$u_{13}^2 + u_{23}^2 + u_{33}^2 = s_{33} \quad \text{so that} \quad u_{33} = \sqrt{s_{33} - u_{13}^2 - u_{23}^2}$$

We now have all the u_{ij} and can write $\mathbf{S} = \mathbf{U}^T\mathbf{U}$.

Having now gone through the computation to find $\mathbf{U}$ we can now substitute (8·3·21) into (8·3·1) to obtain

$$\mathbf{U}^T\mathbf{U}\mathbf{t} = \mathbf{r}$$

This can be broken down into two subproblems by letting $\mathbf{U}\mathbf{t} = \mathbf{y}$ to give

$$\mathbf{U}^T\mathbf{y} = \mathbf{r} \tag{8·3·22}$$

and

$$\mathbf{U}\mathbf{t} = \mathbf{y} \tag{8·3·23}$$

Equation (8·3·22) is a triangular system of algebraic equations that can be solved for $\mathbf{y}$ via forward substitution and (8·3·23) is a triangular system of algebraic equations that can be solved for $\mathbf{t}$ via back substitution.

In the numerical example considered in the previous section,

$$\mathbf{S} = \begin{bmatrix} 4 & -2 \\ -2 & 10 \end{bmatrix}$$

To find $\mathbf{U}$ directly we would compute

$$u_{11} = \sqrt{s_{11}} = \sqrt{4} = 2$$

$$u_{12} = s_{12}/u_{11} = -2/2 = -1$$

$$u_{22} = \sqrt{s_{22} - u_{12}^2} = \sqrt{10 - (-1)^2} = 3$$

Thus we have found

$$\mathbf{U} = \begin{bmatrix} 2 & -1 \\ & 3 \end{bmatrix}$$

The subproblem given by (8·3·22) then becomes

$$\begin{bmatrix} 2 & \\ -1 & 3 \end{bmatrix} \begin{bmatrix} y_1 \\ y_2 \end{bmatrix} = \begin{bmatrix} 18 \\ 36 \end{bmatrix}$$

The first row in the forward substitution process gives $2y_1 = 18$ or $y_1 = 9$. Having found y_1, the second row gives $-1(9) + 3y_2 = 36$ or $y_2 = 15$. Thus,

$$\mathbf{y} = \begin{bmatrix} y_1 \\ y_2 \end{bmatrix} = \begin{bmatrix} 9 \\ 15 \end{bmatrix}$$

The subproblem stated by (8·3·23) then becomes

$$\begin{bmatrix} 2 & -1 \\ & 3 \end{bmatrix} \begin{bmatrix} t_1 \\ t_2 \end{bmatrix} = \begin{bmatrix} 9 \\ 15 \end{bmatrix}$$

Solving by back substitution, the last row gives $3t_2 = 15$ or $t_2 = 5$. Having found t_2, the first row then gives $2t_1 - 5 = 9$ or $t_1 = 7$. Thus

$$\mathbf{t} = \begin{bmatrix} t_1 \\ t_2 \end{bmatrix} = \begin{bmatrix} 7 \\ 5 \end{bmatrix}$$

Two-dimensional conduction problems often use hundreds of nodal points to discretize the area. Computer storage can be a problem in such cases. It is important to be economical in the use of the computer. If a problem uses 100 nodes, $\mathbf{S}$ will be a 100×100 matrix and require 10,000 entries to be completely stored. Since $\mathbf{S}$ is symmetric it is only necessary to store the main diagonal and the upper triangle of $\mathbf{S}$. Any time an entry below the main diagonal is needed it can be found from the symmetric entry in the upper triangle. In a 100-node problem this would reduce the computer storage requirement from 10,000 to 5,050 (about a factor of two)! Furthermore, $\mathbf{S}$ will be banded if the nodal points are numbered in a reasonable manner. It will not be necessary to store entries in the upper triangle that fall outside the band. In a 100-node problem it is reasonable to say that there will be, at most, about $\sqrt{100} = 10$ off-diagonal entries in any row of the upper triangle of $\mathbf{S}$. The *bandwidth* of the upper triangle in this case would be 11. Of the 5050 entries in the upper triangle, 4005 of them are outside of the band and need not be stored. By taking advantage of both the symmetry and band structure of $\mathbf{S}$, the storage requirements can be reduced from 10,000 to 1045 (about an order of magnitude).

Engineers are not usually forced to write their own program to solve algebraic equations. Subroutines have already been written that the engineer can simply use. For the finite-difference solution of conduction problems, Cholesky decomposition and solution subroutines that take advantage of the banded structure of $\mathbf{S}$ are quite appropriate. Appendix I•1 contains listings for two such FORTRAN subroutines, *DBAND* and *SBAND*, for solving $\mathbf{Ax} = \mathbf{b}$ for the kind of $\mathbf{A}$ that arise in finite differences.

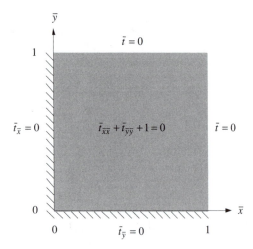

Figure 8•19 *Normalized version of Figure 8•16.*

8•3•4 Numerical results

As a numerical example let us consider the problem discussed in Section 8•3•1 whose finite-difference equations are given by (8·3·9). For numerical computations let us take unit values for k, g''' and L and take the specified nodal temperatures to be each equal to 0. This is equivalent to solving the normalized problem shown in Figure 8•19 where

$$\bar{t} = \frac{t - T_L}{g''' L^2 / k} \quad , \quad \bar{x} = \frac{x}{L} \quad \text{and} \quad \bar{y} = \frac{y}{L}$$

Equation (8.3.9) then becomes

$$\frac{1}{2}
\begin{bmatrix}
1 & -1 & 0 & 0 & & & & & & & & & & & \\
-1 & 4 & -1 & -2 & 0 & & & & & & & & & & \\
0 & -1 & 4 & 0 & -1 & -2 & & & & & & & & & \\
0 & -2 & 0 & 4 & 0 & -2 & 0 & & & & & & & & \\
 & 0 & -1 & 0 & 4 & 0 & 0 & -2 & & & & & & & \\
 & & -2 & -2 & 0 & 8 & 0 & -2 & -2 & & & & & & \\
 & & & 0 & 0 & 0 & 2 & 0 & 0 & 0 & & & & & \\
 & & & & -2 & -2 & 0 & 8 & 0 & 0 & -2 & & & & \\
 & & & & & -2 & 0 & 0 & 4 & 0 & -2 & 0 & & & \\
 & & & & & & 0 & 0 & 0 & 4 & 0 & 0 & 0 & & \\
 & & & & & & & -2 & -2 & 0 & 8 & 0 & -2 & 0 & \\
 & & & & & & & & 0 & 0 & 0 & 4 & 0 & 0 & 0 \\
 & & & & & & & & & 0 & -2 & 0 & 4 & 0 & 0 \\
 & & & & & & & & & & 0 & 0 & 0 & 4 & 0 \\
 & & & & & & & & & & & 0 & 0 & 0 & 1 \\
\end{bmatrix}
\begin{bmatrix} t_1 \\ t_2 \\ t_3 \\ t_4 \\ t_5 \\ t_6 \\ t_7 \\ t_8 \\ t_9 \\ t_{10} \\ t_{11} \\ t_{12} \\ t_{13} \\ t_{14} \\ t_{15} \end{bmatrix}
= \frac{1}{128}
\begin{bmatrix} 1 \\ 4 \\ 4 \\ 4 \\ 4 \\ 8 \\ 0 \\ 8 \\ 4 \\ 0 \\ 8 \\ 0 \\ 4 \\ 0 \\ 0 \end{bmatrix}$$

$$(8\cdot3\cdot24)$$

Because of the symmetry and bandedness of **S**, it will only be necessary to store the main diagonal and upper band of **S**. These are the shaded entries in (8·3·24). This can be done using 15×4 array. We will call this *economy-banded storage* of **S**. We will then write

$$\mathbf{S} = \frac{1}{2}
\begin{bmatrix}
1 & -1 & 0 & 0 \\
4 & -1 & -2 & 0 \\
4 & 0 & -1 & -2 \\
4 & 0 & -2 & 0 \\
4 & 0 & 0 & -2 \\
8 & 0 & -2 & -2 \\
2 & 0 & 0 & 0 \\
8 & 0 & 0 & -2 \\
4 & 0 & -2 & 0 \\
4 & 0 & 0 & 0 \\
8 & 0 & -2 & 0 \\
4 & 0 & 0 & 0 \\
4 & 0 & 0 \\
4 & 0 \\
1
\end{bmatrix}$$

$$(8\cdot3\cdot25)$$

Observe that the main diagonal of (8·3·24) has been stored in the first column of (8·3·25). The first row in the upper band of (8·3·24) has been stored in the first row of (8·3·25). The second row in (8·3·24) has been stored in the second row of (8·3·25) and so forth. There is some storage space in the lower right corner (six entries in this example) of (8·3·25) that do not need to be used. More elaborate storage schemes could be used to save this additional space but these are beyond the scope of our discussion. The entries in row 8 of (8·3·24) are shaded in (8·3·25). The main diagonal entry of row 8 in (8·3·24) is in the first column in row 8 of (8·3·25). The entries in row 8 to the right of the main diagonal in (8·3·24) are in row 8 of (8·3·25). The entries in row 8 to the left of the main diagonal in (8·3·24) are symmetric to the entries in column 8 above the main diagonal in (8·3·24). They can be found in (8·3·25) lying on a 45° line extending upward and to the right from the main diagonal entry as shown in (8·3·25). Thus all of the information in **S** of (8·3·24) can be found in (8·3·25).

In order to solve this system of algebraic equations, we will write a computer program that reads in and stores the entries in **S** (economy-banded storage) and **g** and then calls upon *DBAND* and *SBAND* to obtain the solution. Table 8•1 gives the input data for this example that is required by program *STEADY* that is listed in Table 8•2. The first line of data is a title to identify the problem. The second line gives the number of nodes $N = 15$ and the upper bandwidth *IBW* = 4. Observe that program *STEADY* has been dimensioned so that it can handle up to 50 nodes as long as the upper bandwidth *IBW* does not exceed 10. The DO 2 loop reads in and echo writes the entries in **S** (the next $N = 15$ lines of data with *IBW* = 4 values per line). This is followed by reading and writing the entries in **r** ($N = 15$ values). Program *STEADY* then calls *DBAND* and *SBAND* to solve **St** = **r** for **t**. The second portion of the program computes the boundary heat flows following (8·2·42). The entries in **K** and **g** for this example are shown in (8·3·7). The DO 3 loop reads and echo writes the entries in **K** (the next $N = 15$ lines of data with *IBW* = 4 values per line). This is followed by reading and writing the entries in **g** ($N = 15$ values). The same storage locations have been used for **K** and **g** as were used for **S** and **r** in order to conserve on computer storage space. The DO 4 loop changes **g** to −**g**. The call to subroutine *YAXPB* (described in Appendix I•2) computes

$$\mathbf{q}_o = \mathbf{Kt} + (-\mathbf{g})$$

Finally, the DO 5 loop writes the values of **t** and $\mathbf{q}_o$.

The output from the program is listed in Table 8•3. Observe that the temperatures of nodes 7, 10, 12, 14 and 15 are indeed 0 and that the maximum temperature is at node 1 as expected. The only nonzero boundary heat flows are for nodes 7, 10, 12, 14 and 15. The heat flows for all of the interior nodes (and nodes along a completely adiabatic boundary) are 0 to within the accuracy of the computations (about 3.8×10^{-6} times the smallest heat flow on the boundary). You should also note that the sum of the boundary heat flows is −0.5000 which is equal in magnitude to the energy generation rate within the region.

Table 8•1 *Input data for program STEADY.*

```
STEADY-STATE EXAMPLE, SECTION 8.3
15      4
0.5   -0.5    0.0    0.0
2.0   -0.5   -1.0    0.0
2.0    0.0   -0.5   -1.0
2.0    0.0   -1.0    0.0
2.0    0.0    0.0   -1.0
4.0    0.0   -1.0   -1.0
1.0    0.0    0.0    0.0
4.0    0.0    0.0   -1.0
2.0    0.0   -1.0    0.0
2.0    0.0    0.0    0.0
4.0    0.0   -1.0    0.0
2.0    0.0    0.0    0.0
2.0    0.0    0.0    0.0
2.0    0.0    0.0    0.0
0.5    0.0    0.0    0.0
0.0078125
0.03125
0.03125
0.03125
0.03125
0.06250
0.0
0.06250
0.03125
0.0
0.06250
0.0
0.03125
0.0
0.0
0.5   -0.5    0.0    0.0
2.0   -0.5   -1.0    0.0
2.0    0.0   -0.5   -1.0
2.0    0.0   -1.0    0.0
2.0    0.0   -0.5   -1.0
4.0    0.0   -1.0   -1.0
1.0    0.0    0.0   -0.5
4.0    0.0   -1.0   -1.0
2.0    0.0   -1.0    0.0
2.0    0.0   -0.5    0.0
4.0   -1.0   -1.0    0.0
2.0    0.0   -0.5    0.0
2.0   -1.0    0.0    0.0
2.0   -0.5    0.0    0.0
0.5    0.0    0.0    0.0
0.0078125
0.03125
0.03125
0.03125
0.03125
0.06250
0.015625
0.06250
0.03125
0.03125
0.06250
0.03125
0.03125
0.03125
0.0078125
```

Table 8•2 *Program STEADY for the Solution of Steady-State Finite-Difference Equations.*

```
C**PROGRAM STEADY
      DIMENSION S(50,10),R(50),T(50),Q(50)
      CHARACTER TITLE(20)*4
      DATA IIN/10/,IOUT/11/,NROW/50/,NCOL/10/
      OPEN (UNIT=IIN,FILE='STEADYIN',STATUS='OLD')
      OPEN (UNIT=IOUT,FILE='STEADYOUT',STATUS='UNKNOWN')
      READ (IIN,501,END=999) (TITLE(I),I=1,20)
      WRITE (IOUT,601) (TITLE(I),I=1,20)
      READ (IIN,*) N,IBW
      WRITE (IOUT,602) N,IBW
      WRITE (IOUT,603)
      DO 2 I=1,N
         READ (IIN,*) (S(I,J),J=1,IBW)
         WRITE (IOUT,604) (S(I,J),J=1,IBW)
    2    CONTINUE
      WRITE (IOUT,605)
      READ (IIN,*) (R(I),I=1,N)
      WRITE (IOUT,606) (R(I),I=1,N)
      CALL DBAND(NROW,NCOL,N,IBW,S,NOGO)
         IF (NOGO.EQ.1) GO TO 99
      CALL SBAND(NROW,NCOL,N,IBW,S,R,T)
      WRITE (IOUT,607)
      DO 3 I=1,N
         READ (IIN,*) (S(I,J),J=1,IBW)
         WRITE (IOUT,604) (S(I,J),J=1,IBW)
    3    CONTINUE
      WRITE (IOUT,608)
      READ (IIN,*) (R(I),I=1,N)
      WRITE (IOUT,606) (R(I),I=1,N)
      DO 4 I=1,N
         R(I) = -R(I)
    4    CONTINUE
      CALL YAXPB(NROW,NCOL,N,IBW,S,R,T,Q)
      WRITE (IOUT,609)
      DO 5 I=1,N
         WRITE (IOUT,610) I,T(I),Q(I)
    5    CONTINUE
      GO TO 999
   99 CONTINUE
      WRITE (IOUT,699)
  501 FORMAT (20A4)
  601 FORMAT (' ',20A4)
  602 FORMAT (/' ','N =',I3,5X,'IBW =',I3)
  603 FORMAT (/' ','ENTRIES IN S MATRIX, ECONOMY-BANDED STORAGE')
  604 FORMAT (' ',10E12.4)
  605 FORMAT (/' ','ENTRIES IN R VECTOR')
  606 FORMAT (' ',5E12.4)
  607 FORMAT (/' ','ENTRIES IN K MATRIX, ECONOMY-BANDED STORAGE')
  608 FORMAT (/' ','ENTRIES IN G VECTOR')
  609 FORMAT (/' ','NODAL TEMPERATURE AND HEAT-FLOW RESULTS'/
     1        ' ',5X,'NODE',2X,'TEMPERATURE',2X,'HEAT FLOW')
  610 FORMAT (' ',5X,I3,E13.4,E13.4)
  699 FORMAT (//' ','DBAND FAILED. LOOK FOR ERRORS IN S.')
  999 CONTINUE
      CLOSE (UNIT=IIN,STATUS='KEEP')
      CLOSE (UNIT=IOUT,STATUS='KEEP')
      STOP
      END
```

Table 8•3 *Output from program STEADY.*

```
STEADY-STATE EXAMPLE PROBLEM FOR SECTION 8.3

N = 15    IBW =  4

ENTRIES IN S MATRIX, ECONOMY-BANDED STORAGE
   0.5000E+00  -0.5000E+00   0.0000E+00   0.0000E+00
   0.2000E+01  -0.5000E+00  -0.1000E+01   0.0000E+00
   0.2000E+01   0.0000E+00  -0.5000E+00  -0.1000E+01
   0.2000E+01   0.0000E+00  -0.1000E+01   0.0000E+00
   0.2000E+01   0.0000E+00   0.0000E+00  -0.1000E+01
   0.4000E+01   0.0000E+00  -0.1000E+01  -0.1000E+01
   0.1000E+01   0.0000E+00   0.0000E+00   0.0000E+00
   0.4000E+01   0.0000E+00   0.0000E+00  -0.1000E+01
   0.2000E+01   0.0000E+00  -0.1000E+01   0.0000E+00
   0.2000E+01   0.0000E+00   0.0000E+00   0.0000E+00
   0.4000E+01   0.0000E+00  -0.1000E+01   0.0000E+00
   0.2000E+01   0.0000E+00   0.0000E+00   0.0000E+00
   0.2000E+01   0.0000E+00   0.0000E+00   0.0000E+00
   0.2000E+01   0.0000E+00   0.0000E+00   0.0000E+00
   0.5000E+00   0.0000E+00   0.0000E+00   0.0000E+00


ENTRIES IN R VECTOR
   0.7813E-02   0.3125E-01   0.3125E-01   0.3125E-01   0.3125E-01
   0.6250E-01   0.0000E+00   0.6250E-01   0.3125E-01   0.0000E+00
   0.6250E-01   0.0000E+00   0.3125E-01   0.0000E+00   0.0000E+00


ENTRIES IN K MATRIX, ECONOMY-BANDED STORAGE
   0.5000E+00  -0.5000E+00   0.0000E+00   0.0000E+00
   0.2000E+01  -0.5000E+00  -0.1000E+01   0.0000E+00
   0.2000E+01   0.0000E+00  -0.5000E+00  -0.1000E+01
   0.2000E+01   0.0000E+00  -0.1000E+01   0.0000E+00
   0.2000E+01   0.0000E+00  -0.5000E+00  -0.1000E+01
   0.4000E+01   0.0000E+00  -0.1000E+01  -0.1000E+01
   0.1000E+01   0.0000E+00   0.0000E+00  -0.5000E+00
   0.4000E+01   0.0000E+00  -0.1000E+01  -0.1000E+01
   0.2000E+01   0.0000E+00  -0.1000E+01   0.0000E+00
   0.2000E+01   0.0000E+00  -0.5000E+00   0.0000E+00
   0.4000E+01  -0.1000E+01  -0.1000E+01   0.0000E+00
   0.2000E+01   0.0000E+00  -0.5000E+00   0.0000E+00
   0.2000E+01  -0.1000E+01   0.0000E+00   0.0000E+00
   0.2000E+01  -0.5000E+00   0.0000E+00   0.0000E+00
   0.5000E+00   0.0000E+00   0.0000E+00   0.0000E+00


ENTRIES IN G VECTOR
   0.7813E-02   0.3125E-01   0.3125E-01   0.3125E-01   0.3125E-01
   0.6250E-01   0.1563E-01   0.6250E-01   0.3125E-01   0.3125E-01
   0.6250E-01   0.3125E-01   0.3125E-01   0.3125E-01   0.7813E-02


NODAL TEMPERATURE AND HEAT-FLOW RESULTS
   NODE  TEMPERATURE  HEAT FLOW
      1    0.2911E+00    0.0000E+00
      2    0.2755E+00    0.1490E-07
      3    0.2266E+00   -0.1490E-07
      4    0.2609E+00   -0.2980E-07
      5    0.1381E+00   -0.1490E-07
      6    0.2151E+00   -0.2980E-07
      7    0.0000E+00   -0.8467E-01
      8    0.1317E+00    0.1490E-07
      9    0.1787E+00    0.0000E+00
     10    0.0000E+00   -0.1629E+00
     11    0.1110E+00   -0.1490E-07
     12    0.0000E+00   -0.1422E+00
     13    0.7112E-01    0.0000E+00
     14    0.0000E+00   -0.1024E+00
     15    0.0000E+00   -0.7813E-02
```

The finite-difference solution for temperature is compared to the exact solution of the governing partial differential equation in Table 8•4. The maximum error occurs at node 1 and is 1.22 per cent.

Table 8•4 *Finite-difference temperature solution,* $(t - T_L)/(g'''L^2/k)$, *for 15-node energy-generation example.*

Node	Finite Difference	Exact	Error, Exact – Approx.	Error, % of Node 1
1	0.2911	0.2947	0.0036	1.22
2	0.2755	0.2789	0.0034	1.15
3	0.2266	0.2293	0.0027	0.92
4	0.2609	0.2641	0.0033	1.12
5	0.1381	0.1397	0.0016	0.54
6	0.2151	0.2178	0.0027	0.92
7	0.0000	0.0000	0.0000	0.00
8	0.1317	0.1333	0.0016	0.54
9	0.1787	0.1811	0.0024	0.81
10	0.0000	0.0000	0.0000	0.00
11	0.1110	0.1127	0.0017	0.58
12	0.0000	0.0000	0.0000	0.00
13	0.0711	0.0728	0.0017	0.58
14	0.0000	0.0000	0.0000	0.00
15	0.0000	0.0000	0.0000	0.00

Boundary heat flux (nodal heat flow/nodal system boundary area), rather than nodal heat flow, is tabulated in Table 8•5. The maximum error is the –12.50 per cent that is found for node 15.

Table 8•5 *Finite-difference boundary heat-flux solution,* $q''(L, y)/g'''L$, *for 15-node energy-generation example.*

Node	Finite Difference	Exact	Error, Exact – Approx.	Error, % of $g'''L/2$
7	0.6774	0.6753	–0.0021	–0.42
10	0.6516	0.6488	–0.0028	–0.56
12	0.5688	0.5628	–0.0060	–1.20
14	0.4096	0.3918	–0.0178	–3.56
15	0.0625	0.0000	–0.0625	–12.50

The accuracy of the solution depends upon the number of nodal points used to discretize the region. Figures 8•20 and 8•21 show the increased accuracy that is obtained by going from 3 nodes ($\Delta x = L$) to 6 nodes ($\Delta x = L/2$) to the example we have been considering with 15 nodes ($\Delta x = L/4$). The temperatures shown in Figure 8•20 all fall below the exact solution, but are coming closer as the number of nodes increases. Figure 8•21 shows the improvement in the heat flux as the number of nodes is increased. The steps in the finite-difference heat-flux solution result from assuming that the heat flux is uniform over each nodal boundary area. The area "under" each finite-difference heat flux result is equal to the area "under" the exact curve since both are equal to the energy-generation rate (0.5000) in the region.

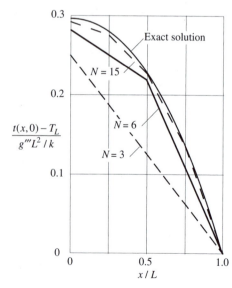

Figure 8•20 *Effect of nodal-point spacing on finite-difference solution along* $y = 0$ *for energy-generation example.*

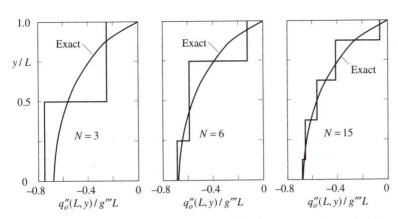

Figure 8•21 *Comparison of finite-difference boundary heat flux to exact solution for energy-generation example.*

The solution of conduction problems via finite differences is relatively straight-forward if the nodal-point spacing is uniform. We have generated the entries in **K**, **g**, **S** and **r** by hand (Table 8•1) and then read them into the computer (Table 8•2) which then called upon subroutines *DBAND* and *SBAND* (Appendix I•1) to solve for the nodal temperatures. Heat flows were then calculated using subroutine *YAXPB* (Appendix I•2). Thus, the computer programming has already been done. You must only provide the necessary matrix entries.

The computation of the entries in **K** and **g** could also be computerized so that you would not have to find them by hand. This is quite a bit of work however (especially for the boundary nodes). Furthermore, one often wants to use irregularly-spaced nodal points which further complicates the finite-difference programming. In such cases it is far easier to use the finite-element method discussed in Chapter 9. Therefore we will leave automating the determination of **K**, **g**, **S** and **r** until Chapter 9.

8•4 TRANSIENT PROBLEMS

In transient problems we must derive and solve a system of ordinary differential equations with an initial condition as given by (8•2•34). In Section 8•4•1 we will look at a specific example to see how to derive the system of finite-difference equations. Sections 8•4•2 and 8•4•3 will each discuss an approximate method of solving the equations. Finally, Section 8•4•4 will look at some numerical results for the example treated in Section 8•4•1.

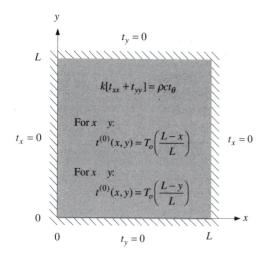

Figure 8•22 *Adiabatic-square example.*

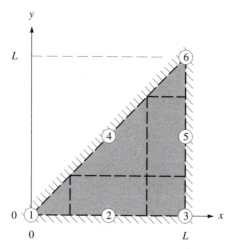

Figure 8•23 *Six-node finite-difference discretization of adiabatic-square example shown in Figure 8•22.*

8•4•1 Discretization examples

As an example of the discretization process for transient problems let us consider Figure 8•22. It is a square region with constant and uniform thermal properties, no energy generation, and completely adiabatic boundaries. The nonuniform initial temperature distribution will change as time goes on and approach a uniform steady-state value.

As usual we want to make as much use as possible of any symmetry that we can. In this example the initial condition is symmetric about the line $y = x$. The boundary conditions are also symmetric about this line. Hence we may discretize as shown in Figure 8•23. Observe that the maximum difference between adjacent nodal-point numbers is two. Hence there will be two diagonals on each side of the main diagonal in the **K** matrix we will derive.

The next step is to construct the system of ordinary differential equations which are given by (8·2·11) as

$$\mathbf{C\dot{t} + Kt = g} \qquad (8\cdot4\cdot1)$$

In this example, **C** and **K** will be 6×6 matrices. **C** will be diagonal as it always is in finite differences. **K** will be symmetric and banded as expected. Let us start by setting all the entries on the diagonal of **C**, within the band in **K**, and in **g** equal to 0. Thus (8·4·1) will begin as

$$
\begin{bmatrix}
0 & & & & & \\
 & 0 & & & & \\
 & & 0 & & & \\
 & & & 0 & & \\
 & & & & 0 & \\
 & & & & & 0
\end{bmatrix}
\begin{bmatrix}
\dot{t}_1 \\ \dot{t}_2 \\ \dot{t}_3 \\ \dot{t}_4 \\ \dot{t}_5 \\ \dot{t}_6
\end{bmatrix}
+
\begin{bmatrix}
0 & 0 & 0 & & & \\
0 & 0 & 0 & 0 & & \\
0 & 0 & 0 & 0 & 0 & \\
 & 0 & 0 & 0 & 0 & 0 \\
 & & 0 & 0 & 0 & 0 \\
 & & & 0 & 0 & 0
\end{bmatrix}
\begin{bmatrix}
t_1 \\ t_2 \\ t_3 \\ t_4 \\ t_5 \\ t_6
\end{bmatrix}
=
\begin{bmatrix}
0 \\ 0 \\ 0 \\ 0 \\ 0 \\ 0
\end{bmatrix}
\qquad (8\cdot4\cdot2)
$$

Now we are ready to write an energy balance on nodal system 1 to determine the entries that will be in row 1 of (8·4·2). For this system the distance between the nodes is $\frac{L}{2}$, the conduction area between nodes 1 and 2 is $\frac{L}{4}$ (for unit depth in the z-direction), and the volume of the system is $\frac{1}{2} \cdot \frac{L}{4} \cdot \frac{L}{4} = \frac{1}{32}L^2$. The energy balance (using approximate rate equations) is then given by

$$\frac{\rho c L^2}{32}\frac{dt_1}{d\theta} + \frac{k(L/4)}{L/2}[t_1 - t_2] = 0$$

When this equation is inserted into row 1 of (8·4·2) we get

$$
\frac{\rho c L^2}{32}
\begin{bmatrix}
1 & & & & & \\
 & 0 & & & & \\
 & & 0 & & & \\
 & & & 0 & & \\
 & & & & 0 & \\
 & & & & & 0
\end{bmatrix}
\begin{bmatrix}
\dot{t}_1 \\ \dot{t}_2 \\ \dot{t}_3 \\ \dot{t}_4 \\ \dot{t}_5 \\ \dot{t}_6
\end{bmatrix}
+
\frac{k}{2}
\begin{bmatrix}
1 & -1 & 0 & & & \\
0 & 0 & 0 & 0 & & \\
0 & 0 & 0 & 0 & 0 & \\
 & 0 & 0 & 0 & 0 & 0 \\
 & & 0 & 0 & 0 & 0 \\
 & & & 0 & 0 & 0
\end{bmatrix}
\begin{bmatrix}
t_1 \\ t_2 \\ t_3 \\ t_4 \\ t_5 \\ t_6
\end{bmatrix}
=
\begin{bmatrix}
0 \\ 0 \\ 0 \\ 0 \\ 0 \\ 0
\end{bmatrix}
$$

$$(8\cdot4\cdot3)$$

Following a similar procedure we may next write an energy balance on node 2 as

$$\frac{\rho c L^2}{8}\frac{dt_2}{d\theta} + \frac{k(L/4)}{L/2}(t_2 - t_1) + \frac{k(L/4)}{L/2}(t_2 - t_3) + \frac{k(L/2)}{L/2}(t_2 - t_4) = 0$$

Upon rearrangement of this equation we get

$$\frac{\rho c L^2}{32}[4]\frac{dt_2}{d\theta} + \frac{k}{2}[-t_1 + 4t_2 - t_3 - 2t_4] = 0$$

Inserting this equation into row 2 of (8·4·3) gives

$$\frac{\rho c L^2}{32}\begin{bmatrix} 1 & & & & & \\ & 4 & & & & \\ & & 0 & & & \\ & & & 0 & & \\ & & & & 0 & \\ & & & & & 0 \end{bmatrix}\begin{bmatrix} \dot{t}_1 \\ \dot{t}_2 \\ \dot{t}_3 \\ \dot{t}_4 \\ \dot{t}_5 \\ \dot{t}_6 \end{bmatrix} + \frac{k}{2}\begin{bmatrix} 1 & -1 & 0 & & & \\ -1 & 4 & -1 & -2 & & \\ 0 & 0 & 0 & 0 & 0 & \\ & 0 & 0 & 0 & 0 & 0 \\ & & 0 & 0 & 0 & 0 \\ & & & 0 & 0 & 0 \end{bmatrix}\begin{bmatrix} t_1 \\ t_2 \\ t_3 \\ t_4 \\ t_5 \\ t_6 \end{bmatrix} = \begin{bmatrix} 0 \\ 0 \\ 0 \\ 0 \\ 0 \\ 0 \end{bmatrix}$$

$$(8·4·4)$$

This process is continued for nodes 3 through 6. The resulting system of equations is given by

$$\frac{\rho c L^2}{32}\begin{bmatrix} 1 & & & & & \\ & 4 & & & & \\ & & 2 & & & \\ & & & 4 & & \\ & & & & 4 & \\ & & & & & 1 \end{bmatrix}\begin{bmatrix} \dot{t}_1 \\ \dot{t}_2 \\ \dot{t}_3 \\ \dot{t}_4 \\ \dot{t}_5 \\ \dot{t}_6 \end{bmatrix} + \frac{k}{2}\begin{bmatrix} 1 & -1 & 0 & & & \\ -1 & 4 & -1 & -2 & & \\ 0 & -1 & 2 & 0 & -1 & \\ & -2 & 0 & 4 & -2 & 0 \\ & & -1 & -2 & 4 & -1 \\ & & & 0 & -1 & 1 \end{bmatrix}\begin{bmatrix} t_1 \\ t_2 \\ t_3 \\ t_4 \\ t_5 \\ t_6 \end{bmatrix} = \begin{bmatrix} 0 \\ 0 \\ 0 \\ 0 \\ 0 \\ 0 \end{bmatrix}$$

$$(8·4·5)$$

In this example there are no specified temperatures. Therefore no modifications will have to be made to (8·4·5). Thus we have $\mathbf{S} = \mathbf{K}$ and $\mathbf{r} = \mathbf{g}$. Equation (8·4·5) is the final system of ordinary differential equations we must solve. Note that the sum of the entries in $\mathbf{C}$ is $\frac{1}{2}\rho c L^2$. Also, the sum of the entries in each row and in each column of $\mathbf{K}$ is 0.

The final step in the discretization process is to determine the initial condition. From the given initial temperature distribution (see Figure 8•22) we see that the initial temperature of node 1 is T_o. Nodes 2 and 4 are initially at $\frac{1}{2}T_o$. Nodes 3, 5, and 6 are initially at 0 temperature. Hence the initial temperature vector $\mathbf{t}^{(0)}$ is given by

$$\mathbf{t}^{(0)} = \begin{bmatrix} t_1 \\ t_2 \\ t_3 \\ t_4 \\ t_5 \\ t_6 \end{bmatrix}^{(0)} = \frac{T_o}{2}\begin{bmatrix} 2 \\ 1 \\ 0 \\ 1 \\ 0 \\ 0 \end{bmatrix} \qquad (8·4·6)$$

Before moving on to the next section to discuss the solution to (8·4·5) it may be instructive to illustrate the construction of the finite-difference equations for the same example, but with a convective boundary along $x = L$ instead of an adiabatic one. In this case the system of equations would have the form given by (8·2·19) as

$$\mathbf{C}\dot{\mathbf{t}} + \mathbf{K}\mathbf{t} + \mathbf{H}\mathbf{t} = \mathbf{g} + \mathbf{h} \qquad (8·4·7)$$

After initializing the matrices and vectors to 0 and writing energy balances on nodes 1 and 2, (8·4·7) is given by

$$\frac{\rho c L^2}{32}\begin{bmatrix}1&&&&&\\&4&&&&\\&&0&&&\\&&&0&&\\&&&&0&\\&&&&&0\end{bmatrix}\begin{bmatrix}\dot t_1\\\dot t_2\\\dot t_3\\\dot t_4\\\dot t_5\\\dot t_6\end{bmatrix}+\frac{k}{2}\begin{bmatrix}1&-1&0&&&\\-1&4&-1&-2&&\\0&0&0&0&0&\\&0&0&0&0&0\\&&0&0&0&0\\&&&0&0&0\end{bmatrix}\begin{bmatrix}t_1\\t_2\\t_3\\t_4\\t_5\\t_6\end{bmatrix}+\begin{bmatrix}0&&&&&\\&0&&&&\\&&0&&&\\&&&0&&\\&&&&0&\\&&&&&0\end{bmatrix}\begin{bmatrix}t_1\\t_2\\t_3\\t_4\\t_5\\t_6\end{bmatrix}=\begin{bmatrix}0\\0\\0\\0\\0\\0\end{bmatrix}+\begin{bmatrix}0\\0\\0\\0\\0\\0\end{bmatrix}$$

$$(8\cdot4\cdot8)$$

Observe that this is essentially the same as (8·4·4) since neither node 1 nor node 2 is on a convection boundary. Notice also that we are expecting **H** to be diagonal since it always is in finite differences.

Now let us write an energy balance on node 3 which is on a convection boundary. The conduction area between nodes 2 and 3 and between nodes 5 and 3 is $\frac{L}{4}$. The distance between nodal points is still $\frac{L}{2}$. The volume of nodal system 3 is $\frac{1}{16}L^2$. The convective heat-transfer area is $\frac{L}{4}$ (for unit depth in the z-direction). The energy balance then becomes

$$h\frac{L}{4}(t_\infty-t_3)=\frac{\rho c L^2}{16}\frac{dt_3}{d\theta}+\frac{k(L/4)}{L/2}(t_3-t_2)+\frac{k(L/4)}{L/2}(t_3-t_5)$$

After rearranging somewhat,

$$\frac{\rho c L^2}{32}[2]\frac{dt_3}{d\theta}+\frac{k}{2}[-t_2+2t_3-t_5]+\frac{hL}{4}t_3=\frac{hL}{4}t_\infty$$

Inserting this equation into row 3 of (8·4·8) as well as the equations for nodes 4, 5, and 6 gives

$$\frac{\rho c L^2}{32}\begin{bmatrix}1&&&&&\\&4&&&&\\&&2&&&\\&&&4&&\\&&&&4&\\&&&&&1\end{bmatrix}\begin{bmatrix}\dot t_1\\\dot t_2\\\dot t_3\\\dot t_4\\\dot t_5\\\dot t_6\end{bmatrix}+\frac{k}{2}\begin{bmatrix}1&-1&0&&&\\-1&4&-1&-2&&\\0&-1&2&0&-1&\\&-2&0&4&-2&0\\&&-1&-2&4&-1\\&&&0&-1&1\end{bmatrix}\begin{bmatrix}t_1\\t_2\\t_3\\t_4\\t_5\\t_6\end{bmatrix}+\frac{hL}{4}\begin{bmatrix}0&&&&&\\&0&&&&\\&&1&&&\\&&&0&&\\&&&&2&\\&&&&&1\end{bmatrix}\begin{bmatrix}t_1\\t_2\\t_3\\t_4\\t_5\\t_6\end{bmatrix}=\begin{bmatrix}0\\0\\0\\0\\0\\0\end{bmatrix}+\frac{hLt_\infty}{4}\begin{bmatrix}0\\0\\1\\0\\2\\1\end{bmatrix}$$

$$(8\cdot4\cdot9)$$

The only nonzero entries in **H** and **h** are in rows 3, 5, and 6 corresponding to nodes on the convective boundary. To program this for the computer, **H** and **h** need not be stored. Their entries could be immediately added to **K** and **g** respectively. However, if you want to compute the boundary heat flows from (8·2·40), you would need to save **K** and **g** separately. This means you would also need to reserve separate storage for **S** (= **K**+**H** plus changes for specified temperatures) and **r** (= **g**+**h**+**q** plus changes for specified temperatures). Since there are no specified temperatures in this example, no further changes need to be made and we now have found

$$\text{Ç}\dot{\mathbf{t}}+\mathbf{St}=\mathbf{r} \qquad\qquad (8\cdot4\cdot10)$$

where Ç = **C**, **S** = **K**+**H** and **r** = **g**+**h** in this example can be found from (8·4·9). In the previous example (completely adiabatic boundaries) Ç = **C**, **S** = **K** and **r** = **g** can be found from (8·4·5).

Although it is possible to obtain the exact solution to (8·4·10), in more-realistic conduction problems having many more nodes this is not yet practical. We will instead discuss the Euler method and the Crank-Nicolson method for obtaining approximate computer solutions of (8·4·10). The exact solution will be discussed later in Section 8•5•1.

8•4•2 Euler approximate solution

The Euler method of solution starts with the given initial condition and moves ahead one step at a time according to the relation

$$\mathbf{t}^{(v+1)} = \mathbf{t}^{(v)} + \dot{\mathbf{t}}^{(v)}\Delta\theta \qquad (8.4.11)$$

This relation says that the temperatures $\mathbf{t}^{(v+1)}$ at the next time step $\theta^{(v+1)}$ are found by starting with the temperatures $\mathbf{t}^{(v)}$ at time $\theta^{(v)}$ and moving ahead in time using the time derivatives $\dot{\mathbf{t}}^{(v)}$ calculated at time $\theta^{(v)}$. This is depicted in Figure 8•24. The slope decreases along the exact path whereas the slope is constant along the Euler path. The time step size, $\Delta\theta = \theta^{(v+1)} - \theta^{(v)}$, is chosen by you (the person carrying out the solution). The bold dot shows $\mathbf{t}^{(v+1)}$ is quite a bit above the value on the exact path at $\theta^{(v+1)}$.

The time derivative in (8.4.11) may be evaluated by rewriting (8.2.34) as

$$\mathbf{\c{C}}\dot{\mathbf{t}} = -\mathbf{St} + \mathbf{r}$$

which at time $\theta^{(v)}$ gives

$$\mathbf{\c{C}}\dot{\mathbf{t}}^{(v)} = -\mathbf{St}^{(v)} + \mathbf{r} \qquad (8.4.12)$$

This is a system of algebraic equations that could be solved for $\dot{\mathbf{t}}^{(v)}$ since $\mathbf{t}^{(v)}$ will be known from either the initial condition (if this is the first step) or from previous calculations (if it is a later step). Rather than actually doing this, let us premultiply (8.4.11) by $\mathbf{\c{C}}$ to get

$$\mathbf{\c{C}}\mathbf{t}^{(v+1)} = \mathbf{\c{C}}\mathbf{t}^{(v)} + \mathbf{\c{C}}\dot{\mathbf{t}}^{(v)}\Delta\theta \qquad (8.4.13)$$

Substitution of (8.4.12) into (8.4.13) will eliminate the time derivative and give

$$\mathbf{\c{C}}\mathbf{t}^{(v+1)} = \mathbf{\c{C}}\mathbf{t}^{(v)} + [-\mathbf{St}^{(v)} + \mathbf{r}]\Delta\theta$$

A slight rearrangement to combine the terms containing $\mathbf{t}^{(v)}$ will give

$$\mathbf{\c{C}}\mathbf{t}^{(v+1)} = (\mathbf{\c{C}} - \mathbf{S}\Delta\theta)\mathbf{t}^{(v)} + \mathbf{r}\Delta\theta \qquad (8.4.14)$$

This result is the basic equation used in carrying out the Euler method of approximating the solution to (8.2.34). At each step in time the solution to find $\mathbf{t}^{(v+1)}$ will consist of two procedures. First, $\mathbf{t}^{(v)}$ is substituted into the right-hand side and the matrix operations are carried out using subroutine *YAXPB* to obtain

$$\mathbf{y}^{(v)} = (\mathbf{\c{C}} - \mathbf{S}\Delta\theta)\mathbf{t}^{(v)} + \mathbf{r}\Delta\theta \qquad (8.4.15)$$

where $\mathbf{A} = \mathbf{\c{C}} - \mathbf{S}\Delta\theta$ and $\mathbf{b} = \mathbf{r}\Delta\theta$ when *YAXPB* is called. Then

$$\mathbf{\c{C}}\mathbf{t}^{(v+1)} = \mathbf{y}^{(v)} \qquad (8.4.16)$$

must be solved for $\mathbf{t}^{(v+1)}$. Since $\mathbf{\c{C}}$ is diagonal, the left-hand side of (8.4.16) it is particularly simple. Equation (8.4.16) can be written out in expanded form as

$$\begin{bmatrix} \varsigma_{11}t_1 \\ \varsigma_{22}t_2 \\ \vdots \\ \varsigma_{NN}t_N \end{bmatrix}^{(v+1)} = \begin{bmatrix} y_1 \\ y_2 \\ \vdots \\ y_N \end{bmatrix}^{(v)} \qquad (8.4.17)$$

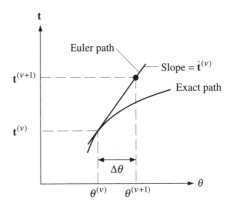

Figure 8•24 *Euler method of moving ahead in time.*

This may be solved *explicitly*[1] for the temperatures $\mathbf{t}^{(v+1)}$ as

$$
\begin{bmatrix} t_1 \\ t_2 \\ \vdots \\ t_N \end{bmatrix}^{(v+1)} = \begin{bmatrix} y_1/\varsigma_{11} \\ y_2/\varsigma_{22} \\ \vdots \\ y_N/\varsigma_{NN} \end{bmatrix}^{(v)}
\tag{8.4.18}
$$

A main program for carrying out the Euler solution is listed in Table 8•6. The first step in the main program is to call on subroutine *INPUT*, Table 8•7, to read in the necessary information [$\mathbf{\c{C}}$, $\mathbf{S}$, $\mathbf{r}$, $\Delta\theta$ and $\mathbf{t}^{(0)}$] and echo write the input data. Some additional data [title, N, IBW, initial time and final time-step number] are also read in. The DO 2 loop in the main program creates $\mathbf{r}\Delta\theta$ and $(\mathbf{\c{C}} - \mathbf{S}\Delta\theta)$ that appear in (8.4.14). Note that $\mathbf{r}\Delta\theta$ is stored in $\mathbf{r}$ and $(\mathbf{\c{C}} - \mathbf{S}\Delta\theta)$ is stored in $\mathbf{S}$ to conserve computer space. The step-by-step solution is then carried out in the DO 4 loop. The call to *YAXPB* in the DO 4 loop carries out the operations shown in (8.4.15) to find $\mathbf{y}^{(v)}$. The DO 3 loop solves for the temperatures as shown in (8.4.18). Some numerical examples will be given in Section 8•4•4.

Table 8•6 *Program EULER for the Euler solution of* $\mathbf{\c{C}}\dot{\mathbf{t}} + \mathbf{S}\mathbf{t} = \mathbf{r}$.

```
C**PROGRAM EULER
      COMMON C(50),S(50,10),R(50),T(50),Y(50)
      COMMON N,IBW,TIME,DTIME,NUMAX
      DATA IIN/10/,IOUT/11/,NROW/50/,NCOL/10/
      OPEN (UNIT=IIN,FILE='EULERIN',STATUS='OLD')
      OPEN (UNIT=IOUT,FILE='EULEROUT',STATUS='UNKNOWN')
      CALL INPUT
      WRITE (IOUT,601)
      DO 2 I=1,N
        R(I) = R(I)*DTIME
        DO 1 J=1,IBW
          S(I,J) = -S(I,J)*DTIME
1         CONTINUE
        S(I,1) = C(I) + S(I,1)
2       CONTINUE
      WRITE (IOUT,602)
      DO 4 NU=1,NUMAX
        TIME = TIME + DTIME
        WRITE (IOUT,603) NU,TIME
        CALL YAXPB(NROW,NCOL,N,IBW,S,R,T,Y)
        DO 3 I=1,N
          T(I) = Y(I)/C(I)
3         CONTINUE
        WRITE (IOUT,604) (T(I),I=1,N)
4       CONTINUE
601   FORMAT (/' ','EULER APPROXIMATION WILL BE USED')
602   FORMAT (/' ','TRANSIENT TEMPERATURE VARIATION')
603   FORMAT (/' ','NU =',I5,5X,'TIME =',E12.4)
604   FORMAT (' ',10E12.4)
      CLOSE (UNIT=IIN,STATUS='KEEP')
      CLOSE (UNIT=IOUT,STATUS='KEEP')
      STOP
      END
```

---

[1] A method is said to be *explicit* when the unknowns can be solved for directly without having to simultaneously solve a system of algebraic equations.

Table 8•7 *Subroutine INPUT called by program EULER (Table 8•6).*

```
      SUBROUTINE INPUT
      COMMON C(50),S(50,10),R(50),T(50),Y(50)
      COMMON N,IBW,TIME,DTIME,NUMAX
      CHARACTER TITLE(20)*4
      DATA IIN/10/,IOUT/11/
      READ (IIN,501,END=999) (TITLE(I),I=1,20)
      WRITE (IOUT,601) (TITLE(I),I=1,20)
      READ (IIN,*) N,IBW
      WRITE (IOUT,602) N,IBW
      WRITE (IOUT,603)
      READ (IIN,*) (C(I),I=1,N)
      WRITE (IOUT,604) (C(I),I=1,N)
      WRITE (IOUT,605)
      DO 1 I=1,N
         READ (IIN,*) (S(I,J),J=1,IBW)
         WRITE (IOUT,604) (S(I,J),J=1,IBW)
    1    CONTINUE
      WRITE (IOUT,606)
      READ (IIN,*) (R(I),I=1,N)
      WRITE (IOUT,604) (R(I),I=1,N)
      READ (IIN,*) TIME,DTIME,NUMAX
      WRITE (IOUT,607) TIME,DTIME,NUMAX
      WRITE (IOUT,608)
      READ (IIN,*) (T(I),I=1,N)
      WRITE (IOUT,604) (T(I),I=1,N)
  501 FORMAT (20A4)
  601 FORMAT (' ',20A4)
  602 FORMAT (/' ','N =',I3,5X,'IBW =',I3)
  603 FORMAT (/' ','MAIN-DIAGONAL ENTRIES IN C MATRIX')
  604 FORMAT (' ',10E12.4)
  605 FORMAT (/' ','ENTRIES IN S MATRIX, ECONOMY-BANDED STORAGE')
  606 FORMAT (/' ','ENTRIES IN R VECTOR')
  607 FORMAT (/' ','TIME =',E11.4,5X,'DTIME =',E11.4,5X,'NUMAX =',I4)
  608 FORMAT (/' ','INITIAL TEMPERATURE DISTRIBUTION')
  999 CONTINUE
      RETURN
      END
```

As the Euler procedure (8·4·14) continues (*i.e.*, as v increases) a steady-state solution $\mathbf{t}^{(\infty)}$ should be approached. Taking one more time step should also give $\mathbf{t}^{(\infty)}$. For the steady-state condition[1] we can substitute $\mathbf{t}^{(\infty)}$ for both $\mathbf{t}^{(v+1)}$ and $\mathbf{t}^{(v)}$ in (8·4·14) to obtain

$$\mathbf{Ç}\mathbf{t}^{(\infty)} = (\mathbf{Ç} - \mathbf{S}\Delta\theta)\mathbf{t}^{(\infty)} + \mathbf{r}\Delta\theta$$

The terms involving $\mathbf{Ç}$ cancel and we may divide by $\Delta\theta$ and rearrange to obtain

$$\mathbf{S}\mathbf{t}^{(\infty)} = \mathbf{r}$$

Comparison of this result with (8·3·1) shows that $\mathbf{t}^{(\infty)}$ will be the correct steady-state solution of the problem. Since $\Delta\theta$ has canceled, $\mathbf{t}^{(\infty)}$ will be independent of the size of the time step.

---

[1] As will be shown later, if $\Delta\theta$ is too large, unstable, numerically induced oscillations will appear in the Euler solution and a steady-state condition will never be approached.

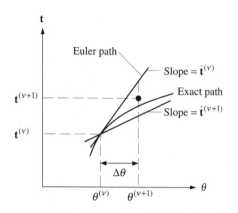

Figure 8•25 *Crank-Nicolson method of moving ahead in time.*

8•4•3 Crank-Nicolson approximate solution

Figure 8•25 shows the slopes at $\theta^{(v)}$ and at $\theta^{(v+1)}$. If you move ahead using $\dot{\mathbf{t}}^{(v)}$ then $\mathbf{t}^{(v+1)}$ is too high. If you move ahead using $\dot{\mathbf{t}}^{(v+1)}$ then $\mathbf{t}^{(v+1)}$ is too low. The Crank-Nicolson method uses the arithmetic average of these two time derivatives. Since this is a better approximation to the time derivative, it gives an improved solution. The bold dot is closer to the exact path in Figure 8•25 than it was in Figure 8•24. The Crank-Nicolson solution moves ahead in time using the equation

$$\mathbf{t}^{(v+1)} = \mathbf{t}^{(v)} + \left[\frac{\dot{\mathbf{t}}^{(v)} + \dot{\mathbf{t}}^{(v+1)}}{2} \right] \Delta\theta \tag{8·4·19}$$

Premultiplication of this equation by $\mathbf{\zeta}$ gives

$$\mathbf{\zeta t}^{(v+1)} = \mathbf{\zeta t}^{(v)} + \left[\mathbf{\zeta \dot{t}}^{(v)} + \mathbf{\zeta \dot{t}}^{(v+1)} \right] \frac{\Delta\theta}{2}$$

The time derivatives may be eliminated by substituting (8·4·12) for $\mathbf{\zeta \dot{t}}^{(v)}$ and a similar expression for $\mathbf{\zeta \dot{t}}^{(v+1)}$ to give

$$\mathbf{\zeta t}^{(v+1)} = \mathbf{\zeta t}^{(v)} + \left[(-\mathbf{S t}^{(v)} + \mathbf{r}) + (-\mathbf{S t}^{(v+1)} + \mathbf{r}) \right] \frac{\Delta\theta}{2}$$

The unknown $\mathbf{t}^{(v+1)}$ that appears on the right-hand side of this equation can be moved to the left-hand side to give, after some slight rearrangement of terms,

$$\left[\mathbf{\zeta} + \mathbf{S}\frac{\Delta\theta}{2} \right] \mathbf{t}^{(v+1)} = \left[\mathbf{\zeta} - \mathbf{S}\frac{\Delta\theta}{2} \right] \mathbf{t}^{(v)} + \mathbf{r}\Delta\theta \tag{8·4·20}$$

As in the Euler method, $\mathbf{t}^{(v)}$ will be known from previous computation (or from the initial condition if this is the first step in time) so the right-hand-side matrix operations may be carried out. The right-hand-side operations are very similar to those required by the Euler method in (8·4·14). Once the right-hand side has been computed, however, the computation of $\mathbf{t}^{(v+1)}$ is considerably more complicated in the Crank-Nicolson method than in the Euler method because the left-hand matrix $\mathbf{\zeta} + \mathbf{S}\frac{\Delta\theta}{2}$ is not diagonal as it was for the Euler method. Thus we will have a system of algebraic equations to solve at every step in time. This is called an *implicit* solution.

We may simplify (8·4·20) by defining

$$\mathbf{A} = \mathbf{\zeta} + \mathbf{S}\frac{\Delta\theta}{2} \tag{8·4·21}$$

$$\mathbf{B} = \mathbf{\zeta} - \mathbf{S}\frac{\Delta\theta}{2} \tag{8·4·22}$$

and

$$\mathbf{b} = \mathbf{r}\Delta\theta \tag{8·4·23}$$

then (8·4·20) becomes

$$\mathbf{A t}^{(v+1)} = \mathbf{B t}^{(v)} + \mathbf{b} \tag{8·4·24}$$

Since $\mathbf{A}$, $\mathbf{B}$ and $\mathbf{b}$ are constants, they need to be computed only once, at the beginning of the problem. The first step of the solution is to update the right-hand side of (8·4·24) using subroutine *YAXPB* from Appendix I•2 to give

$$\mathbf{y}^{(v)} = \mathbf{B t}^{(v)} + \mathbf{b} \tag{8·4·25}$$

The second step is to solve

$$At^{(v+1)} = y^{(v)} \tag{8.4.26}$$

This can be done using subroutines *DBAND* and *SBAND* described in Appendix I•1. Since **A** is a constant, its decomposition with *DBAND* needs to be done only once and then saved. From then on only *SBAND* needs to be used to solve (8·4·26).

A main program for carrying out the Crank-Nicolson solution is listed in Table 8•8. In the Crank-Nicolson program the dimensioning for Ç is C(50,10) whereas in the Euler program we had C(50). The reason for this is that the Crank-Nicolson method requires enough storage for **A** and **B** in (8·4·24). If we had unlimited storage space we could store Ç, **S**, **A** and **B** separately by dimensioning C(50), S(50,10), A(50,10) and B(50,10). This provides storage for 1550 entries. By only dimensioning C(50,10) and S(50,10) and then storing **A** and **B** by over-writing them into Ç and **S** which are no longer needed, the storage requirement has been reduced to 1000 entries, a saving of 550 entries. Since storage is often critical, it is important not to be too extravagant. Double up on storage when possible.

Table 8•8 *Program CRANK for Crank-Nicolson solution of* $\dot{\text{Ç}} + \text{St} = \text{r}$.

```
C**PROGRAM CRANK
      COMMON C(50,10),S(50,10),R(50),T(50),Y(50)
      COMMON N,IBW,TIME,DTIME,NUMAX
      DATA IIN/10/,IOUT/11/,NROW/50/,NCOL/10/
      OPEN (UNIT=IIN,FILE='CRANKIN',STATUS='OLD')
      OPEN (UNIT=IOUT,FILE='CRANKOUT',STATUS='UNKNOWN')
      CALL INPUT
      WRITE (IOUT,601)
      DT2 = DTIME/2.0
      DO 2 I=1,N
         R(I) = R(I)*DTIME
         SAVE = C(I,1)
         DO 1 J=1,IBW
            C(I,J) = S(I,J)*DT2
            S(I,J) = -C(I,J)
1           CONTINUE
         C(I,1) = SAVE + C(I,1)
         S(I,1) = SAVE + S(I,1)
2     CONTINUE
      CALL DBAND(NROW,NCOL,N,IBW,C,NOGO)
      IF (NOGO.EQ.1) GO TO 99
      WRITE (IOUT,602)
      DO 3 NU=1,NUMAX
         TIME = TIME + DTIME
         WRITE (IOUT,603) NU,TIME
         CALL YAXPB(NROW,NCOL,N,IBW,S,R,T,Y)
         CALL SBAND(NROW,NCOL,N,IBW,C,Y,T)
         WRITE (IOUT,604) (T(I),I=1,N)
3     CONTINUE
      GO TO 999
99    CONTINUE
      WRITE (IOUT,605)
601   FORMAT (/' ','CRANK-NICOLSON APPROXIMATION WILL BE USED')
602   FORMAT (/' ','TRANSIENT TEMPERATURE VARIATION')
603   FORMAT (/' ','NU =',I5,5X,'TIME =',E12.4)
604   FORMAT (' ',10E12.4)
605   FORMAT (//' ','DBAND FAILED. LOOK FOR INPUT ERRORS.')
999   CONTINUE
      CLOSE (UNIT=IIN,STATUS='KEEP')
      CLOSE (UNIT=IOUT,STATUS='KEEP')
      STOP
      END
```

The first step in the main program is to call subroutine *INPUT* listed in Table 8•9 to read in and echo write all the necessary input data [$Ç$, S, r, $\Delta\theta$ and $t^{(0)}$]. The only difference between the Crank-Nicolson input subroutine in Table 8•9 and the Euler input subroutine in Table 8•6 is the storage of $Ç$. The creation of A, B and b and their subsequent storage in $C(50,10)$, $S(50,10)$ and $R(50)$ is carried out in the DO 2 loop. The decomposition of A is then carried out by *DBAND*. After calling *DBAND*, $Ç$ will contain the decomposition of A. The step-by-step transient solution is carried out in the DO 3 loop. The call to *YAXPB* finds $y^{(v)}$ in (8·4·25). The call to *SBAND* solves (8·4·26) for $t^{(v+1)}$.

Table 8•9 *Subroutine INPUT called by program CRANK (Table 8•8).*

```
SUBROUTINE INPUT
COMMON C(50,10),S(50,10),R(50),T(50),Y(50)
COMMON N,IBW,TIME,DTIME,NUMAX
CHARACTER TITLE(20)*4
DATA IIN/10/,IOUT/11/
READ (IIN,501,END=999) (TITLE(I),I=1,20)
WRITE (IOUT,601) (TITLE(I),I=1,20)
READ (IIN,*) N,IBW
WRITE (IOUT,602) N,IBW
WRITE (IOUT,603)
READ (IIN,*) (C(I,1),I=1,N)
WRITE (IOUT,604) (C(I,1),I=1,N)
WRITE (IOUT,605)
DO 1 I=1,N
   READ (IIN,*) (S(I,J),J=1,IBW)
   WRITE (IOUT,604) (S(I,J),J=1,IBW)
1     CONTINUE
WRITE (IOUT,606)
READ (IIN,*) (R(I),I=1,N)
WRITE (IOUT,604) (R(I),I=1,N)
READ (IIN,*) TIME,DTIME,NUMAX
WRITE (IOUT,607) TIME,DTIME,NUMAX
WRITE (IOUT,608)
READ (IIN,*) (T(I),I=1,N)
WRITE (IOUT,604) (T(I),I=1,N)
501 FORMAT (20A4)
601 FORMAT (' ',20A4)
602 FORMAT (/' ','N =',I3,5X,'IBW =',I3)
603 FORMAT (/' ','MAIN-DIAGONAL ENTRIES IN C MATRIX')
604 FORMAT (' ',10E12.4)
605 FORMAT (/' ','ENTRIES IN S MATRIX, ECONOMY-BANDED STORAGE')
606 FORMAT (/' ','ENTRIES IN R VECTOR')
607 FORMAT (/' ','TIME =',E11.4,5X,'DTIME =',E11.4,5X,'NUMAX =',I4)
608 FORMAT (/' ','INITIAL TEMPERATURE DISTRIBUTION')
999 CONTINUE
RETURN
END
```

As the Crank-Nicolson procedure (8·4·20) continues, a steady-state solution $t^{(\infty)}$ will be approached and (8·4·20) will give

$$\left[Ç+S\frac{\Delta\theta}{2}\right]t^{(\infty)} = \left[Ç-S\frac{\Delta\theta}{2}\right]t^{(\infty)} + r\Delta\theta$$

The terms involving $Ç$ cancel and we may divide by $\Delta\theta$ and rearrange to obtain $St^{(\infty)} = r$. Comparison of this result with (8·3·1) shows that $t^{(\infty)}$ is the correct steady-state solution of the problem. Since $\Delta\theta$ has canceled, $t^{(\infty)}$ will be independent of the size of the time step.

8•4•4 Numerical results

Let us consider the solution to (8·4·5) with the initial condition given by (8·4·6). For numerical computations, let us take unit values for k, ρ, c, L and T_o. This is equivalent to solving the normalized problem shown in Figure 8•26 where

$$\bar{t} = \frac{t}{T_o}, \quad \bar{x} = \frac{x}{L}, \quad \bar{y} = \frac{y}{L} \quad \text{and} \quad \bar{\theta} = \frac{k\theta}{\rho c L^2}$$

Equation (8·4·5) then becomes

$$\frac{1}{32}\begin{bmatrix} 1 & & & & & \\ & 4 & & & & \\ & & 2 & & & \\ & & & 4 & & \\ & & & & 4 & \\ & & & & & 1 \end{bmatrix}\begin{bmatrix} \dot{t}_1 \\ \dot{t}_2 \\ \dot{t}_3 \\ \dot{t}_4 \\ \dot{t}_5 \\ \dot{t}_6 \end{bmatrix} + \frac{1}{2}\begin{bmatrix} 1 & -1 & 0 & & & \\ -1 & 4 & -1 & -2 & & \\ 0 & -1 & 2 & 0 & -1 & \\ & -2 & 0 & 4 & -2 & 0 \\ & & -1 & -2 & 4 & -1 \\ & & & 0 & -1 & 1 \end{bmatrix}\begin{bmatrix} t_1 \\ t_2 \\ t_3 \\ t_4 \\ t_5 \\ t_6 \end{bmatrix} = \begin{bmatrix} 0 \\ 0 \\ 0 \\ 0 \\ 0 \\ 0 \end{bmatrix}$$

$$(8\cdot4\cdot27)$$

with the initial condition (8·4·6) given by

$$\mathbf{t}^{(0)} = \begin{bmatrix} t_1 \\ t_2 \\ t_3 \\ t_4 \\ t_5 \\ t_6 \end{bmatrix}^{(0)} = \frac{1}{2}\begin{bmatrix} 2 \\ 1 \\ 0 \\ 1 \\ 0 \\ 0 \end{bmatrix} \qquad (8\cdot4\cdot28)$$

We must decide upon a choice for $\Delta\theta$ to use in the numerical solution of (8·4·27) and the number of time steps to take. In the discussion of response times in Chapter 7 it was seen that L^2/α can be expected to be a significant time for a plane-wall transient. A significant length to use for Figure 8•23 might be $\frac{L}{2}$ rather than L. Hence, for the problem we are considering, $\frac{1}{4}L^2/\alpha$ should be a significant time. Or, for unit values of k, ρ, c and L, the significant time would be 0.25. The choice of $\Delta\theta$ should be convenient for plotting the solution. For convenience let us take $\Delta\theta =$ 0.02. This would require 10 or 15 steps in time (a reasonable number of points to compute and plot) to reach what we expect to be a physically significant time (around $\theta = 0.25$). The number and size of the time steps can be adjusted after seeing the results of the computer solution.

First let us try the Euler solution since it is the easiest to compute. The output from the Euler program listed in tables 8•6 and 8•7 is given in Table 8•10. The temperature-time histories for two corners of Figure 8•23, (0, 0) and (L, L), are compared to the exact solution of the governing partial differential equation in Figure 8•27. The inadequacy of the solution shown in Figure 8•27 is primarily due to the limited number of nodal points that were used to discretize the problem.

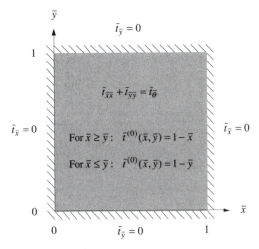

Figure 8•26 *Normalized version of Figure 8•22.*

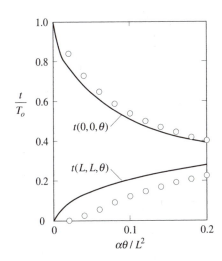

Figure 8•27 *Euler 6-node finite-difference solution to adiabatic-square problem with* $\Delta\theta = 0.02 L^2/\alpha$.

Table 8•10 *Output from program EULER.*

```
EULER EXAMPLE FOR PROBLEM IN SECTION 8.4 (6 NODES)

N =  6     IBW =  3

MAIN-DIAGONAL ENTRIES IN C MATRIX
   0.3125E-01  0.1250E+00  0.6250E-01  0.1250E+00  0.1250E+00  0.3125E-01

ENTRIES IN S MATRIX, ECONOMY-BANDED STORAGE
   0.5000E+00 -0.5000E+00  0.0000E+00
   0.2000E+01 -0.5000E+00 -0.1000E+01
   0.1000E+01  0.0000E+00 -0.5000E+00
   0.2000E+01 -0.1000E+01  0.0000E+00
   0.2000E+01 -0.5000E+00  0.0000E+00
   0.5000E+00  0.0000E+00  0.0000E+00

ENTRIES IN R VECTOR
   0.0000E+00  0.0000E+00  0.0000E+00  0.0000E+00  0.0000E+00  0.0000E+00

TIME = 0.0000E+00    DTIME = 0.2000E-01    NUMAX =  10

INITIAL TEMPERATURE DISTRIBUTION
   0.1000E+01  0.5000E+00  0.0000E+00  0.5000E+00  0.0000E+00  0.0000E+00

EULER APPROXIMATION WILL BE USED

TRANSIENT TEMPERATURE VARIATION

NU =   1     TIME =  0.2000E-01
   0.8400E+00  0.5000E+00  0.8000E-01  0.4200E+00  0.8000E-01  0.0000E+00

NU =   2     TIME =  0.4000E-01
   0.7312E+00  0.4808E+00  0.1472E+00  0.3784E+00  0.1280E+00  0.2560E-01

NU =   3     TIME =  0.6000E-01
   0.6511E+00  0.4578E+00  0.1975E+00  0.3547E+00  0.1614E+00  0.5837E-01

NU =   4     TIME =  0.8000E-01
   0.5892E+00  0.4359E+00  0.2334E+00  0.3403E+00  0.1870E+00  0.9134E-01

NU =   5     TIME =  0.1000E+00
   0.5402E+00  0.4167E+00  0.2584E+00  0.3311E+00  0.2076E+00  0.1219E+00

NU =   6     TIME =  0.1200E+00
   0.5006E+00  0.4002E+00  0.2756E+00  0.3250E+00  0.2245E+00  0.1493E+00

NU =   7     TIME =  0.1400E+00
   0.4685E+00  0.3862E+00  0.2873E+00  0.3210E+00  0.2387E+00  0.1734E+00

NU =   8     TIME =  0.1600E+00
   0.4422E+00  0.3745E+00  0.2954E+00  0.3182E+00  0.2505E+00  0.1943E+00

NU =   9     TIME =  0.1800E+00
   0.4205E+00  0.3645E+00  0.3008E+00  0.3164E+00  0.2604E+00  0.2123E+00

NU =  10     TIME =  0.2000E+00
   0.4026E+00  0.3562E+00  0.3046E+00  0.3151E+00  0.2688E+00  0.2277E+00
```

To improve the solution, let us reduce the spacing between nodal points by a factor of two. This would discretize the region with 15 nodes as shown in Figure 8•18. The resulting solution, using the same time step of $\Delta\theta = 0.02$, is shown in Figure 8•28. Observe that the computer solution has unstable oscillations. These *numerically-induced oscillations* can be eliminated by taking sufficiently smaller time steps.

Reducing the time step by a factor of two gives the results shown in Figure 8•29. We now have a reasonable solution that compares very well with the exact solution of the governing partial differential equation. It is much better than the solution shown in Figure 8•27. It should be pointed out, however, that reducing Δx by a factor of two has required us to reduce $\Delta\theta$ and hence take more time steps to get a reasonable solution. Not only has the computational effort at each time step been increased by going from 6 to 15 nodes, but it now takes 20 time steps instead of 10 to reach $\theta = 0.20$.

Now let us look at some Crank-Nicolson solutions. Since a 6-node Euler solution seemed inadequate in comparison to the exact solution to the governing partial differential equation as shown in Figure 8•27, we will only consider 15-node Crank-Nicolson solutions. For $\Delta\theta = 0.02$ we obtain the solution shown in Figure 8•30. This should be compared to the 15-node Euler solution using the same time-step size shown in Figure 8•28. The unstable oscillations are not present in the Crank-Nicolson solution and a reasonable solution has been obtained. In fact, the Crank-Nicolson solution with $\Delta\theta = 0.02$ is about as good as the Euler solution with $\Delta\theta = 0.01$ shown in Figure 8•29. Thus, although the Crank-Nicolson solution requires more computational effort at each time step (since the solution is implicit rather than explicit), not as many time steps are required.

We have seen that as more nodes are used in the spatial discretization, the time step must be reduced to avoid undesirable, numerically-induced oscillations in the Euler solution. The largest time step for which an Euler solution will be stable is called the *critical time step* $\Delta\theta_c$. Section 8•5 discusses numerically-induced oscillations and the critical-time-step problem. It will be shown there that the Euler solution will have unstable oscillations of $\Delta\theta > \Delta\theta_c$, stable oscillations if $\frac{1}{2}\Delta\theta_c < \Delta\theta < \Delta\theta_c$ and no oscillations at all if $\Delta\theta < \frac{1}{2}\Delta\theta_c$. Crank-Nicolson solutions can also exhibit numerically-induced oscillations if $\Delta\theta > \Delta\theta_c$, but they will always be stable and have decreasing amplitude as time goes on.

It will be shown in Section 8•5•4 that an estimate $\Delta\theta_{c,est}$ of the critical time step can be computed from

$$\Delta\theta_{c,est} = \frac{2}{\underset{\substack{i=1 \\ i\neq i_s}}{\overset{N}{\text{Max}}} \left[\dfrac{1}{\varsigma_{ii}} \displaystyle\sum_{j=1}^{N} |s_{ij}| \right]} \tag{8.4.29}$$

where i_s denotes a specified-temperature row. In the denominator of this equation, each row of $\mathbf{\varsigma}$ and $\mathbf{S}$ (except for specified-temperature rows) is inspected to find the maximum value of $\varsigma_{ii}^{-1}\Sigma_j |s_{ij}|$. It will be shown in Section 8•5•3 that if $\Delta\theta \leq \Delta\theta_{c,est}$ an Euler solution will be stable (but it may still have numerically-induced oscillations) and a Crank-Nicolson solution will not have any numerically-induced oscillations.

For (8.4.5) each row produces the same value in (8.4.29) and gives $\Delta\theta_{c,est} = \rho c L^2 / 16k$. For unit values of k, ρ, c and L, this gives $\Delta\theta_{c,est} = 0.0625$. Observe that the Euler solution in Figure 8•27 ($\Delta\theta = 0.02$) is well within this estimate and is stable. If the 15-node system of equations is examined using (8.4.29) one finds that $\Delta\theta_{c,est} = \rho c L^2 / 64k$ or, for unit values, $\Delta\theta_{c,est} = 0.015625$. The Euler solution in Figure 8•28 has a time step ($\Delta\theta = 0.02$) that exceeds this estimate and is unstable. The time step in Figure 8•29 ($\Delta\theta = 0.01$) is less than this estimate and is stable. The Crank-Nicolson solution in Figure 8•30 ($\Delta\theta = 0.02$) exceeds this estimate but the oscillations, if any, are not very noticeable.

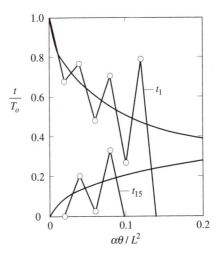

Figure 8•28 *Euler 15-node finite-difference solution to adiabatic-square problem with* $\Delta\theta = 0.02L^2/\alpha$.

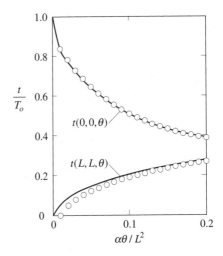

Figure 8•29 *Euler 15-node finite-difference solution to adiabatic-square problem with* $\Delta\theta = 0.01L^2/\alpha$.

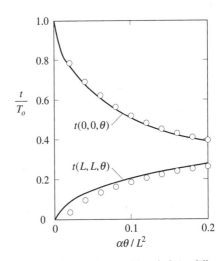

Figure 8•30 *Crank-Nicolson 15-node finite-difference solution to adiabatic-square problem with* $\Delta\theta = 0.02L^2/\alpha$.

8•5 ANALYSIS OF TRANSIENT SOLUTIONS

In Section 8•4 we went right from the discrete equation

$$\mathbf{Ç\dot{t} + St = r} \tag{8.5.1}$$

whose initial condition is given by

$$\mathbf{t}(0) = \mathbf{t}^{(0)} \tag{8.5.2}$$

to a step-by-step computer solution as would be done in practice. In this section we will examine these transient solutions in more detail. We will first obtain the exact solution to (8.5.1) in Section 8•5•1 and compare it to the exact solution of the governing partial differential equation. Next, in Section 8•5•2, we will cast the approximate computer solutions (Euler and Crank-Nicolson) in a form that will be compared in Section 8•5•3 to the exact solution obtained in Section 8•5•1. This comparison will explain why oscillations can occur in the computer solutions. The critical time step estimate given by (8.4.29) is then derived in Section 8•5•4. Finally, Section 8•5•5 discusses the factors that influence the critical time step.

8•5•1 Exact solution of $\mathbf{Ç\dot{t} + St = r}$

In practice the exact solution of (8.5.1) is too expensive to carry out because of the large number of nodes usually involved. Instead the Euler and Crank-Nicolson approximations described in sections 8•4•2 and 8•4•3 are commonly used. It is instructive, however, to examine the exact solution to (8.5.1) because it can be used to give insights into the behavior of the approximate solutions.

Since (8.5.1) is nonhomogeneous we will follow the partial-solution idea developed in Section 3•2•1 and write

$$\mathbf{t}(\theta) = \mathbf{a} + \mathbf{b}\theta + \mathbf{u}(\theta) \tag{8.5.3}$$

where, in most cases, $\mathbf{a}$ is the steady-state temperature distribution, $\mathbf{u}(\theta)$ is a transient solution that decays to zero as θ increases, and $\mathbf{b}\theta$ is a diverging part that is only needed if there is a "run-away" solution due to a continual accumulation (or depletion) of energy within the system.

Substitution of (8.5.3) into (8.5.1) gives

$$\mathbf{Ç(b + \dot{u}) + S(a + b\theta + u) = r}$$

This equation will be satisfied if we choose $\mathbf{a}$, $\mathbf{b}$ and $\mathbf{u}$ to satisfy:

$$\mathbf{Ç\dot{u} + Su = 0} \tag{8.5.4}$$

$$\mathbf{Sb = 0} \tag{8.5.5}$$

$$\mathbf{Çb + Sa = r} \qquad \text{or} \qquad \mathbf{Sa = r - Çb} \tag{8.5.6}$$

The initial condition for (8.5.4) is obtained by substituting (8.5.3) into (8.5.2) to give

$$\mathbf{a + u}(0) = \mathbf{t}^{(0)}$$

Thus,

$$\mathbf{u}(0) = \mathbf{u}^{(0)} = \mathbf{t}^{(0)} - \mathbf{a} \tag{8.5.7}$$

Since $\mathbf{a}$ will usually be the steady-state solution, $\mathbf{u}(0)$ is the difference between the initial temperature distribution and the steady-state temperature distribution.

The values of $\mathbf{a}$ and $\mathbf{b}$ will depend on whether or not $\det(\mathbf{S}) = 0$. The most common case is when there is either a convective boundary or a specified-temperature boundary. In this case, $\det(\mathbf{S}) \neq 0$. Equation (8·5·5) will then give

$$\mathbf{b} = \mathbf{0} \qquad (8\cdot5\cdot8)$$

and (8·5·6) will reduce to

$$\mathbf{Sa} = \mathbf{r} \qquad (8\cdot5\cdot9)$$

which can be solved for $\mathbf{a}$. Observe that in this case $\mathbf{a}$ is the steady-state solution of (8·5·1).

Only when there is neither a convective boundary nor a specified temperature along the boundary will we have $\det(\mathbf{S}) = 0$. Since in this case $\mathbf{S}$ will also be such that $\Sigma_j s_{ij} = 0$ (where Σ_j denotes a sum over j), it can be reasoned that (8·5·5) will be satisfied if the entries in $\mathbf{b}$ are equal to one another. This can be written as

$$\mathbf{b} = b\mathbf{1} \qquad (8\cdot5\cdot10)$$

where $\mathbf{1}$ is a column vector filled with ones and b is a constant that must now be determined. Substitution of $\mathbf{b} = b\mathbf{1}$ into (8·5·6), recognizing that $\mathbf{Ç} = \mathbf{C}$ when there is no specified temperature, gives

$$\mathbf{Sa} = \mathbf{r} - b\mathbf{C}\mathbf{1} = \mathbf{r} - b\mathbf{c} \qquad (8\cdot5\cdot11)$$

where it is easily shown that the entry in row i of $\mathbf{c} = \mathbf{C}\mathbf{1}$ is $\Sigma_j c_{ij}$. Since $\mathbf{S}$ is symmetric, it is also true that $\Sigma_i s_{ij} = 0$, and the addition of rows 2 through N to row 1 in (8·5·11) will give

$$
\begin{bmatrix}
0 & 0 & 0 & 0 & 0 & 0 & 0 & 0 & 0 & 0 \\
s_{21} & s_{22} & & & \cdots & & & & & s_{2N} \\
& & & & & & & & & \\
\vdots & & & \ddots & & & & \vdots & & \\
& & & & & & & & & \\
s_{N1} & & & & \cdots & & & & & s_{NN}
\end{bmatrix}
\begin{bmatrix}
a_1 \\ a_2 \\ \\ \vdots \\ \\ a_N
\end{bmatrix}
=
\begin{bmatrix}
\Sigma_i r_i - b\Sigma_i c_i \\ r_2 - bc_2 \\ \\ \vdots \\ \\ r_N - bc_N
\end{bmatrix}
\qquad (8\cdot5\cdot12)
$$

Since the left-hand side of row 1 is 0, the right-hand side must also be 0. Thus,

$$0 = \Sigma_i r_i - b\Sigma_i c_i$$

Upon solving for b and recalling that $r_i = g_i + q_i$ and $c_i = \Sigma_j c_{ij}$ we obtain

$$b = \frac{\Sigma_i r_i}{\Sigma_i c_i} = \frac{\Sigma_i g_i + \Sigma_i q_i}{\Sigma_i \Sigma_j c_{ij}} \qquad (8\cdot5\cdot13)$$

For $\det(\mathbf{S}) = 0$, we can only have specified heat flux conditions around the entire boundary. In this case, $\Sigma_i r_i$ will be the algebraic sum of all internal generation of energy within the solid plus the heat inflows (outflows are negative inflows) from outside the solid. Whenever $\Sigma_i r_i > 0$ (or < 0), there will be a continual net accumulation (or depletion) of energy in the solid. We will then have $\mathbf{b} = b\mathbf{1} \neq \mathbf{0}$ and (8·5·3) will contain a term that increases (or decreases) linearly with time. The denominator of (8·5·13) contains the total thermal capacitance in the problem. A small capacitance means b will be large and the linear term in (8·5·3) will change rapidly with time.

Since the first row in (8·5·12) has been used to determine b, there are only $N-1$ rows (equations) left to determine the N entries in **a**. The best we will be able to do is to find $N-1$ of the entries in **a**. The remaining entry is arbitrary. Rather than reducing the number of equations from N to $N-1$, we will insert s_{11} back as the diagonal term in row 1 which will then give $s_{11}a_1 = 0$. This arbitrary choice gives $a_1 = 0$. The system of equations (8·5·12), with s_{11} inserted, is no longer symmetric since the off-diagonal entries in the first row are all 0 whereas the off-diagonal entries in the first column are not 0. To regain symmetry we will transfer each of the terms multiplying a_1 (first column in rows 2 through N) from the left-hand side to the right-hand side. Since we have conveniently chosen $a_1 = 0$, there are no changes that need to be made to the right-hand side. We can now set each of the off-diagonal entries in the first column of rows 2 through N equal to 0 to regain symmetry. We will then have

$$
\begin{bmatrix}
s_{11} & 0 & 0 & 0 & 0 & \cdots & 0 & 0 & 0 & 0 \\
0 & s_{22} & & & \cdots & & & & s_{2N} \\
0 \\
0 \\
0 \\
\vdots & \vdots & & & \ddots & & & \vdots \\
0 \\
0 \\
0 \\
0 & s_{N2} & & & \cdots & & & & s_{NN}
\end{bmatrix}
\begin{bmatrix}
a_1 \\
a_2 \\
\vdots \\
\vdots \\
a_N
\end{bmatrix}
=
\begin{bmatrix}
0 \\
r_2 - bc_2 \\
\vdots \\
\vdots \\
r_N - bc_N
\end{bmatrix}
$$

or

$$\mathbf{S'a} = (\mathbf{r} - b\mathbf{c})' \tag{8·5·14}$$

Since $\det(\mathbf{S'})$ will not be 0, (8·5·14) can be solved without trouble. The solution will be one of the solutions to (8·5·11)—the one having $a_1 = 0$.

In summary, when $\det(\mathbf{S}) \neq 0$, **a** and **b** are determined by (8·5·8) and (8·5·9). When $\det(\mathbf{S}) = 0$, **a** and **b** are determined from (8·5·10), (8·5·13) and (8·5·14).

Now we will have to find the solution to (8·5·4) whose initial condition is given by (8·5·7). If (8·5·4) were simply a single ordinary differential equation we could obtain a solution by assuming an exponential form of the solution as discussed in Appendix A•2•3. The analogous step for our system of equations (8·5·4) is to assume the solution has the form

$$
\mathbf{u}(\theta) = \mathbf{x}\exp(-\lambda\theta) =
\begin{bmatrix}
x_1 \\
x_2 \\
\vdots \\
x_i \\
\vdots \\
x_N
\end{bmatrix}
\exp(-\lambda\theta) =
\begin{bmatrix}
x_1\exp(-\lambda\theta) \\
x_2\exp(-\lambda\theta) \\
\vdots \\
x_i\exp(-\lambda\theta) \\
\vdots \\
x_N\exp(-\lambda\theta)
\end{bmatrix}
\tag{8·5·15}
$$

This is equivalent to assuming that each entry in **u** is a constant (contained in **x**) times a decaying exponential.

Now we must substitute (8·5·15) into (8·5·4) to find out the restrictions on $\mathbf{x}$ and λ. Upon substitution we obtain

$$-\lambda \mathbf{\varsigma x} \exp(-\lambda\theta) + \mathbf{Sx} \exp(-\lambda\theta) = \mathbf{0}$$

The exponentials cancel and we find that λ and $\mathbf{x}$ must satisfy

$$\mathbf{Sx} = \lambda \mathbf{\varsigma x} \tag{8·5·16}$$

This is a *generalized eigenproblem*. The situation is somewhat analogous to Chapter 3 where separation of variables led to an ordinary differential equation with boundary conditions to be solved for eigenvalues and the corresponding eigenfunctions. In finite differences, spatial discretization has replaced the partial differential equation and boundary conditions by a system of ordinary differential equations whose exact solution involves a set of algebraic equations (8·5·16) to be solved for the eigenvalues λ and the corresponding eigenvectors $\mathbf{x}$.

Now we need to learn about the generalized eigenproblem that must be solved. Equation (8·5·16) may be rewritten as

$$(\mathbf{S} - \lambda\mathbf{\varsigma})\mathbf{x} = \mathbf{0} \tag{8·5·17}$$

This is seen to be a system of algebraic equations with zero right-hand sides. In order for such a system to have a solution, the determinant of the coefficients must be 0. That is, we must have

$$\det(\mathbf{S} - \lambda\mathbf{\varsigma}) = 0 \tag{8·5·18}$$

Since there are N equations, there will be N values of λ for which this determinant will be zero. That is, there will be N eigenvalues. Eigenvalue λ_i can then be substituted into (8·5·17) to find eigenvector $\mathbf{x}_i$. That is, $\mathbf{x}_i$ is found from

$$(\mathbf{S} - \lambda_i\mathbf{\varsigma})\mathbf{x}_i = \mathbf{0} \tag{8·5·19}$$

Any multiple of $\mathbf{x}_i$ will also satisfy (8·5·19) since the equations are homogeneous.

As in the separation of variables solutions in Chapter 3 where the eigenfunctions were orthogonal to each other, the eigenvectors of (8·5·16) will also be orthogonal to each other. To show this let us write (8·5·16) using λ_i and $\mathbf{x}_i$ and then premultiply both sides by $\mathbf{x}_j^T$ to give

$$\mathbf{x}_j^T\mathbf{Sx}_i = \lambda_i\mathbf{x}_j^T\mathbf{\varsigma x}_i \tag{8·5·20}$$

Similarly, eigenvalue λ_j and eigenvector $\mathbf{x}_j$ must also satisfy

$$\mathbf{x}_i^T\mathbf{Sx}_j = \lambda_j\mathbf{x}_i^T\mathbf{\varsigma x}_j \tag{8·5·21}$$

Subtracting (8·5·21) from (8·5·20) gives

$$\mathbf{x}_j^T\mathbf{Sx}_i - \mathbf{x}_i^T\mathbf{Sx}_j = \lambda_i\mathbf{x}_j^T\mathbf{\varsigma x}_i - \lambda_j\mathbf{x}_i^T\mathbf{\varsigma x}_j \tag{8·5·22}$$

Since $\mathbf{x}_i^T\mathbf{Sx}_j$ is a scalar quantity it is equal to its own transpose. That is, we can write

$$\mathbf{x}_i^T\mathbf{Sx}_j = (\mathbf{x}_i^T\mathbf{Sx}_j)^T$$

Two applications of (8·1·9), $(\mathbf{AB})^T = \mathbf{B}^T\mathbf{A}^T$, to the right-hand side gives

$$\mathbf{x}_i^T\mathbf{Sx}_j = \mathbf{x}_j^T(\mathbf{x}_i^T\mathbf{S})^T = \mathbf{x}_j^T\mathbf{S}^T\mathbf{x}_i$$

Since $\mathbf{S}$ is symmetric, $\mathbf{S}^T = \mathbf{S}$, this becomes

$$\mathbf{x}_i^T \mathbf{S} \mathbf{x}_j = \mathbf{x}_j^T \mathbf{S} \mathbf{x}_i$$

A similar proof also holds to give

$$\mathbf{x}_i^T \mathbf{\mathsf{C}} \mathbf{x}_j = \mathbf{x}_j^T \mathbf{\mathsf{C}} \mathbf{x}_i$$

Substitution of these last two results into (8·5·22) gives

$$0 = (\lambda_i - \lambda_j)\mathbf{x}_i^T \mathbf{\mathsf{C}} \mathbf{x}_j \qquad (8\cdot5\cdot23)$$

Thus, when $i \neq j$ we must have $\mathbf{x}_i^T \mathbf{C} \mathbf{x}_j = 0$ and when $i = j$, $\mathbf{x}_i^T \mathbf{\mathsf{C}} \mathbf{x}_j = \mathbf{x}_j^T \mathbf{\mathsf{C}} \mathbf{x}_j$ will be a positive value which, by suitable normalization of $\mathbf{x}_j$, can be set equal to 1. This orthogonality relationship can be written as

$$\mathbf{x}_i^T \mathbf{\mathsf{C}} \mathbf{x}_j = \delta_{ij} \qquad (8\cdot5\cdot24)$$

where δ_{ij} is the *Kronecker delta* which is 1 when $i = j$ and 0 when $i \neq j$. We say that the eigenvectors $\mathbf{x}_i$ and $\mathbf{x}_j$ are orthogonal to each other with respect to the matrix $\mathbf{\mathsf{C}}$. Equation (8·5·23) is analogous to (3·1·14) given at the end of the orthogonality discussion in Chapter 3.

Upon substituting (8·5·24) into (8·5·21) we obtain

$$\mathbf{x}_i^T \mathbf{S} \mathbf{x}_j = \delta_{ij} \lambda_j \qquad (8\cdot5\cdot25)$$

When $i = j$ this gives $\mathbf{x}_j^T \mathbf{S} \mathbf{x}_j = \lambda_j$ and when $i \neq j$ this gives $\mathbf{x}_i^T \mathbf{S} \mathbf{x}_j = 0$.

Example Determine the eigenvalues and eigenvectors of $\mathbf{S}\mathbf{x} = \lambda \mathbf{\mathsf{C}} \mathbf{x}$ for

$$\mathbf{S} = \begin{bmatrix} 3 & -2 \\ -2 & 4 \end{bmatrix} \text{ and } \mathbf{\mathsf{C}} = \begin{bmatrix} 1 & 0 \\ 0 & 2 \end{bmatrix}$$

Solution: Substitution of $\mathbf{S}$ and $\mathbf{\mathsf{C}}$ into (8·5·18) gives

$$\det(\mathbf{S} - \lambda \mathbf{\mathsf{C}}) = \begin{vmatrix} 3-\lambda & -2 \\ -2 & 4-2\lambda \end{vmatrix} = (3-\lambda)(4-2\lambda) - (-2)(-2) = 0$$

This is a quadratic equation which may be solved for λ to find $\lambda_1 = 1$ and $\lambda_2 = 4$. We will always number the eigenvalues so that $\lambda_1 \leq \lambda_2 \leq \cdots \leq \lambda_N$. To find $\mathbf{x}_1$ we will substitute $\lambda_1 = 1$ and $\mathbf{S}$ and $\mathbf{\mathsf{C}}$ into (8·5·19) to obtain

$$\left[\begin{bmatrix} 3 & -2 \\ -2 & 4 \end{bmatrix} - 1 \begin{bmatrix} 1 & 0 \\ 0 & 2 \end{bmatrix} \right] \begin{bmatrix} x_1 \\ x_2 \end{bmatrix}_1 = \begin{bmatrix} 2 & -2 \\ -2 & 2 \end{bmatrix} \begin{bmatrix} x_1 \\ x_2 \end{bmatrix}_1 = \begin{bmatrix} 0 \\ 0 \end{bmatrix}$$

This relation is satisfied as long as $x_2 = x_1$. Hence we may write

$$\mathbf{x}_1 = x_1 \begin{bmatrix} 1 \\ 1 \end{bmatrix}$$

Next we will normalize $\mathbf{x}_1$ by choosing the value for x_1 so that (8·5·24) is satisfied. Substitution of $\mathbf{x}_1$ and $\mathbf{\mathsf{C}}$ into (8·5·24) gives

$$x_1 \begin{bmatrix} 1 & 1 \end{bmatrix} \begin{bmatrix} 1 & 0 \\ 0 & 2 \end{bmatrix} x_1 \begin{bmatrix} 1 \\ 1 \end{bmatrix} = 1$$

Upon carrying out the matrix operations we find that $3x_1^2 = 1$, or that $x_1 = \frac{1}{\sqrt{3}}$. Thus, the normalized $\mathbf{x}_1$ is given by

$$\mathbf{x}_1 = \frac{1}{\sqrt{3}} \begin{bmatrix} 1 \\ 1 \end{bmatrix}$$

Next we will substitute $\lambda_2 = 4$ and $\mathbf{S}$ and $\mathbf{\mathsf{C}}$ into (8·5·19) to give

$$\begin{bmatrix} -1 & -2 \\ -2 & -4 \end{bmatrix} \begin{bmatrix} x_1 \\ x_2 \end{bmatrix}_2 = \begin{bmatrix} 0 \\ 0 \end{bmatrix}$$

This relation will be satisfied as long as $x_1 = -2x_2$. Hence we may write

$$\mathbf{x}_2 = x_2 \begin{bmatrix} -2 \\ 1 \end{bmatrix}$$

Next we will normalize $\mathbf{x}_2$ by choosing the value of x_2 so that (8·5·24) is satisfied. Substitution of $\mathbf{x}_2$ into (8·5·24) gives

$$x_2 \begin{bmatrix} -2 & 1 \end{bmatrix} \begin{bmatrix} 1 & 0 \\ 0 & 2 \end{bmatrix} x_2 \begin{bmatrix} -2 \\ 1 \end{bmatrix} = 1$$

Upon carrying out the matrix multiplications we find that $6x_2^2 = 1$, or that $x_2 = \frac{1}{\sqrt{6}}$. Thus, the normalized $\mathbf{x}_2$ is given by

$$\mathbf{x}_2 = \frac{1}{\sqrt{6}} \begin{bmatrix} -2 \\ 1 \end{bmatrix}$$

This completes the solution of the example eigenproblem since we have now found the two eigenvalues and their corresponding eigenvectors.

We are now ready to return to our solution of (8·5·4) which we have assumed to have the form given by (8·5·15). We found that $\mathbf{x}$ and λ in (8·5·15) must satisfy the eigenproblem given by (8·5·16). We therefore have a set of N eigenvalues λ_1, λ_2, ..., λ_N and a corresponding N eigenvectors $\mathbf{x}'_1$, $\mathbf{x}'_2$, ..., $\mathbf{x}'_N$ that will satisfy (8·5·16). Consequently we will generalize our assumed solution (8·5·15) by writing

$$\mathbf{u}(\theta) = \sum_{j=1}^{N} d_j \mathbf{x}_j \exp(-\lambda_j \theta) \tag{8·5·26}$$

where the d_j are a set of N arbitrary constants d_1, d_2, ..., d_N that will be determined by forcing (8·5·26) to satisfy the initial condition for $\mathbf{u}(\theta)$ given by (8·5·7). Setting $\theta = 0$ in (8·5·26) gives

$$\mathbf{u}(0) = \sum_{j=1}^{N} d_j \mathbf{x}_j \tag{8·5·27}$$

The coefficients d_j will now be evaluated by using orthogonality in a manner similar to what we did for separation of variables in Chapter 3. Premultiplication of (8·5·27) by $\mathbf{x}_i^T \mathbf{\c C}$ gives

$$\mathbf{x}_i^T \mathbf{\c C} \mathbf{u}(0) = \sum_{j=1}^{N} d_j \mathbf{x}_i^T \mathbf{\c C} \mathbf{x}_j$$

From (8·5·24) we see that each term in the right-hand side summation is 0 except for when $j = i$ when it is d_i. Thus only one term in the sum, $j = i$, will remain and this relation reduces to

$$\mathbf{x}_i^T \mathbf{\c C} \mathbf{u}(0) = d_i$$

We can use (8·5·7) to write $\mathbf{u}(0)$ in terms of the initial temperature values, replace the dummy index i by j and rearrange to obtain

$$d_j = \mathbf{x}_j^T \mathbf{\c C} \mathbf{u}^{(0)} = \mathbf{x}_j^T \mathbf{\c C} (\mathbf{t}^{(0)} - \mathbf{a}) \tag{8·5·28}$$

This determines the constants d_j in (8·5·26).

If we define a *modified eigenvector* $\mathbf{x}'_j$ as

$$\mathbf{x}'_j = d_j \mathbf{x}_j \tag{8.5.29}$$

we may then write the solution (8.5.26) as

$$\mathbf{u}(\theta) = \sum_{j=1}^{N} \mathbf{x}'_j \exp(-\lambda_j \theta) \tag{8.5.30}$$

Upon writing out the sum in (8.5.30),

$$\mathbf{u}(\theta) = \mathbf{x}'_1 \exp(-\lambda_1 \theta) + \mathbf{x}'_2 \exp(-\lambda_2 \theta) + \cdots + \mathbf{x}'_N \exp(-\lambda_N \theta)$$

It is often convenient to write this sum of products using matrix notation. To do this we will place the N eigenvectors $\mathbf{x}'_1$, $\mathbf{x}'_2$, ..., $\mathbf{x}'_N$ in a square $N \times N$ *eigenvector matrix* $\mathbf{X}'$ as given by

$$\mathbf{X}' = \begin{bmatrix} \mathbf{x}'_1 & \mathbf{x}'_2 & \cdots & \mathbf{x}'_j & \cdots & \mathbf{x}'_N \end{bmatrix} = \begin{bmatrix} x'_{11} & x'_{12} & & x'_{1j} & & x'_{1N} \\ x'_{21} & x'_{22} & & x'_{2j} & & x'_{2N} \\ \vdots & \vdots & & \vdots & & \vdots \\ x'_{i1} & x'_{i2} & \cdots & x'_{ij} & \cdots & x'_{iN} \\ \vdots & \vdots & & \vdots & & \vdots \\ x'_{N1} & x'_{N2} & & x'_{Nj} & & x'_{NN} \end{bmatrix} \tag{8.5.31}$$

The first modified eigenvector $\mathbf{x}'_1$ has been placed in the first column of $\mathbf{X}'$, the second modified eigenvector $\mathbf{x}'_2$ is placed in the second column, and so forth. Thus, entry x'_{ij} in $\mathbf{X}'$ would be entry i in eigenvector j. We will also define an $N \times 1$ *eigenvalue vector* $\boldsymbol{\lambda}$ containing the N eigenvalues λ_1, λ_2, ..., λ_N and an $N \times 1$ vector $\exp(-\boldsymbol{\lambda}\theta)$ containing the N decaying exponentials in (8.5.30) as

Table 8•11 *Values of* $\mathbf{a}$ *and* $\mathbf{b}$ *in (8.5.34).*

det($\mathbf{S}$)	$\mathbf{a}$	$\mathbf{b}$
$\neq 0$	$\mathbf{Sa} = \mathbf{r}$	$\mathbf{0}$
$= 0$	Any solution to $\mathbf{Sa} = \mathbf{r} - \mathbf{Cb}$	$\dfrac{\sum_{i=1}^{N} r_i}{\sum_{i=1}^{N}\sum_{j=1}^{N} c_{ij}} \mathbf{1}$

$$\boldsymbol{\lambda} = \begin{bmatrix} \lambda_1 \\ \lambda_2 \\ \vdots \\ \lambda_i \\ \vdots \\ \lambda_N \end{bmatrix} \qquad \exp(-\boldsymbol{\lambda}\theta) = \begin{bmatrix} \exp(-\lambda_1 \theta) \\ \exp(-\lambda_2 \theta) \\ \vdots \\ \exp(-\lambda_i \theta) \\ \vdots \\ \exp(-\lambda_N \theta) \end{bmatrix} \tag{8.5.32}$$

We can then write (8.5.30) in terms of $\boldsymbol{\lambda}$ and $\mathbf{X}'$ as

$$\mathbf{u}(\theta) = \mathbf{X}' \exp(-\boldsymbol{\lambda}\theta) \tag{8.5.33}$$

where $\mathbf{X}'$ contains the modified eigenvectors $\mathbf{x}'_1$, $\mathbf{x}'_2$, ..., $\mathbf{x}'_N$. This is the solution of the homogeneous problem.

The complete solution is found by substituting (8.5.30) or (8.5.33) into (8.5.3) to obtain

$$\mathbf{t}(\theta) = \mathbf{a} + \mathbf{b}\theta + \sum_{j=1}^{N} \mathbf{x}'_j \exp(-\lambda_j \theta) = \mathbf{a} + \mathbf{b}\theta + \mathbf{X}' \exp(-\boldsymbol{\lambda}\theta) \tag{8.5.34}$$

Table 8•12 *Values of* λ_j *and* $\mathbf{x}'_j$ *in (8.5.34).*

$\mathbf{Sx}_j = \lambda_j \mathbf{Cx}_j$	$d_j = \mathbf{x}_j^T \mathbf{C}(\mathbf{t}^{(0)} - \mathbf{a})$
	$\mathbf{x}'_j = d_j \mathbf{x}_j$

where $\mathbf{a}$, $\mathbf{b}$, λ_j and $\mathbf{x}'_j$ are summarized in tables 8•11 and 8•12.

As an example of obtaining an exact transient solution to (8·5·1) and (8·5·2), let us consider (8·4·5) and (8·4·6). The first step is to obtain the eigenvalues and eigenvectors that satisfy (8·5·16). Although this could be done by hand, it is rather tedious when there are more than three or four rows in **C** and **S**. Even the writing of a computer subroutine is a task that is best left to the experts. With the assistance of a borrowed eigenproblem subroutine (a typical engineering approach to using the computer), it is found that the solution to (8·5·16) is given by

$$
\mathbf{X} = \frac{\sqrt{2}}{\sqrt{\rho c L^2}}
\begin{bmatrix}
1 & 2 & 2 & \sqrt{2} & 2 & 1 \\
1 & 1 & 0 & 0 & -1 & -1 \\
1 & 0 & -2 & \sqrt{2} & 0 & 1 \\
1 & 0 & 0 & -\sqrt{2} & 0 & 1 \\
1 & -1 & 0 & 0 & 1 & -1 \\
1 & -2 & 2 & \sqrt{2} & -2 & 1
\end{bmatrix}
\quad \text{and} \quad
\boldsymbol{\lambda} = \frac{\alpha}{L^2}
\begin{bmatrix}
0 \\ 8 \\ 16 \\ 16 \\ 24 \\ 32
\end{bmatrix}
\qquad (8·5·35)
$$

Since there is neither a convective nor a specified-temperature boundary condition in this example, the determinant of **S** will be 0. In this case **a** and **b** are determined from (8·5·10), (8·5·13), and (8·5·14). The result is that **a** = **0** and **b** = **0** in this example. Calculation of d_j from (8·5·28) and $\mathbf{x}'_j$ from (8·5·29) then gives the modified eigenvectors as

$$
\mathbf{X}' = \frac{T_o}{16}
\begin{bmatrix}
5 & 8 & 4 & -2 & 0 & 1 \\
5 & 4 & 0 & 0 & 0 & -1 \\
5 & 0 & -4 & -2 & 0 & 1 \\
5 & 0 & 0 & 2 & 0 & 1 \\
5 & -4 & 0 & 0 & 0 & -1 \\
5 & -8 & 4 & -2 & 0 & 1
\end{bmatrix}.
\qquad (8·5·36)
$$

Substitution of **a**, **b**, **X**′ and **λ** into (8·5·34) then gives the exact solution of the 6-node, finite-difference equations as

$$
\mathbf{t}(\theta) = \frac{T_o}{16}
\begin{bmatrix}
5 & 8 & 4 & -2 & 0 & 1 \\
5 & 4 & 0 & 0 & 0 & -1 \\
5 & 0 & -4 & -2 & 0 & 1 \\
5 & 0 & 0 & 2 & 0 & 1 \\
5 & -4 & 0 & 0 & 0 & -1 \\
5 & -8 & 4 & -2 & 0 & 1
\end{bmatrix}
\begin{bmatrix}
1 \\
\exp(-8\alpha\theta/L^2) \\
\exp(-16\alpha\theta/L^2) \\
\exp(-16\alpha\theta/L^2) \\
\exp(-24\alpha\theta/L^2) \\
\exp(-32\alpha\theta/L^2)
\end{bmatrix}
\qquad (8·5·37)
$$

This 6-node result (8·5·37) and the results for $N = 3$ and 15 nodes are compared to the exact solution ($N = \infty$) of the governing partial differential equation in Figure 8•31 for the corners at (0,0) and (L, L). The differences between these solutions are entirely due to the number of nodes in the spatial discretization since time was not discretized. Note the improvement in the finite-difference solutions as the number of nodes is increased. Another difference that can be pointed out is that the steady-state solution of (8·5·37) is that each nodal temperature approaches $\frac{5}{16}T_0 = 0.3125T_o$ whereas the exact value is $\frac{1}{3}T_0 = 0.3333T_o$. The exact value is determined by the initial internal energy in the region. The initial internal energy has not been modeled exactly by using only 6 nodal points. Notice also that the fifth modified eigenvector $\mathbf{x}'_5$ in (8·5·37) is **0**. In this case $\exp(-\lambda_5\theta) = \exp(-24\alpha\theta/L^2)$ drops out of the solution because its coefficient $\mathbf{x}'_5 = \mathbf{0}$.

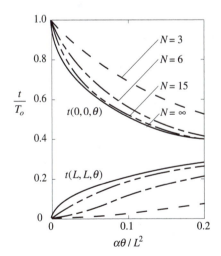

Figure 8•31 *Comparison of the exact finite-difference solutions to the partial-differential-equation solution of the adiabatic-square problem.*

For the 15-node example the modified eigenvectors and the eigenvalues are given by

$$\mathbf{X}' = \frac{T_o}{64}\begin{bmatrix} 21 & 27.3 & 13.7 & -8 & 0 & 4.7 & 4 & -2 & 0 & 0 & 0 & 0 & 2.3 & 0 & 1 \\ 21 & 23.3 & 9.7 & -4 & 0 & 0.7 & 0 & 0 & 0 & 0 & 0 & 0 & -1.7 & 0 & -1 \\ 21 & 13.7 & 0 & 0 & 0 & 2.3 & -4 & -2 & 0 & 0 & 0 & 0 & 0 & 0 & 1 \\ 21 & 19.3 & 6.8 & 0 & 0 & -3.3 & 0 & 2 & 0 & 0 & 0 & 0 & 1.2 & 0 & 1 \\ 21 & 4 & -9.7 & -4 & 0 & 4 & 0 & 0 & 0 & 0 & 0 & 0 & 1.7 & 0 & -1 \\ 21 & 9.7 & 0 & 4 & 0 & -1.7 & 0 & 0 & 0 & 0 & 0 & 0 & 0 & 0 & -1 \\ 21 & 0 & -13.7 & -8 & 0 & 0 & 4 & -2 & 0 & 0 & 0 & 0 & -2.3 & 0 & 1 \\ 21 & 0 & -6.8 & 0 & 0 & 0 & 0 & 2 & 0 & 0 & 0 & 0 & -1.2 & 0 & 1 \\ 21 & 0 & 0 & 8 & 0 & 0 & 4 & -2 & 0 & 0 & 0 & 0 & 0 & 0 & 1 \\ 21 & -4 & -9.7 & -4 & 0 & -4 & 0 & 0 & 0 & 0 & 0 & 0 & 1.7 & 0 & -1 \\ 21 & -9.7 & 0 & 4 & 0 & 1.7 & 0 & 0 & 0 & 0 & 0 & 0 & 0 & 0 & -1 \\ 21 & -13.7 & 0 & 0 & 0 & -2.3 & -4 & -2 & 0 & 0 & 0 & 0 & 0 & 0 & 1 \\ 21 & -19.3 & 6.8 & 0 & 0 & 3.3 & 0 & 2 & 0 & 0 & 0 & 0 & 1.2 & 0 & 1 \\ 21 & -23.3 & 9.7 & -4 & 0 & -0.7 & 0 & 0 & 0 & 0 & 0 & 0 & -1.7 & 0 & -1 \\ 21 & -27.3 & 13.7 & -8 & 0 & -4.7 & 4 & -2 & 0 & 0 & 0 & 0 & 2.3 & 0 & 1 \end{bmatrix} \quad \text{and} \quad \boldsymbol{\lambda} = \frac{\alpha}{L^2}\begin{bmatrix} 0 \\ 9.37 \\ 18.75 \\ 32 \\ 41.37 \\ 54.63 \\ 64 \\ 64 \\ 64 \\ 73.37 \\ 86.63 \\ 96 \\ 109.3 \\ 118.6 \\ 128 \end{bmatrix}$$

$$(8\cdot5\cdot38)$$

Notice that the entries in $\mathbf{X}'$ generally get smaller as you go from $\mathbf{x}'_1$ on the left toward $\mathbf{x}'_{15}$ on the right. The magnitude of the entries in $\mathbf{x}'_{15}$ is a little less than five percent ($1/21$) of the entries in $\mathbf{x}'_1$. The entries in $\mathbf{X}'$ are the coefficients of the exponential terms in the solution (8·5·34). Entry x'_{ij} is the coefficient of $\exp(-\lambda_j\theta)$ in $t_i(\theta)$. The higher eigenvalues are less important because they decay rapidly for $\theta > 0$ and because their coefficients (toward the right in $\mathbf{X}'$) are small. We observed this behavior in Chapter 3 when looking at analytical solutions to partial differential equations.

The analytical solution of the governing partial differential equation (as found using separation of variables) can be put into the form

$$t(x,y,\theta) = \sum_{j=1}^{\infty} a_j(x,y)\exp(-\lambda_j\theta) \qquad (8\cdot5\cdot39)$$

This solution has an infinite number of eigenvalues λ_j and corresponding coefficients $a_j(x,y)$. At the location of node 1 $(x,y) = (0,0)$ the analytical solution (8·5·39) gives

$$t(0,0,\theta) = \sum_{j=1}^{\infty} a_j(0,0)\exp(-\lambda_j\theta) \qquad (8\cdot5\cdot40)$$

The first ten eigenvalues, 0, π^2, $2\pi^2$, ..., $16\pi^2$, are listed in Table 8•13. The first ten values of $a_j(0,0)$ are listed in Table 8•14.

The exact finite-difference solution (8·5·34) is an approximation to the analytical solution (8·5·39). If (8·5·34) is written for node 1 we obtain

$$t_1(\theta) = \sum_{j=1}^{N} x'_{1j}\exp(-\lambda_j\theta) \qquad (8\cdot5\cdot41)$$

Comparison of (8·5·41) with (8·5·40) shows that x'_{1j} is an approximation to $a_j(0,0)$. The values of x'_{1j} are the entries in the first row of $\mathbf{X}'$. Tables 8•13 and 8•14 give results of the exact finite-difference solutions for $N = 3, 6, 15$ and 45. Table 8•13 lists the first ten finite-difference eigenvalues.

Note that the finite-difference values approach the exact values as N increases. Table 8•14 lists the first ten values of x'_{1j}. The values of x'_{1j} approach the exact values $a_j(0,0)$ as N increases.

Since $\lambda_1 = 0$, the first term in (8·5·40) and in (8·5·41) is the steady-state solution. The steady-state solution (Table 8•14, $j = 1$) approaches the exact value of $1/3$ as the number of nodes increases. The terms that have negative coefficients in the solution of the partial differential equation (Table 8•14, $j = 1$ and 10) are also negative in the 45-node finite-difference solution.

Table 8•13 *Finite-difference approximations to the eigenvalues $L^2 \lambda_j / \alpha$ in (8·5·39).*

j	$N = 3$	6	15	45	Exact ($N = \infty$)
1	0	0	0	0	0
2	4	8	9.37	9.74	$9.87 = \pi^2$
3	8	16	18.75	19.49	$19.74 = 2\pi^2$
4		16	32	37.49	$39.48 = 4\pi^2$
5		24	41.37	47.23	$49.35 = 5\pi^2$
6		32	54.63	74.98	$78.96 = 8\pi^2$
7			64	79.02	$88.83 = 9\pi^2$
8			64	88.76	$98.70 = 10\pi^2$
9			64	116.5	$128.30 = 13\pi^2$
10			73.37	128.0	$157.91 = 16\pi^2$

Table 8•14 *Finite-difference approximations x'_{1j} to the coefficients $a_j(0,0)$ in (8·5·39).*

j	$N = 3$	6	15	45	Exact ($N = \infty$)
1	0.2500	0.3125	0.3281	0.3320	$0.3333 = 1/3$
2	0.5000	0.5000	0.4268	0.4105	$0.4053 = 4/\pi^2$
3	0.2500	0.2500	0.2134	0.2053	$0.2026 = 2/\pi^2$
4		−0.1250	−0.1250	−0.1067	$-0.1013 = -1/\pi^2$
5		0.0000	0.0000	0.0000	0
6		0.0625	0.0732	0.0534	$0.0507 = 1/2\pi^2$
7			−0.0045	0.0506	$0.0450 = 4/9\pi^2$
8			0.0256	0.0000	0
9			0.0101	0.0000	0
10			0.0000	−0.0312	$-0.0253 = -1/4\pi^2$

8•5•2 Euler and Crank-Nicolson solutions of $\mathbf{C\dot{t}} + \mathbf{St} = \mathbf{r}$

The Euler method (8·4·14) and the Crank-Nicolson method (8·4·20) may each be written in the same general form

$$\mathbf{At}^{(v+1)} = \mathbf{Bt}^{(v)} + \mathbf{r}\Delta\theta \tag{8·5·42}$$

where $\mathbf{A}$ and $\mathbf{B}$ are given in Table 8•15.

Table 8•15 $\mathbf{A}$ *and* $\mathbf{B}$ *for (8·5·42).*

Method	$\mathbf{A}$	$\mathbf{B}$
Euler	$\mathbf{C}$	$\mathbf{C} - \mathbf{S}\Delta\theta$
Crank-Nicolson	$\mathbf{C} + \mathbf{S}\dfrac{\Delta\theta}{2}$	$\mathbf{C} - \mathbf{S}\dfrac{\Delta\theta}{2}$

Insight into the nature of the solution to (8·5·42) may be obtained by assuming the solution to have the form

$$\mathbf{t}^{(v)} = \mathbf{a} + \mathbf{b}\theta^{(v)} + \mathbf{u}^{(v)} \tag{8·5·43}$$

Observe the similarity of this assumed form of the approximate solution to the assumed form of the exact solution given by (8·5·3). At time $\theta^{(v+1)} = \theta^{(v)} + \Delta\theta$ (8·5·43) becomes

$$\mathbf{t}^{(v+1)} = \mathbf{a} + \mathbf{b}\theta^{(v+1)} + \mathbf{u}^{(v+1)} = \mathbf{a} + \mathbf{b}(\theta^{(v)} + \Delta\theta) + \mathbf{u}^{(v+1)} \tag{8·5·44}$$

Substituting (8·5·43) and (8·5·44) into (8·5·42) gives

$$\mathbf{A}\left[\mathbf{a} + \mathbf{b}(\theta^{(v)} + \Delta\theta) + \mathbf{u}^{(v+1)}\right] = \mathbf{B}\left[\mathbf{a} + \mathbf{b}\theta^{(v)} + \mathbf{u}^{(v)}\right] + \mathbf{r}\Delta\theta$$

This may be rearranged to give

$$(\mathbf{A} - \mathbf{B})\mathbf{a} + (\mathbf{Ab} - \mathbf{r})\Delta\theta + (\mathbf{A} - \mathbf{B})\mathbf{b}\theta^{(v)} + \mathbf{Au}^{(v+1)} = \mathbf{Bu}^{(v)} \tag{8·5·45}$$

Observing from Table 8•15 that $\mathbf{A} - \mathbf{B} = \mathbf{S}\Delta\theta$ for both the Euler and the Crank-Nicolson methods, (8·5·45) becomes

$$[\mathbf{Sa} + \mathbf{Ab} - \mathbf{r}]\Delta\theta + \mathbf{Sb}^{(v)}\theta^{(v)}\Delta\theta + \mathbf{Au}^{(v+1)} = \mathbf{Bu}^{(v)}$$

This equation will be satisfied if the transient solution $\mathbf{u}$ satisfies

$$\mathbf{Au}^{(v+1)} = \mathbf{Bu}^{(v)} \tag{8·5·46}$$

and $\mathbf{b}$ and $\mathbf{a}$ satisfy

$$\mathbf{Sb} = \mathbf{0} \tag{8·5·47}$$

and

$$\mathbf{Sa} + \mathbf{Ab} = \mathbf{r} \qquad \text{or} \qquad \mathbf{Sa} = \mathbf{r} - \mathbf{Ab} \tag{8·5·48}$$

The initial condition for (8·5·46) is obtained by setting $v = 0$ and $\theta^{(v)} = \theta^{(0)} = 0$ in (8·5·43) to give

$$\mathbf{t}^{(0)} = \mathbf{a} + \mathbf{u}^{(0)}$$

or

$$\mathbf{u}^{(0)} = \mathbf{t}^{(0)} - \mathbf{a} \tag{8·5·49}$$

The vector $\mathbf{b}$ must satisfy (8·5·47) which happens to be identical to (8·5·5) which $\mathbf{b}$ had to satisfy in the exact solution. As in the exact solution, the solution for **b** will depend on the value of $\det(S)$ and there are two cases to consider:

1. If $\det(S) \neq 0$, the solution to (8·5·47) is **b** = **0** and (8·5·48) reduces to $S a = r$ which can be solved for **a**, the steady-state solution. These results are identical to (8·5·8) and (8·5·9) for the exact solution.

2. If $\det(S) = 0$, the solution to (8·5·47) is $\mathbf{b} = b\mathbf{1}$ and (8·5·48) becomes

$$S a = r - bA\mathbf{1} \tag{8·5·50}$$

For the Euler method, Table 8•15 gives $A = \text{Ç}$. For $\det(S) = 0$ it is also true that $S = K$ and $\text{Ç} = C$. Substituting $A = C$ into (8·5·50) gives

$$S a = r - bC\mathbf{1} \tag{8·5·51}$$

For the Crank-Nicolson method, Table 8•15 gives

$$A = \text{Ç} + S\frac{\Delta\theta}{2} = C + S\frac{\Delta\theta}{2}$$

Then, upon substituting this into (8·5·50) we obtain

$$S a = r - b\left[C + S\frac{\Delta\theta}{2}\right]\mathbf{1} = r - bC\mathbf{1} - b\frac{\Delta\theta}{2}S\mathbf{1}$$

Since $S\mathbf{1} = K\mathbf{1} = 0$ for $\det(S) = 0$, this also reduces to (8·5·51). These results, $\mathbf{b} = b\mathbf{1}$ and (8·5·51), are identical to (8·5·10) and (8·5·11) for the exact solution.

Therefore the values of **a** and **b** in the approximate solution (8·5·43) are identical to the values of **a** and **b** in the exact solution (8·5·3) summarized in Table 8•11. The only difference between the Euler and Crank-Nicolson approximate solutions and the exact solution is the solution for $\mathbf{u}(\theta)$.

We must now look for the solution to (8·5·46) with the initial condition given by (8·5·49). We can look at the exact solution to get a clue of how to proceed. There (8·5·15) we assumed

$$\mathbf{u}(\theta) = \mathbf{x}\exp(-\lambda\theta)$$

Evaluation of this assumed exact solution at $\theta = \theta^{(v+1)}$ would be

$$\mathbf{u}(\theta^{(v+1)}) = \mathbf{x}\exp(-\lambda\theta^{(v+1)})$$

Upon letting $\mathbf{u}(\theta^{(v+1)}) = \mathbf{u}^{(v+1)}$ and since $\theta^{(v+1)} = (v+1)\Delta\theta$ we may write the exact solution as

$$\mathbf{u}^{(v+1)} = \mathbf{x}\exp[-\lambda(v+1)\Delta\theta] = \mathbf{x}[\exp(-\lambda\Delta\theta)]^{v+1}$$

Thus we will assume that the solution to (8·5·46) has the form given by

$$\mathbf{u}^{(v+1)} = \mathbf{z}\mu^{v+1} \tag{8·5·52}$$

We can think of **z** as an approximation to **x** and μ as an approximation to $\exp(-\lambda\theta)$. At time step v we may write

$$\mathbf{u}^{(v)} = \mathbf{z}\mu^{v} \tag{8·5·53}$$

Substitution of (8·5·52) and (8·5·53) into (8·5·46) gives

$$A\mathbf{z}\mu^{v+1} = B\mathbf{z}\mu^{v}$$

Upon dividing by μ^{v} and rearranging,

$$B\mathbf{z} = \mu A\mathbf{z} \tag{8·5·54}$$

This is a generalized eigenproblem for μ and **z**.

The eigenproblem (8·5·54) may be rewritten as

$$[\mathbf{B} - \mu\mathbf{A}]\mathbf{z} = \mathbf{0} \tag{8·5·55}$$

It will be instructive to look at this eigenproblem for both the Euler method and the Crank-Nicolson method. Rather than actually computing μ and $\mathbf{z}$, we will relate them to the λ and $\mathbf{x}$ obtained in the exact solution (8·5·17).

Euler method Substituting the values of $\mathbf{A}$ and $\mathbf{B}$ from Table 8•15 for the Euler method into (8·5·55) gives

$$[\mathbf{Ç} - \mathbf{S}\Delta\theta - \mu\mathbf{Ç}]\mathbf{z} = \mathbf{0}$$

Combining the $\mathbf{Ç}$ terms, dividing by $-\Delta\theta$ and rearranging gives

$$\left[\mathbf{S} - \frac{1-\mu}{\Delta\theta}\mathbf{Ç}\right]\mathbf{z} = \mathbf{0}$$

In the exact solution of Section 5•1•1 we have already solved the generalized eigenproblem given by (8·5·17) as $(\mathbf{S} - \lambda\mathbf{Ç})\mathbf{x} = \mathbf{0}$. By comparison of these two eigenproblems we can say that

$$\frac{1-\mu}{\Delta\theta} = \lambda \qquad \text{and} \qquad \mathbf{z} = \mathbf{x}$$

Upon rearranging to solve for μ,

$$\mu = 1 - \lambda\Delta\theta \qquad\qquad \text{[Euler]} \tag{8·5·56}$$

The eigenvalues μ of the Euler method are related to the eigenvalues λ of the exact solution to the discrete finite-difference equations by (8·5·56). The eigenvectors $\mathbf{z}$ of the Euler method are identical to the eigenvectors $\mathbf{x}$ of the exact solution to the discrete finite-difference equations.

Crank-Nicolson method Substituting the values of $\mathbf{A}$ and $\mathbf{B}$ from Table 8•15 for the Crank-Nicolson method into (8·5·55) gives

$$\left[\mathbf{Ç} - \mathbf{S}\frac{\Delta\theta}{2} - \mu\left(\mathbf{Ç} + \mathbf{S}\frac{\Delta\theta}{2}\right)\right]\mathbf{z} = \mathbf{0}$$

Rearranging,

$$\left[(1-\mu)\mathbf{Ç} - \mathbf{S}\frac{\Delta\theta}{2}(1+\mu)\right]\mathbf{z} = \mathbf{0}$$

Dividing by $-\frac{\Delta\theta}{2}(1+\mu)$ and rearranging,

$$\left[\mathbf{S} - \frac{2(1-\mu)}{\Delta\theta(1+\mu)}\mathbf{Ç}\right]\mathbf{z} = \mathbf{0}$$

Comparison of this eigenproblem to the one given by (8·5·17) gives

$$\frac{2(1-\mu)}{\Delta\theta(1+\mu)} = \lambda \qquad \text{and} \qquad \mathbf{z} = \mathbf{x}$$

Upon rearranging to solve for μ,

$$\mu = \frac{2 - \lambda\Delta\theta}{2 + \lambda\Delta\theta} \qquad\qquad \text{[Crank-Nicolson]} \tag{8·5·57}$$

The eigenvalues μ of the Crank-Nicolson method are related to the eigenvalues λ of the exact solution to the discrete finite-difference equations by (8·5·57). The eigenvectors $\mathbf{z}$ of the Crank-Nicolson method are identical to the eigenvectors $\mathbf{x}$ of the exact solution to the discrete finite-difference equations.

Summary The Euler and Crank-Nicolson eigenvalues μ are related to the exact eigenvalues as given by (8·5·56) and (8·5·57). The Euler and Crank-Nicolson eigenvectors $\mathbf{z}$ are both identical to the exact eigenvectors $\mathbf{x}$.

Since the problem (8·5·46) we are solving is linear and homogeneous, we will write the solution (8·5·52) as a sum over each of the eigensolutions. Following (8·5·26), we will write the general solution as

$$\mathbf{u}^{(v+1)} = \sum_{j=1}^{N} d_j \mathbf{x}_j \mu_j^{v+1} \tag{8·5·58}$$

where we have used the result that the eigenvectors $\mathbf{z}$ are identical to the eigenvectors $\mathbf{x}$ obtained for the exact solution in Section 8•5•1. As we did for (8·5·26), we will choose the constants d_j to make the solution (8·5·58) satisfy the initial condition (8·5·49). Setting $v = -1$ in (8·5·58) gives

$$\mathbf{u}^{(0)} = \sum_{j=1}^{N} d_j \mathbf{x}_j \mu_j^0 = \sum_{j=1}^{N} d_j \mathbf{x}_j$$

Substituting this result into the initial condition (8·5·49) gives

$$\sum_{j=1}^{N} d_j \mathbf{x}_j = \mathbf{t}^{(0)} - \mathbf{a} \tag{8·5·59}$$

To evaluate the d_j we will use the orthogonality property (8·5·24) of the eigenvectors discussed in Section 8•5•1. Premultiplying (8·5·59) by $\mathbf{x}_i^T \mathbf{C}$ gives

$$\sum_{j=1}^{N} d_j \mathbf{x}_i^T \mathbf{C} \mathbf{x}_j = \mathbf{x}_i^T \mathbf{C}(\mathbf{t}^{(0)} - \mathbf{a})$$

From (8·5·24) we see that $\mathbf{x}_i^T \mathbf{C} \mathbf{x}_j$ is 0 for $j \neq i$ and is 1 for $j = i$. Therefore each term in the summation is 0 except when $j = i$. Only one term, $j = i$, will remain in the sum and we may then write

$$d_i = \mathbf{x}_i^T \mathbf{C}(\mathbf{t}^{(0)} - \mathbf{a}) \tag{8·5·60}$$

The eigenvectors $\mathbf{x}_i$ are obtained from the eigenproblem given by (8·5·16). Following (8·5·29), we will now define a set of modified eigenvectors $\mathbf{x}_j' = d_j \mathbf{x}_j$, where d_j is obtained from (8·5·60). The solution (8·5·58) may then be written as

$$\mathbf{u}^{(v+1)} = \sum_{j=1}^{N} \mathbf{x}_j' \mu_j^{v+1} \tag{8·5·61}$$

As in the exact solution (8·5·30) it is often convenient to write this sum of products using matrix notation. Upon writing out the sum in (8·5·61),

$$\mathbf{u}^{(v+1)} = \mathbf{x}_1' \mu_1^{v+1} + \mathbf{x}_2' \mu_2^{v+1} + \cdots + \mathbf{x}_N' \mu_N^{v+1}$$

The eigenvector matrix $\mathbf{X}'$ has already been defined by (8·5·31) and can be introduced here. Following the definition of $\boldsymbol{\lambda}$ (8·5·32) we will now define an $N \times 1$ eigenvalue vector $\boldsymbol{\mu}$ containing the N eigenvalues μ_1, μ_2, ..., μ_N and an $N \times 1$ vector $\boldsymbol{\mu}^{v+1}$ containing μ_1^{v+1}, μ_2^{v+1}, ..., μ_N^{v+1} as

$$
\boldsymbol{\mu} = \begin{bmatrix} \mu_1 \\ \mu_2 \\ \vdots \\ \mu_i \\ \vdots \\ \mu_N \end{bmatrix}
\qquad
\boldsymbol{\mu}^{v+1} = \begin{bmatrix} \mu_1^{v+1} \\ \mu_2^{v+1} \\ \vdots \\ \mu_i^{v+1} \\ \vdots \\ \mu_N^{v+1} \end{bmatrix}
\tag{8·5·62}
$$

We can then write (8·5·61) in terms of $\boldsymbol{\mu}$ and $\mathbf{X}'$ as

$$
\mathbf{u}^{(v+1)} = \mathbf{X}'\boldsymbol{\mu}^{v+1}
\tag{8·5·63}
$$

where $\mathbf{X}'$ contains the modified eigenvectors $\mathbf{x}_1'$, $\mathbf{x}_2'$, ..., $\mathbf{x}_N'$.

The complete solution to (8·5·42) for $\mathbf{t}^{(v+1)}$ is then found by substituting (8·5·61) or (8·5·63) into (8·5·44) to give

$$
\mathbf{t}^{(v+1)} = \mathbf{a} + \mathbf{b}\theta^{(v+1)} + \sum_{j=1}^{N} \mathbf{x}_j'\mu_j^{v+1} = \mathbf{a} + \mathbf{b}\theta^{(v+1)} + \mathbf{X}'\boldsymbol{\mu}^{v+1}
\tag{8·5·64}
$$

The vectors $\mathbf{a}$, $\mathbf{b}$ and $\mathbf{x}_j'$ are still as given in Table 8•12 for the exact solution of the finite-difference equations. The constants μ_j are obtained by solving the exact eigenproblem (8·5·16) for λ_j. Then, for the Euler method μ_j is found from (8·5·56) and for the Crank-Nicolson method μ_j is found from (8·5·57).

Equation (8·5·64) is the step-by-step numerical solution to (8·5·42) in a form that will be very helpful for understanding numerical stability and oscillation. Computation of the eigenvalues and eigenvectors in (8·5·64) takes too much effort in large problems and thus either the Euler or Crank-Nicolson solution discussed in sections 8•4•2 and 8•4•3 would be used in practice. It will be instructive, however, to return to the examples discussed in Section 8•4•4 and discuss them in the context of (8·5·64).

For the $\mathbf{\zeta} = \mathbf{C}$, $\mathbf{S}$ and $\mathbf{r}$ given in (8·4·5), $\mathbf{a}$ and $\mathbf{b}$ are both zero, $\boldsymbol{\lambda}$ and $\mathbf{X}$ are given by (8·5·35), and the modified eigenvector matrix $\mathbf{X}'$ is given by (8·5·36). For the 6-node Euler example, $\boldsymbol{\mu}$ can be obtained using (8·5·56) with $\Delta\theta = 0.02 L^2 / \alpha$ to give

$$
\boldsymbol{\mu} = \begin{bmatrix} \mu_1 \\ \mu_2 \\ \mu_3 \\ \mu_4 \\ \mu_5 \\ \mu_6 \end{bmatrix} = \begin{bmatrix} 1.00 \\ 0.84 \\ 0.68 \\ 0.68 \\ 0.52 \\ 0.36 \end{bmatrix}
$$

Since all entries in $\boldsymbol{\mu}$ are between 0 and 1, each μ_i^{v+1} will steadily decay toward 0 (except for μ_1^{v+1} which stays at 1) as v increases. Figure 8•32 shows how μ_2^{v+1} and μ_6^{v+1} vary with time step v. The effect of the largest eigenvalue μ_N is essentially zero for $v \geq 5$. This behavior of $\boldsymbol{\mu}^{v+1}$ corresponds to the smooth response of the finite-difference solution shown in Figure 8•27.

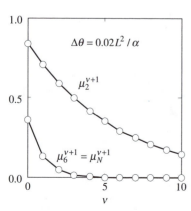

Figure 8•32 *Behavior of terms involving μ_2 and μ_6 in 6-node finite-difference Euler solution.*

For the 15-node finite-difference Euler solution shown in Figure 8•28, the vector $\boldsymbol{\mu}$ is computed from the λ_i in (8·5·38) and (8·5·56) to be

$$\boldsymbol{\mu} = \begin{bmatrix} \mu_1 \\ \mu_2 \\ \mu_3 \\ \mu_4 \\ \mu_5 \\ \mu_6 \\ \mu_7 \\ \mu_8 \\ \mu_9 \\ \mu_{10} \\ \mu_{11} \\ \mu_{12} \\ \mu_{13} \\ \mu_{14} \\ \mu_{15} \end{bmatrix} = \begin{bmatrix} 1 \\ 0.8125 \\ 0.6250 \\ 0.3600 \\ 0.1726 \\ -0.0926 \\ -0.2800 \\ -0.2800 \\ -0.2800 \\ -0.4674 \\ -0.7326 \\ -0.9200 \\ -1.186 \\ -1.372 \\ -1.560 \end{bmatrix}$$

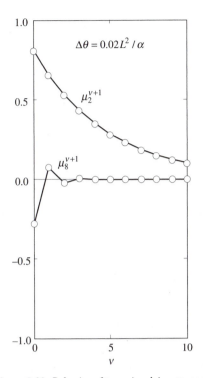

Figure 8•33 *Behavior of terms involving μ_2 and μ_8 in 15-node finite-difference Euler solution.*

Observe that $\boldsymbol{\mu}$ contains 10 negative entries. These entries in $\boldsymbol{\mu}^{v+1}$ will alternate in sign as v increases. Figure 8•33 shows how μ_2^{v+1} and μ_8^{v+1} vary with v. The damped oscillations of μ_8^{v+1} are basically zero for $v \geq 3$. The three entries in $\boldsymbol{\mu}$ that are less than -1 will grow in amplitude as v increases as shown in Figure 8•34 for μ_{15}^{v+1}. This explains the unstable oscillatory behavior seen in Figure 8•28. Reducing $\Delta\theta$ by a factor of two reduces the negative entries in $\boldsymbol{\mu}$ from 10 to 3, with $\mu_{15} = -0.28$. This produces the smooth finite-difference solution shown in Figure 8•29.

For the 15-node finite-difference Crank-Nicolson solution shown in Figure 8•30 the vector $\boldsymbol{\mu}$ is computed from (8·5·38) and (8·5·57) to be

$$\boldsymbol{\mu} = \begin{bmatrix} 1 \\ 0.8286 \\ 0.6842 \\ 0.5152 \\ 0.4147 \\ 0.2934 \\ 0.2195 \\ 0.2195 \\ 0.2195 \\ 0.1536 \\ 0.0716 \\ 0.0204 \\ -0.0444 \\ -0.0851 \\ -0.1228 \end{bmatrix}$$

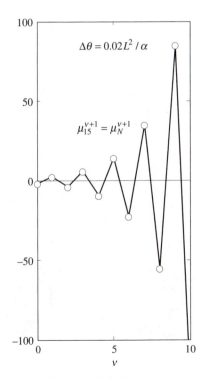

Figure 8•34 *Behavior of term involving $\mu_{15} = \mu_N$ in 15-node finite-difference Euler solution.*

Notice that $\boldsymbol{\mu}$ contains three negative entries, but the absolute values of these entries are less than 1. These entries in $\boldsymbol{\mu}^{v+1}$ will oscillate as v increases but will decay in amplitude. In fact, since their absolute values are so small, they will decay rapidly toward zero. These oscillations are not even noticeable in Figure 8•30.

8•5•3 Comparison of solutions to Çt + St = r

The exact finite-difference solution at $\theta = \theta^{(v+1)} = (v+1)\Delta\theta$ is found from (8·5·34) as

$$\mathbf{t}^{(v+1)} = \mathbf{a} + \mathbf{b}\theta^{(v+1)} + \mathbf{X}'\exp[-\boldsymbol{\lambda}(v+1)\Delta\theta]$$

Upon rearrangement of the exponential this becomes

$$\mathbf{t}^{(v+1)} = \mathbf{a} + \mathbf{b}\theta^{(v+1)} + \mathbf{X}'[\exp(-\boldsymbol{\lambda}\Delta\theta)]^{v+1} \qquad (8.5.65)$$

The approximate Euler and Crank-Nicolson solutions (8·5·64) are both given by

$$\mathbf{t}^{(v+1)} = \mathbf{a} + \mathbf{b}\theta^{(v+1)} + \sum_{j=1}^{N}\mathbf{x}'_j\mu_j^{v+1} = \mathbf{a} + \mathbf{b}\theta^{(v+1)} + \mathbf{X}'\boldsymbol{\mu}^{v+1} \qquad (8.5.66)$$

Comparison of the approximate solutions (8·5·66) with the exact solution (8·5·65) shows that they are identical with the exception that $\boldsymbol{\mu}$ in the approximate solution (8·5·66) is an approximation to $\exp(-\boldsymbol{\lambda}\Delta\theta)$ in the exact solution (8·5·65). All three solutions may then be written as (8·5·66) with the entries in $\boldsymbol{\mu}$ as given in Table 8•16.

Table 8•16 *Values of μ_i for (8·5·66).*

Method	μ_i	
Exact	$\exp(-\lambda_i\Delta\theta)$	(8.5.67)
Euler	$1 - \lambda_i\Delta\theta$	(8.5.56)
Crank-Nicolson	$\dfrac{2 - \lambda_i\Delta\theta}{2 + \lambda_i\Delta\theta}$	(8.5.57)

The series representation for $\exp(-\lambda_i\Delta\theta)$ is given by

$$\exp(-\lambda_i\Delta\theta) = 1 - \lambda_i\Delta\theta + \frac{(\lambda_i\Delta\theta)^2}{2!} - \frac{(\lambda_i\Delta\theta)^3}{3!} + \cdots \qquad (8.5.68)$$

Note that μ_i, as given by (8·5·56) for the Euler method, agrees with the first two terms in the series (8·5·68) for $\exp(-\lambda_i\Delta\theta)$. For the Crank-Nicolson method, if the division in (8·5·57) is carried out (long hand) one finds that

$$\mu_i = 1 - \lambda_i\Delta\theta + \frac{(\lambda_i\Delta\theta)^2}{2} - \frac{(\lambda_i\Delta\theta)^3}{4} + \cdots \qquad (8.5.69)$$

Note that μ_i, as given by (8·5·69) for the Crank-Nicolson method, agrees with the first three terms in the series (8·5·68) for $\exp(-\lambda_i\Delta\theta)$. This is better agreement than the Euler method.

The three curves of μ_i as a function of $\lambda_i\Delta\theta$ are shown in Figure 8•35. Observe that both the Euler and the Crank-Nicolson approximations are very close to the exact curve if the time step $\Delta\theta$ is very small. Euler and Crank-Nicolson both become worse as $\Delta\theta$ increases. The Euler method is far more in error than the Crank-Nicolson method. This is not unexpected since the Euler method of approximating the time derivatives to move ahead in time, (8·4·11), is not as good as the Crank-Nicolson procedure, (8·4·19). Also, the Crank-Nicolson result (8·5·69) agrees with the first three terms of the exact solution (8·5·68) whereas the Euler result (8·5·56) only agrees with the first two terms of the exact solution (8·5·68).

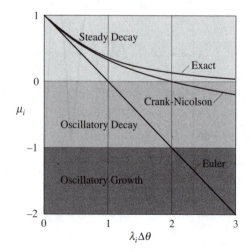

Figure 8•35 *Comparison of Euler and Crank-Nicolson values of μ_i to exact values.*

The behavior of the exact, Euler and Crank-Nicolson solutions can be examined if the solution (8·5·66) is written out in terms of the individual eigenvectors and eigenvalues as

$$\mathbf{t}^{(v+1)} = \mathbf{a} + \mathbf{b}\theta^{(v+1)} + \mathbf{x}'_1\mu_1^{v+1} + \mathbf{x}'_2\mu_2^{v+1} + \cdots + \mathbf{x}'_i\mu_i^{v+1} + \cdots + \mathbf{x}'_N\mu_N^{v+1}$$

$$(8\cdot5\cdot70)$$

The values of $\mathbf{a}$, $\mathbf{b}$, $\mathbf{x}'_1$, $\mathbf{x}'_2$, ..., $\mathbf{x}'_N$ are constants and are the same for all three solutions (exact, Euler and Crank-Nicolson). The different behavior of the each of the three solutions depends only on the values of the μ_i. From (8·5·67) and as shown in Figure 8•35, the values of μ_i for the exact solution are between 1 and 0 for all values of $\Delta\theta$. From (8·5·56) and as shown in Figure 8•35, the values of μ_i for the Euler solution can run from 1 to large negative values as $\Delta\theta$ increases. From (8·5·57) the values of μ_i for Crank-Nicolson run from 1 at $\Delta\theta = 0$ and approach -1 as $\Delta\theta$ increases.

The exact solution of the finite difference equations has N eigenvalues $\lambda_1 \leq \lambda_2 \leq \cdots \leq \lambda_i \leq \cdots \leq \lambda_N$. Each of these eigenvalues has a corresponding eigenvalue $\mu_1 \geq \mu_2 \geq \cdots \geq \mu_i \geq \cdots \geq \mu_N$ in (8·5·70). The relations between λ_i and μ_i are given in Table 8•16 and shown in Figure 8•35. A value of μ_i (for any i) between 1 and 0 produces a stable decay in the solution (8·5·70) because μ_i^{v+1} will decrease steadily as v increases. A value of μ_i between 0 and -1 will produce stable oscillations in (8·5·70) because μ_i^{v+1} will alternate sign and decrease in amplitude as v increases. Values of μ_i less than -1 will produce unstable oscillations in the solution (8·5·70) because μ_i^{v+1} will alternate sign and increase in amplitude as v increases. The Euler method can exhibit all three types of behavior whereas Crank-Nicolson values of μ_i will never be less than -1 and hence a Crank-Nicolson solution may have oscillations but will never be unstable.

The severity of the oscillations depends upon the value of $\Delta\theta$. As $\Delta\theta$ increases each $\lambda_i\Delta\theta$ will increase and each μ_i will decrease. As soon as $\lambda_N\Delta\theta$ increases to the point where $\mu_N < 0$ stable oscillations will appear in the solution. As $\Delta\theta$ continues to increase, more and more of the μ_i will become negative and the severity of the oscillations will increase. If $\Delta\theta$ increases to the point where $\mu_N < -1$, the oscillations will become unstable. As $\Delta\theta$ continues to increase, more and more of the μ_i will drop below -1 and the severity of the unstable oscillations will increase.

It should also be pointed out that $\lambda_i\Delta\theta = 2$ is the point beyond which the Euler oscillations will become unstable. In order to prevent this undesirable behavior $\Delta\theta$ must be chosen small enough so that $\lambda_N\Delta\theta \leq 2$. This "critical value" of $\Delta\theta$ is called the *critical time step* $\Delta\theta_c$ and is given by

$$\Delta\theta_c = \frac{2}{\lambda_N} \tag{8\cdot5\cdot71}$$

where λ_N is the maximum eigenvalue of (8·5·16). It should also be observed from Figure 8•35 that $\Delta\theta_c$ is also the time step beyond which oscillations will begin to creep into the Crank-Nicolson solution. Also, if $\Delta\theta < \Delta\theta_c/2$ no oscillations will occur in the Euler solution.

It is possible that $\mathbf{x}'_N$ in (8·5·70) could be zero. In this case λ_N and μ_N would drop out of the solution and λ_{N-1} would be used in (8·5·71) to find $\Delta\theta_c$. As discussed in the next section, modified eigenvectors associated with specified temperature nodes will be zero and thus drop out of the solution.

8·5·4 Critical-time-step estimate

It is helpful to have a knowledge of the critical time step $\Delta\theta_c$ so you can pick a time step $\Delta\theta$ that will avoid undesirable oscillations in the solution. The value of $\Delta\theta_c$ may be calculated from the maximum eigenvalue λ_N using (8·5·71). Unfortunately, for the large problems often found in conduction analyses (*e.g.*, several hundred nodes or more), it is expensive to actually calculate the maximum eigenvalue required in (8·5·71). In practice one relies on an estimate of the maximum eigenvalue $\lambda_{N,est}$ to obtain an estimate of the critical time step $\Delta\theta_{c,est}$ given by

$$\Delta\theta_{c,est} = \frac{2}{\lambda_{N,est}} \tag{8·5·72}$$

As long as $\Delta\theta \leq \Delta\theta_{c,est} \leq \Delta\theta_c$ unstable oscillations will not appear in an Euler solution. To ensure that $\Delta\theta_{c,est} \leq \Delta\theta_c$ we want $\lambda_{N,est} \geq \lambda_N$.

We may substitute λ_N and the corresponding eigenvector $\mathbf{x}_N$ into the eigenproblem (8·5·17) to obtain

$$[\mathbf{S} - \lambda_N \mathbf{\varsigma}]\mathbf{x}_N = \mathbf{0}$$

Upon writing out a few of the entries in this equation,

$$
\begin{bmatrix}
s_{11} - \lambda_N \varsigma_{11} & s_{12} & \cdots & s_{1i} & \cdots & s_{1N} \\
s_{21} & s_{22} - \lambda_N \varsigma_{22} & \cdots & s_{2i} & \cdots & s_{2N} \\
\vdots & \vdots & & \vdots & & \vdots \\
s_{i1} & s_{i2} & \cdots & s_{ii} - \lambda_N \varsigma_{ii} & \cdots & s_{iN} \\
\vdots & \vdots & & \vdots & & \vdots \\
s_{N1} & s_{N2} & \cdots & s_{Ni} & \cdots & s_{NN} - \lambda_N \varsigma_{NN}
\end{bmatrix}
\begin{bmatrix}
x_{1N} \\ x_{2N} \\ \vdots \\ x_{iN} \\ \vdots \\ x_{NN}
\end{bmatrix}
=
\begin{bmatrix}
0 \\ 0 \\ \vdots \\ 0 \\ \vdots \\ 0
\end{bmatrix}
\tag{8·5·73}
$$

We will assume that x_{iN} has the largest magnitude of the entries in $\mathbf{x}_N$. We can then learn something about λ_N by looking at row i. Multiplying out the matrices to obtain the equation for row i,

$$s_{i1}x_{1N} + s_{i2}x_{2N} + \cdots + (s_{ii} - \lambda_N \varsigma_{ii})x_{iN} + \cdots + s_{iN}x_{NN} = 0$$

Upon solving for λ_N,

$$\lambda_N = \frac{1}{\varsigma_{ii}}\left[s_{i1}\frac{x_{1N}}{x_{iN}} + s_{i2}\frac{x_{2N}}{x_{iN}} + \cdots + s_{ii} + \cdots + s_{iN}\frac{x_{NN}}{x_{iN}} \right]$$

Since we are looking for an upper bound on λ_N, we may write

$$\lambda_N \leq \frac{1}{\varsigma_{ii}}\sum_{j=1}^{N}\left| s_{ij}\frac{x_{jN}}{x_{iN}} \right| \tag{8·5·74}$$

Since we have assumed that x_{iN} has the largest magnitude in $\mathbf{x}_N$, the largest magnitude that any of the ratios in (8·5·74) can have is 1. Upon setting each of the ratios in (8·5·74) equal to 1 gives

$$\lambda_N \le \frac{1}{\varsigma_{ii}} \sum_{j=1}^{N} |s_{ij}| \tag{8.5.75}$$

This result was found by considering only row i which was assumed to have the largest magnitude in $\mathbf{x}_N$. Since we have not calculated $\mathbf{x}_N$, we don't really know which row in $\mathbf{x}_N$ has the largest magnitude. To be safe we will check every row, using (8.5.75) to estimate λ_N for $i = 1$ to N and take the maximum of these N values as $\lambda_{N,est}$. That is, we will take

$$\lambda_{N,est} = \underset{i=1}{\overset{N}{\text{Max}}} \left[\frac{1}{\varsigma_{ii}} \sum_{j=1}^{N} |s_{ij}| \right] \tag{8.5.76}$$

Effect of specified temperatures When row i is a specified-temperature row, not necessarily the row containing the maximum value of x_{iN}, it follows from the discussion about specified-temperature nodes in Section 8•2•2 leading to (8.2.27) that $\mathbf{S}$ and $\mathbf{C}$ will have only zeros as the off-diagonal entries in row i and in column i. Thus (8.5.73) will be

$$\begin{bmatrix} s_{11} - \lambda_N \varsigma_{11} & s_{12} & \cdots & 0 & \cdots & s_{1N} \\ s_{21} & s_{22} - \lambda_N \varsigma_{22} & \cdots & 0 & \cdots & s_{2N} \\ \vdots & \vdots & & \vdots & & \vdots \\ & & & 0 & & \\ 0 & 0 & \cdots 0 & s_{ii} - \lambda_N \varsigma_{ii} \ 0 & \cdots & 0 \\ & & & 0 & & \\ \vdots & \vdots & & \vdots & & \vdots \\ s_{N1} & s_{N2} & \cdots & 0 & \cdots & s_{NN} - \lambda_N \varsigma_{NN} \end{bmatrix} \begin{bmatrix} x_{1N} \\ x_{2N} \\ \vdots \\ x_{iN} \\ \vdots \\ x_{NN} \end{bmatrix} = \begin{bmatrix} 0 \\ 0 \\ \vdots \\ 0 \\ \vdots \\ 0 \end{bmatrix}$$

$$\tag{8.5.77}$$

The eigenvalue λ_N in (8.5.77) is obtained by setting the determinant of the coefficient matrix equal to zero.

The determinant of an $N \times N$ matrix $\mathbf{A}$ is defined by the equation

$$\det(\mathbf{A}) = |\mathbf{A}| = \sum_{i=1}^{N} (-1)^{i+j} a_{ij} M_{ij} \tag{8.5.78}$$

where M_{ij} is the minor determinant of a_{ij} formed from $\mathbf{A}$ by deleting row i and column j. The evaluation in (8.5.78) may be carried out in terms of any column j. As an example, the evaluation of a 3×3 determinant in terms of the entries in column 1 is given by

$$\det(\mathbf{A}) = \begin{vmatrix} a_{11} & a_{12} & a_{13} \\ a_{21} & a_{22} & a_{23} \\ a_{31} & a_{32} & a_{33} \end{vmatrix} = a_{11} \begin{vmatrix} a_{22} & a_{23} \\ a_{32} & a_{33} \end{vmatrix} - a_{21} \begin{vmatrix} a_{12} & a_{13} \\ a_{32} & a_{33} \end{vmatrix} + a_{31} \begin{vmatrix} a_{12} & a_{13} \\ a_{22} & a_{23} \end{vmatrix}$$

Note that if each of entries in column 1 is zero (*i.e.*, $a_{11} = a_{21} = a_{31} = 0$) the value of the determinant will be 0.

The determinant of the coefficient matrix in (8.5.77) will be 0 if $\lambda_N = s_{ii}/\varsigma_{ii}$ because $s_{ii} - \lambda_N \varsigma_{ii}$ will then be 0 and therefore every entry in column i will be 0. Thus $\lambda_N = s_{ii}/\varsigma_{ii}$ will actually be one of the eigenvalues, not just an estimate. If, after checking all of the rows, $\lambda_N = s_{ii}/\varsigma_{ii}$ is the maximum value obtained, it will be the exact value of λ_N. In this case the eigenvector $\mathbf{x}_N$ in (8·5·77) will be satisfied by

$$\mathbf{x}_N = \begin{bmatrix} 0 \\ 0 \\ \vdots \\ 0 \\ x_{iN} \\ 0 \\ \vdots \\ 0 \end{bmatrix} \tag{8·5·79}$$

The exact, Euler and Crank-Nicolson solutions are each given by (8·5·66) as

$$\mathbf{t}^{(v+1)} = \mathbf{a} + \mathbf{b}\theta^{(v+1)} + \sum_{j=1}^{N} d_j \mathbf{x}_j \mu_j^{v+1} \tag{8·5·80}$$

where d_j may be found from (8·5·28) or (8·5·60) as

$$d_j = \mathbf{x}_j^T \mathbf{\varsigma}(\mathbf{t}^{(0)} - \mathbf{a})$$

For $j = N$,

$$d_N = \mathbf{x}_N^T \mathbf{\varsigma}(\mathbf{t}^{(0)} - \mathbf{a}) \tag{8·5·81}$$

We can obtain $\mathbf{x}_N^T$ from (8·5·79). As discussed in Section 8•5•1, when there is a specified temperature, $\mathbf{a}$ will be the steady-state solution. If $t_i = T_i$ is specified then $t_i^{(0)} = T_i$ and $a_i = T_i$. Thus the entry in row i of $(\mathbf{t}^{(0)} - \mathbf{a})$ will be $T_i - T_i = 0$. We may thus write (8·5·81) as

$$d_N = \begin{bmatrix} 0 & 0 & \cdots & 0 & x_{iN} & 0 & \cdots & 0 \end{bmatrix} \begin{bmatrix} \varsigma_{11} & & & & \\ & \varsigma_{22} & & & \\ & & \varsigma_{ii} & & \\ & & & & \\ & & & & \varsigma_{NN} \end{bmatrix} \begin{bmatrix} t_1^{(0)} - a_1 \\ t_2^{(0)} - a_2 \\ 0 \\ t_N^{(0)} - a_N \end{bmatrix}$$

Multiplying to find $\mathbf{x}_N^T \mathbf{\mathsf{C}}$ and then $\mathbf{x}_N^T \mathbf{\mathsf{C}}(\mathbf{t}^{(0)} - \mathbf{a})$,

$$d_N = \begin{bmatrix} 0 & \cdots & 0 & \varsigma_{ii} x_{iN} & 0 & \cdots & 0 \end{bmatrix} \begin{bmatrix} t_1^{(0)} - a_1 \\ \\ t_2^{(0)} - a_2 \\ \\ \\ 0 \\ \\ \\ \\ \\ \\ \\ \\ \\ t_N^{(0)} - a_N \end{bmatrix} = 0$$

Since $d_N = 0$, μ_N will not appear in the solution (8·5·80). Therefore the eigenvalue estimate $\lambda_N = s_{ii} / \varsigma_{ii}$ we obtained from specified-temperature row i will not affect the solution and can be disregarded. This means that in using (8·5·76) we do not need to consider any specified-temperature row i_s. We then modify (8·5·76) to convey this fact and write

$$\lambda_{N,est} = \underset{\substack{i=1 \\ i \neq i_s}}{\overset{N}{\text{Max}}} \left[\frac{1}{\varsigma_{ii}} \sum_{j=1}^{N} | s_{ij} | \right] \tag{8·5·82}$$

Substituting (8·5·82) into (8·5·72) gives

$$\Delta\theta_{c,est} = \frac{2}{\underset{\substack{i=1 \\ i \neq i_s}}{\overset{N}{\text{Max}}} \left[\dfrac{1}{\varsigma_{ii}} \sum_{j=1}^{N} | s_{ij} | \right]} \tag{8·5·83}$$

This is the result that was stated without proof as (8·4·29). It is a convenient way to examine the $\mathbf{S}$ and $\mathbf{\mathsf{C}}$ matrices to arrive at an estimate of the critical time step. It is much faster to do this than to actually compute the maximum eigenvalue.

Example Determine $\Delta\theta_{c,est}$ for the transient discussed in Section 8•4•4 whose finite-difference equation (8·4·5) is given by

$$\frac{\rho c L^2}{32} \begin{bmatrix} 1 & & & & & \\ & 4 & & & & \\ & & 2 & & & \\ & & & 4 & & \\ & & & & 4 & \\ & & & & & 1 \end{bmatrix} \begin{bmatrix} \dot{t}_1 \\ \dot{t}_2 \\ \dot{t}_3 \\ \dot{t}_4 \\ \dot{t}_5 \\ \dot{t}_6 \end{bmatrix} + \frac{k}{2} \begin{bmatrix} 1 & -1 & 0 & & & \\ -1 & 4 & -1 & -2 & & \\ 0 & -1 & 2 & 0 & -1 & \\ & -2 & 0 & 4 & -2 & 0 \\ & & -1 & -2 & 4 & -1 \\ & & & 0 & -1 & 1 \end{bmatrix} \begin{bmatrix} t_1 \\ t_2 \\ t_3 \\ t_4 \\ t_5 \\ t_6 \end{bmatrix} = \begin{bmatrix} 0 \\ 0 \\ 0 \\ 0 \\ 0 \\ 0 \end{bmatrix}$$

Solution: In this problem there are no specified temperatures so we must check all six rows. Application of (8·5·82) to row 1 gives

$$\text{Row 1:} \quad \frac{1}{\varsigma_{11}} \sum_{j=1}^{N} | s_{1j} | = \frac{1}{\rho c L^2 / 32} \left[\frac{k}{2} + \frac{k}{2} \right] = 32 \frac{k}{\rho c L^2} = 32 \frac{\alpha}{L^2}$$

Row 2: $\dfrac{1}{\varsigma_{22}}\displaystyle\sum_{j=1}^{N}|s_{2j}| = \dfrac{1}{4\rho c L^2/32}\left[\dfrac{k}{2}+\dfrac{4k}{2}+\dfrac{k}{2}+\dfrac{2k}{2}\right] = 32\dfrac{k}{\rho c L^2} = 32\dfrac{\alpha}{L^2}$

Row 3: $\dfrac{1}{\varsigma_{33}}\displaystyle\sum_{j=1}^{N}|s_{3j}| = \dfrac{1}{2\rho c L^2/32}\left[\dfrac{k}{2}+\dfrac{2k}{2}+\dfrac{k}{2}\right] = 32\dfrac{k}{\rho c L^2} = 32\dfrac{\alpha}{L^2}$

Row 4: $\dfrac{1}{\varsigma_{44}}\displaystyle\sum_{j=1}^{N}|s_{4j}| = \dfrac{1}{4\rho c L^2/32}\left[\dfrac{2k}{2}+\dfrac{4k}{2}+\dfrac{2k}{2}\right] = 32\dfrac{k}{\rho c L^2} = 32\dfrac{\alpha}{L^2}$

Row 5: $\dfrac{1}{\varsigma_{55}}\displaystyle\sum_{j=1}^{N}|s_{5j}| = \dfrac{1}{4\rho c L^2/32}\left[\dfrac{k}{2}+\dfrac{2k}{2}+\dfrac{4k}{2}+\dfrac{k}{2}\right] = 32\dfrac{k}{\rho c L^2} = 32\dfrac{\alpha}{L^2}$

Row 6: $\dfrac{1}{\varsigma_{66}}\displaystyle\sum_{j=1}^{N}|s_{6j}| = \dfrac{1}{\rho c L^2/32}\left[\dfrac{k}{2}+\dfrac{k}{2}\right] = 32\dfrac{k}{\rho c L^2} = 32\dfrac{\alpha}{L^2}$

In this example each row gives the same value. Thus $\lambda_{N,est} = 32\alpha/L^2$. The exact value of λ_N given by (8·5·35) also happens to be $32\alpha/L^2$. This agreement between the row estimates and the exact value will not happen in general. Substitution into (8·5·83) gives

$$\Delta\theta_{c,est} = \dfrac{2}{32\alpha/L^2} = 0.0625L^2/\alpha$$

In formulating the finite-difference system of equations (8·4·5) the spacing between the nodal points was $\Delta x = L/2$. In the above example, if L is replaced by $2\Delta x$ in $\Delta\theta_{c,est}$ one obtains

$$\Delta\theta_{c,est} = \dfrac{2}{32\alpha/L^2} = \dfrac{2(2\Delta x)^2}{32\alpha} = \dfrac{(\Delta x)^2}{4\alpha}$$

Upon normalizing by multiplying both sides by $\alpha/(\Delta x)^2$ we obtain,

$$\dfrac{\alpha\Delta\theta_{c,est}}{(\Delta x)^2} = \dfrac{1}{4} \tag{8·5·84}$$

This fairly famous result is for uniformly spaced nodes in two-dimensional problems but it should not be trusted if there is a convection boundary.

To illustrate the effect of convection, application of (8·5·82) to (8·4·9) shows that row 6 gives the largest estimate of λ_N and then from (8·5·83)

$$\dfrac{\alpha\Delta\theta_{c,est}}{(\Delta x)^2} = \dfrac{1}{4\left[1+\dfrac{1}{4}\dfrac{hL}{k}\right]} \tag{8·5·85}$$

rather than (8·5·84). For a Biot number of $hL/k = 4$, (8·5·85) gives $\Delta\theta_{c,est}$ of one-half of the value given by (8·5·84). Convection has reduced $\Delta\theta_{c,est}$. Furthermore, as h increases $\Delta\theta_{c,est}$ decreases.

Observe that if Δx is halved $\Delta\theta_{c,est}$ decreases by a factor of four. Decreasing nodal spacing not only increases the number of equations to solve, it requires you to take much smaller time steps to avoid unwanted numerical oscillations. In the above example $\Delta\theta = 0.02L^2/\alpha$ is less than $\Delta\theta_{c,est} = 0.0625L^2/\alpha$ and there are no unstable oscillations in Figure 8•23. If the nodal spacing is halved (which gives the 15-node formulation shown in Figure 8•24), then $\Delta\theta_{c,est} = 0.0625L^2/4\alpha = 0.015625L^2/\alpha$. A time step of $\Delta\theta = 0.02L^2/\alpha$ exceeds $\Delta\theta_{c,est}$ and unstable oscillations are quite apparent in Figure 8•24.

8•6 SUPPLEMENTARY PROBLEMS

At this point we have seen how to use a regular system of nodal points to solve problems in complicated regions with some relatively simple boundary conditions. In this section we will extend our capabilities to handle even more complicated problems.

8•6•1 Irregular nodal-point locations

In discretizing the spatial problem in Section 8•2 we restricted ourselves to regularly spaced nodal points. To get a fine spacing of nodal points in one portion of the problem it is thus necessary to accept a fine spacing in other portions of the region where such fine spacing might not really be needed. This adds to the work involved in solving the problem. It would be helpful if we could use irregularly spaced nodal points as shown in Figure 8•36. Figures 8•36 and 8•5 both have 26 nodes along the boundary. By allowing irregularly spaced nodes, the boundary approximation in Figure 8•36 is much better than in Figure 8•5. Observe that the total number of nodes in Figure 8•36 is only 49 as compared to 69 in Figure 8•5. We have not been forced into using as many interior nodes in order to get a good approximation of the boundary.

Figure 8•37 shows a typical interior node i surrounded by five adjacent nodes j, k, ℓ, m and n. Nodal system i is constructed by drawing the perpendicular bisectors to the line segments joining node i to the adjacent nodes. The approximate rate equations for conduction out of nodal system i, generation within system i and energy storage within system i are given by

$$q_{ij} = \frac{k_{ij}A_{ij}}{d_{ij}}(t_i - t_j) \quad , \quad (g'''V)_i \quad \text{and} \quad (\rho cV)_i \frac{dt_i}{d\theta}$$

Everything mentioned in Section 8•2 remains the same except that now every nodal system may be different. Some interior nodal systems may be connected to five surrounding nodes, but others may be connected to three, four, six or other numbers of nodes. Conduction areas will not be uniform. Volumes of the nodal systems will not be uniform. There will be many more calculations to make, but this can be done. We will still get a banded, symmetric set of coupled ordinary differential equations to solve.

Although, with a computer, this might look like a simple way to handle irregularly spaced nodal points, there are several headaches. The area A_{ij} will depend on the location of nodes k and n. Figure 8•38 shows that A_{ij} gets smaller if nodes k and n are moved closer to node j. It is therefore more difficult to program the computer to calculate A_{ij} when the nodes are irregularly spaced. To carry this to an extreme, A_{ij} will fall outside the polygon surrounding node i if nodes k and n are moved to the positions shown in Figure 8•39. In this case we would not connect node j to node i but rather connect node i only to nodes k, ℓ, m and n. Thus we cannot be completely arbitrary about the way we connect nodal points together to form the nodal systems.

In summary, although irregularly spaced nodal points can be handled by finite differences, it is very cumbersome even with the computer. Doing so by hand is out of the question when you might be dealing with hundreds or thousands of nodes. The finite-element method discussed in Chapter 9 is far superior in handling irregularly spaced nodal points.

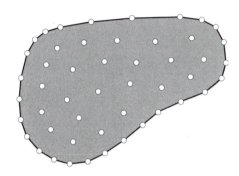

Figure 8•36 *Approximation of irregular region using irregularly spaced nodal points.*

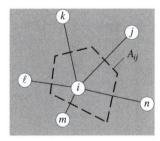

Figure 8•37 *Typical interior node i in an irregular pattern of nodal points.*

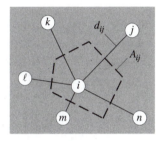

Figure 8•38 *A_{ij} is decreased as nodes k and n move toward node j.*

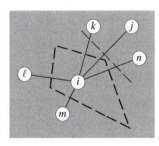

Figure 8•39 *Nodes k and n can be located so that A_{ij} falls outside the polygon surrounding node i.*

8·6·2 Time-dependent specified nodal temperatures

When nodal temperatures are specified functions of time, the discussion of specified temperatures in Section 8•2•2 must be modified. Suppose that the temperature at node i is a specified function of time as given by

$$t_i = T_i(\theta) \tag{8·6·1}$$

In this case (8·2·27) will not give the desired result (8·6·1) because the capacitance term is not zero. Instead we will replace (8·2·27) by

$$c_{ii}\frac{dt_i}{d\theta} + \kappa_{ii}t_i = \kappa_{ii}T_i(\theta) + c_{ii}\frac{dT_i(\theta)}{d\theta} \tag{8·6·2}$$

As in Section 8•2•2, κ_{ii} is really the diagonal entry in row i of $(\mathbf{K}+\mathbf{H})$, not $\mathbf{K}$. When $T_i(\theta)$ is constant the time derivative in the capacitance term on the right-hand side of (8·6·2) is zero and (8·6·2) reduces to (8·2·27). The capacitance term is included on the right-hand side of (8·6·2) to produce a differential equation whose solution will be $t_i = T_i(\theta)$ when $t_i(0) = T_i(0)$.

Upon embedding (8·6·2) in the global matrix equation and indicating the modifications to other equations required to maintain symmetry,

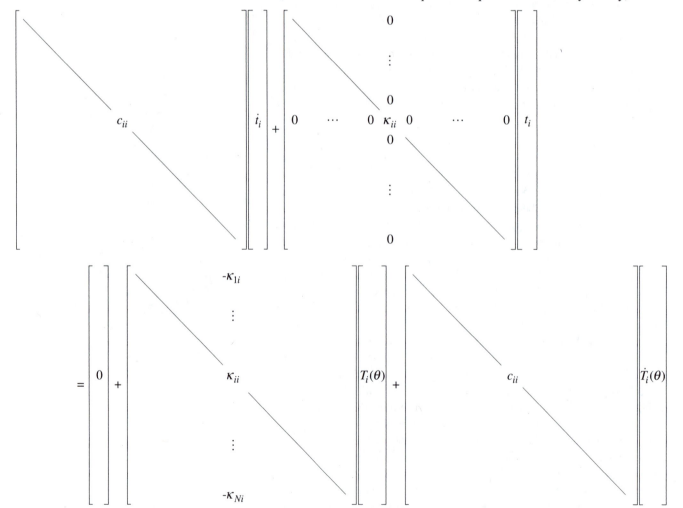

When this is done for all specified-temperature nodes we may still write

$$\mathbf{Ç}\dot{\mathbf{t}} + \mathbf{St} = \mathbf{r}$$

where $\mathbf{Ç}$ and $\mathbf{S}$ are the same matrices as one would obtain if the specified temperatures had been constants, but $\mathbf{r}$ is now time dependent. We can write $\mathbf{r}$ as

$$\mathbf{r} = \mathbf{r}' + \mathbf{K}'\mathbf{t}_s + \mathbf{C}'\dot{\mathbf{t}}_s \tag{8.6.3}$$

where $\mathbf{r}' = (\mathbf{g} + \mathbf{h} + \mathbf{q})$ with zeros in the specified-temperature rows. The matrices $\mathbf{K}'$ and $\mathbf{C}'$ contain the entries from modifications to $(\mathbf{K} + \mathbf{H})$ and $\mathbf{C}$, respectively, that are required to move the specified temperatures from the left-hand side to the right-hand side and to keep $\mathbf{Ç}$ and $\mathbf{S}$ symmetric. The vector $\mathbf{t}_s$ and its time derivative $\dot{\mathbf{t}}_s$ contain zeros except for the specified-temperature rows which contain the known values and time derivatives of the specified temperatures.

When $\mathbf{r}$ is time dependent, the Crank-Nicolson method discussed in Section 8•4•3 must be modified. Instead of (8·4·20) Crank-Nicolson will now give

$$\left[\mathbf{Ç} + \mathbf{S}\frac{\Delta\theta}{2}\right]\mathbf{t}^{(v+1)} = \left[\mathbf{Ç} - \mathbf{S}\frac{\Delta\theta}{2}\right]\mathbf{t}^{(v)} + \left[\mathbf{r}^{(v)} + \mathbf{r}^{(v+1)}\right]\frac{\Delta\theta}{2}$$

which, instead of (8·4·24), we may rewrite as

$$\mathbf{A}\mathbf{t}^{(v+1)} = \mathbf{B}\mathbf{t}^{(v)} + \mathbf{b}^{(v+1)} \tag{8.6.4}$$

where

$$\mathbf{A} = \mathbf{Ç} + \mathbf{S}\frac{\Delta\theta}{2} \qquad \mathbf{B} = \mathbf{Ç} - \mathbf{S}\frac{\Delta\theta}{2} \qquad \mathbf{b}^{(v+1)} = \left[\mathbf{r}^{(v)} + \mathbf{r}^{(v+1)}\right]\frac{\Delta\theta}{2} \tag{8.6.5}$$

The only difference between (8·6·4) and (8·4·24) is that $\mathbf{b}$ is no longer constant but must be evaluated at each time step. When the entries in $\mathbf{t}_s$ and $\dot{\mathbf{t}}_s$ are both given for each time step, $\mathbf{r}$ can be evaluated directly from (8·6·3). When only $\mathbf{t}_s$ is given at each time step we will resort to an approximation to replace $\dot{\mathbf{t}}_s$. With the help of (8·6·3), we may write

$$\mathbf{b}^{(v+1)} = \left[\mathbf{r}' + \mathbf{K}'\mathbf{t}_s^{(v)} + \mathbf{C}'\dot{\mathbf{t}}_s^{(v)} + \mathbf{r}' + \mathbf{K}'\mathbf{t}_s^{(v+1)} + \mathbf{C}'\dot{\mathbf{t}}_s^{(v+1)}\right]\frac{\Delta\theta}{2}$$

Rearranging,

$$\mathbf{b}^{(v+1)} = \mathbf{r}'\Delta\theta + \mathbf{K}'\frac{\Delta\theta}{2}\left\{\mathbf{t}_s^{(v)} + \mathbf{t}_s^{(v+1)}\right\} + \mathbf{C}'\left\{\frac{\dot{\mathbf{t}}_s^{(v)} + \dot{\mathbf{t}}_s^{(v+1)}}{2}\right\}\Delta\theta \tag{8.6.6}$$

To avoid having to evaluate $\dot{\mathbf{t}}_s^{(v)}$ and $\dot{\mathbf{t}}_s^{(v+1)}$ we will write

$$\frac{\dot{\mathbf{t}}_s^{(v)} + \dot{\mathbf{t}}_s^{(v+1)}}{2} = \frac{\mathbf{t}_s^{(v+1)} - \mathbf{t}_s^{(v)}}{\Delta\theta} \tag{8.6.7}$$

The left-hand side of (8·6·7) is the average derivative during $\Delta\theta$. The right-hand side is the derivative if $\mathbf{t}_s$ varied linearly during $\Delta\theta$. Substituting (8·6·7) into (8·6·6),

$$\mathbf{b}^{(v+1)} = \mathbf{r}'\Delta\theta + \mathbf{K}'\frac{\Delta\theta}{2}\left\{\mathbf{t}_s^{(v)} + \mathbf{t}_s^{(v+1)}\right\} + \mathbf{C}'\left\{\mathbf{t}_s^{(v+1)} - \mathbf{t}_s^{(v)}\right\}$$

Or, after slight rearrangement,

$$\mathbf{b}^{(v+1)} = \mathbf{r}'\Delta\theta + \left[\mathbf{C}' + \mathbf{K}'\frac{\Delta\theta}{2}\right]\mathbf{t}_s^{(v+1)} - \left[\mathbf{C}' - \mathbf{K}'\frac{\Delta\theta}{2}\right]\mathbf{t}_s^{(v)} \tag{8.6.8}$$

The computation of $\mathbf{b}^{(v+1)}$ from (8·6·8) is more complicated than the calculation of $\mathbf{b} = \mathbf{r}\Delta\theta$ when the specified temperatures are constant. Furthermore, $\mathbf{b}^{(v+1)}$ must be calculated at each time step rather than only once at the start. More computer storage space is needed to find $\mathbf{b}^{(v+1)}$ but the computations can be carried out never-the-less. The solution for $\mathbf{t}^{(v+1)}$ given by (8·6·4) will contain the exact values of the specified temperatures at $\theta^{(v+1)}$.

8·6·3 Variable Ç, S and r

Entries in **Ç** depend on the ρc product which may vary with temperature. Entries in **S** depend upon thermal conductivity k which may also depend on temperature. When there is a convection boundary, entries in **r** will depend on t_∞. For a specified-heat-flux boundary, entries in **r** will depend on q_s''. Both t_∞ and q_s'' may be known functions of time. When there is energy generation, entries in **r** will depend upon g''' which may depend on temperature and/or time. Problems such as these require modifications to what we have seen so far. When **Ç**, **S** and **r** are variables, the system of differential equations (8·2·34) being solved may be written at time $\theta^{(v)}$ as

$$\mathbf{Ç}^{(v)}\dot{\mathbf{t}}^{(v)} + \mathbf{S}^{(v)}\mathbf{t}^{(v)} = \mathbf{r}^{(v)} \tag{8·6·9}$$

The easiest situation to treat is when **Ç** and **S** are constants and **r** depends only on time. This occurs when k and ρc do not depend upon temperature and g''', t_∞ and q_s'' are known functions of time. The Crank-Nicolson method is given by (8·4·19) as

$$\mathbf{t}^{(v+1)} = \mathbf{t}^{(v)} + \left[\frac{\dot{\mathbf{t}}^{(v)} + \dot{\mathbf{t}}^{(v+1)}}{2}\right]\Delta\theta \tag{8·6·10}$$

Premultiplying (8·6·10) by **Ç** gives

$$\mathbf{Ç}\mathbf{t}^{(v+1)} = \mathbf{Ç}\mathbf{t}^{(v)} + \left[\mathbf{Ç}\dot{\mathbf{t}}^{(v)} + \mathbf{Ç}\dot{\mathbf{t}}^{(v+1)}\right]\frac{\Delta\theta}{2} \tag{8·6·11}$$

We can eliminate the time derivatives in (8·6·11) using (8·6·9) to evaluate **Ç**ṫ at $\theta^{(v)}$ and at $\theta^{(v+1)}$. Thus, (8·6·11) may be written as

$$\mathbf{Ç}\mathbf{t}^{(v+1)} = \mathbf{Ç}\mathbf{t}^{(v)} + \left[(-\mathbf{S}\mathbf{t}^{(v)} + \mathbf{r}^{(v)}) + (-\mathbf{S}\mathbf{t}^{(v+1)} + \mathbf{r}^{(v+1)})\right]\frac{\Delta\theta}{2}$$

where **r** has superscripts because it is time dependent. Rearranging terms,

$$\left[\mathbf{Ç} + \mathbf{S}\frac{\Delta\theta}{2}\right]\mathbf{t}^{(v+1)} = \left[\mathbf{Ç} - \mathbf{S}\frac{\Delta\theta}{2}\right]\mathbf{t}^{(v)} + \left[\mathbf{r}^{(v)} + \mathbf{r}^{(v+1)}\right]\frac{\Delta\theta}{2} \tag{8·6·12}$$

The only difference between (8·6·12) and (8·4·20) is that **r** must be continually updated as the transient goes on. This is not a major problem since **r** is a known function of time.

The next easiest case is a steady-state problem where **S** and **r** depend upon temperature. For a steady-state problem, (8·6·9) simplifies to give

$$\mathbf{S}\mathbf{t} = \mathbf{r} \tag{8·6·13}$$

We are facing an iterative solution since **S** and **r** depend upon **t** and **t** cannot be found until **S** and **r** are known. We will start the iteration by using some appropriate average property values to predict the entries in **S** and **r**. Then we can solve (8·6·13) to predict **t**. The predicted **t** can be used to correct **S** and **r**. Then we can solve (8·6·13) again but with the corrected **S** and **r** to correct **t**. The corrected **t** can then be used to correct **S** and **r** again. Then (8·6·13) can be used to correct **t** again. This process can be continued until the change in **t** is acceptably small. If we use $\mathbf{t}^{(p)}$ to denote the current predicted value of **t** and $\mathbf{t}^{(p+1)}$ to denote the corrected or next predicted value of **t**, we evaluate $\mathbf{t}^{(p+1)}$ by repeatedly solving

$$\mathbf{S}^{(p)}\mathbf{t}^{(p+1)} = \mathbf{r}^{(p)} \tag{8·6·14}$$

where $\mathbf{S}^{(p)}$ and $\mathbf{r}^{(p)}$ are evaluated using $\mathbf{t}^{(p)}$.

The most difficult case is when **Ç**, **S** and **r** are all functions of **t**. One way to handle this problem for one time step is to assume $\mathbf{Ç}^{(v+1)} \approx \mathbf{Ç}^{(v)}$,

$\mathbf{S}^{(v+1)} \approx \mathbf{S}^{(v)}$ and $\mathbf{r}^{(v+1)} \approx \mathbf{r}^{(v)}$. The Crank-Nicolson scheme (8·4·20) can then be written as

$$\left[\mathbf{\c{C}}^{(v)} + \mathbf{S}^{(v)}\frac{\Delta\theta}{2}\right]\mathbf{t}^{(v+1)} = \left[\mathbf{\c{C}}^{(v)} - \mathbf{S}^{(v)}\frac{\Delta\theta}{2}\right]\mathbf{t}^{(v)} + \mathbf{r}^{(v)}\Delta\theta \qquad (8\cdot6\cdot15)$$

This system of algebraic equations can be solved for $\mathbf{t}^{(v+1)}$ which then can be used to find $\mathbf{\c{C}}^{(v+1)}$, $\mathbf{S}^{(v+1)}$ and $\mathbf{r}^{(v+1)}$. To move ahead in time, we can increase the superscripts in (8·6·15) and calculate $\mathbf{t}^{(v+2)}$ from

$$\left[\mathbf{\c{C}}^{(v+1)} + \mathbf{S}^{(v+1)}\frac{\Delta\theta}{2}\right]\mathbf{t}^{(v+2)} = \left[\mathbf{\c{C}}^{(v+1)} - \mathbf{S}^{(v+1)}\frac{\Delta\theta}{2}\right]\mathbf{t}^{(v+1)} + \mathbf{r}^{(v+1)}\Delta\theta$$

This process can be repeated for succeeding time steps.

There are a lot of computations in the scheme represented by (8·6·15). We must recalculate $\mathbf{\c{C}}$, $\mathbf{S}$, $\mathbf{r}$ and the matrix coefficients of $\mathbf{t}^{(v)}$ and $\mathbf{t}^{(v+1)}$ at each time step. Also, since (8·6·15) is a system of algebraic equations to solve and the coefficient matrix of $\mathbf{t}^{(v+1)}$ is different for each time step, this matrix must be redecomposed at each time step. If the calculations at each time step are too much work, the Euler method is another alternative for moving ahead in time. Application of the Euler method (8·4·14) gives

$$\mathbf{\c{C}}^{(v)}\mathbf{t}^{(v+1)} = (\mathbf{\c{C}}^{(v)} - \mathbf{S}^{(v)}\Delta\theta)\mathbf{t}^{(v)} + \mathbf{r}^{(v)}\Delta\theta \qquad (8\cdot6\cdot16)$$

instead of (8·6·15). The quantities $\mathbf{\c{C}}$, $\mathbf{S}$, $\mathbf{r}$ and the matrix coefficient of $\mathbf{t}^{(v)}$ must still be recalculated at each time step. However, because $\mathbf{\c{C}}$ is diagonal, (8·6·16) can be solved explicitly for $\mathbf{t}^{(v+1)}$. Decomposition of $\mathbf{\c{C}}$ is not needed. Also, recall that Euler is less stable than Crank-Nicolson and might force you to take smaller times steps for a reasonable solution.

If you can afford increased computation and want improved accuracy, there is a predictor-corrector scheme that can be used. The initial condition $\mathbf{t}^{(0)}$ can be used to evaluate $\mathbf{\c{C}}^{(0)}$, $\mathbf{S}^{(0)}$ and $\mathbf{r}^{(0)}$. The initial time derivatives $\dot{\mathbf{t}}^{(0)}$ can be found from (8·6·9) by solving

$$\mathbf{\c{C}}^{(0)}\dot{\mathbf{t}}^{(0)} = -\mathbf{S}^{(0)}\mathbf{t}^{(0)} + \mathbf{r}^{(0)}$$

Starting with the initial values of $\mathbf{t}^{(0)}$ and $\dot{\mathbf{t}}^{(0)}$, you can step along in time by first predicting temperatures at $\theta^{(v+1)}$ using the Euler equation (8·4·11)

$$\mathbf{t}_{pred}^{(v+1)} = \mathbf{t}^{(v)} + \dot{\mathbf{t}}^{(v)}\Delta\theta \qquad (8\cdot6\cdot17)$$

The predicted temperatures $\mathbf{t}_{pred}^{(v+1)}$ can be used to calculate $\mathbf{\c{C}}_{pred}^{(v+1)}$, $\mathbf{S}_{pred}^{(v+1)}$ and $\mathbf{r}_{pred}^{(v+1)}$. The predicted time derivatives $\dot{\mathbf{t}}_{pred}^{(v+1)}$ can then be calculated from (8·6·9) by solving

$$\mathbf{\c{C}}_{pred}^{(v+1)}\dot{\mathbf{t}}_{pred}^{(v+1)} = -\mathbf{S}_{pred}^{(v+1)}\mathbf{t}_{pred}^{(v+1)} + \mathbf{r}_{pred}^{(v+1)} \qquad (8\cdot6\cdot18)$$

Temperatures can be corrected using Crank-Nicolson (8·6·10) as given by

$$\mathbf{t}_{corr}^{(v+1)} = \mathbf{t}^{(v)} + \left[\dot{\mathbf{t}}^{(v)} + \dot{\mathbf{t}}_{pred}^{(v+1)}\right]\frac{\Delta\theta}{2} \qquad (8\cdot6\cdot19)$$

The corrected temperatures $\mathbf{t}_{corr}^{(v+1)}$ can then be used to calculate $\mathbf{\c{C}}_{corr}^{(v+1)}$, $\mathbf{S}_{corr}^{(v+1)}$ and $\mathbf{r}_{corr}^{(v+1)}$. The corrected time derivatives $\dot{\mathbf{t}}_{corr}^{(v+1)}$ can be calculated from (8·6·9) by solving

$$\mathbf{\c{C}}_{corr}^{(v+1)}\dot{\mathbf{t}}_{corr}^{(v+1)} = -\mathbf{S}_{corr}^{(v+1)}\mathbf{t}_{corr}^{(v+1)} + \mathbf{r}_{corr}^{(v+1)} \qquad (8\cdot6\cdot20)$$

The steps from (8·6·19) through (8·6·20) may be repeated until successive values of $\mathbf{t}_{corr}^{(v+1)}$ are acceptably close. If more than one or two iterations are required you may want to reduce the step size $\Delta\theta$.

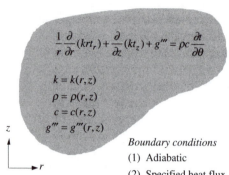

$$\frac{1}{r}\frac{\partial}{\partial r}(krt_r) + \frac{\partial}{\partial z}(kt_z) + g''' = \rho c\frac{\partial t}{\partial \theta}$$

$$k = k(r,z)$$
$$\rho = \rho(r,z)$$
$$c = c(r,z)$$
$$g''' = g'''(r,z)$$

Boundary conditions
(1) Adiabatic
(2) Specified heat flux

Initial condition   (3) Convection
$$t(r,z,0) = t^{(0)}(r,z)$$ (4) Specified temperature

Figure 8•40 *A general two-dimensional, transient conduction problem in cylindrical coordinates.*

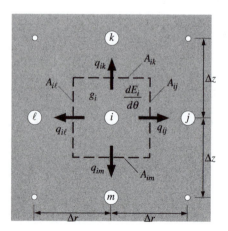

Figure 8•41 *Placement of nodal points in a square pattern showing approximation of region boundaries.*

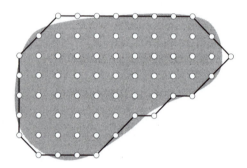

Figure 8•42 *Interior node i not at or adjacent to r = 0.*

8•6•4 Cylindrical coordinates

Some geometries are best described by using cylindrical coordinates. One example might be a cylindrical pressure vessel. If the pressure vessel is long, the temperature of the cylindrical walls might only be a function of radial position r and time θ. However, the temperature in the thick-walled ends of the pressure vessel will be a function of radial position r, axial position z and time θ. We will assume that temperature does not vary with angular position ϕ around the z-axis. Since only two space coordinates are needed, this is still a two-dimensional problem.

In this section we will consider the general class of problems pictured in Figure 8•40. The solid is formed by revolving the shaded area shown in Figure 8•40 about the z-axis. The cross-sectional geometry is not very simple and the material properties can vary with position. The governing partial differential equation is a simplification of (1·1·11) to eliminate dependence on ϕ. The finite-difference analysis of this problem will be quite similar to the discussion in Section 8•2 using x,y coordinates.

The spatial problem can be discretized by using a set of regularly spaced nodal points as shown in Figure 8•41. The spacing in the r-direction is uniform and equal to the spacing in the z-direction. A rectangular pattern could almost as easily have been chosen. An irregular pattern, however, would complicate the analysis considerably. A finer pattern of nodes would improve the approximation of the boundary.

Interior nodes (not at or adjacent to r = 0) An interior nodal point i that is not at or adjacent to $r = 0$ is shown in Figure 8•42. Node i is surrounded by nodal points j, k, ℓ and m. The dashed boundary around node i is the cross-section of a ring around the z-axis. The radial position of node i is r_i, the thickness of the ring in the r-direction is Δr and the height of the ring in the z-direction is Δz. The areas of the top and bottom surfaces of the ring are given by

$$A_{ik} = A_{im} = \pi\left[r_i + \frac{\Delta r}{2}\right]^2 - \pi\left[r_i - \frac{\Delta r}{2}\right]^2$$

Upon squaring the bracketed terms and simplifying,

$$A_{ik} = A_{im} = \pi\left[r_i^2 + r_i\Delta r + \frac{(\Delta r)^2}{4}\right] - \pi\left[r_i^2 - r_i\Delta r + \frac{(\Delta r)^2}{4}\right] = 2\pi r_i\Delta r \quad (8\cdot6\cdot21)$$

The volume V_i of the ring is the product of the area of the base times the height of the ring as given by

$$V_i = A_{im}\Delta z = 2\pi r_i\Delta r\Delta z \qquad (8\cdot6\cdot22)$$

Volume V_i is also the product of $2\pi r_i$ times the cross-sectional area $\Delta r\Delta z$ of the ring.

An energy balance on system i is given by

$$g_i = \frac{dE_i}{d\theta} + q_{im} + q_{i\ell} + q_{ij} + q_{ik} \qquad (8\cdot6\cdot23)$$

The rate equations for generation and energy storage are given by

$$g_i = g'''V_i \qquad \qquad \frac{dE_i}{d\theta} = \rho c V_i\frac{dt_i}{d\theta} \qquad (8\cdot6\cdot24)$$

The axial conduction terms q_{im} and q_{ik} are given by

$$q_{im} = \frac{kA_{im}}{\Delta z}(t_i - t_m) \qquad q_{ik} = \frac{kA_{ik}}{\Delta z}(t_i - t_k) \qquad (8\cdot6\cdot25)$$

The radial conduction terms q_{ij} and $q_{i\ell}$ will be obtained from the analysis of steady-state one-dimensional conduction in a thick-walled cylinder. The heat-transfer rate q_{ij} through a cylinder of inside radius r_i, outside radius r_j, inside temperature t_i, outside temperature t_j and length Δz is given by[1]

$$q_{ij} = \frac{t_i - t_j}{\dfrac{1}{2\pi k \Delta z} \ln(r_j / r_i)} = \frac{2\pi k \Delta z}{\ln(r_j / r_i)}(t_i - t_j) \qquad (8 \cdot 6 \cdot 26)$$

Similarly,

$$q_{i\ell} = \frac{t_i - t_\ell}{\dfrac{1}{2\pi k \Delta z} \ln(r_i / r_\ell)} = \frac{2\pi k \Delta z}{\ln(r_i / r_\ell)}(t_i - t_\ell)$$

Substituting the six rate equations into the energy balance (8·6·23) gives

$$g''' V_i = \rho c V_i \frac{dt_i}{d\theta} + \frac{k A_{im}}{\Delta z}(t_i - t_m) + \frac{2\pi k \Delta z}{\ln(r_i / r_\ell)}(t_i - t_\ell) + \frac{2\pi k \Delta z}{\ln(r_j / r_i)}(t_i - t_j)$$
$$+ \frac{k A_{ik}}{\Delta z}(t_i - t_k)$$

Upon rearranging,

$$g''' V_i = \rho c V_i \frac{dt_i}{d\theta} + \left[-\frac{k A_{im}}{\Delta z} t_m - \frac{2\pi k \Delta z}{\ln(r_i / r_\ell)} t_\ell + \left(\frac{k A_{im}}{\Delta z} + \frac{2\pi k \Delta z}{\ln(r_i / r_\ell)} \right. \right.$$
$$\left. \left. + \frac{2\pi k \Delta z}{\ln(r_j / r_i)} + \frac{k A_{ik}}{\Delta z} \right) t_i - \frac{2\pi k \Delta z}{\ln(r_j / r_i)} t_j - \frac{k A_{ik}}{\Delta z} t_k \right]$$
$$(8 \cdot 6 \cdot 27)$$

Next we will simplify the notation by defining

$$g_i = g''' V_i , \qquad c_{ii} = \rho c V_i , \qquad \kappa_{ii} = \frac{k A_{im}}{\Delta z} + \frac{2\pi k \Delta z}{\ln(r_i / r_\ell)} + \frac{2\pi k \Delta z}{\ln(r_j / r_i)} + \frac{k A_{ik}}{\Delta z}$$

$$\kappa_{im} = -\frac{k A_{im}}{\Delta z} , \qquad \kappa_{i\ell} = -\frac{2\pi k \Delta z}{\ln(r_i / r_\ell)} , \qquad \kappa_{ij} = -\frac{2\pi k \Delta z}{\ln(r_j / r_i)} , \qquad \kappa_{ik} = -\frac{k A_{ik}}{\Delta z}$$
$$(8 \cdot 6 \cdot 28)$$

Upon introducing (8·6·28) into (8·6·27) and switching left- and right-hand sides we obtain

$$c_{ii} \frac{dt_i}{d\theta} + \left[\kappa_{im} t_m + \kappa_{i\ell} t_\ell + \kappa_{ii} t_i + \kappa_{ij} t_j + \kappa_{ik} t_k \right] = g_i \qquad (8 \cdot 6 \cdot 29)$$

This result is typical of any interior node that is not at or adjacent to $r = 0$. It is the same as we obtained in Section 8•2•1 for x, y coordinates. This is one of a system of coupled, first-order, ordinary differential equations. Matrix notation can be used and the matrices will be similar to those we derived for x, y coordinates. That is, the capacitance matrix will be diagonal and the conduction matrix will be banded and symmetric. The only difference comes in the calculation of the coefficients in the differential equation (8·6·28) which are now found using (8·6·21), (8·6·22) and (8·6·28).

---

[1] When r_i and r_j are large and Δr is small, $\ln(r_j / r_i) \approx \Delta r / r_i$ and we may write

$$q_{ij} = \frac{2\pi k \Delta z}{\ln(r_j / r_i)}(t_i - t_j) = \frac{2\pi k \Delta z}{\Delta r / r_i}(t_i - t_j) = \frac{k 2\pi r_i \Delta z}{\Delta r}(t_i - t_j) = \frac{k A_{ij}}{\Delta r}(t_i - t_j)$$

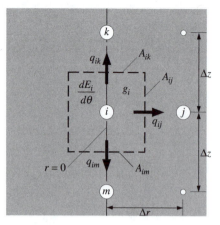

Figure 8•43 *Interior node i at r = 0.*

Interior nodes (at or adjacent to r = 0) An interior nodal point i at $r = 0$ is shown in Figure 8•43. The dashed boundary around node i designates a solid cylinder of diameter Δr. The conduction areas of this cylinder are

$$A_{ik} = A_{im} = \pi(\Delta r / 2)^2 \qquad \text{and} \qquad A_{ij} = \pi \Delta r \Delta z \qquad (8 \cdot 6 \cdot 30)$$

The volume of the cylinder is

$$V_i = \pi(\Delta r / 2)^2 \Delta z \qquad (8 \cdot 6 \cdot 31)$$

An energy balance on system i is given by

$$g_i = \frac{dE_i}{d\theta} + q_{im} + q_{ij} + q_{ik} \qquad (8 \cdot 6 \cdot 32)$$

There is no $q_{i\ell}$ in Figure 8•43, as there was in Figure 8•42, because q_{ij} includes the radial heat flow around the entire cylinder. The rate equations for g_i and $dE_i / d\theta$ are still given by (8·6·24) except that now V_i is given by (8·6·31) rather than (8·6·22). The rate equations for q_{im} and q_{ik} are still given by (8·6·25) except that now A_{ik} and A_{im} are given by (8·6·30) rather than (8·6·21). The rate equation for q_{ij} given by (8·6·26) cannot be used here since, for $r_i = 0$, it gives $q_{ij} = 0$. We must find another rate equation for q_{ij}.

To determine a rate equation for q_{ij} we will consider one-dimensional, steady-state conduction in an infinitely long, solid cylinder with uniform energy generation. For $t = t_i$ at $r = 0$ the analytical solution is given by

$$t(r) = t_i - \frac{g'''}{4k} r^2$$

The temperature t_j at $r = r_j$ is then given by

$$t_j = t_i - \frac{g'''}{4k} r_j^2 \qquad (8 \cdot 6 \cdot 33)$$

The heat-transfer rate q_{ij} at $r = r_j / 2 = \Delta r / 2$, for no axial conduction and a cylinder length of Δz, may be found from an energy balance as

$$q_{ij} = (g''' V)_i = g''' \pi \left(\frac{\Delta r}{2} \right)^2 \Delta z = \frac{1}{4} g''' \pi (\Delta r)^2 \Delta z \qquad (8 \cdot 6 \cdot 34)$$

Upon solving (8·6·33) for g''' and substituting it into (8·6·34) gives

$$q_{ij} = \frac{1}{4} \left[\frac{4k}{r_j^2} (t_i - t_j) \right] \pi (\Delta r)^2 \Delta z = \frac{k}{r_j^2} \pi (\Delta r)^2 \Delta z (t_i - t_j)$$

Since $\Delta r = r_j - r_i$ and $r_i = 0$, the rate equation for q_{ij} reduces to

$$q_{ij} = k \pi \Delta z (t_i - t_j) = \frac{k \pi \Delta r \Delta z}{\Delta r} (t_i - t_j) = \frac{k A_{ij}}{\Delta r} (t_i - t_j) \qquad (8 \cdot 6 \cdot 35)$$

This rate equation will also be used for nodes adjacent to nodes at $r = 0$.

Substituting the rate equations (8·6·24), (8·6·25) and (8·6·35) into the energy balance (8·6·32) gives

$$g''' V_i = \rho c V_i \frac{dt_i}{d\theta} + \frac{k A_{im}}{\Delta z} (t_i - t_m) + \frac{k A_{ij}}{\Delta r} (t_i - t_j) + \frac{k A_{ik}}{\Delta z} (t_i - t_k) \qquad (8 \cdot 6 \cdot 36)$$

where A_{im}, A_{ik} and V_i are now given by (8·6·30) and (8·6·31). With appropriate definitions of c_{ii}, κ_{im}, κ_{ii}, κ_{ij}, κ_{ik} and g_i, (8·6·36) can be written in the form given by (8·6·29).

Boundary nodes Boundary nodes that are not located at $r = 0$ are handled following the previous discussion in this section and the discussion for x, y coordinates in Section 8·2·2. The cross-sections of ring nodal systems on the boundary are still as shown in Figure 8·12. Volumes of these systems are $2\pi r_c$ times shaded area in Figure 8·12 where r_c is the centroid of the area about the z-axis. Surface areas swept out by a line segment are $2\pi r_c$ times the length of the boundary line segment where r_c is the radial distance to the midpoint of the line segment.

When boundary node i is at $r_i = 0$, seven possible nodal systems are shown in Figure 8·44. For $r_i = 0$, the radial conduction rate equation is given by (8·6·35) as

$$q_{ij} = \frac{kA_{ij}}{\Delta r}(t_i - t_j)$$

where for figures 8·44(a), (d) and (e) the axial length of the nodal system for radial conduction is $\Delta z / 2$ so that $A_{ij} = \pi \Delta r \Delta z / 2$ and for figures 8·44(c) and (f) the length of the nodal system is Δz so that $A_{ij} = \pi \Delta r \Delta z$. The rate equations for axial conduction q_{ik} and q_{im} are given by

$$q_{ik} = \frac{kA_{ik}}{\Delta z}(t_i - t_k) \quad \text{and} \quad q_{im} = \frac{kA_{im}}{\Delta z}(t_i - t_m)$$

where $A_{ik} = A_{im} = \pi(\Delta r / 2)^2$.

As an example of a boundary node at $r = 0$, consider the system shown in Figure 8·44(c). An enlarged view is shown in Figure 8·45. The energy terms include a convection boundary. The energy balance is given by

$$g_i + q_{c,i} = \frac{dE_i}{d\theta} + q_{ij} + q_{ik}$$

Upon substituting the rate equations we get

$$g'''V_i + h_{c,i}S_i(t_{\infty,i} - t_i) = \rho c V_i \frac{dt_i}{d\theta} + \frac{kA_{ij}}{\Delta r}(t_i - t_j) + \frac{kA_{ik}}{\Delta z}(t_i - t_k)$$

where the ambient temperature outside of node i is $t_{\infty,i}$ and is assumed to be *given*. The volume and areas of this system are given by

$$V_i = \frac{5}{24}\pi(\Delta r)^2 \Delta z , \quad A_{ij} = \pi \Delta r \Delta z , \quad A_{ik} = \pi(\Delta r / 2)^2$$

$$S_i = 2\pi \frac{\Delta r}{4}\sqrt{\left(\frac{\Delta r}{2}\right)^2 + \left(\frac{\Delta z}{2}\right)^2} = \pi \frac{\Delta r}{4}\sqrt{(\Delta r)^2 + (\Delta z)^2}$$

Upon rearranging the energy-balance equation and introducing notation similar to (8·6·28), we may write

$$c_{ii}\frac{dt_i}{d\theta} + \left[(\kappa_{ii} + h_{ii})t_i + \kappa_{ij}t_j + \kappa_{ik}t_k\right] = g_i + h_i \tag{8·6·37}$$

where the convection conductance h_{ii} is given by

$$h_{ii} = h_{c,i}S_i$$

and h_i is given by

$$h_i = h_{c,i}S_i t_{\infty,i}$$

The differential equation (8·6·37) can be included with the differential equations obtained for interior nodes.

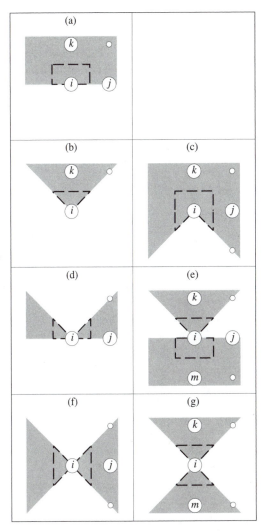

Figure 8·44 *Systems for node i at $r_i = 0$.*

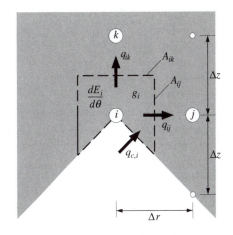

Figure 8·45 *Convection boundary node i at $r_i = 0$.*

Summary The finite-difference formulation for r, z-coordinate problems is close to the analysis presented in Section 8•2 for x, y-coordinate problems. The result is still a system of first-order ordinary differential equations that can be written as

$$\mathbf{C\dot{t}} + (\mathbf{K} + \mathbf{H})\mathbf{t} = \mathbf{g} + \mathbf{q} + \mathbf{h}$$

The entries in $\mathbf{C}$, $\mathbf{K}$, $\mathbf{H}$, $\mathbf{g}$, $\mathbf{q}$ and $\mathbf{h}$ are calculated in a slightly different manner to account for the variable conduction area. Modifications for specified-temperature nodes follow the discussion in Section 8•2•2. Programs *STEADY*, *EULER* and *CRANK* can still be used. Subroutines *DBAND*, *SBAND* and *YAXPB* are still applicable. The critical time step can still be estimated using (8·4·29).

8•7   SUMMARY REMARKS ON FINITE DIFFERENCES

This chapter has developed the fundamental ideas of the finite-difference method of finding approximate solutions to two-dimensional heat-conduction problems. With the aid of the digital computer the finite-difference method is capable of handling almost any problem (if you are willing to do the computations). The basic ideas were presented in sections 8•2 to 8•5. These were then extended in Section 8•6 to incorporate some of the complications that arise in practice. The extension of this chapter to three-dimensional problems would follow the same pattern as developed in this chapter.

One of the headaches that must be faced in handling practical problems is that computational times can become quite long (even on the computer). This is especially true in two- and three-dimensional transient problems. Bandwidths can be quite large so that implicit methods become costly. At the same time, however, the boundary conditions and irregular nodal spacing can cause severe restrictions in the allowable time-step size in the Euler method because of stability considerations. There are many alterations which people are forced to make to save on computational time or to improve stability limits. The particular modifications that must be made for a specific problem are usually determined by personal experience and trial and error. Textbooks and research papers can only serve as guides to suggest possible alternatives. Some of the references at the end of this chapter contain introductions to some of these techniques.

The example computer programs given in sections 8•3 and 8•4 assumed that the entries in $\mathbf{Ç}$, $\mathbf{S}$ and $\mathbf{r}$ had been derived by hand and simply needed to be read into the computer. The creation of $\mathbf{Ç}$, $\mathbf{S}$ and $\mathbf{r}$ is not a small task in general (especially when the nodal-point spacing is irregular). Section 8•6•1 hinted at some of the difficulties you would encounter in trying to do this. Rather than attempt to develop a general finite-difference subroutine to form $\mathbf{Ç}$, $\mathbf{S}$ and $\mathbf{r}$ it is much better to go on the Chapter 9 and study finite elements. Finite elements provides a computationally-convenient way to form $\mathbf{Ç}$, $\mathbf{S}$ and $\mathbf{r}$ for irregular problems.

SELECTED REFERENCES

1. Dahlquist, G., Björck, A. and N. Anderson: *Numerical Methods*, Prentice-Hall, Inc., Englewood Cliffs, NJ, 1974.

2. Forsythe, G. E. and C. B. Moler: *Computer Solution of Linear Algebraic Systems*, Prentice-Hall, Inc., Englewood Cliffs, NJ, 1967.

3. Fox, L.: *Numerical Solution of Ordinary and Partial Differential Equations*, Pergamon Press, New York, 1962.

4. Ketter, R. L. and S. P. Prawel: *Modern Methods of Engineering Computation*, McGraw-Hill Book Company, New York, 1969.

5. Noble, B.: *Applied Linear Algebra*, Prentice-Hall, Inc., Englewood Cliffs, NJ, 1969.

6. Smith, G. D.: *Numerical Solution of Partial Differential Equations*, 2nd ed., Oxford University Press, London, 1978.

7. Stiefel, E. L.: *An Introduction to Numerical Mathematics*, Academic Press, Inc., New York, 1963.

EXERCISES

8·1 Carry out the following matrix operations:

(a) $\begin{bmatrix} 2 & -2 & 0 \\ -2 & 4 & -2 \\ 0 & -2 & 2 \end{bmatrix}\begin{bmatrix} 3 \\ 2 \\ 1 \end{bmatrix} + \begin{bmatrix} -3 \\ 2 \\ 4 \end{bmatrix}$

(b) $\begin{bmatrix} 1 & 0 & 0 \\ 0 & 2 & -1 \\ 0 & -1 & 1 \end{bmatrix}\begin{bmatrix} 2 & -1 & 0 \\ -1 & 3 & -2 \\ 0 & -2 & 2 \end{bmatrix}$

(c) $\begin{bmatrix} 2 & -2 & 0 \\ -2 & 4 & -2 \\ 0 & -2 & 2 \end{bmatrix} + \begin{bmatrix} 1 & -1 & 0 \\ -1 & 2 & -1 \\ 0 & -1 & 1 \end{bmatrix}$

8·2 Determine $\mathbf{\Lambda}^3$ for

$$\mathbf{\Lambda} = \begin{bmatrix} 1 & 0 & 0 \\ 0 & 2 & 0 \\ 0 & 0 & 3 \end{bmatrix}$$

8·3 For

$$\mathbf{A} = \begin{bmatrix} 2 & -1 & 0 \\ -1 & 3 & -2 \\ 0 & -2 & 2 \end{bmatrix} \quad \text{and} \quad \mathbf{B} = \begin{bmatrix} 4 & -2 & -1 \\ -2 & 5 & -2 \\ -1 & -2 & 3 \end{bmatrix}$$

(a) Show that $\mathbf{AB} \neq \mathbf{BA}$.
(b) Show that $(\mathbf{BA})^T = \mathbf{AB}$. Is this true in general? If not, why is it true here?

8·4 Determine $\mathbf{C}^{-1}$ when $\mathbf{C}$ is diagonal as given by

$$\mathbf{C} = \begin{bmatrix} c_{11} & 0 & 0 \\ 0 & c_{22} & 0 \\ 0 & 0 & c_{33} \end{bmatrix}$$

8·5 Determine $\mathbf{C}^{-1}$ for

$$\mathbf{C} = \begin{bmatrix} 2 & 1 & 1 \\ 1 & 2 & 1 \\ 1 & 1 & 2 \end{bmatrix}$$

8·6 Determine $\det(\mathbf{S})$ for

$$\mathbf{S} = \begin{bmatrix} 2 & -2 & 0 \\ -2 & 4 & -2 \\ 0 & -2 & 3 \end{bmatrix}$$

8·7 Determine $\det(\mathbf{S})$ for

$$\mathbf{S} = \begin{bmatrix} 2 & -2 & 0 \\ -2 & 4 & -2 \\ 0 & -2 & 2 \end{bmatrix}$$

8·8 Determine $\det(\mathbf{A})$ for

$$\mathbf{A} = \begin{bmatrix} 1 & 2 & 1 \\ 3 & 2 & 2 \\ 2 & 1 & 3 \end{bmatrix}$$

If $\mathbf{A}'$ is formed from $\mathbf{A}$ by adding α times column 2 to column 1, show that $\det(\mathbf{A}') = \det(\mathbf{A})$.

8·9 Show that the approximation given by (8·1·11) is exact for a point midway between x_i and x_j when the temperature profile between x_i and x_j is parabolic.

8·10 Determine the steady-state, finite-difference equation for node i on the boundary between two different materials. The thermal conductivity for material 1 is k_1 and for material 2 it is k_2. There is no energy generation in either material.

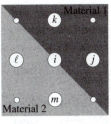

Solve for t_i in terms of t_j, t_k, t_ℓ and t_m and simplify. Does your result seem reasonable? Why or why not?

8•11 Determine the finite-difference differential equation for node i on the boundary between two different materials. Material 1 has properties k_1, ρ_1, c_1, g_1'''. Material 2 has properties k_2, ρ_2, c_2, g_2'''. The nodal-point spacing $d_{ij} = d$ is uniform.

8•12 A portion of a two-dimensional composite region is shown in the sketch. Material 1 has properties k_1, ρ_1, c_1, g_1'''. Material 2 has properties k_2, ρ_2, c_2, g_2'''.

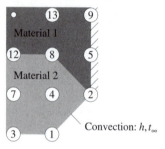

The nodes are uniformly spaced a distance d apart. There are 25 nodes in total.
(a) Determine the differential equation for nodal system 2.
(b) Determine $\mathbf{C}^{(2)}$, $\mathbf{K}^{(2)}$, $\mathbf{H}^{(2)}$, $\mathbf{g}^{(2)}$, $\mathbf{q}^{(2)}$ and $\mathbf{h}^{(2)}$.

8•13 Number the nodes in Figure 8•5 to obtain the minimum bandwidth. Determine the value of the bandwidth.

8•14 A 9-node portion of a region discretized using finite differences is shown in the sketch. Economy-banded storage for $\mathbf{K}$ is dimensioned to handle up to 25 nodes with a bandwidth of 5.

```
•     •     •
8     9     12

•     •     •
3     5     7

•     •     •
1     2     4
```

(a) Where will the nonzero entries in $\mathbf{K}$ for nodal system 5 be found?
(b) Is the dimensioning of $\mathbf{K}$ adequate for the nodes as numbered? If not, could the problem be corrected without redimensioning $\mathbf{K}$? If so, how?

8•15 Determine the **LU** decomposition of **S** for

$$\mathbf{S} = \begin{bmatrix} 2 & -2 & 0 \\ -2 & 4 & -2 \\ 0 & -2 & 3 \end{bmatrix}$$

and then find $\mathbf{t}$ in $\mathbf{LUt} = \mathbf{r}$ where $\mathbf{r}^T = \begin{bmatrix} 2 & 4 & 1 \end{bmatrix}$.

8•16 Determine the Cholesky decomposition of **S** for

$$\mathbf{S} = \begin{bmatrix} 4 & 2 & 0 \\ 2 & 10 & 12 \\ 0 & 12 & 17 \end{bmatrix}$$

8•17 Determine the Cholesky decomposition of **S** for

$$\mathbf{S} = \begin{bmatrix} \dfrac{k}{L} & -\dfrac{k}{L} \\ -\dfrac{k}{L} & \dfrac{k}{L}+h \end{bmatrix}$$

8•18 Consider the steady-state uniform generation of energy in a plane wall. The surface at $x = 0$ is adiabatic and the surface at $x = L$ is at t_L. For a 3-node, finite-difference formulation where node 1 is at $x = 0$, node 2 is at $x = L/2$ and node 3 is at $x = L$,
(a) Determine $\mathbf{K}$ and $\mathbf{g}$.
(b) Modify $\mathbf{K}$ and $\mathbf{g}$ to obtain $\mathbf{S}$ and $\mathbf{r}$.
(c) For k, g''' and $L = 1$ and $t_L = 0$, determine the steady-state solution.
(d) Compare the finite-difference solution with the exact solution of the governing differential equation.
(e) Determine $\mathbf{q}_o$ (per unit area) using $\mathbf{q}_o = \mathbf{Kt} - \mathbf{g}$ and compare the result to the exact value.

8•19 A two-dimensional, square region has a specified temperature on all four sides. The model is given by

$$t_{xx} + t_{yy} = 0$$

$$t(0, y) = t_0$$
$$t(L, y) = t_0$$
$$t(x, 0) = t_0$$
$$t(x, L) = t_0 + T\sin(\pi x / L)$$

Using a uniformly spaced 15-node, finite-difference model, determine a solution for temperature. Carry out the solution in symbols as far as you feel is practical. Then, for $k = 2$, $L = 1$, $t_0 = 0$ and $T = 1$, determine the solution and compare it to the exact solution found in Exercise 3•44.

8•20 For a 3-node, finite difference formulation (one node at each corner) of the problem shown in Figure 8•18,
(a) Determine $\mathbf{K}$, $\mathbf{g}$, $\mathbf{S}$ and $\mathbf{r}$.
For k, g''' and $L = 1$ and $T_L = 0$,
(b) Determine $\mathbf{t}$ and compare the result to Figure 8•20.
(c) Determine $\mathbf{q}_o$ using $\mathbf{q}_o = \mathbf{Kt} - \mathbf{g}$ and then determine the heat fluxes. Compare the results to Figure 8•21.

8•21 Consider steady-state uniform energy generation g''' in a square $(L \times L)$ region [conductivity $= k$]. The boundaries along $x = 0$ and $y = 0$ are adiabatic. The boundaries along $x = L$ and $y = L$ are maintained at $t = t_L$. Consider a three-node, finite-difference model where node 1 is at $(0,0)$, node 2 is at $(L,0)$ and node 3 is at (L,L).
(a) Determine **K** and **g**.
(b) Determine **S** and **r**.
(c) Determine **t** and $\mathbf{q}_o$.
For $k = 2$, $g''' = 8$ and $L = 1$ and $t_L = 0$,
(d) Determine **t** and $\mathbf{q}_o$.

8•22 Consider steady-state uniform energy generation g''' in a square $(L \times L)$ region [conductivity $= k$]. The boundaries along $x = 0$ and $y = 0$ are adiabatic. The boundaries along $x = L$ and $y = L$ are cooled [heat-transfer coefficient $= h$] by an ambient at $t = t_\infty$. Consider a three-node, finite-difference model where node 1 is at $(0,0)$, node 2 is at $(L,0)$ and node 3 is at (L,L).
(a) Determine **K**, **H**, **g** and **h**.
(b) Determine **S** and **r**.
For $k = 2$, $g''' = 8$, $h = 4$, $L = 1$ and $t_\infty = 0$,
(c) Determine **t**.
(d) Determine $\mathbf{q}_o$ from **K**, **g** and **t**.
(e) Determine $\mathbf{q}_o$ from **H**, **h** and **t**. Compare to part (d).

8•23 Compare **t** and $\mathbf{q}_o$ from exercises 8•21 and 8•22.

8•24 A thin rod (length $= L$, cross-sectional area $= A$, perimeter $= p$, conductivity $= k$) is convectively cooled (heat-transfer coefficient $= h$, ambient temperature $= t_\infty$). Each end of the rod is held at a temperature of t_0. Using four uniformly spaced nodes in one half of the rod,
(a) Determine **S** and **r**.
For $L = 1$, $A = 1/64$, $p = 1/2$, $k = 96$, $h = 24$, $t_\infty = 0$ and $t_0 = 1$,
(b) Determine **t**.
(c) Determine the heat-transfer rate from the rod.
(d) Compare parts (b) and (c) to the exact values.

8•25 A thin, square plate-fin (conductivity $= k$, thickness $= \delta$, length and width $= L$) is convectively cooled on both faces (heat-transfer coefficient $= h$, ambient temperature $= t_\infty$). Each of the four edges of the plate-fin is maintained at a temperature of t_0. The fin is at steady-state with $g''' = 0$. For a finite-difference formulation, using six uniformly spaced nodes (taking maximum advantage of symmetry),
(a) Determine **K**, **H**, **h**, **S** and **r** in terms of symbols.
For $L = 2$ ft, $\delta = 1$ in, $k = 120$ Btu/hr-ft-F, $h = 16$ Btu/hr-ft^2-F, $t_0 = 100$ F and $t_\infty = 0$ F,
(b) Determine the numerical values in **K**, **H**, **h**, **S** and **r**.
(c) Determine the temperature [F] at the center of the plate.
(d) Determine the heat-transfer rate [Btu/hr] from the plate to the ambient by two different methods. Do the two values agree?

8•26 Obtain a 6-node, finite-difference solution for uniform energy generation in an equilateral triangle. The length of each of the three sides is L. Each side is maintained at a temperature t_o. A uniform rectangular arrangement of nodal points will probably be most convenient. Make use of symmetry to get maximum mileage out of the six nodes. For $k = 2\sqrt{3}$, $g''' = 128\sqrt{3}$, $L = 1$ and $t_o = 0$, determine the maximum temperature.

8•27 Consider uniform energy generation in the region shown.

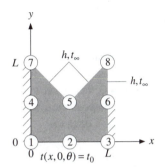

Take the initial temperature to be uniformly at $t^{(0)}$. Using the eight uniformly spaced nodal points shown,
(a) Derive **C**, **K**, **H**, **g** and **h**. The entries should be in terms of k, g''', L, h, ρ, c and t_∞.
For k, g''', L, h and $t_0 = 1$ and $t_\infty = 0$,
(b) Modify **K** + **H** and **g** + **h** to obtain **S** and **r**.
(c) Determine the steady-state **t**.
(d) Show that an overall energy balance is satisfied.

8•28 A one-dimensional plane wall [k, ρ, and c] has an adiabatic surface at $x = 0$. The initial wall temperature is uniformly at $t^{(0)}$. The surface at $x = L$ is suddenly changed to a new specified value t_L at $\theta = 0$ and maintained at this value for all time. The generation rate is zero. Consider a 3-node, finite-difference formulation where node 1 is at $x = 0$, node 2 is at $x = L/2$ and node 3 is at $x = L$.
(a) Determine **C**, **K** and **g** in terms of k, ρ, c and L.
(b) Determine **Ç**, **S**, **r** and $\mathbf{t}^{(0)}$.
(c) Determine $\Delta\theta_{c,est}$.
For k, ρ, c, L and $t^{(0)} = 1$ and $t_L = 0$,
(d) Determine and plot the Euler solution, with $\Delta\theta = \Delta\theta_{c,est}$, up to $\theta = 1$.
(e) Determine and plot the Crank-Nicolson solution, with $\Delta\theta = \Delta\theta_{c,est}$, up to $\theta = 1$.

8•29 Consider the same physical problem as in Exercise 8•28, but use a 4-node, finite-difference formulation with node 1 at $x = 0$, node 2 at $x = L/3$, node 3 at $x = 2L/3$ and node 4 at $x = L$.
(a) Determine **Ç**, **S**, **r** and $\mathbf{t}^{(0)}$.
(b) Determine $\Delta\theta_{c,est}$ and compare it to the 3-node value determined in Exercise 8•28.

8•30 Consider the same physical problem as in Exercise 8•28 but use a 3-node, finite-difference formulation with node 1 at $x = 0$, node 2 at $x = 3L/4$ and node 3 at $x = L$.

(a) Determine $\mathbf{\c{C}}$, $\mathbf{S}$, $\mathbf{r}$ and $\mathbf{t}^{(0)}$.

(b) Determine $\Delta\theta_{c,est}$.

For k, ρ, c, L and $t^{(0)} = 1$ and $t_L = 0$,

(c) Determine the Euler solution with $\Delta\theta = \frac{1}{8}$ until $\theta = 1$.

(d) Does the solution in (c) behave as you would expect?

8•31 For the plane-wall transient considered in exercises 8•28 and 8•30, the exact solution of the partial differential equation is given by (3·1·30). For k, ρ, c, L and $t^{(0)} = 1$ and $t_L = 0$,

(a) Evaluate the exact solution and make two separate plots of $t(0,\theta)$ as a function of θ for θ running from 0 to 1.

(b) On the first plot from part (a), superimpose the Euler solutions from exercises 8•28 and 8•30. Compare.

(c) On the other plot from part (a), superimpose the Crank-Nicolson solution from Exercise 8•28. Compare.

8•32 A one-dimensional plane wall $[k,\ \rho,\ \text{and}\ c]$ is initially at a uniform temperature $t^{(0)}$. The surface at $x = 0$ is adiabatic. The surface at $x = L$ is exposed to an ambient temperature that is suddenly changed from $t^{(0)}$ to a new value t_∞ at $\theta = 0$. The heat-transfer coefficient at $x = L$ is h. Consider a 3-node, finite-difference formulation where node 1 is at $x = 0$, node 2 is at $x = L/2$ and node 3 is at $x = L$.

(a) Derive $\mathbf{\c{C}}$, $\mathbf{S}$ and $\mathbf{r}$.

(b) Determine $\Delta\theta_{c,est}$ and compare it to the value obtained in Exercise 8•28.

8•33 A one-dimensional plane wall $[k,\ \rho,\ \text{and}\ c]$ is initially at a uniform temperature $t^{(0)}$. The surface at $x = 0$ is adiabatic. The surface at $x = L$ is suddenly exposed to a constant heat-flux input q_o'' at $\theta = 0$. Consider a 2-node, finite-difference formulation where node 1 is at $x = 0$ and node 2 is at $x = L$.

(a) Derive $\mathbf{\c{C}}$, $\mathbf{S}$ and $\mathbf{r}$.

(b) Determine $\Delta\theta_{c,est}$.

For k, ρ, c, L and $q_o'' = 1$ and $t^{(0)} = 0$,

(c) For $\Delta\theta = 1/2$, determine and plot the Euler solution for five time steps. Discuss.

(d) For $\Delta\theta = 1/4$, determine and plot the Euler solution for four time steps. Compare to the exact solution (3·1·99) for large θ.

8•34 For a 3-node, finite-difference formulation (one node at each corner) of the problem discussed in sections 8•4•1 and 8•4•4,

(a) Determine $\mathbf{\c{C}}$, $\mathbf{S}$ and $\mathbf{r}$.

(b) Determine the Euler and Crank-Nicolson solutions for $\Delta\theta = \frac{1}{2}\Delta\theta_{c,est}$, $\Delta\theta_{c,est}$ and $2\Delta\theta_{c,est}$. Plot and discuss.

(c) Determine and plot the Crank-Nicolson solution for 20 steps with $\Delta\theta = 20\Delta\theta_{c,est}$. Discuss.

8•35 A one-dimensional, composite, plane-wall is initially at a uniform temperature $t^{(0)}$. At $\theta = 0$ energy generation starts and the temperature begins to change toward a new steady-state profile. The surface at $x = 0$ is adiabatic while the surface at $x = L$ is maintained at $t^{(0)}$. The region for $0 \le x \le L/2$ is made of material 1 and the region for $0.5 \le x \le 1$ is made of material 2. Consider a 3-node, finite-difference formulation where node 1 is at $x = 0$, node 2 is at $x = L/2$ and node 3 is at $x = L$.

(a) Determine $\mathbf{C}$, $\mathbf{K}$ and $\mathbf{g}$.

(b) Determine $\mathbf{\c{C}}$, $\mathbf{S}$, $\mathbf{r}$ and $\mathbf{t}^{(0)}$.

(c) Determine $\mathbf{q}_o$.

For $L = 1$, $t^{(0)} = 1$ and the following property values,

Material	k	ρ	c	g'''
1	2	3	3	3
2	1	1	1	1

(d) Substitute numerical values into $\mathbf{\c{C}}$, $\mathbf{S}$, $\mathbf{r}$, $\mathbf{t}^{(0)}$ and $\mathbf{q}_o$.

(e) Determine $\Delta\theta_{c,est}$.

(f) For the Euler method with $\Delta\theta = 1/2$, determine $\mathbf{\c{C}}$, $(\mathbf{\c{C}} - \mathbf{S}\Delta\theta)$ and $\mathbf{r}\Delta\theta$.

(g) Using your favorite language or software, calculate $\mathbf{t}^{(v)}$ and $\mathbf{q}_o^{(v)}$ for the Euler method of part (f).

(h) Plot the Euler solution for t_1 and t_2 out to $\theta = 5$. Do they behave as you would expect? Discuss.

(i) Plot the Euler solution for q_{o3} out to $\theta = 5$. Does it behave as you would expect? Discuss.

(j) Determine the Euler solutions $\mathbf{t}^{(\infty)}$ and $\mathbf{q}_o^{(\infty)}$.

(k) From the governing ordinary differential equation and the boundary conditions, determine the exact steady-state solutions for t_1, t_2 and q_{o3}. Compare to part (j).

8•36 One face of a plane wall initially at $t^{(0)}$ is suddenly exposed to a step change in temperature that then decays exponentially back to the initial value again. The other face is insulated. The mathematical description is given by

$$kt_{xx} = \rho c t_\theta \qquad t_x(0,\theta) = 0$$

$$t(L,\theta) = t^{(0)} + T\exp(-a\theta)$$

$$t(x,0) = t^{(0)}$$

Consider a 3-node, finite-difference formulation with node 1 at $x = 0$, node 2 at $x = L/2$ and node 3 at $x = L$.

(a) Determine $\mathbf{C}$, $\mathbf{K}$ and $\mathbf{g}$.

(b) Determine $\mathbf{\c{C}}$, $\mathbf{S}$, $\mathbf{r}$ and $\mathbf{t}^{(0)}$ to account for the specified time-dependent temperature $t(L,\theta)$.

(c) Determine $\mathbf{q}_o$.

For k, ρ, c, a, L and $T = 1$ and $t^{(0)} = 0$,

(d) Determine the matrices needed for a Crank-Nicolson solution.

(e) Using your favorite language or software, and taking $\Delta\theta = 0.1$, calculate $\mathbf{t}^{(v)}$ and $\mathbf{q}_o^{(v)}$ out to $\theta = 1$.

(f) Plot t_1, t_2 and t_3 as a function of θ.

(g) Plot q_{o3} as a function of θ.

8•37 A 2-node, finite-difference model of a plane wall (adiabatic at one surface, convective at the other surface) is given by:

$$\begin{bmatrix} 3 & 0 \\ 0 & 3 \end{bmatrix}\dot{\mathbf{t}} + \begin{bmatrix} 1 & -1 \\ -1 & 2 \end{bmatrix}\mathbf{t} = \begin{bmatrix} 0 \\ 1 \end{bmatrix} \quad \text{with} \quad \mathbf{t}(0) = \begin{bmatrix} 0 \\ 0 \end{bmatrix}$$

(a) Determine $\dot{\mathbf{t}}(0)$.
(b) Determine $\Delta\theta_{c,est}$.

8•38 A square region of uniform thermal properties [k, ρ and c] has a uniform initial temperature of $t^{(0)}$. Along $x = L$ and along $y = L$ the temperature is suddenly changed to a new specified value t_L at $\theta = 0$ and maintained at this value for all time. The boundaries along $x = 0$ and along $y = 0$ are adiabatic. For a 6-node, finite-difference formulation where the nodes are arranged as shown in Figure 8•23,

(a) Determine $\mathbf{C}$, $\mathbf{K}$ and $\mathbf{g}$.
(b) Determine $\mathbf{Ç}$, $\mathbf{S}$, $\mathbf{r}$ and $\mathbf{t}^{(0)}$.
(c) Determine $\Delta\theta_{c,est}$.
For k, ρ, c, L and $t^{(0)} = 1$ and $t_L = 0$,
(d) Determine the numerical values in $\mathbf{Ç}$, $\mathbf{S}$, $\mathbf{r}$ and $\mathbf{t}^{(0)}$.
(e) For an Euler solution with values of $\Delta\theta = 0.05$ and 0.10, calculate $\mathbf{t}$ out to $\theta = 0.5$. Plot. Discuss.
(f) For a Crank-Nicolson solution with values of $\Delta\theta = 0.05$ and 0.10, calculate $\mathbf{t}$ out to $\theta = 0.5$. Plot. Discuss.
(g) For a Crank-Nicolson solution with $\Delta\theta = 5.0$, calculate $\mathbf{t}$ out to $\theta = 250$. Plot. Discuss.

8•39 A one-dimensional plane wall [k, ρ, c and g'''] is initially at a uniform temperature $t^{(0)}$. The surface at $x = 0$ is adiabatic. The surface at $x = L$ is exposed to an ambient temperature that is suddenly changed from $t^{(0)}$ to a new value t_∞ at $\theta = 0$. The heat-transfer coefficient at $x = L$ is h. For a 3-node, finite-difference formulation where node 1 is at $x = 0$, node 2 is at $x = L/2$ and node 3 is at $x = L$,

(a) Determine $\mathbf{C}$, $\mathbf{K}$, $\mathbf{H}$, $\mathbf{g}$, $\mathbf{h}$, $\mathbf{q}$, $\mathbf{Ç}$, $\mathbf{S}$, $\mathbf{r}$, $\mathbf{t}^{(0)}$ and $\mathbf{q}_o$.
For k, ρ, c, g''', h and $L = 1$ and $t^{(0)} = t_\infty = 0$,
(b) Determine the values in $\mathbf{Ç}$, $\mathbf{S}$, $\mathbf{r}$, $\mathbf{t}^{(0)}$ and $\mathbf{q}_o$.
Determine and plot t_1, t_2, t_3 and q_{o3} for $\theta = 0$ to 4 for
(c) An Euler solution with $\Delta\theta = 0.125$. If oscillations are a problem, reduce $\Delta\theta$ by a factor of two.
(d) A Crank-Nicolson solution with $\Delta\theta = 0.125$.

8•40 Find the eigenvalues and eigenvectors of $\mathbf{Sx} = \lambda\mathbf{Çx}$ for

$$\mathbf{S} = \begin{bmatrix} 2 & -2 \\ -2 & 6 \end{bmatrix} \quad \text{and} \quad \mathbf{Ç} = \begin{bmatrix} 1 & 0 \\ 0 & 2 \end{bmatrix}$$

Normalize the eigenvectors so that $\mathbf{x}_i^T\mathbf{Çx}_j = \delta_{ij}$. Check your results by verifying that $\mathbf{x}_i^T\mathbf{Sx}_j = \delta_{ij}\lambda_j$.

8•41 For the $\mathbf{x}_3$ and $\mathbf{x}_4$ in (8•5•35), show that $\mathbf{x}_3^T\mathbf{Çx}_4 = 0$.

8•42 For the plane-wall transient of Exercise 8•28,

(a) Determine λ_1, λ_2, λ_3, $\mathbf{x}_1$, $\mathbf{x}_2$ and $\mathbf{x}_3$. Normalize the eigenvectors so that $\mathbf{x}_i^T\mathbf{Çx}_j = \delta_{ij}$.
(b) Determine the exact solution of $\mathbf{Çṫ} + \mathbf{St} = \mathbf{r}$.
For k, ρ, c, L and $t^{(0)} = 1$ and $t_L = 0$,
(c) Determine and plot the solution in part (b) for t_1 and t_2.
(d) On another graph, plot t_1 from part (c), $t(0,\theta)$ from Exercise 8•31(a) and t_1 from Exercise 8•28(e).
(e) Determine the exact value of $\Delta\theta_c$ based on (8•5•71). Compare $\Delta\theta_c$ to $\Delta\theta_{c,est}$ found in Exercise 8•28(c).
(f) Determine a relation between $t_1^{(0)}$ and $t_2^{(0)}$ that will eliminate λ_3 from the solution.

8•43 For the plane-wall transient of Exercise 8•30,

(a) Determine λ_1, λ_2, λ_3, $\mathbf{x}_1$, $\mathbf{x}_2$ and $\mathbf{x}_3$.
(b) Determine the exact solution of $\mathbf{Çṫ} + \mathbf{St} = \mathbf{r}$.
(c) Determine and plot the solution in part (b) for t_1 and t_2.

8•44 For the plane-wall transient of Exercise 8•33,

(a) Determine λ_1, λ_2, $\mathbf{x}_1$ and $\mathbf{x}_2$.
(b) Determine the exact solution of $\mathbf{Çṫ} + \mathbf{St} = \mathbf{r}$.
(c) Determine and plot the solution in part (b) for t_1 and t_2. Compare to the exact solution (3•1•99) for large θ.

8•45 For the two-dimensional transient of Exercise 8•34,

(a) Determine λ_1, λ_2, λ_3, $\mathbf{x}_1$, $\mathbf{x}_2$ and $\mathbf{x}_3$.
(b) Determine the exact solution of $\mathbf{Çṫ} + \mathbf{St} = \mathbf{r}$.
(c) Determine and plot the solution in part (b) for t_1, t_2 and t_3. Compare the solution to the Crank-Nicolson result for $\Delta\theta = \frac{1}{2}\Delta\theta_{c,est}$ found in Exercise 8•34.

8•46 For the system of equations derived for Exercise 8•38,

(a) Determine λ_1 through λ_6 and $\mathbf{x}_1$ through $\mathbf{x}_6$.
(b) Determine the exact solution of $\mathbf{Çṫ} + \mathbf{St} = \mathbf{r}$.
(c) Determine the exact value of $\Delta\theta_c$ based on (8•5•71) and compare it to $\Delta\theta_{c,est}$ found in Exercise 8•38.

8•47 A one-dimensional plane wall [k, ρ and c] has an adiabatic surface at $x = 0$. Initially the wall temperature is uniformly at $t^{(0)}$. The surface at $x = L$ is suddenly changed to a new value t_L at $\theta = 0$. The cross-sectional area is A. Consider a three-node, finite-difference model where node 1 is at $x = 0$, node 2 is at $x = L/2$ and node 3 is at $x = L$.

(a) Determine $\mathbf{C}$, $\mathbf{K}$, $\mathbf{Ç}$, $\mathbf{S}$, $\mathbf{r}$ and $\mathbf{t}^{(0)}$.
(b) Determine $\Delta\theta_{c,est}$ and $\Delta\theta_c$. Compare.
(c) Determine the exact analytical solution to the finite-difference model derived in part (a).
For k, ρ, c, L and $A = 1$, $t^{(0)} = 0$ and $t_L = 1$,
(d) Determine the numerical values for t_1 and t_2 in the result obtained in part (c) for $\theta = 0$ to 2.
(e) Illustrate the critical-time-step theory as applied to this problem by plotting the Euler solution for t_1 for $\Delta\theta_{c,est} < \Delta\theta < \Delta\theta_c$ and for $\Delta\theta$ slightly greater than $\Delta\theta_c$. Also, overlay t_1 from part (d) for comparison. Discuss.

8•48 A one-dimensional, plane-wall [k, ρ and c] has an adiabatic surface at $x = 0$. Initially the wall temperature is uniformly at $t^{(0)}$. The surface at $x = L$ is suddenly changed to a new value t_L at $\theta = 0$. The cross-sectional area is A. The exact 3-node finite-difference solution for k, ρ, c, L and $A = 1$, $t^{(0)} = 0$ and $t_L = 1$ was worked out in Exercise 8•47 for node 1 at $x = 0$, node 2 at $x = L/2$ and node 3 at $x = L$. The exact solution to the governing partial differential equation is given by

$$t(x,\theta) = t_L + 2(t^{(0)} - t_L) \sum_{n=1}^{\infty} \frac{\sin(\lambda_n L)}{\lambda_n L} \cos(\lambda_n x) \exp(-\lambda_n^2 \alpha \theta)$$

where the eigenvalues are given by $\lambda_n L = (2n-1)\pi/2$. Express the exact solution to the partial differential equation at $x = 0$, $L/2$ and L using matrix notation. That is, write it in the form $\mathbf{t}(\theta) = \mathbf{a} + \mathbf{X}' \exp(-\boldsymbol{\lambda}\theta)$. Use only the first three terms in the infinite series. Your final answer should contain numerical values in $\mathbf{a}$, $\mathbf{X}'$ and $\boldsymbol{\lambda}$. Compare to Exercise 8•47.

8•49 A one-dimensional, plane-wall [k, ρ and c] has an adiabatic surface at $x = 0$. Initially the wall temperature is uniformly at $t^{(0)}$. The surface at $x = L$ is suddenly changed to a new value t_L at $\theta = 0$. The cross-sectional area is A. Exercise 8•47 considers a three-node model where node 1 is at $x = 0$, node 2 is at $x = L/2$ and node 3 is at $x = L$. To improve accuracy, add a node at $x = 3L/4$. Now node 1 is at $x = 0$, node 2 is at $x = L/2$, node 3 is at $x = 3L/4$ and node 4 is at $x = L$.

(a) Determine $\mathbf{C}$, $\mathbf{K}$, $\mathbf{Ç}$, $\mathbf{S}$, $\mathbf{r}$ and $\mathbf{t}^{(0)}$.
(b) Determine and compare $\Delta\theta_{c,est}$ and $\Delta\theta_c$. Compare $\Delta\theta_c$ for the four-node model to $\Delta\theta_c$ for the three-node model.

For k, ρ, c, L and $A = 1$, $t^{(0)} = 0$ and $t_L = 1$,

(c) Illustrate the critical-time-step theory as applied to this problem by plotting the Euler solution for t_1 for $\Delta\theta_{c,est} < \Delta\theta < \Delta\theta_c$ and for $\Delta\theta$ slightly greater than $\Delta\theta_c$.

8•50 A one-dimensional plane wall [k, ρ and c] has an adiabatic surface at $x = 0$. Initially the wall temperature is uniformly at $t^{(0)}$. The surface at $x = L$ is suddenly changed to a new value t_L at $\theta = 0$. The cross-sectional area is A. Consider a three-node, finite-difference model where node 1 is at $x = 0$, node 2 is at $x = 3L/4$ and node 3 is at $x = L$.

(a) Determine $\mathbf{C}$, $\mathbf{K}$, $\mathbf{Ç}$, $\mathbf{S}$, $\mathbf{r}$ and $\mathbf{t}^{(0)}$.
(b) Determine $\Delta\theta_{c,est}$ and $\Delta\theta_c$. Compare.
(c) Determine the exact analytical solution to the finite-difference model derived in part (a).

For k, ρ, c, L and $A = 1$, $t^{(0)} = 0$ and $t_L = 1$,

(d) Determine the numerical values for t_1 and t_2 in the result obtained in part (c) for $\theta = 0$ to 2.
(e) Illustrate the critical-time-step theory as applied to this problem by plotting the Euler solution for t_1 for $\Delta\theta_{c,est} < \Delta\theta < \Delta\theta_c$ and for $\Delta\theta$ slightly greater than $\Delta\theta_c$. Also, overlay t_1 from part (d) for comparison. Discuss.

8•51 A one-dimensional plane wall [k, ρ and c] has an adiabatic surface at $x = 0$. Initially the wall temperature is uniformly at $t^{(0)}$. The surface at $x = L$ is suddenly changed to a new value t_L at $\theta = 0$. A five-node, finite-difference model, where nodes 1 through 5 are at $x = 0$, $L/4$, $L/2$, $3L/4$ and L, respectively, gives:

$$\mathbf{Ç} = \frac{\rho c L}{8} \begin{bmatrix} 1 & & & & \\ & 2 & & & \\ & & 2 & & \\ & & & 2 & \\ & & & & 1 \end{bmatrix} \quad \mathbf{S} = \frac{4k}{L} \begin{bmatrix} 1 & -1 & & & \\ -1 & 2 & -1 & & \\ & -1 & 2 & -1 & \\ & & -1 & 2 & 0 \\ & & & 0 & 1 \end{bmatrix}$$

$$\mathbf{r} = \frac{4k}{L} \begin{bmatrix} 0 \\ 0 \\ 0 \\ t_L \\ t_L \end{bmatrix} \quad\quad\quad\quad \mathbf{t}^{(0)} = \begin{bmatrix} t^{(0)} \\ t^{(0)} \\ t^{(0)} \\ t^{(0)} \\ t_L \end{bmatrix}$$

For k, ρ, c and $L = 1$ and for $t^{(0)} = 0$ and $t_L = 1$, the eigenvalues and eigenvectors of this model are given by

$$\boldsymbol{\lambda} = \begin{bmatrix} 2.44 \\ 19.75 \\ 32.00 \\ 44.25 \\ 61.56 \end{bmatrix} \quad \mathbf{X} = \begin{bmatrix} -1.414 & 1.414 & 0 & 1.414 & -1.414 \\ -1.307 & 0.541 & 0 & -0.541 & 1.307 \\ -1.000 & -1.000 & 0 & -1.000 & -1.000 \\ -0.541 & -1.307 & 0 & 1.307 & 0.541 \\ 0.000 & 0.000 & \sqrt{8} & 0.000 & 0.000 \end{bmatrix}$$

$$\mathbf{X}' = \begin{bmatrix} -1.257 & 0.374 & 0 & -0.167 & 0.050 \\ -1.161 & 0.143 & 0 & 0.064 & -0.046 \\ -0.889 & -0.265 & 0 & 0.118 & 0.035 \\ -0.481 & -0.346 & 0 & -0.154 & -0.019 \\ 0.000 & 0.000 & 0 & 0.000 & 0.000 \end{bmatrix}$$

The exact solution to the governing partial differential equation is given by

$$t(x,\theta) = t_L + 2(t^{(0)} - t_L) \sum_{n=1}^{\infty} \frac{\sin(\lambda_n L)}{\lambda_n L} \cos(\lambda_n x) \exp(-\lambda_n^2 \alpha \theta)$$

where the eigenvalues are given by $\lambda_n L = (2n-1)\pi/2$.

For k, ρ, c and $L = 1$ and for $t^{(0)} = 0$ and $t_L = 1$,

(a) Evaluate the exact finite-difference solutions for nodes 1 and 4 for $0 \le \theta \le 1.0$.
(b) For comparison with the temperatures of nodes 1 and 4, evaluate the solution of the partial-differential equation for $t(0,\theta)$ and $t(3L/4,\theta)$ for $0 \le \theta \le 1.0$.
(c) Plot the solutions found in parts (a) and (b) on the same graph. Compare.
(d) Compare the first two eigenvalues in parts (a) and (b).
(e) Plot the first two eigenfunctions for the solution of the partial differential equation as functions of x on the same graph. For comparison, plot the appropriate data points from the eigenvectors in $\mathbf{X}'$ on the same graph. Compare.

8•52 A one-dimensional plane wall [k, ρ and c] has an adiabatic surface at $x = 0$. The surface at $x = L$ is suddenly changed to a new value t_L at $\theta = 0$. A five-node, finite-difference model, where nodes 1 through 5 are at $x = 0$, $L/4$, $L/2$, $3L/4$ and L, respectively, gives:

$$\mathsf{C} = \frac{\rho c L}{8} \begin{bmatrix} 1 & & & & \\ & 2 & & & \\ & & 2 & & \\ & & & 2 & \\ & & & & 1 \end{bmatrix} \qquad \mathsf{S} = \frac{4k}{L} \begin{bmatrix} 1 & -1 & & & \\ -1 & 2 & -1 & & \\ & -1 & 2 & -1 & \\ & & -1 & 2 & 0 \\ & & & 0 & 1 \end{bmatrix}$$

$$\mathbf{r} = \frac{4k}{L} \begin{bmatrix} 0 \\ 0 \\ 0 \\ t_L \\ t_L \end{bmatrix} \qquad \mathbf{t}^{(0)} = \begin{bmatrix} t_1^{(0)} \\ t_2^{(0)} \\ t_3^{(0)} \\ t_4^{(0)} \\ t_L \end{bmatrix}$$

For k, ρ, c and $L = 1$ and for $t_1^{(0)}$, $t_2^{(0)}$, $t_3^{(0)}$, $t_4^{(0)} = 0$ and $t_L = 1$, the eigenvalues and eigenvectors of this model are given by

$$\boldsymbol{\lambda} = \begin{bmatrix} 2.44 \\ 19.75 \\ 32.00 \\ 44.25 \\ 61.56 \end{bmatrix} \quad \mathbf{X} = \begin{bmatrix} -1.414 & 1.414 & 0 & 1.414 & -1.414 \\ -1.307 & 0.541 & 0 & -0.541 & 1.307 \\ -1.000 & -1.000 & 0 & -1.000 & -1.000 \\ -0.541 & -1.307 & 0 & 1.307 & 0.541 \\ 0.000 & 0.000 & \sqrt{8} & 0.000 & 0.000 \end{bmatrix}$$

$$\mathbf{X}' = \begin{bmatrix} -1.257 & 0.374 & 0 & -0.167 & 0.050 \\ -1.161 & 0.143 & 0 & 0.064 & -0.046 \\ -0.889 & -0.265 & 0 & 0.118 & 0.035 \\ -0.481 & -0.346 & 0 & -0.154 & -0.019 \\ 0.000 & 0.000 & 0 & 0.000 & 0.000 \end{bmatrix}$$

For k, ρ, c and $L = 1$ and for $t_1^{(0)}$, $t_2^{(0)}$, $t_3^{(0)}$, $t_4^{(0)} = 0$ and $t_L = 1$,
(a) Evaluate the exact finite-difference solutions for nodes 1 and 4 for $0 \le \theta \le 1.0$.
(b) Determine $\Delta\theta_{c,est}$ and $\Delta\theta_c$.
(c) For comparison with part (a), determine the Euler solution using $\Delta\theta = 0.02$.
(d) Plot the solutions found in parts (a) and (c) on the same graph.
(e) On another graph, for comparison to part (d), plot the solution found in part (a) and the Euler solution for nodes 1 and 4 for $\Delta\theta = 0.04$.
(f) Determine the Euler solution using $\Delta\theta = 0.02$ out to $\theta = 0.08$, then change $\Delta\theta$ to 0.04 and continue the Euler solution to $\theta = 1.00$.
(g) Plot the solution found in part (f) to compare with the previous solutions.
(h) Discuss and compare the solutions shown in parts (d), (e) and (g). Give reasons for the behavior you see.

8•53 For transient conduction the standard finite-difference analysis assumes the temperature varies linearly between adjacent nodes in approximating the conduction heat flows but is uniform within each nodal system when modeling the energy storage. For a one-dimensional, plane wall it might be a better approximation to assume the temperature also varies linearly between adjacent nodes in modeling energy storage. For adjacent, uniformly spaced nodes i, j and k having a linear temperature profile between nodes,
(a) Show that the average temperature $\langle t_j \rangle$ of nodal system j is given by

$$\langle t_j \rangle = \frac{t_i + 6t_j + t_k}{8}$$

Consider a one-dimensional, plane wall [k, ρ and c] with an adiabatic surface at $x = 0$. The initial temperature is uniformly at $t^{(0)}$. The surface at $x = L$ is suddenly changed to a new value t_L at $\theta = 0$. Consider a three-node, finite-difference model where node 1 is at $x = 0$, node 2 is at $x = L/2$ and node 3 is at $x = L$. Using the improved model for energy storage,
(b) Determine $\mathbf{C}$, $\mathbf{K}$, C, S, $\mathbf{r}$ and $\mathbf{t}^{(0)}$.
(c) Determine $\dot{\mathbf{t}}^{(0)}$.
(d) Determine $\Delta\theta_c$.
(e) Compare your results for $\dot{\mathbf{t}}^{(0)}$ and $\Delta\theta_c$ to the results for the standard energy-storage model.
(f) Determine and plot the Euler solution for t_1 and t_2 out to a value of θ where $t_2 \approx 0.5$. For $\Delta\theta < \Delta\theta_c/2$ all numerically induced oscillations (even stable ones) will be eliminated. To increase accuracy take $\Delta\theta \le \Delta\theta_c/4$. Superimpose the slopes obtained in part (c).

8•54 A two-dimensional, square ($L \times L$) region [k, ρ and c] has adiabatic boundaries along $x = 0$, $y = 0$ and $y = L$. Initially, the temperature within the region is uniformly at $t^{(0)}$. At $\theta = 0$ the temperature along $x = L$ is suddenly changed to vary linearly from t_0 at $y = 0$ to t_L at $y = L$. The generation rate is zero. The three adiabatic boundaries remain adiabatic. Consider a 4-node, finite-difference model where node 1 is at $(0,0)$, node 2 is at $(L,0)$, node 3 is at $(0, L)$ and node 4 is at (L, L).
(a) Determine $\mathbf{C}$, $\mathbf{K}$, C, S, $\mathbf{r}$ and $\mathbf{t}^{(0)}$
(b) Determine the steady-state solution.
(c) Determine $\Delta\theta_{c,est}$.
(d) Determine $\Delta\theta_c$ and compare it to $\Delta\theta_{c,est}$.
(e) Determine the exact, transient solution of the finite-difference model.
For k, ρ, c and $L = 1$ and $t^{(0)} = 0$, $t_0 = 0$ and $t_L = 1$,
(f) Evaluate the exact, transient finite-difference solution for θ going from 0 to 1.
(g) Determine the Euler solution, using $\Delta\theta = 0.2$, for θ going from 0 to 1.
(h) Superimpose the exact solution from part (f) and the Euler solution from part (g) for t_1 and t_3 on the same plot. Discuss.

CHAPTER

9

FINITE ELEMENTS

9•0 INTRODUCTION

One of the most important applications of conduction analyses is to provide information for thermal-stress calculations. It is therefore important for the heat-transfer engineer and the stress engineer to have compatible ways of analyzing problems. The finite-element method has developed over the years as one of the most useful ways to attack stress problems numerically [1, 2]. The heat-transfer engineer who works with the stress group often finds it is essential to adapt to using finite-element methods to solve the thermal problem in order to work effectively with the stress group.

In addition to the cooperative benefits of using the finite-element method for solving conduction problems, there are also some technical advantages. One of the biggest headaches in applying the finite-difference method occurs when either the arrangement of the nodes or the geometry is irregular. Some of these problems were mentioned in Section 8•6•1. The finite-element method simplifies many of these problems.

This chapter is intended to provide the framework for understanding the finite-element method. Section 9•1 discusses some of the mathematics that will be needed. Discretization of the spatial problem is considered in Section 9•2. The steady-state and transient examples in Sections 9•3 and 9•4 are the same ones that were discussed using finite differences in Sections 8•3 and 8•4 respectively. Comparisons with finite differences are made to enable you to better appreciate the two methods. The analysis of the transient solutions in Section 9•5 gives further comparisons with the finite-difference method.

One of the most appealing features of the finite-element method is the ease with which it can be programmed for the computer so that irregular problems can be readily treated. $FEHT^1$ is a program that can be used to illustrate the theory presented in Chapter 9.

The discussion through Section 9•5 is for two-dimensional problems using linear triangular elements. Extensions to other problems and other elements are indicated in Section 9•6.

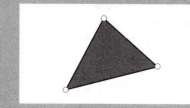

1 Klein, S. A., W. A. Beckman and G. E. Myers: *FEHT – Finite Element Analysis*, F-Chart Software < http : //www.fchart.com/ >

9•1 FUNDAMENTAL CONCEPTS

The governing partial differential equation we will consider is the two-dimensional, transient heat equation given by

$$\frac{\partial}{\partial x}(kt_x) + \frac{\partial}{\partial y}(kt_y) + g''' = \rho c \frac{\partial t}{\partial \theta} \qquad (9\cdot1\cdot1)$$

We want to satisfy this partial differential equation in region A with one or the other of the following boundary conditions on the boundary B:

$$q_o'' = -k\frac{\partial t}{\partial n} = q_s'' + h(t_\infty - t) \qquad (9\cdot1\cdot2)$$

where n is the normal direction into the region, or

$$t = T_s \qquad (9\cdot1\cdot3)$$

where T_s is a specified boundary temperature.

For a transient problem we must also satisfy the initial condition:

$$t(x,y,0) = t^{(0)}(x,y) \qquad (9\cdot1\cdot4)$$

The parameters k, ρ, c, g''', h, t_∞, q_s'' and T_s will be allowed to be functions of position, but not time. This is the same problem we treated by finite differences in Chapter 8.

In the finite-element method of approximating the solution to this problem we will again derive a system of ordinary differential equations to solve. During the derivation it will be necessary to understand the ideas of matrix calculus discussed in Section 9•1•1. The finite-element theory derived in Section 9•2 will be based on the Galerkin weighted-residual method given in Section 9•1•2. The triangular coordinates discussed in Section 9•1•3 are a convenient way to deal with the triangles that will be used in the finite-element theory.

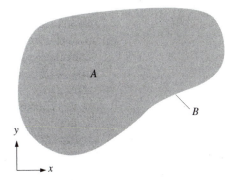

Figure 9•1 *General two-dimensional region.*

9•1•1 Matrix calculus

In the development of the finite-element method you will need to know how to differentiate and integrate matrices. These calculus operations will be illustrated by small examples.

If $\mathbf{A}$ is an $N \times N$ matrix whose entries are functions of x, we say that $\mathbf{A}$ is a function of x. The derivative of $\mathbf{A}(x)$ with respect to x will be an $N \times N$ matrix whose entries are the derivatives of the corresponding entries in $\mathbf{A}(x)$. Consider a 2×2 matrix $\mathbf{A}$ whose entries are functions of x. We may write

$$\mathbf{A}(x) = \begin{bmatrix} a_{11}(x) & a_{12}(x) \\ a_{21}(x) & a_{22}(x) \end{bmatrix}$$

The derivative of $\mathbf{A}$ is then defined to be

$$\frac{d\mathbf{A}}{dx} = \begin{bmatrix} \dfrac{da_{11}}{dx} & \dfrac{da_{12}}{dx} \\ \dfrac{da_{21}}{dx} & \dfrac{da_{22}}{dx} \end{bmatrix}$$

If $\mathbf{A}$ is an $N \times N$ matrix whose entries are functions of x and y, we say that $\mathbf{A}$ is a function of x and y. The partial derivative of $\mathbf{A}(x,y)$ with respect to x or y will be an $N \times N$ matrix whose entries are the partial derivatives with respect to x or y of the corresponding entries in $\mathbf{A}(x,y)$. Consider a 2×2 matrix $\mathbf{A}$ whose entries are functions of x and y. We may write

$$\mathbf{A}(x,y) = \begin{bmatrix} a_{11}(x,y) & a_{12}(x,y) \\ a_{21}(x,y) & a_{22}(x,y) \end{bmatrix}$$

The partial derivatives of $\mathbf{A}$ with respect to x and with respect to y are then defined to be

$$\frac{\partial \mathbf{A}}{\partial x} = \mathbf{A}_x(x,y) = \begin{bmatrix} \dfrac{\partial a_{11}}{\partial x} & \dfrac{\partial a_{12}}{\partial x} \\ \dfrac{\partial a_{21}}{\partial x} & \dfrac{\partial a_{22}}{\partial x} \end{bmatrix} \quad \text{and} \quad \frac{\partial \mathbf{A}}{\partial y} = \mathbf{A}_y(x,y) = \begin{bmatrix} \dfrac{\partial a_{11}}{\partial y} & \dfrac{\partial a_{12}}{\partial y} \\ \dfrac{\partial a_{21}}{\partial y} & \dfrac{\partial a_{22}}{\partial y} \end{bmatrix}$$

The integral of $\mathbf{A}(x)$ will be an $N \times N$ matrix whose entries are the integrals of the corresponding entries in $\mathbf{A}(x)$. For a 2×2 matrix $\mathbf{A}$ whose entries are functions of x we may write

$$\int \mathbf{A}(x)dx = \begin{bmatrix} \int a_{11}dx & \int a_{12}dx \\ \int a_{21}dx & \int a_{22}dx \end{bmatrix}$$

The limits of integration are the same on all integrals.

The integral of $\mathbf{A}(x,y)$ will be an $N \times N$ matrix whose entries are the integrals of the corresponding entries in $\mathbf{A}(x,y)$. For a 2×2 matrix $\mathbf{A}$ whose entries are functions of x and y. We may write

$$\iint \mathbf{A}(x,y)dxdy = \begin{bmatrix} \iint a_{11}dxdy & \iint a_{12}dxdy \\ \iint a_{21}dxdy & \iint a_{22}dxdy \end{bmatrix}$$

The limits of integration are the same on all integrals.

9•1•2 Galerkin weighted-residual method

It will be convenient to rearrange the governing partial differential equation (9·1·1) as

$$\rho c \frac{\partial t}{\partial \theta} + \left[\frac{\partial}{\partial x}(-kt_x) + \frac{\partial}{\partial y}(-kt_y) \right] - g''' = 0 \qquad (9 \cdot 1 \cdot 5)$$

The value of the left-hand side obtained by substituting $t = t(x, y, \theta)$ is called the *residual*. If the exact solution is substituted into the left-hand side the residual will be 0 for all (x, y, θ). If an approximate solution is substituted into the left-hand side the residual will, in general, not be 0. The value of the residual will depend upon position and time (x, y, θ).

In the *method of weighted residuals* (9·1·5) is multiplied by a *weighting function* $f(x, y)$ and then integrated over the region of interest A to give

$$\iint_A f(x, y) \left\{ \rho c \frac{\partial t}{\partial \theta} + \left[\frac{\partial}{\partial x}(-kt_x) + \frac{\partial}{\partial y}(-kt_y) \right] - g''' \right\} dx\, dy = 0$$

The exact solution $t(x, y, \theta)$ will make the integrand identically zero for all (x, y, θ) in region A since the term within the braces is always zero for the exact solution. The method of weighted residuals finds an approximate solution that will make the integral (a weighted average of the residuals) equal zero even though the integrand will not be zero. That is, a weighted average of the residual will be zero rather than the residual itself.

Rewriting this integral as the sum of three integrals,

$$\iint_A f(x, y) \rho c \frac{\partial t}{\partial \theta} dx\, dy + \iint_A f(x, y) \left[\frac{\partial}{\partial x}(-kt_x) + \frac{\partial}{\partial y}(-kt_y) \right] dx\, dy$$
$$- \iint_A f(x, y) g''' dx\, dy = 0 \qquad (9 \cdot 1 \cdot 6)$$

The two conduction integrals (those involving k) can be integrated by parts. The first one may be integrated first over x to give

$$\iint_A f(x, y) \frac{\partial}{\partial x}(-kt_x) dx\, dy = \int \left[\int_{x_W}^{x_E} f(x, y) \frac{\partial}{\partial x}(-kt_x) dx \right] dy$$

where x_W denotes the "western" boundary of A and x_E is on the "eastern" boundary as show in Figure 9•2. Integrating by parts,

$$\iint_A f(x, y) \frac{\partial}{\partial x}(-kt_x) dx\, dy = \int \left[\left[f \cdot (-kt_x) \right]_{x_W}^{x_E} - \int_{x_W}^{x_E} f_x \cdot (-kt_x) dx \right] dy$$

The right-hand side may be rewritten to give

$$\iint_A f(x, y) \frac{\partial}{\partial x}(-kt_x) dx\, dy = \int \left[f \cdot (-kt_x) \right]_{x_W}^{x_E} dy + \iint_A f_x kt_x\, dx\, dy$$

The boundary integrals along the eastern and western boundaries may be written separately to obtain

$$\iint_A f(x, y) \frac{\partial}{\partial x}(-kt_x) dx\, dy = \int_E f \cdot (-kt_x) dy - \int_W f \cdot (-kt_x) dy + \iint_A f_x kt_x\, dx\, dy$$

If we now note that the heat flux in the x direction is given by $q''_x = -kt_x$, we may rewrite the boundary integrals to give

$$\iint_A f(x, y) \frac{\partial}{\partial x}(-kt_x) dx\, dy = \int_E f q''_x\, dy - \int_W f q''_x\, dy + \iint_A f_x kt_x\, dx\, dy$$

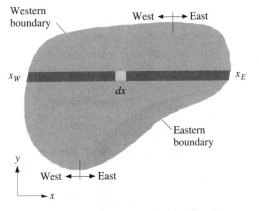

Figure 9•2 *Parts integration in the x-direction.*

A similar analysis will show that the second conduction integral in (9·1·6) may be integrated by parts to give

$$\iint\limits_{A} f(x,y)\frac{\partial}{\partial y}(-kt_y)\,dx\,dy = \int\limits_{N} fq_y''\,dx - \int\limits_{S} fq_y''\,dx + \iint\limits_{A} f_y kt_y\,dx\,dy$$

Substituting both of these integrated conduction integrals into (9·1·6) and moving the boundary integrals to the right-hand side we obtain

$$\iint\limits_{A} f\rho c\frac{\partial t}{\partial \theta}\,dx\,dy + \iint\limits_{A}\left[f_x kt_x + f_y kt_y\right]dx\,dy - \iint\limits_{A} fg'''\,dx\,dy$$
$$= -\int\limits_{E} fq_x''\,dy + \int\limits_{W} fq_x''\,dy - \int\limits_{N} fq_y''\,dx + \int\limits_{S} fq_y''\,dx$$

The integral along the eastern boundary may be broken into two parts, a northeast part and a southeast part. Similarly the other boundary integrals may each be written in two parts. We then may write

$$\iint\limits_{A} f\rho c\frac{\partial t}{\partial \theta}\,dx\,dy + \iint\limits_{A}\left[f_x kt_x + f_y kt_y\right]dx\,dy - \iint\limits_{A} fg'''\,dx\,dy$$
$$= -\int\limits_{NE} fq_x''\,dy - \int\limits_{SE} fq_x''\,dy + \int\limits_{NW} fq_x''\,dy + \int\limits_{SW} fq_x''\,dy$$
$$- \int\limits_{NE} fq_y''\,dx - \int\limits_{NW} fq_y''\,dx + \int\limits_{SE} fq_y''\,dx + \int\limits_{SW} fq_y''\,dx$$

The pairs of *NE*, *NW*, *SW* and *SE* integrals may be combined to give

$$\iint\limits_{A} f\rho c\frac{\partial t}{\partial \theta}\,dx\,dy + \iint\limits_{A}\left[f_x kt_x + f_y kt_y\right]dx\,dy - \iint\limits_{A} fg'''\,dx\,dy$$
$$= \int\limits_{NE} f(-q_x''\,dy - q_y''\,dx) + \int\limits_{NW} f(q_x''\,dy - q_y''\,dx)$$
$$+ \int\limits_{SW} f(q_x''\,dy + q_y''\,dx) + \int\limits_{SE} f(-q_x''\,dy + q_y''\,dx)$$

Now let us look at the northeast boundary integral. Figure 9•3 shows a differential system along a portion of the northeast boundary. The term $q_x''\,dy$ is the conduction rate into the system in the x direction; $q_y''\,dx$ is the conduction rate into the system in the y direction; $q_o''\,ds$ is the rate of heat transfer into the system from outside the region. An energy balance gives

$$q_x''\,dy + q_y''\,dx + q_o''\,ds = 0$$

Rearranging,

$$-q_x''\,dy - q_y''\,dx = q_o''\,ds$$

We see that the parenthesized term in the northeast integral is equal to the rate of heat transfer into the boundary from outside the conduction region. Similar analyses may be made for the northwest, southwest, and southeast boundary integrals. We may then write

$$\iint\limits_{A} f\rho c\frac{\partial t}{\partial \theta}\,dx\,dy + \iint\limits_{A}\left[f_x kt_x + f_y kt_y\right]dx\,dy - \iint\limits_{A} fg'''\,dx\,dy$$
$$= \int\limits_{NE} fq_o''\,ds + \int\limits_{NW} fq_o''\,ds + \int\limits_{SW} fq_o''\,ds + \int\limits_{SE} fq_o''\,ds$$

The four boundary integrals may be combined into one integral around the entire boundary. Thus we may write

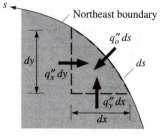

Figure 9•3 *Energy balance along the northeast boundary.*

$$\iint\limits_{A} f\rho c\,\frac{\partial t}{\partial \theta}\,dx\,dy + \iint\limits_{A}\left[f_x k t_x + f_y k t_y \right]dx\,dy - \iint\limits_{A} fg'''\,dx\,dy = \int\limits_{B} fq_o''\,ds \qquad (9\cdot1\cdot7)$$

where q_o'' is the heat flux into region A along boundary B from outside the region.

The exact temperature distribution will satisfy this equation at every point (x,y) in region A and at every time θ. For problems simple enough to obtain an exact solution by separation of variables, we would first assume the variables could be separated and write $t = W(x,y)\Theta(\theta)$ to separate the time variable from the space variables. We would find an infinite number of these solutions which satisfy the partial differential equation and its boundary conditions. These solutions would then be appropriately added to also satisfy the initial condition. Here we are going to assume that the temperature can be approximated by only N terms as

$$t(x,y,\theta) = w_1(x,y)t_1(\theta) + \cdots + w_i(x,y)t_i(\theta) + \cdots + w_N(x,y)t_N(\theta) \qquad (9\cdot1\cdot8)$$

We may write this using matrix notation as

$$t(x,y,\theta) = \begin{bmatrix} w_1 & \cdots & w_i & \cdots & w_N \end{bmatrix}\begin{bmatrix} t_1 \\ \vdots \\ t_i \\ \vdots \\ t_N \end{bmatrix}$$

$$= \mathbf{w}^T(x,y)\mathbf{t}(\theta) \qquad (9\cdot1\cdot9)$$

where $\mathbf{w}^T(x,y)$ contains the N spatial functions $w_i(x,y)$ and $\mathbf{t}(\theta)$ contains the N time variables $t_i(\theta)$.

Partial derivatives of temperature with respect to x and y are given by

$$t_x = \mathbf{w}_x^T\mathbf{t} \qquad\text{and}\qquad t_y = \mathbf{w}_y^T\mathbf{t} \qquad (9\cdot1\cdot10)$$

The partial derivative of temperature with respect to θ is given by

$$t_\theta = \mathbf{w}^T\dot{\mathbf{t}} \qquad (9\cdot1\cdot11)$$

Substitution of the assumed temperature distribution and its derivatives, $(9\cdot1\cdot9)$, $(9\cdot1\cdot10)$, and $(9\cdot1\cdot11)$, into $(9\cdot1\cdot7)$ gives

$$\left[\iint\limits_{A} f\rho c\mathbf{w}^T(x,y)\,dx\,dy\right]\dot{\mathbf{t}} + \left[\iint\limits_{A}\left[f_x k\mathbf{w}_x^T + f_y k\mathbf{w}_y^T \right]dx\,dy\right]\mathbf{t} - \iint\limits_{A} fg'''\,dx\,dy$$

$$= \int\limits_{B} fq_o''\,ds \qquad (9\cdot1\cdot12)$$

Note that $\mathbf{t}$ and $\dot{\mathbf{t}}$ have been removed from the integrals since they are only functions of θ, not of x or y. This is now a single ordinary differential equation relating $t_1(\theta)$, ..., $t_i(\theta)$, ..., $t_N(\theta)$ and their first derivatives. There is a different differential equation for each independent $f(x,y)$.

Since there are N dependent variables $t_1(\theta)$, ..., $t_i(\theta)$, ..., $t_N(\theta)$ we will need N independent ordinary differential equations to obtain a unique solution. To arrive at a set of N differential equations we will use N independent functions $f_1(x,y)$, ..., $f_i(x,y)$, ..., $f_N(x,y)$. The Galerkin technique takes each $f_i(x,y) = w_i(x,y)$. In matrix notation we will write

$$\mathbf{f} = \begin{bmatrix} f_1 \\ \vdots \\ f_i \\ \vdots \\ f_N \end{bmatrix} = \begin{bmatrix} w_1 \\ \vdots \\ w_i \\ \vdots \\ w_N \end{bmatrix} = \mathbf{w}$$

Then, upon replacing f by $\mathbf{f} = \mathbf{w}$, f_x by $\mathbf{f}_x = \mathbf{w}_x$, and f_y by $\mathbf{f}_y = \mathbf{w}_y$ in (9·1·12), we obtain

$$\left[\iint_A \mathbf{w}\rho c\mathbf{w}^T \, dx\, dy \right]\dot{\mathbf{t}} + \left[\iint_A \left(\mathbf{w}_x k\mathbf{w}_x^T + \mathbf{w}_y k\mathbf{w}_y^T \right) dx\, dy \right]\mathbf{t} - \iint_A \mathbf{w}g''' \, dx\, dy$$
$$= \int_B \mathbf{w}q_o'' \, ds$$

We now have a system of N ordinary differential equations each relating $t_1(\theta)$, ..., $t_i(\theta)$, ..., $t_N(\theta)$ and their first derivatives. This system of ordinary differential equations may be written using matrix notation as

$$\mathbf{C}\dot{\mathbf{t}} + \mathbf{K}\mathbf{t} - \mathbf{g} = \mathbf{q}_o \tag{9·1·13}$$

where

$$\mathbf{C} = \iint_A \mathbf{w}\rho c\mathbf{w}^T \, dx\, dy \tag{9·1·14}$$

$$\mathbf{K} = \iint_A \left(\mathbf{w}_x k\mathbf{w}_x^T + \mathbf{w}_y k\mathbf{w}_y^T \right) dx\, dy \tag{9·1·15}$$

$$\mathbf{g} = \iint_A \mathbf{w}g''' \, dx\, dy \tag{9·1·16}$$

$$\mathbf{q}_o = \int_B \mathbf{w}q_o'' \, ds \tag{9·1·17}$$

Since the products $\mathbf{w}\mathbf{w}^T$, $\mathbf{w}_x\mathbf{w}_x^T$, and $\mathbf{w}_y\mathbf{w}_y^T$ are $N \times N$ matrices, $\mathbf{C}$ and $\mathbf{K}$ will be $N \times N$ matrices. Since $\mathbf{w}$ is an $N \times 1$ vector, $\mathbf{g}$ and $\mathbf{q}_o$ will also be $N \times 1$ vectors.

In the Galerkin weighted-residual method reasonable assumptions are made for the $w_i(x,y)$ contained in $\mathbf{w} = \mathbf{w}(x,y)$. The entries in $\mathbf{C}$, $\mathbf{K}$, $\mathbf{g}$ and $\mathbf{q}_o$ are then obtained. The system of ordinary differential equations is then solved for the $t_i(\theta)$ contained in $\mathbf{t} = \mathbf{t}(\theta)$. Substitution of the $w_i(x,y)$ and the $t_i(\theta)$ into (9·1·8) then gives an approximate solution for $t(x,y,\theta)$.

9•1•3 Triangular coordinates

In the derivation of the finite-element method in the next section we will consider temperature distributions in triangular regions. It will be helpful to use a special coordinate system to describe the location of a point within such a triangular region.

Nodes i, j and k, located at positions (x_i, y_i), (x_j, y_j) and (x_k, y_k), respectively, may be connected together by straight lines to form a triangle as shown in Figure 9•4. Such a triangle is called an *element*. The area A of this triangular element may be written in terms of the (x, y) coordinates of its three nodes. This area may be found by calculating the area of the trapezoid below side ik and then subtracting the areas of the trapezoids below sides ij and jk to give

$$A = \frac{y_i + y_k}{2}(x_k - x_i) - \frac{y_i + y_j}{2}(x_j - x_i) - \frac{y_j + y_k}{2}(x_k - x_j)$$

To simplify notation we will define $x_{ij} = x_j - x_i$ and $x_{jk} = x_k - x_j$. We may then write

$$x_k - x_i = (x_k - x_j) + (x_j - x_i) = x_{jk} + x_{ij}$$

Upon making these substitutions and factoring out the 2, the area may be written as

$$A = \frac{1}{2}\Big[(y_i + y_k)(x_{jk} + x_{ij}) - (y_i + y_j)x_{ij} - (y_j + y_k)x_{jk}\Big]$$

Rearranging gives

$$A = \frac{1}{2}\Big[(y_i + y_k - y_j - y_k)x_{jk} + (y_i + y_k - y_i - y_j)x_{ij}\Big]$$

Upon canceling y_k from the first term and y_i from the second term and defining $y_{ij} = y_j - y_i$ and $y_{jk} = y_k - y_j$ we may write the area as

$$A = \frac{1}{2}\Big[x_{ij}y_{jk} - x_{jk}y_{ij}\Big] = \frac{1}{2}b_{ijk} \tag{9·1·18}$$

where we have defined

$$b_{ijk} = x_{ij}y_{jk} - x_{jk}y_{ij} \tag{9·1·19}$$

A point p located at position (x, y) within the triangular element will divide the triangle into three separate areas as shown in Figure 9•5. The area "attached" to the side opposite node i will be called A_i. Similarly, the areas opposite nodes j and k will be called A_j and A_k respectively. The areas of these smaller triangles may be found in the same way we found the total area. For area A_i we may add the trapezoidal area under line segments jp and pk and subtract the area under jk to obtain

$$A_i = \frac{y + y_j}{2}(x - x_j) + \frac{y_k + y}{2}(x_k - x) - \frac{y_j + y_k}{2}(x_k - x_j)$$

Upon factoring out the 2 and removing parentheses,

$$A_i = \frac{1}{2}\Big[xy + y_j x - x_j y - x_j y_j + x_k y_k + x_k y - xy_k$$
$$-x_k y_j - x_k y_k + x_j y_j + x_j y_k\Big]$$

Six of these terms will cancel and the rest may be rearranged to give

$$A_i = \frac{1}{2}\Big[(x_j y_k - x_k y_j) + x(y_j - y_k) + (x_k - x_j)y\Big]$$

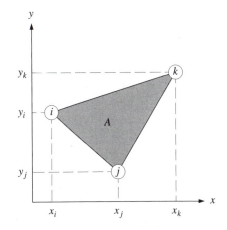

Figure 9•4 *Triangular element.*

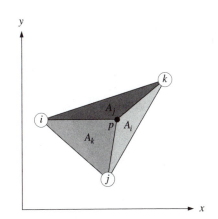

Figure 9•5 *Point p located within triangular element.*

Introducing $x_{jk} = x_k - x_j$ and $y_{jk} = y_k - y_j$ gives

$$A_i = \frac{1}{2}\left[(x_j y_k - x_k y_j) - x y_{jk} + x_{jk} y\right] \tag{9·1·20}$$

A *triangular coordinate* ξ_i is defined as the ratio A_i / A. Upon substituting (9·1·18) and (9·1·20) for A and A_i we obtain

$$\xi_i = \frac{A_i}{A} = \frac{1}{b_{ijk}}\left[(x_j y_k - x_k y_j) - x y_{jk} + x_{jk} y\right]$$

When point p is at node i area A_i will equal area A and ξ_i will equal 1. Whenever point p lies along side jk (opposite node i), area A_i will be 0 and hence ξ_i will also be 0. Whenever point p moves along a line parallel to side jk, area A_i will be constant (since the base and height of A_i will be constant) and hence ξ_i will also be constant. Thus, lines of constant ξ_i are parallel to the side opposite node i. The value of ξ_i varies linearly from 1 at node i to 0 along the side opposite node i.

We may also define triangular coordinates ξ_j and ξ_k and show that

$$\xi_j = \frac{A_j}{A} = \frac{1}{b_{ijk}}\left[(x_k y_i - x_i y_k) + x y_{ik} - x_{ik} y\right]$$

$$\xi_k = \frac{A_k}{A} = \frac{1}{b_{ijk}}\left[(x_i y_j - x_j y_i) - x y_{ij} + x_{ij} y\right]$$

The value of ξ_j varies linearly from 1 at node j to 0 along the side opposite node j and is constant along lines parallel to the side opposite node j. The value of ξ_k varies linearly from 1 at node k to 0 along the side opposite node k and is constant along lines parallel to the side opposite node k.

The three equations relating ξ_i, ξ_j and ξ_k to x and y may be written using matrix notation as

$$\begin{bmatrix} \xi_i \\ \xi_j \\ \xi_k \end{bmatrix} = \frac{1}{b_{ijk}} \begin{bmatrix} (x_j y_k - x_k y_j) & -y_{jk} & x_{jk} \\ -(x_i y_k - x_k y_i) & y_{ik} & -x_{ik} \\ (x_i y_j - x_j y_i) & -y_{ij} & x_{ij} \end{bmatrix} \begin{bmatrix} 1 \\ x \\ y \end{bmatrix} \tag{9·1·21}$$

If you add these three equations together you will obtain $\xi_i + \xi_j + \xi_k = 1$. This is to be expected since $A_i + A_j + A_k = A$. Therefore only two of the three triangular coordinates are independent.

The location of a point at (x, y) within the triangular element is fixed by specifying any two triangular coordinates. For example, if ξ_i and ξ_j are given, (x, y) must be at the intersection of the lines of constant ξ_i and ξ_j as shown in Figure 9•7.

During the finite-element development in the next section we will need partial derivatives of ξ_i, ξ_j and ξ_k with respect to x and y. Differentiating (9·1·21) gives

$$\frac{\partial}{\partial x}\begin{bmatrix} \xi_i \\ \xi_j \\ \xi_k \end{bmatrix} = \frac{1}{b_{ijk}}\begin{bmatrix} -y_{jk} \\ y_{ik} \\ -y_{ij} \end{bmatrix} \quad \text{and} \quad \frac{\partial}{\partial y}\begin{bmatrix} \xi_i \\ \xi_j \\ \xi_k \end{bmatrix} = \frac{1}{b_{ijk}}\begin{bmatrix} x_{jk} \\ -x_{ik} \\ x_{ij} \end{bmatrix} \tag{9·1·22}$$

We will have need to evaluate integrals of powers of ξ_i, ξ_j and ξ_k over a triangular area. These integrals may be evaluated from [3]:

$$\int_A \xi_i^a \xi_j^b \xi_k^c \, dA = \frac{a!\,b!\,c!}{(a+b+c+2)!}\,2A \tag{9·1·23}$$

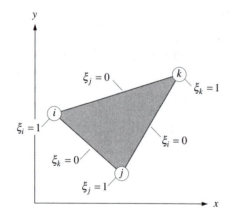

Figure 9·6 *Triangular-coordinate values.*

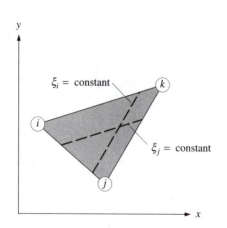

Figure 9·7 *Point within triangular element is fixed by two triangular coordinates.*

9•2 DISCRETIZING THE SPATIAL PROBLEM

The formulation of a two-dimensional conduction problem considers a system $\Delta x \times \Delta y$, writes an energy balance, and uses the rate equations. In an exact formulation, leading to a partial differential equation, the system is shrunk to zero size (*i.e.*, Δx and Δy approach 0 in the limit). In the finite-difference formulation of Chapter 8, Δx and Δy remain finite and finite-difference approximations of the rate equations are used. The finite-element formulation in this book starts with the exact formulation (the partial differential equation) and uses the Galerkin weighted-residual method as discussed in Section 9•1•2 to approximate the solution to the partial differential equation and its boundary conditions.

The major results of the Galerkin weighted-residual method are that the temperature distribution is approximated by (9·1·8) or (9·1·9) and the partial differential equation and its boundary conditions are replaced by a system of ordinary differential equations (9·1·13).

The spatial problem will be discretized by choosing a set of N nodal points as shown in Figure 9•8 for $N = 49$. These nodal points may be placed at any convenient locations you desire. They do not need to be placed in a regular pattern as one usually tries to do in the finite-difference method. The fact that irregularly spaced nodal points are no more difficult to handle than regularly spaced nodal points is what makes the finite-element method so attractive in handling two-dimensional problems.

In placing the nodal points one usually looks at the boundary first. A sufficient number of nodes are used to give an adequate approximation of the boundary. The nodes along the boundary are connected by straight-line segments to form a polygon. The polygon approximation of the boundary in Figure 9•8 is much better than the finite-difference approximation shown in Figure 8•4 using the same number of boundary nodes (26). The approximate finite-element polygon boundary in Figure 9•8 can hardly be distinguished from the exact boundary. This is an advantage of the finite-element method over the finite-difference method that occurs due to the ease with which irregularly spaced nodes can be used to approximate the boundaries.

Next the interior nodes are placed. These nodes can be more densely located in regions where steep temperature gradients are expected or where more accuracy is desired. Each nodal point is given a number (1, 2, 3, ..., 49 in this example) as shown in Figure 9•8. It is important in large problems to number the nodes to keep the bandwidth small as was mentioned in connection with finite differences.

Finally, the nodal points are connected to form a network of triangular elements as shown in Figure 9•9. Each element is also given a number (**1**, **2**, **3**, ..., **70** in this example). The numbering of the elements is not important insofar as the bandwidth is concerned. Generally, equilateral elements are felt to be better than long, thin elements, but any triangular shape will do.

In the approximate temperature distribution (9·1·8) we will take $t_i(\theta)$ to be the temperature at node i. There will be N of these nodal temperatures.

The function $w_i(x,y)$ is an *interpolation function* to help find the temperature at points inside the elements that surround node i. In order that (9·1·8) give $t(x,y,\theta) = t_i(\theta)$ at node i, we will insist that

$$w_i(x,y) = 1 \text{ at node } i$$

and

$$w_i(x,y) = 0 \text{ at all other nodes.}$$

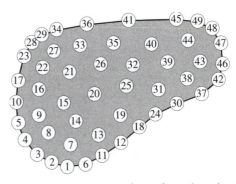

Figure 9•8 *Approximation of irregular region using 49 irregularly spaced nodal points.*

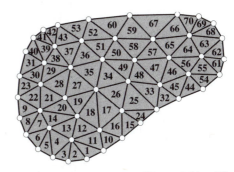

Figure 9•9 *Discretization of Figure 9•8 into 70 triangular elements.*

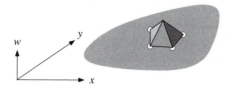

Figure 9•10 *Pyramid function.*

We will assume that $w_i(x, y)$ varies linearly from 1 at node i to 0 at nodes connected directly to node i and is 0 everywhere outside of the elements around node i. Since the sketch of $w_i(x, y)$ shown in Figure 9•10 looks like a pyramid, this function is sometimes called a *pyramid function*. Within any triangular element surrounding node i the value of $w_i(x, y)$ may be most easily expressed using the area coordinate ξ_i as

$$w_i(x, y) = \xi_i$$

Consider element e to be the element formed by connecting nodes i, j and k. The temperature at a point (x, y) located within element e will be called $t^{(e)}(x, y, \theta)$. To evaluate $t^{(e)}(x, y, \theta)$ from (9·1·8) we must evaluate each of the N pyramid functions at (x, y). The only nonzero pyramid functions will be $w_i(x, y)$, $w_j(x, y)$ and $w_k(x, y)$. In matrix form we may then write

$$t^{(e)}(x, y, \theta) = \begin{bmatrix} & w_i & w_j & w_k & \end{bmatrix} \begin{bmatrix} t_i \\ t_j \\ \\ t_k \\ \\ \end{bmatrix}$$

where only the nonzero values in $\mathbf{w}^T(x, y)$ are shown. The values of w_i, w_j and w_k appear in columns i, j and k respectively.

Since within this triangle, $w_i(x, y) = \xi_i$, $w_j(x, y) = \xi_j$, and $w_k(x, y) = \xi_k$ we may write

$$\mathbf{w}^T = \begin{bmatrix} & w_i & w_j & w_k & \end{bmatrix}$$

$$= \begin{bmatrix} & \xi_i & \xi_j & \xi_k & \end{bmatrix} \tag{9·2·1}$$

Upon multiplying $\mathbf{w}^T\mathbf{t}$ we find that the temperature within element e is given by

$$t^{(e)} = \xi_i t_i + \xi_j t_j + \xi_k t_k$$

This may be written in matrix notation as

$$t^{(e)} = \begin{bmatrix} t_i & t_j & t_k \end{bmatrix} \begin{bmatrix} \xi_i \\ \xi_j \\ \xi_k \end{bmatrix}$$

Upon substituting (9·1·21) we obtain

$$t^{(e)} = \begin{bmatrix} t_i & t_j & t_k \end{bmatrix} \frac{1}{b_{ijk}} \begin{bmatrix} (x_j y_k - x_k y_j) & -y_{jk} & x_{jk} \\ -(x_i y_k - x_k y_i) & y_{ik} & -x_{ik} \\ (x_i y_j - x_j y_i) & -y_{ij} & x_{ij} \end{bmatrix} \begin{bmatrix} 1 \\ x \\ y \end{bmatrix} \tag{9·2·2}$$

Upon multiplying the 1×3 row vector of temperatures by the 3×3 matrix of nodal-position values we obtain a 1×3 row vector containing the three nodal temperatures and the nodal position values. If, for shorthand, we call the entries in this product c_1, c_2 and c_3, we may write

$$t^{(e)} = \begin{bmatrix} c_1 & c_2 & c_3 \end{bmatrix} \begin{bmatrix} 1 \\ x \\ y \end{bmatrix}$$

Upon multiplying further we obtain

$$t^{(e)} = c_1 + c_2 x + c_3 y$$

We chose the $w_i(x,y)$ so that $t^{(e)}(x_i, y_i, \theta) = t_i(\theta)$. We now see that the temperature within an element varies linearly between its three nodal-point temperatures.

For heat-flux computation within an element we will need the partial derivatives of $t^{(e)}$ in both the x- and the y-directions. Upon taking the partial derivative of (9·2·2) with respect to x,

$$t_x^{(e)} = \begin{bmatrix} t_i & t_j & t_k \end{bmatrix} \frac{1}{b_{ijk}} \begin{bmatrix} (x_j y_k - x_k y_j) & -y_{jk} & x_{jk} \\ -(x_i y_k - x_k y_i) & y_{ik} & -x_{ik} \\ (x_i y_j - x_j y_i) & -y_{ij} & x_{ij} \end{bmatrix} \begin{bmatrix} 0 \\ 1 \\ 0 \end{bmatrix}$$

The partial derivative of (9·2·2) with respect to y gives,

$$t_y^{(e)} = \begin{bmatrix} t_i & t_j & t_k \end{bmatrix} \frac{1}{b_{ijk}} \begin{bmatrix} (x_j y_k - x_k y_j) & -y_{jk} & x_{jk} \\ -(x_i y_k - x_k y_i) & y_{ik} & -x_{ik} \\ (x_i y_j - x_j y_i) & -y_{ij} & x_{ij} \end{bmatrix} \begin{bmatrix} 0 \\ 0 \\ 1 \end{bmatrix}$$

Since we have taken the temperature profile within an element to be linear, the partial derivatives of temperature within an element are constants.

Upon multiplying out the matrices we obtain

$$t_x^{(e)} = \frac{-y_{jk} t_i + y_{ik} t_j - y_{ij} t_k}{b_{ijk}}$$

and

$$t_y^{(e)} = \frac{x_{jk} t_i - x_{ik} t_j + x_{ij} t_k}{b_{ijk}}$$

These two expressions may be expressed in one matrix relation by writing

$$\begin{bmatrix} t_x \\ t_y \end{bmatrix}^{(e)} = \frac{1}{b_{ijk}^{(e)}} \begin{bmatrix} -y_{jk} & y_{ik} & -y_{ij} \\ x_{jk} & -x_{ik} & x_{ij} \end{bmatrix}^{(e)} \begin{bmatrix} t_i \\ t_j \\ t_k \end{bmatrix}^{(e)} \tag{9·2·3}$$

The value of $b_{ijk}^{(e)}$ in (9·2·2) and (9·2·3) is given by (9·1·19).

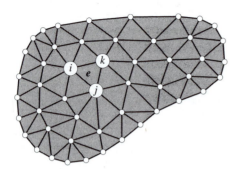

Figure 9•11 *Typical interior element.*

9•2•1 Interior elements

The next step in the formulation is to evaluate **C**, **K** and **g** as given by (9·1·14), (9·1·15) and (9·1·16), respectively. Each of these involves a double integral over the entire area A. Discretization has replaced the actual area by a polygon whose boundary is "acceptably close" to the actual boundary. We will evaluate the area integrals by writing the total integral as the sum of integrals over each of the triangular elements. If e is the element number and there are a total of NE elements, we will write

$$\mathbf{C} = \iint_A \mathbf{w}\rho c \mathbf{w}^T \, dx \, dy = \sum_{e=1}^{NE} \iint_{A^{(e)}} \mathbf{w}\rho c \mathbf{w}^T \, dx \, dy$$

$$\mathbf{K} = \iint_A \left(\mathbf{w}_x k \mathbf{w}_x^T + \mathbf{w}_y k \mathbf{w}_y^T \right) dx \, dy = \sum_{e=1}^{NE} \iint_{A^{(e)}} \left(\mathbf{w}_x k \mathbf{w}_x^T + \mathbf{w}_y k \mathbf{w}_y^T \right) dx \, dy$$

$$\mathbf{g} = \iint_A \mathbf{w} g''' \, dx \, dy = \sum_{e=1}^{NE} \iint_{A^{(e)}} \mathbf{w} g''' \, dx \, dy$$

The values of k, ρ, c and g''' will be assumed to be uniform within an element but they can, in general, be different for each element. These uniform values may be factored out of the element integrals to give

$$\mathbf{C} = \sum_{e=1}^{NE} \rho^{(e)} c^{(e)} \iint_{A^{(e)}} \mathbf{w}\mathbf{w}^T \, dx \, dy$$

$$\mathbf{K} = \sum_{e=1}^{NE} k^{(e)} \iint_{A^{(e)}} \left(\mathbf{w}_x \mathbf{w}_x^T + \mathbf{w}_y \mathbf{w}_y^T \right) dx \, dy$$

$$\mathbf{g} = \sum_{e=1}^{NE} g'''^{(e)} \iint_{A^{(e)}} \mathbf{w} \, dx \, dy$$

where $k^{(e)}$, $\rho^{(e)}$, $c^{(e)}$ and $g'''^{(e)}$ are the values of k, ρ, c and g''' for element e.

An *element capacitance matrix* $\mathbf{C}^{(e)}$, an *element conduction matrix* $\mathbf{K}^{(e)}$, and an *element generation vector* $\mathbf{g}^{(e)}$ are convenient to define as

$$\mathbf{C}^{(e)} = \rho^{(e)} c^{(e)} \iint_{A^{(e)}} \mathbf{w}\mathbf{w}^T \, dx \, dy \tag{9·2·4}$$

$$\mathbf{K}^{(e)} = k^{(e)} \iint_{A^{(e)}} \left(\mathbf{w}_x \mathbf{w}_x^T + \mathbf{w}_y \mathbf{w}_y^T \right) dx \, dy \tag{9·2·5}$$

$$\mathbf{g}^{(e)} = g'''^{(e)} \iint_{A^{(e)}} \mathbf{w} \, dx \, dy \tag{9·2·6}$$

The *global capacitance matrix* **C**, *global conduction matrix* **K** and *global generation vector* **g** are the sums of $\mathbf{C}^{(e)}$, $\mathbf{K}^{(e)}$ and $\mathbf{g}^{(e)}$ as given by

$$\mathbf{C} = \sum_{e=1}^{NE} \mathbf{C}^{(e)} \tag{9·2·7}$$

$$\mathbf{K} = \sum_{e=1}^{NE} \mathbf{K}^{(e)} \tag{9·2·8}$$

$$\mathbf{g} = \sum_{e=1}^{NE} \mathbf{g}^{(e)} \tag{9·2·9}$$

Next we must learn how to find $\mathbf{C}^{(e)}$, $\mathbf{K}^{(e)}$ and $\mathbf{g}^{(e)}$ so we can construct **C**, **K** and **g**.

Capacitance matrix To find $\mathbf{C}^{(e)}$ we must first determine $\mathbf{w}\mathbf{w}^T$. Since $\mathbf{w}$ is $N \times 1$ and $\mathbf{w}^T$ is $1 \times N$, the product $\mathbf{w}\mathbf{w}^T$ will be $N \times N$. We can find $\mathbf{w}$ and $\mathbf{w}^T$ from (9·2·1). The product will then give

$$\mathbf{w}\mathbf{w}^T = \begin{bmatrix} \xi_i \\ \xi_j \\ \xi_k \end{bmatrix} \begin{bmatrix} \xi_i & \xi_j & \xi_k \end{bmatrix} = \begin{bmatrix} \xi_i\xi_i & \xi_i\xi_j & \xi_i\xi_k & \text{Row } i \\ \xi_j\xi_i & \xi_j\xi_j & \xi_j\xi_k & \text{Row } j \\ \xi_k\xi_i & \xi_k\xi_j & \xi_k\xi_k & \text{Row } k \\ \text{Column } i & \text{Column } j & \text{Column } k & \end{bmatrix}$$

All entries in $\mathbf{w}\mathbf{w}^T$ are 0 except for those at the intersection of rows i, j and k and columns i, j and k. Note that $\mathbf{w}\mathbf{w}^T$ is a symmetric matrix.

The next step in finding $\mathbf{C}^{(e)}$ is to evaluate the integral of $\mathbf{w}\mathbf{w}^T$. Substitution into (9·1·23) gives

$$\int_{A^{(e)}} \xi_i^2 \, dA = \frac{2!0!0!}{(2+0+0+2)!} 2A^{(e)} = \frac{A^{(e)}}{6}$$

$$\int_{A^{(e)}} \xi_i\xi_j \, dA = \frac{1!1!0!}{(1+1+0+2)!} 2A^{(e)} = \frac{A^{(e)}}{12}$$

The first integral is typical of the three terms on the main diagonal. The second integral is typical of the off-diagonal terms. Upon factoring out $\frac{1}{12}A^{(e)}$, the element capacitance matrix $\mathbf{C}^{(e)}$ in (9·2·4) is then given by

$$\mathbf{C}^{(e)} = \frac{\rho^{(e)}c^{(e)}A^{(e)}}{12} \begin{bmatrix} 2 & 1 & 1 & \text{Row } i \\ 1 & 2 & 1 & \text{Row } j \\ 1 & 1 & 2 & \text{Row } k \\ \text{Column } i & \text{Column } j & \text{Column } k & \end{bmatrix} \qquad (9\cdot2\cdot10)$$

Only the nonzero terms in $\mathbf{C}^{(e)}$ are shown. The sum of these nine entries in $\mathbf{C}^{(e)}$ is $\rho^{(e)}c^{(e)}A^{(e)}$, the thermal capacitance of element e. This thermal capacitance is distributed to nine positions in the global capacitance matrix.

From (9·2·7), the global capacitance matrix $\mathbf{C}$ is found by summing the element capacitance matrices $\mathbf{C}^{(e)}$ for all of the elements. Although $\mathbf{C}$ will be symmetric, it will not be diagonal as it was in finite differences. For reasonable numbering of the nodes, $\mathbf{C}$ will be banded. The banded $\mathbf{C}$ will require more computer storage space and it will alter transient behavior.

Conduction matrix To determine the element conduction matrix $\mathbf{K}^{(e)}$ in (9·2·5) we will need $\mathbf{w}_x^T$ and $\mathbf{w}_y^T$. These can be found by differentiating (9·2·1) to give

$$\mathbf{w}_x^T = \begin{bmatrix} \dfrac{\partial \xi_i}{\partial x} & \dfrac{\partial \xi_j}{\partial x} & \dfrac{\partial \xi_k}{\partial x} \end{bmatrix}$$

$$\mathbf{w}_y^T = \begin{bmatrix} \dfrac{\partial \xi_i}{\partial y} & \dfrac{\partial \xi_j}{\partial y} & \dfrac{\partial \xi_k}{\partial y} \end{bmatrix}$$

The six partial derivatives on the right-hand sides may be obtained from (9·1·22). Thus,

$$\mathbf{w}_x^T = \frac{1}{b_{ijk}} \begin{bmatrix} -y_{jk} & y_{ik} & -y_{ij} \end{bmatrix}$$

$$\mathbf{w}_y^T = \frac{1}{b_{ijk}} \begin{bmatrix} x_{jk} & -x_{ik} & x_{ij} \end{bmatrix}$$

Next we can find $\mathbf{w}_x \mathbf{w}_x^T$ as

$$\mathbf{w}_x \mathbf{w}_x^T = \frac{1}{b_{ijk}} \begin{bmatrix} -y_{jk} \\ y_{ik} \\ -y_{ij} \end{bmatrix} \frac{1}{b_{ijk}} \begin{bmatrix} -y_{jk} & y_{ik} & -y_{ij} \end{bmatrix} = \frac{1}{b_{ijk}^2} \begin{bmatrix} y_{jk}y_{jk} & -y_{jk}y_{ik} & y_{jk}y_{ij} & \text{Row } i \\ -y_{ik}y_{jk} & y_{ik}y_{ik} & -y_{ik}y_{ij} & \text{Row } j \\ y_{ij}y_{jk} & -y_{ij}y_{ik} & y_{ij}y_{ij} & \text{Row } k \\ \text{Column } i & \text{Column } j & \text{Column } k & \end{bmatrix}$$

Similarly, $\mathbf{w}_y \mathbf{w}_y^T$ is given by

$$\mathbf{w}_y \mathbf{w}_y^T = \frac{1}{b_{ijk}} \begin{bmatrix} x_{jk} \\ -x_{ik} \\ x_{ij} \end{bmatrix} \frac{1}{b_{ijk}} \begin{bmatrix} x_{jk} & -x_{ik} & x_{ij} \end{bmatrix} = \frac{1}{b_{ijk}^2} \begin{bmatrix} x_{jk}x_{jk} & -x_{jk}x_{ik} & x_{jk}x_{ij} & \text{Row } i \\ -x_{ik}x_{jk} & x_{ik}x_{ik} & -x_{ik}x_{ij} & \text{Row } j \\ x_{ij}x_{jk} & -x_{ij}x_{ik} & x_{ij}x_{ij} & \text{Row } k \\ \text{Column } i & \text{Column } j & \text{Column } k & \end{bmatrix}$$

It is important to note that both $\mathbf{w}_x \mathbf{w}_x^T$ and $\mathbf{w}_y \mathbf{w}_y^T$ are symmetric matrices. Also, they are both constants, independent of x and y within the element e. Therefore they can be factored out of the integral in (9·2·5). The integral that remains simply gives the area of the element. Thus, upon adding $\mathbf{w}_x \mathbf{w}_x^T + \mathbf{w}_y \mathbf{w}_y^T$ and then multiplying by $k^{(e)}A^{(e)}$, the element conduction matrix may be written as

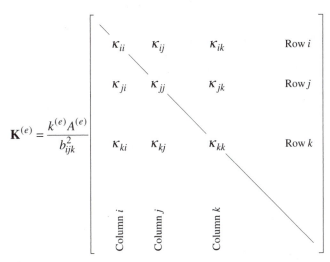

$$\mathbf{K}^{(e)} = \frac{k^{(e)}A^{(e)}}{b_{ijk}^2}$$

where

$$\kappa_{ii} = x_{jk}x_{jk} + y_{jk}y_{jk} \qquad \kappa_{ij} = -(x_{jk}x_{ik} + y_{jk}y_{ik}) \qquad \kappa_{ik} = x_{jk}x_{ij} + y_{jk}y_{ij}$$

$$\kappa_{jj} = x_{ik}x_{ik} + y_{ik}y_{ik} \qquad \kappa_{jk} = -(x_{ik}x_{ij} + y_{ik}y_{ij})$$

$$\kappa_{kk} = x_{ij}x_{ij} + y_{ij}y_{ij}$$

From (9·1·18) we see that $A^{(e)} = \frac{1}{2}b_{ijk}$ where b_{ijk} is given by (9·1·19). Actually this is true only for nodes i, j and k going around the element in a counter-clockwise direction. If i, j and k are in a clockwise order then $A^{(e)} = -\frac{1}{2}b_{ijk}$. Thus, to be more general, we should write

$$A^{(e)} = \frac{1}{2}\left|b_{ijk}\right| = \frac{1}{2}\left|x_{ij}y_{jk} - x_{jk}y_{ij}\right| \qquad (9\cdot2\cdot11)$$

No matter which way the nodes are numbered, $b_{ijk}^2 = 4A^{(e)^2}$. The element conduction matrix may then be written as

$$\mathbf{K}^{(e)} = \frac{k^{(e)}}{4A^{(e)}}
\begin{bmatrix}
\kappa_{ii} & \kappa_{ij} & \kappa_{ik} & & \text{Row } i \\
\kappa_{ji} & \kappa_{jj} & \kappa_{jk} & & \text{Row } j \\
\kappa_{ki} & \kappa_{kj} & \kappa_{kk} & & \text{Row } k \\
& & & & \\
\text{Column } i & \text{Column } j & \text{Column } k & &
\end{bmatrix}$$

$$(9\cdot2\cdot12)$$

The element conduction matrix is an $N \times N$ symmetric matrix with at most nine nonzero entries. For an element having nodes i, j and k, these nonzero entries in $\mathbf{K}^{(e)}$ are at the intersections of rows and column i, j and k. Another useful fact is that the sum of the entries in any one row in $\mathbf{K}^{(e)}$ is always 0. Since $\mathbf{K}^{(e)}$ is symmetric, the sum of the entries in any one column is also 0.

From (9·2·8), the global conduction matrix $\mathbf{K}$ is found by summing the element conduction matrices $\mathbf{K}^{(e)}$ for all of the elements. Note that $\mathbf{K}$ will be symmetric and, with reasonable nodal-point numbering, banded.

Generation vector   To determine the element generation vector $\mathbf{g}^{(e)}$ in (9·2·6) we will need to integrate $\mathbf{w}$ over the element area. For element e whose nodes are i, j and k, we can find $\mathbf{w}$ in terms of triangular coordinates from (9·2·1). Substitution into (9·2·6) then gives

$$\mathbf{g}^{(e)} = g'''^{(e)} \int_{A^{(e)}} \begin{bmatrix} \xi_i \\ \\ \xi_j \\ \\ \xi_k \\ \\ \\ \\ \\ \\ \\ \end{bmatrix} dA$$

The three integrals may be evaluated from (9·1·23). For row i,

$$\int_{A^{(e)}} \xi_i \, dA = \frac{1!0!0!}{(1+0+0+2)!} 2A^{(e)} = \frac{1}{3} A^{(e)}$$

The same result is obtained for the two other nonzero integrals. Upon substituting the integrals into the equation for $\mathbf{g}^{(e)}$ and factoring $\frac{1}{3} A^{(e)}$ we obtain

$$\mathbf{g}^{(e)} = \frac{g'''^{(e)} A^{(e)}}{3} \begin{bmatrix} 1 & \text{Row } i \\ \\ 1 & \text{Row } j \\ \\ \\ \\ 1 & \text{Row } k \\ \\ \\ \\ \\ \\ \\ \\ \\ \end{bmatrix}$$

$$(9·2·13)$$

Only the nonzero terms are shown. They appear in rows i, j and k. The sum of the entries in $\mathbf{g}^{(e)}$ is $g'''^{(e)} A^{(e)}$, the total rate of energy generation in element e. This generation rate is equally distributed in each row.

From (9·2·9), the global generation vector $\mathbf{g}$ is found by summing the element capacitance vectors $\mathbf{g}^{(e)}$ for all of the elements.

Summary   Each element contributes to the global $\mathbf{C}$, $\mathbf{K}$ and $\mathbf{g}$. In practice one does not save space for each of the element matrices. The contributions from each element are added directly into $\mathbf{C}$, $\mathbf{K}$ and $\mathbf{g}$. In finite elements $\mathbf{C}$, $\mathbf{K}$ and $\mathbf{g}$ are not constructed one row at a time as they were in finite differences. Several different elements can contribute to the same row. All elements connected to node i contribute to row i.

9•2•2 Boundary segments

In the previous section we learned how to evaluate $\mathbf{C}$, $\mathbf{K}$ and $\mathbf{g}$ in (9·1·13). We must now find $\mathbf{q}_o$ in (9·1·13) which is given by (9·1·17) as an integral around the boundary. The discretization has replaced the actual boundary by a set of line segments. We will evaluate the boundary integral by writing the total integral as the sum of integrals over each of the boundary segments. If b is the boundary-segment number and there are a total of NB boundary segments, we will write (9·1·17) as

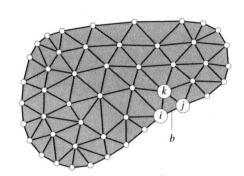

Figure 9•12 *Typical boundary segment b.*

$$\mathbf{q}_o = \int_B \mathbf{w} q_o'' \, ds = \sum_{b=1}^{NB} \int_{B^{(b)}} \mathbf{w} q_o'' \, ds$$

There are two types of boundary segments we must consider–specified heat flux (zero or nonzero) and convection. At this stage we will consider boundary segments having a specified temperature as being adiabatic. We will write the summation over the NB boundary segments in two parts, NB_q and NB_h, which correspond to the two boundary conditions. If we let b_q be the boundary-segment number for boundary segments having a specified heat flux and b_h be the boundary-segment number for boundary segments having a convection boundary we may then write

$$\mathbf{q}_o = \sum_{b_q=1}^{NB_q} \int_{B^{(b_q)}} \mathbf{w} q_o'' \, ds + \sum_{b_h=1}^{NB_h} \int_{B^{(b_h)}} \mathbf{w} q_o'' \, ds$$

where NB_q is the total number of boundary segments with a specified heat flux and NB_h is the total number of boundary segments with a convection boundary.

For specified-heat-flux boundaries we will let $q_o'' = q_s''$. For convection boundaries the heat flux into the boundary segment from outside the region is given by $q_o'' = h(t_\infty - t)$. Upon making these substitutions and writing the convection integral in two parts,

$$\mathbf{q}_o = \sum_{b_q=1}^{NB_q} \int_{B^{(b_q)}} \mathbf{w} q_s'' \, ds + \sum_{b_h=1}^{NB_h} \int_{B^{(b_h)}} \mathbf{w} h t_\infty \, ds - \sum_{b_h=1}^{NB_h} \int_{B^{(b_h)}} \mathbf{w} h t \, ds$$

From (9·1·9), the temperature in the third integral may be written as $t = \mathbf{w}^T \mathbf{t}$. Since $\mathbf{t}$ is only a function of θ it may be factored out of the integral and we may write

$$\mathbf{q}_o = \sum_{b_q=1}^{NB_q} \int_{B^{(b_q)}} \mathbf{w} q_s'' \, ds + \sum_{b_h=1}^{NB_h} \int_{B^{(b_h)}} \mathbf{w} h t_\infty \, ds - \left[\sum_{b_h=1}^{NB_h} \int_{B^{(b_h)}} \mathbf{w} h \mathbf{w}^T \, ds \right] \mathbf{t}$$

We will now define a *boundary heat-flow vector* $\mathbf{q}^{(b_q)}$, a *boundary convection vector* $\mathbf{h}^{(b_h)}$ and a *boundary convection matrix* $\mathbf{H}^{(b_h)}$ as

$$\mathbf{q}^{(b_q)} = \int_{B^{(b_q)}} \mathbf{w} q_s'' \, ds \qquad \mathbf{h}^{(b_h)} = \int_{B^{(b_h)}} \mathbf{w} h t_\infty \, ds \qquad \mathbf{H}^{(b_h)} = \int_{B^{(b_h)}} \mathbf{w} h \mathbf{w}^T \, ds \qquad (9\cdot2\cdot14)$$

Thus, we may write

$$\mathbf{q}_o = \sum_{b_q=1}^{NB_q} \mathbf{q}^{(b_q)} + \sum_{b_h=1}^{NB_h} \mathbf{h}^{(b_h)} - \left[\sum_{b_h=1}^{NB_h} \mathbf{H}^{(b_h)} \right] \mathbf{t} = \mathbf{q} + \mathbf{h} - \mathbf{H} \mathbf{t} \qquad (9\cdot2\cdot15)$$

where the sums of $\mathbf{q}^{(b_q)}$, $\mathbf{h}^{(b_h)}$ and $\mathbf{H}^{(b_h)}$ have been defined as $\mathbf{q}$, $\mathbf{h}$ and $\mathbf{H}$. Now we must learn how to evaluate $\mathbf{q}^{(b_q)}$, $\mathbf{h}^{(b_h)}$ and $\mathbf{H}^{(b_h)}$ so that we may construct the global quantities $\mathbf{q}$, $\mathbf{h}$ and $\mathbf{H}$.

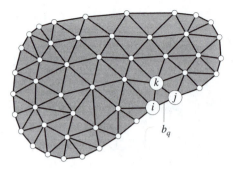

Figure 9•13 *Typical specified-heat-flux boundary segment b_q.*

Specified heat flux For each specified-heat-flux boundary segment we will assume that q_s'' is uniform. Upon factoring q_s'' out of the integral for $\mathbf{q}^{(b_q)}$ given in (9·2·14) we may write

$$\mathbf{q}^{(b_q)} = q_s''^{(b_q)} \int_{B^{(b_q)}} \mathbf{w}\, ds$$

We must now integrate $\mathbf{w}$ along boundary segment b_q. Within an element $\mathbf{w}^T$ is given by (9·2·1). If we take boundary segment b_q to run from node i to node j (*i.e.*, opposite node k), ξ_k will be 0 along the boundary segment and from (9·2·1) we may write

$$\mathbf{w}^T = \begin{bmatrix} \xi_i & \xi_j & \end{bmatrix}$$

Furthermore, as can be seen from Figure 9•6, ξ_j runs from 0 at node i to 1 at node j. We will also note that $ds = s_{ij}^{(b_q)}\, d\xi_j$ along the boundary from node i to node j. Thus, we may write

$$\mathbf{q}^{(b_q)} = q_s''^{(b_q)} s_{ij}^{(b_q)} \int_{\xi_j=0}^{1} \begin{bmatrix} \xi_i \\ \\ \xi_j \\ \\ \end{bmatrix} d\xi_j$$

Noting that $\xi_i = 1 - \xi_j$, we can carry out the integrations to obtain

$$\int_{\xi_j=0}^{1} \xi_i\, d\xi_j = \int_{\xi_j=0}^{1} (1-\xi_j)\, d\xi_j = \frac{1}{2} \qquad \text{and} \qquad \int_{\xi_j=0}^{1} \xi_j\, d\xi_j = \frac{1}{2}$$

Thus, upon factoring out the 2 we obtain

$$\mathbf{q}^{(b_q)} = \frac{q_s''^{(b_q)} s_{ij}^{(b_q)}}{2} \begin{bmatrix} 1 & \text{Row } i \\ \\ 1 & \text{Row } j \\ \\ \\ \\ \end{bmatrix}$$

$$(9\cdot2\cdot16)$$

The sum of the two entries in $\mathbf{q}^{(b_q)}$ is equal to $q_s''^{(b_q)} s_{ij}^{(b_q)}$. This is the total heat flow into boundary segment b_q. Half of this heat flow is assigned to row i and the other half is assigned to row j.

Convection For each convection boundary segment we will take h and t_∞ to be constant. We may then factor h and t_∞ from the integral for $\mathbf{h}^{(b_h)}$ given in (9·2·14) and write

$$\mathbf{h}^{(b_h)} = h^{(b_h)} t_\infty^{(b_h)} \int_{B^{(b_h)}} \mathbf{w}\, ds$$

where $h^{(b_h)}$ and $t_\infty^{(b_h)}$ are the values of h and t_∞ for boundary segment b_h. In general, $h^{(b_h)}$ and $t_\infty^{(b_h)}$ can be different for each boundary segment.

To evaluate $\mathbf{h}^{(b_h)}$ we will use (9·2·1) to give $\mathbf{w}$ (recalling that $\xi_k = 0$ along the segment from node i to node j) and let $ds = s_{ij}^{(b_h)} d\xi_j$. Thus,

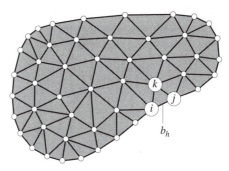

Figure 9•14 *Typical convection boundary segment b_h.*

$$\mathbf{h}^{(b_h)} = h^{(b_h)} t_\infty^{(b_h)} s_{ij}^{(b_h)} \int_{\xi_j=0}^{1} \begin{bmatrix} \xi_i \\ \\ \xi_j \\ \\ \\ \\ \end{bmatrix} d\xi_j$$

Noting that $\xi_i = 1 - \xi_j$, we can carry out the integrations to obtain

$$\int_{\xi_j=0}^{1} \xi_i\, d\xi_j = \int_{\xi_j=0}^{1} (1 - \xi_j)\, d\xi_j = \frac{1}{2} \qquad \text{and} \qquad \int_{\xi_j=0}^{1} \xi_j\, d\xi_j = \frac{1}{2}$$

Therefore,

$$\mathbf{h}^{(b_h)} = \frac{h^{(b_h)} t_\infty^{(b_h)} s_{ij}^{(b_h)}}{2} \begin{bmatrix} 1 & \text{Row } i \\ \\ 1 & \text{Row } j \\ \\ \\ \\ \\ \\ \end{bmatrix}$$

$$(9\cdot2\cdot17)$$

To find $\mathbf{H}^{(b_h)}$ we will again take h to be constant along boundary segment b_h and then factor h from the integral for $\mathbf{H}^{(b_h)}$ given in (9·2·14) to obtain

$$\mathbf{H}^{(b_h)} = h^{(b_h)} \int_{B^{(b_h)}} \mathbf{w}\mathbf{w}^T\, ds$$

We can obtain $\mathbf{w}\mathbf{w}^T$ from (9·2·1). Since $\xi_k = 0$ along the boundary segment between nodes i and j, we may write

$$\mathbf{w}\mathbf{w}^T = \begin{bmatrix} \xi_i \\ \xi_j \end{bmatrix} \begin{bmatrix} \xi_i & \xi_j \end{bmatrix} = \begin{bmatrix} \xi_i\xi_i & \xi_i\xi_j & & \text{Row } i \\ \xi_j\xi_i & \xi_j\xi_j & & \text{Row } j \\ & & & \\ \text{Column } i & \text{Column } j & & \end{bmatrix}$$

Again noting that $\xi_i = 1 - \xi_j$ and $ds = s_{ij}^{(b_h)}d\xi_j$, we can carry out the integrations for ξ_j running from 0 to 1 to obtain

$$\int_{B^{(b_h)}} \xi_i\xi_i\, ds = s_{ij}^{(b_h)} \int_{\xi_j=0}^{1} \xi_i\xi_i\, d\xi_j = s_{ij}^{(b_h)} \int_{\xi_j=0}^{1} (1-\xi_j)^2\, d\xi_j = \frac{1}{3}s_{ij}^{(b_h)}$$

$$\int_{B^{(b_h)}} \xi_i\xi_j\, ds = s_{ij}^{(b_h)} \int_{\xi_j=0}^{1} \xi_i\xi_j\, d\xi_j = s_{ij}^{(b_h)} \int_{\xi_j=0}^{1} (1-\xi_j)\xi_j\, d\xi_j = \frac{1}{6}s_{ij}^{(b_h)}$$

$$\int_{B^{(b_h)}} \xi_j\xi_j\, ds = s_{ij}^{(b_h)} \int_{\xi_j=0}^{1} \xi_j^2\, d\xi_j = \frac{1}{3}s_{ij}^{(b_h)}$$

We may then write

$$\mathbf{H}^{(b_h)} = \frac{h^{(b_h)} s_{ij}^{(b_h)}}{6} \begin{bmatrix} 2 & 1 & & \text{Row } i \\ 1 & 2 & & \text{Row } j \\ & & & \\ \text{Column } i & \text{Column } j & & \end{bmatrix} \qquad (9\cdot2\cdot18)$$

The global $\mathbf{H}$ may now be constructed by adding each of the $\mathbf{H}^{(b_h)}$. Note that $\mathbf{H}$ will be symmetric since each $\mathbf{H}^{(b_h)}$ is symmetric. Also, since $\mathbf{H}^{(b_h)}$ is not diagonal, $\mathbf{H}$ will not be diagonal as it was for finite differences.

Upon carrying out the matrix operations, one can show that the entry in row i of $\mathbf{h}^{(b_h)} - \mathbf{H}^{(b_h)}\mathbf{t}$ is given by

$$\frac{h^{(b_h)} s_{ij}^{(b_h)}}{2}\left[t_\infty - \frac{2t_i + t_j}{3} \right]$$

This is the convective heat-transfer rate into one-half of the boundary segment. The surface temperature is a weighted average of the nodal temperatures t_i and t_j at the ends of the segment.

Specified temperature We are going to handle specified-temperature nodes in finite elements just the same way as we did for finite differences. After adding the contributions from each element and each boundary segment, assuming all boundary segments are either specified heat flux (+, 0 or −) or convective, we have a system of ordinary differential equations that may be written as

$$\mathbf{C}\dot{\mathbf{t}} + (\mathbf{K} + \mathbf{H})\mathbf{t} = \mathbf{g} + \mathbf{q} + \mathbf{h} \qquad\qquad (9 \cdot 2 \cdot 19)$$

An energy interpretation of this equation will be given in the next section. In expanded form (9·2·19) may be written as

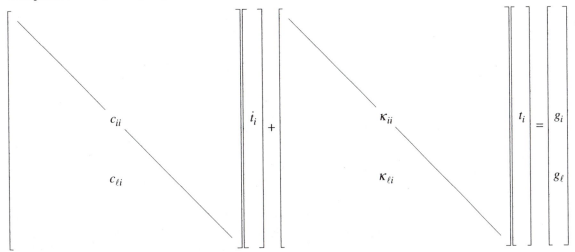

where now κ denotes an entry in $\mathbf{K} + \mathbf{H}$ rather than $\mathbf{K}$ and g denotes an entry in $\mathbf{g} + \mathbf{q} + \mathbf{h}$ rather than just $\mathbf{g}$. Although $\mathbf{C}$ and $\mathbf{K}$ are both banded, only the diagonal terms c_{ii} and κ_{ii} are shown in row i. In row ℓ only the entries $c_{\ell i}$ and $\kappa_{\ell i}$ which multiply $\dot{t}_i$ and t_i are shown.

When the temperature at node i is specified to be T_i we will replace the differential equation in row i by

$$c_{ii}\frac{dt_i}{d\theta} + \kappa_{ii}t_i = \kappa_{ii}T_i$$

When the initial value of t_i is T_i the solution to this equation is $t_i = T_i$. Substituting this differential equation into row i of the global system of differential equations we obtain

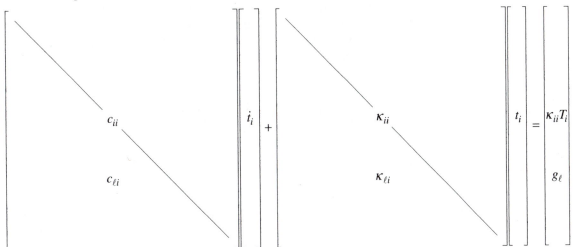

where now all of the off-diagonal terms in row i in $\mathbf{C}$ and $\mathbf{K}$ are 0. To regain the symmetry of $\mathbf{C}$ and $\mathbf{K}$ we will have to set all of the off-diagonal entries in column i in **C** and **K** equal to 0. We will do this by transferring these terms to the right-hand side of the system of equations. For row ℓ, the term $c_{\ell i}\dot{t}_i$ will be 0 since we are only considering problems where $t_i = T_i$ is a constant and therefore $\dot{t}_i = 0$. Thus, we can set $c_{\ell i} = 0$ without modifying the right-hand side. The term $\kappa_{\ell i}t_i = \kappa_{\ell i}T_i$ must be transferred to the right-hand side before setting $\kappa_{\ell i} = 0$ on the left-hand side. After making these modifications, the system of differential equations becomes

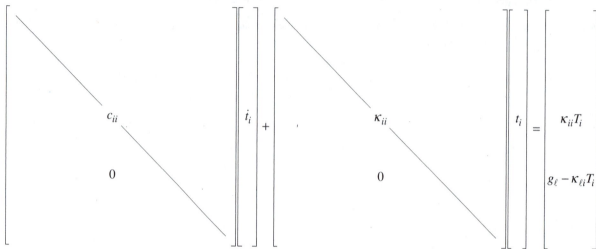

The nonzero terms in row ℓ of **C** and **K** (remember **K** is really **K + H**) are not shown. Similar transfers must be made until each off-diagonal term in column i on the left-hand side is 0. These transfers must be made for each column that multiplies a specified temperature.

Alternatively, we could have handled a specified temperature for row i by multiplying c_{ii} and κ_{ii} by a large constant (*e.g.*, 10^8) and replaced the right-hand side of row i by $10^8\kappa_{ii}T_i$. The off-diagonal entries in row i and column i would remain unchanged.

In addition to the modifications of **C**, **K + H** and **g + q + h**, the initial condition $\mathbf{t}^{(0)}$ would have to be modified to replace the initial value in the specified-temperature row by the value of the specified temperature. If this change in $\mathbf{t}^{(0)}$ is not made the solution will not give $t_i = T_i$.

A major difference between finite elements and finite differences is that **C** is banded in finite elements, not diagonal as it was in finite differences. This means that specified-temperature nodes require modification of the off-diagonal terms in the finite-element **C** whereas in finite differences no changes in **C** were required to accommodate specified temperatures. We will let **Ç** denote the modified **C**. The modified system of equations may be written as

$$\mathbf{Ç}\dot{\mathbf{t}} + \mathbf{S}\mathbf{t} = \mathbf{r} \qquad\qquad (9\cdot2\cdot20)$$

where **Ç**, **S** and **r** are **C**, **K + H** and **g + q + h** as modified for specified temperatures.

9•2•3 Heat-flow computation

Once (9·2·20) has been solved for the nodal temperatures it is often of interest to compute the heat fluxes within the solid or the heat flows at the boundary of the solid. A typical element is shown in Figure 9•15. The element heat flux components q_x'' and q_y'' are shown. The origin of these heat-flux arrows is shown at the centroid of the triangular element. The vector sum of q_x'' and q_y'' gives the direction and magnitude of the element heat flux. The isotherms in the triangular element are perpendicular to the element heat flux arrow.

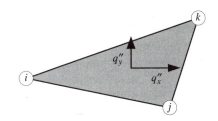

Figure 9•15 *Element heat-flux vectors.*

We will write the element heat flux as a 2×1 vector containing the x- and y-components of the heat flux. That is, we will write

$$\begin{bmatrix} q_x'' \\ q_y'' \end{bmatrix}^{(e)} = \begin{bmatrix} -kt_x \\ -kt_y \end{bmatrix}^{(e)}$$

Upon factoring out the conductivity,

$$\begin{bmatrix} q_x'' \\ q_y'' \end{bmatrix}^{(e)} = -k^{(e)} \begin{bmatrix} t_x \\ t_y \end{bmatrix}^{(e)}$$

The element temperature gradient vector may be obtained from (9·2·3). Thus,

$$\begin{bmatrix} q_x'' \\ q_y'' \end{bmatrix}^{(e)} = -\frac{k^{(e)}}{b_{ijk}^{(e)}} \begin{bmatrix} -y_{jk} & y_{ik} & -y_{ij} \\ x_{jk} & -x_{ik} & x_{ij} \end{bmatrix}^{(e)} \begin{bmatrix} t_i \\ t_j \\ t_k \end{bmatrix}^{(e)} \qquad (9\cdot2\cdot21)$$

This equation allows us to compute the x- and y-components of the heat flux within an element.

Another helpful quantity is the heat flux in the β direction where the β-axis makes an angle β with the x-axis as shown in Figure 9•16. As seen from the figure,

$$q_\beta''^{(e)} = q_x''^{(e)} \cos(\beta) + q_y''^{(e)} \sin(\beta)$$

or, using matrix notation,

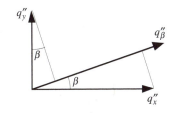

Figure 9•16 *Element heat-flux in the b direction.*

$$q_\beta''^{(e)} = \begin{bmatrix} \cos(\beta) & \sin(\beta) \end{bmatrix} \begin{bmatrix} q_x'' \\ q_y'' \end{bmatrix}^{(e)} \qquad (9\cdot2\cdot22)$$

The heat flux into an element normal to the side of the element opposite to node i will be called $q_{ni}''^{(e)}$. Figure 9•17 shows this heat flux vector for a typical element. The origin of the normal heat-flux arrow is shown at the midpoint of the side opposite node i. From the figure we see that $\beta = \pi - \phi$. Then, using some trigonometric identities, we see that

$$\cos(\beta) = \cos(\pi - \phi) = \cos(\pi)\cos(\phi) + \sin(\pi)\sin(\phi) = -\cos(\phi)$$

and

$$\sin(\beta) = \sin(\pi - \phi) = \sin(\pi)\cos(\phi) - \cos(\pi)\sin(\phi) = \sin(\phi)$$

Furthermore, as also seen from Figure 9•17,

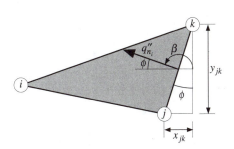

Figure 9•17 *Element heat flux normal to side opposite node i.*

$$\cos(\phi) = \frac{y_{jk}}{s_{jk}} \qquad \text{and} \qquad \sin(\phi) = \frac{x_{jk}}{s_{jk}}$$

where s_{jk} is the length of the side opposite node i. We may then write

$$\cos(\beta) = -\frac{y_{jk}}{s_{jk}} \qquad \text{and} \qquad \sin(\beta) = \frac{x_{jk}}{s_{jk}}$$

Substituting these expressions into (9·2·22) gives

$$q_{ni}''^{(e)} = \left[-\frac{y_{jk}}{s_{jk}} \quad \frac{x_{jk}}{s_{jk}} \right]^{(e)} \left[\begin{matrix} q_x'' \\ q_y'' \end{matrix} \right]^{(e)}$$

The heat flow per unit depth (into the page) across the side opposite node i is then given by

$$q_{ni}'^{(e)} = s_{jk} q_{ni}''^{(e)} = \left[-y_{jk} \quad x_{jk} \right]^{(e)} \left[\begin{matrix} q_x'' \\ q_y'' \end{matrix} \right]^{(e)} \tag{9·2·23}$$

Similar analyses will show that the heat flows per unit depth normal to the other two sides are given by

$$q_{nj}'^{(e)} = \left[y_{ik} \quad -x_{ik} \right]^{(e)} \left[\begin{matrix} q_x'' \\ q_y'' \end{matrix} \right]^{(e)} \tag{9·2·24}$$

and

$$q_{nk}'^{(e)} = \left[-y_{ij} \quad x_{ij} \right]^{(e)} \left[\begin{matrix} q_x'' \\ q_y'' \end{matrix} \right]^{(e)} \tag{9·2·25}$$

Equations (9·2·23), (9·2·24) and (9·2·25) may be combined by writing

$$\left[\begin{matrix} q_{ni}' \\ q_{nj}' \\ q_{nk}' \end{matrix} \right]^{(e)} = \left[\begin{matrix} -y_{jk} & x_{jk} \\ y_{ik} & -x_{ik} \\ -y_{ij} & x_{ij} \end{matrix} \right]^{(e)} \left[\begin{matrix} q_x'' \\ q_y'' \end{matrix} \right]^{(e)} \tag{9·2·26}$$

This result has been obtained for Figure 9•17 in which nodes i, j and k progressed in a counter-clockwise direction around the element. A similar analysis for clockwise nodal-point numbering will show that

$$\left[\begin{matrix} q_{ni}' \\ q_{nj}' \\ q_{nk}' \end{matrix} \right]^{(e)} = - \left[\begin{matrix} -y_{jk} & x_{jk} \\ y_{ik} & -x_{ik} \\ -y_{ij} & x_{ij} \end{matrix} \right]^{(e)} \left[\begin{matrix} q_x'' \\ q_y'' \end{matrix} \right]^{(e)} \tag{9·2·27}$$

The only difference between (9·2·26) and (9·2·27) is the minus sign that appears on the right-hand side of (9·2·27). A similar situation occurs in computing the area of the element as discussed regarding (9·2·11). Since $b_{ijk}^{(e)}$ is positive for counter-clockwise numbering and negative for clockwise numbering we can write

$$\frac{b_{ijk}^{(e)}}{\left| b_{ijk}^{(e)} \right|} = \pm 1$$

where +1 is for counter-clockwise and −1 is for clockwise numbering. Therefore, (9·2·26) and (9·2·27) can be combined by writing

$$\left[\begin{matrix} q_{ni}' \\ q_{nj}' \\ q_{nk}' \end{matrix} \right]^{(e)} = \frac{b_{ijk}^{(e)}}{\left| b_{ijk}^{(e)} \right|} \left[\begin{matrix} -y_{jk} & x_{jk} \\ y_{ik} & -x_{ik} \\ -y_{ij} & x_{ij} \end{matrix} \right]^{(e)} \left[\begin{matrix} q_x'' \\ q_y'' \end{matrix} \right]^{(e)} \tag{9·2·28}$$

This equation holds for either counter-clockwise or clockwise numbering of the nodal points.

Substituting (9·2·21) into (9·2·28) gives

$$
\begin{bmatrix} q'_{ni} \\ q'_{nj} \\ q'_{nk} \end{bmatrix}^{(e)} = \frac{b^{(e)}_{ijk}}{\left| b^{(e)}_{ijk} \right|} \begin{bmatrix} -y_{jk} & x_{jk} \\ y_{ik} & -x_{ik} \\ -y_{ij} & x_{ij} \end{bmatrix}^{(e)} \frac{-k^{(e)}}{b^{(e)}_{ijk}} \begin{bmatrix} -y_{jk} & y_{ik} & -y_{ij} \\ x_{jk} & -x_{ik} & x_{ij} \end{bmatrix}^{(e)} \begin{bmatrix} t_i \\ t_j \\ t_k \end{bmatrix}^{(e)}
$$

Upon canceling $b^{(e)}_{ijk}$, replacing $\left| b^{(e)}_{ijk} \right|$ by $2A^{(e)}$ following (9·2·11), and multiplying the first two matrices on the right-hand side will give

$$
\begin{bmatrix} q'_{ni} \\ q'_{nj} \\ q'_{nk} \end{bmatrix}^{(e)} = \frac{-k^{(e)}}{2A^{(e)}} \begin{bmatrix} \kappa_{ii} & \kappa_{ij} & \kappa_{ik} \\ \kappa_{ji} & \kappa_{jj} & \kappa_{jk} \\ \kappa_{ki} & \kappa_{kj} & \kappa_{kk} \end{bmatrix}^{(e)} \begin{bmatrix} t_i \\ t_j \\ t_k \end{bmatrix}^{(e)}
$$

where the 3×3 matrix is symmetric and

$$\kappa_{ii} = x_{jk}x_{jk} + y_{jk}y_{jk} \qquad \kappa_{ij} = -(x_{jk}x_{ik} + y_{jk}y_{ik}) \qquad \kappa_{ik} = x_{jk}x_{ij} + y_{jk}y_{ij}$$

$$\kappa_{jj} = x_{ik}x_{ik} + y_{ik}y_{ik} \qquad \kappa_{jk} = -(x_{ik}x_{ij} + y_{ik}y_{ij})$$

$$\kappa_{kk} = x_{ij}x_{ij} + y_{ij}y_{ij}$$

The entries in the 3×3 matrix are seen to be –2 times the entries in $\mathbf{K}^{(e)}$ as given by (9·2·12).

It should be pointed out that the heat flow across segment jk in Figure 9•18 as computed from the temperature distribution in element e, $q'^{(e)}_{ni}$, will not (in general) be the same as the heat flow computed using the temperature distribution in element f, $q'^{(f)}_{n\ell}$. There will be a discontinuity in heat flow at the element boundaries. This discontinuity will become smaller as element size is reduced and the linear temperature approximation within the elements becomes more adequate.

It is also of interest to subdivide element e into three quadrilaterals as shown in Figure 9•19. These quadrilaterals are constructed by drawing lines from each node [i, j and k] to the midpoints [i', j' and k'] of the opposite sides. These three lines will intersect at the centroid [c] of the element and divide the triangle into three quadrilaterals [i, j and k]. The area of each of these quadrilaterals is $\frac{1}{3} A^{(e)}$. Furthermore, the distance between c and i' will be $\frac{1}{3}$ of the distance between i and i'. That is, $s_{ci'} = \frac{1}{3} s_{ii'}$. Similarly, $s_{cj'} = \frac{1}{3} s_{jj'}$ and $s_{ck'} = \frac{1}{3} s_{kk'}$. Next we will determine the heat flow $q'^{(e)}_{i}$ leaving quadrilateral i and entering quadrilaterals j and k.

Consider the subelement triangle $j'jk$ shown in Figure 9•20 along with quadrilateral k. Equation (9·2·28) can be applied to this triangle by simply replacing i by j' to obtain the heat flow leaving quadrilateral i and entering quadrilateral k in Figure 9•19 as

$$
q'^{(e)}_{i \to k} = \frac{1}{3} \frac{b^{(e)}_{j'jk}}{\left| b^{(e)}_{j'jk} \right|} \begin{bmatrix} -y_{j'j} & x_{j'j} \end{bmatrix}^{(e)} \begin{bmatrix} q''_x \\ q''_y \end{bmatrix}^{(e)}
$$

where the $\frac{1}{3}$ accounts for the fact that $s_{cj'} = \frac{1}{3} s_{jj'}$. It should be observed that $b^{(e)}_{j'jk}$ will have the same sign as $b^{(e)}_{ijk}$ and hence we may write

$$
q'^{(e)}_{i \to k} = \frac{1}{3} \frac{b^{(e)}_{ijk}}{\left| b^{(e)}_{ijk} \right|} \begin{bmatrix} -y_{j'j} & x_{j'j} \end{bmatrix}^{(e)} \begin{bmatrix} q''_x \\ q''_y \end{bmatrix}^{(e)} \tag{9.2.29}
$$

A similar derivation will show that the heat flow leaving quadrilateral i and entering quadrilateral j is given by

$$
q'^{(e)}_{i \to j} = \frac{1}{3} \frac{b^{(e)}_{ijk}}{\left| b^{(e)}_{ijk} \right|} \begin{bmatrix} y_{k'k} & -x_{k'k} \end{bmatrix}^{(e)} \begin{bmatrix} q''_x \\ q''_y \end{bmatrix}^{(e)} \tag{9.2.30}
$$

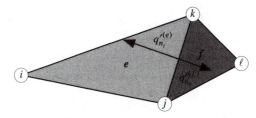

Figure 9•18 *Heat flow across boundary between two adjacent elements.*

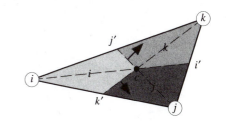

Figure 9•19 *Subdivision of a triangular element into three quadrilaterals.*

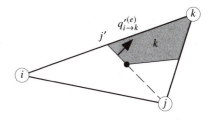

Figure 9•20 *Subelement triangle $j'jk$.*

Addition of (9·2·29) and (9·2·30) gives the heat flow leaving quadrilateral i and entering the remainder of the triangular element. That is,

$$q_i'^{(e)} = \frac{1}{3}\frac{b_{ijk}^{(e)}}{\left|b_{ijk}^{(e)}\right|}\left[(-y_{j'j}+y_{k'k}) \quad (x_{j'j}-x_{k'k})\right]^{(e)}\begin{bmatrix}q_x''\\q_y''\end{bmatrix}^{(e)}$$

This result may be simplified by observing that

$$x_{j'j} - x_{k'k} = x_j - x_{j'} - (x_k - x_{k'})$$

$$= x_j - \frac{x_i+x_k}{2} - x_k + \frac{x_i+x_j}{2} = \frac{3}{2}(x_j - x_k) = -\frac{3}{2}x_{jk}$$

and similarly,

$$-y_{j'j} + y_{k'k} = \frac{3}{2}y_{jk}$$

Thus, the heat flow leaving quadrilateral i and entering the remainder of the element, Figure 9·21(a), is given by

$$q_i'^{(e)} = \frac{1}{2}\frac{b_{ijk}^{(e)}}{\left|b_{ijk}^{(e)}\right|}\left[y_{jk} \quad -x_{jk}\right]^{(e)}\begin{bmatrix}q_x''\\q_y''\end{bmatrix}^{(e)} \tag{9·2·31}$$

Similarly, the heat flow leaving quadrilateral j, Figure 9·21(b), is given by

$$q_j'^{(e)} = \frac{1}{2}\frac{b_{ijk}^{(e)}}{\left|b_{ijk}^{(e)}\right|}\left[-y_{ik} \quad x_{ik}\right]^{(e)}\begin{bmatrix}q_x''\\q_y''\end{bmatrix}^{(e)} \tag{9·2·32}$$

and the heat flow leaving quadrilateral k, Figure 9·21(c), is given by

$$q_k'^{(e)} = \frac{1}{2}\frac{b_{ijk}^{(e)}}{\left|b_{ijk}^{(e)}\right|}\left[y_{ij} \quad -x_{ij}\right]^{(e)}\begin{bmatrix}q_x''\\q_y''\end{bmatrix}^{(e)} \tag{9·2·33}$$

We may combine (9·2·31), (9·2·32), and (9·2·33) into one equation as

$$\begin{bmatrix}q_i'\\q_j'\\q_k'\end{bmatrix}^{(e)} = \frac{1}{2}\frac{b_{ijk}^{(e)}}{\left|b_{ijk}^{(e)}\right|}\begin{bmatrix}y_{jk} & -x_{jk}\\-y_{ik} & x_{ik}\\y_{ij} & -x_{ij}\end{bmatrix}^{(e)}\begin{bmatrix}q_x''\\q_y''\end{bmatrix}^{(e)} \tag{9·2·34}$$

Upon substituting (9·2·21) for the heat fluxes we obtain

$$\begin{bmatrix}q_i'\\q_j'\\q_k'\end{bmatrix}^{(e)} = \frac{1}{2}\frac{b_{ijk}^{(e)}}{\left|b_{ijk}^{(e)}\right|}\begin{bmatrix}y_{jk} & -x_{jk}\\-y_{ik} & x_{ik}\\y_{ij} & -x_{ij}\end{bmatrix}^{(e)}\frac{-k^{(e)}}{b_{ijk}^{(e)}}\begin{bmatrix}-y_{jk} & y_{ik} & -y_{ij}\\x_{jk} & -x_{ik} & x_{ij}\end{bmatrix}^{(e)}\begin{bmatrix}t_i\\t_j\\t_k\end{bmatrix}^{(e)}$$

Upon canceling $b_{ijk}^{(e)}$, replacing $\left|b_{ijk}^{(e)}\right|$ by $2A^{(e)}$ following (9·2·11), and multiplying the first two matrices on the right-hand side, we obtain

$$\begin{bmatrix}q_i'\\q_j'\\q_k'\end{bmatrix}^{(e)} = \frac{k^{(e)}}{4A^{(e)}}\begin{bmatrix}\kappa_{ii} & \kappa_{ij} & \kappa_{ik}\\\kappa_{ji} & \kappa_{jj} & \kappa_{jk}\\\kappa_{ki} & \kappa_{kj} & \kappa_{kk}\end{bmatrix}^{(e)}\begin{bmatrix}t_i\\t_j\\t_k\end{bmatrix}^{(e)}$$

where the 3×3 matrix is symmetric and

$$\kappa_{ii} = x_{jk}x_{jk}+y_{jk}y_{jk} \quad \kappa_{ij} = -(x_{jk}x_{ik}+y_{jk}y_{ik}) \quad \kappa_{ik} = x_{jk}x_{ij}+y_{jk}y_{ij}$$

$$\kappa_{jj} = x_{ik}x_{ik}+y_{ik}y_{ik} \quad \kappa_{jk} = -(x_{ik}x_{ij}+y_{ik}y_{ij})$$

$$\kappa_{kk} = x_{ij}x_{ij}+y_{ij}y_{ij}$$

(a)
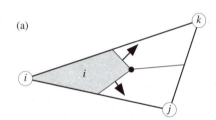

(b)
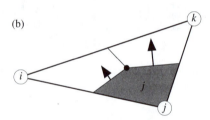

(c)
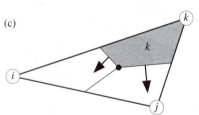

Figure 9·21 *Conduction out of quadrilaterals i, j and k in a triangular element.*

The entries in the 3×3 matrix are identical to the nonzero entries in $\mathbf{K}^{(e)}$ as given by (9·2·12). Upon expanding to global matrix size, we may write

$$
\begin{bmatrix} q_i' \\ q_j' \\ q_k' \end{bmatrix}^{(e)} = \frac{k^{(e)}}{4A^{(e)}}
\begin{bmatrix}
\kappa_{ii} & \kappa_{ij} & \kappa_{ik} & & \text{Row } i \\
\kappa_{ji} & \kappa_{jj} & \kappa_{jk} & & \text{Row } j \\
\kappa_{ki} & \kappa_{kj} & \kappa_{kk} & & \text{Row } k \\
& & & & \\
\text{Column } i & \text{Column } j & \text{Column } k & &
\end{bmatrix}^{(e)}
\begin{bmatrix} t_i \\ t_j \\ t_k \end{bmatrix}
$$

The $N \times N$ matrix on the right-hand side is the element conduction matrix as given by (9·2·12). Thus we may write

$$ \mathbf{q}'^{(e)} = \mathbf{K}^{(e)}\mathbf{t} $$

If we now sum over each of the elements we obtain

$$ \sum_{e=1}^{NE} \mathbf{q}'^{(e)} = \sum_{e=1}^{NE} \mathbf{K}^{(e)}\mathbf{t} $$

The sum of the $\mathbf{K}^{(e)}$ on the right-hand side is the global conduction matrix $\mathbf{K}$. The sum on the left-hand side will be defined to be a *global nodal heat-conduction vector*. Thus,

$$ \mathbf{q}' = \mathbf{K}\mathbf{t} $$

A physical interpretation of $\mathbf{K}\mathbf{t}$ may be seen by considering the construction of $\mathbf{q}'$. The contribution from each element is given by

$$
\mathbf{q}'^{(e)} =
\begin{bmatrix}
q_i' \\
q_j' \\
q_k'
\end{bmatrix}^{(e)}
$$

The contribution to row j in $\mathbf{K}\mathbf{t}$ from element e is $q_j'^{(e)}$ where $q_j'^{(e)}$ is the heat flow from quadrilateral j in Figure 9•21 into quadrilaterals i and k. The sum of all the entries in row j then represents the heat flow out of the set of quadrilaterals surrounding node j as shown in Figure 9•22. In other words, row j in $\mathbf{K}\mathbf{t}$ contains the total heat conducted out of *finite-element nodal system* j which consists of all the quadrilaterals surrounding node j

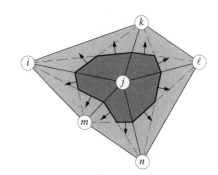

Figure 9•22 *Nodal system j surrounding node j.*

as shown in Figure 9•22. Thus, **Kt** is a vector containing the energy conducted out of each of the N nodal systems into surrounding nodal systems! You should recall that in the discussion of (8·2·33) for finite differences, **Kt** contained the conduction rates out of each of the N finite-difference nodal systems. We have now seen that a similar interpretation may be given to the finite-element **Kt**.

Examination of (9·2·9) and (9·2·13) shows that **g** is a vector containing the generation rates within each of the N finite-element nodal systems. This is the same interpretation for **g** as in finite differences.

A physical interpretation of **Cṫ** may also be found. The contribution to this vector from element e may be seen from (9·2·7) and (9·2·10) to be

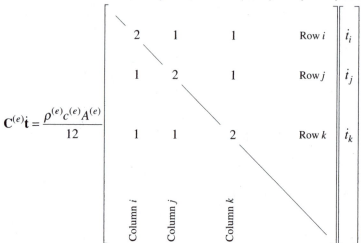

The total contribution of element e to row i is given by

$$\frac{\rho^{(e)}c^{(e)}A^{(e)}}{12}(2\dot{t}_i + \dot{t}_j + \dot{t}_k) \qquad (9·2·36)$$

Similarly the total contribution from element e to row j is given by

$$\frac{\rho^{(e)}c^{(e)}A^{(e)}}{12}(\dot{t}_i + 2\dot{t}_j + \dot{t}_k) \qquad (9·2·37)$$

And for row k,

$$\frac{\rho^{(e)}c^{(e)}A^{(e)}}{12}(\dot{t}_i + \dot{t}_j + 2\dot{t}_k) \qquad (9·2·38)$$

Upon adding these three contributions, we find that the total contribution of element e to **Cṫ** is given by

$$\frac{\rho^{(e)}c^{(e)}A^{(e)}}{12}(4\dot{t}_i + 4\dot{t}_j + 4\dot{t}_k) = \rho^{(e)}c^{(e)}A^{(e)}\frac{d}{d\theta}\left(\frac{t_i + t_j + t_k}{3}\right)$$

This is the energy storage rate in element e based on the time rate of change of the average temperature of the element. The contribution to row j given by (9·2·37) may be rewritten as

$$\frac{\rho^{(e)}c^{(e)}A^{(e)}}{3}\frac{d}{d\theta}\left(\frac{t_i + 2t_j + t_k}{4}\right)$$

This may be interpreted as the rate of energy storage in quadrilateral j in element e with the average temperature of the quadrilateral given by

$$\frac{t_i + 2t_j + t_k}{4}$$

If we now sum over every element, row j in $\mathbf{C}\dot{\mathbf{t}}$ will contain the energy storage in finite-element nodal system j shown in Figure 9·22. This is the same interpretation of $\mathbf{C}\dot{\mathbf{t}}$ as in finite differences.

In the discussion of the boundary-segment heat-flow vector (9·2·16) we saw that row i contained the specified heat flow into the half of the boundary segment next to node i. The entry in row i in the boundary-segment heat-flow vector for the boundary segment on the other side of node i would give the specified heat flow into the half of the boundary segment attached to row i. Together the sum of the two boundary-segment heat-flow vectors with entries in row i would give the total specified heat flow into finite-element nodal system for node i. The sum of the boundary-segment heat-flow vectors for all of the boundary segments will contain the specified heat flows into each of the N finite-element nodal systems. It will contain zeros for all of the interior nodes and for all of the boundary nodes along an adiabatic boundary.

Based on the discussion of (9·2·17) and (9·2·18), the physical interpretation of $\mathbf{h} - \mathbf{Ht}$ is that it is a vector containing the convective heat inflows to each of the N nodes. It will contain zeros for all of the interior nodes and for all of the boundary nodes not along a convective boundary.

We can now give an energy interpretation to (9·2·19) which was obtained by a Galerkin weighted-residual approximation of the governing partial differential equation and its boundary conditions. Upon rearranging (9·2·19) we can write

$$\mathbf{g} + \mathbf{q} + (\mathbf{h} - \mathbf{Ht}) = \mathbf{C}\dot{\mathbf{t}} + \mathbf{Kt} \tag{9·2·39}$$

Although the numerical values in $\mathbf{g}$, $\mathbf{q}$, $\mathbf{h}$, $\mathbf{H}$, $\mathbf{C}$ and $\mathbf{K}$ will not, in general, be the same in this equation as in (8·2·30) which was obtained by finite differences, it has the same physical interpretation. The left-hand side gives the sum of the energy generation plus the specified heat inflow plus the convection inflow to each of the N finite-element nodal systems. The right-hand side gives the energy-storage rate in each finite-element nodal system plus the conduction out of each nodal system into the surrounding nodal systems. Thus (9·2·39) represents an energy balance on each of the N finite-element nodal systems. As in finite differences, the boundaries of nodal systems with specified nodal temperatures have been assumed to be adiabatic. In finite differences the energy-balance interpretation for (8·2·30) was more obvious since the equation was derived directly from energy balances on the nodal systems. In finite elements, the derivation of (9·2·39) was different, but the same energy-balance interpretation can be given. The finite-element nodal systems are similar to the one shown in Figure 9·22.

As in finite differences (8·2·37), the boundary heat inflows may be computed from

$$\mathbf{q}_o = \mathbf{C}\dot{\mathbf{t}} + \mathbf{Kt} - \mathbf{g} \tag{9·2·40}$$

The temperature vector $\mathbf{t}$ comes from the solution of (9·2·20). Once $\mathbf{t}$ has been found, $\dot{\mathbf{t}}$ may be obtained by rearranging (9·2·20) as

$$\mathbf{Ç}\dot{\mathbf{t}} = \mathbf{r} - \mathbf{St} \tag{9·2·41}$$

and then solving for $\dot{\mathbf{t}}$. In finite elements the computations to find $\dot{\mathbf{t}}$ are more involved than for finite differences. Since $\mathbf{Ç}$ is not diagonal in finite elements, (9·2·41) is a system of algebraic equations that must be solved simultaneously for $\dot{\mathbf{t}}$. In finite differences $\mathbf{Ç}$ was diagonal and the solution for $\dot{\mathbf{t}}$ was explicit. For steady-state problems, $\dot{\mathbf{t}} = \mathbf{0}$.

9•2•4 Program *FEHT*

The graphics capability of the Macintosh and Windows operating systems has permitted the writing of program *FEHT*.[1] The program is based on the finite-element theory developed in this chapter for solving two-dimensional conduction problems. The program is capable of solving both steady-state and transient (Euler and Crank-Nicolson) problems. *FEHT* calls on the three general-purpose subroutines *DBAND*, *SBAND* and *YAXPB* that were also used in Chapter 8 and which are listed in Appendix I. The main objective of mentioning *FEHT* is that it is easy to use to illustrate the theory in this chapter.

In using *FEHT* the discretization and data input are done by drawing on the computer screen by pointing, clicking and dragging a mouse. A two-dimensional region may be drawn and discretized into triangular elements in minutes. As an example, the region shown in Figure 9•8 was copied to the clipboard and pasted into *FEHT* as a template and used as a pattern to copy and discretize into triangular elements. The result, as output from *FEHT*, is shown in Figure 9•23. Material properties and boundary conditions have also been specified. In this example, the eight darkened nodes along the southern and western boundary have been specified to have a temperature of 100. The four darkened nodes along the eastern boundary have been specified to have a temperature of 0. The two portions of the boundary that do not lie along the specified-temperature portions of the boundary have been specified to be adiabatic.

The method of discretizing the region may be shown by asking *FEHT* to display the nodal-point numbers as shown in Figure 9•24. The outline of the boundary was drawn by first clicking the mouse at the location of node 1, then clicking at nodes 2, 3, …, 26 and then node 1 again to complete the outline. Material properties for the region were then specified. The seven boundary segments between nodes 1 and 8 were then clicked upon and specified to have a temperature of 100. The three boundary segments between nodes 12 and 15 were specified to have a temperature of 0. The four boundary segments between nodes 8 and 12 and the 12 boundary segments between 15 and 1 were then specified to be adiabatic. The first interior element line was established by clicking on a boundary node (*e.g.*, node 3) and then clicking at the location of node 27. This gives the element line from node 3 to node 27. Clicking again on node 27 and then at the location of node 28 gives the element line between nodes 27 and 28 and locates node 28. Clicking on node 27 and then node 4 (both previously established) draws the element line between the two nodes but does not establish a new node number. It does, however, form a triangular element with corners at nodes 3, 4 and 27. By following the node numbers in Figure 9•24 you can see the order in which new nodes were established.

Once the region has been discretized and the properties and boundary conditions specified, the calculations for a steady-state problem can be run. Nodal temperatures can then be displayed at the node locations instead of the node numbers as shown in Figure 9•25. For a dense mesh of nodal points the display of temperatures as in Figure 9•25 becomes too cluttered to be of value. In this case, however, the temperature at any location in the solid may be read in the information strip by pointing to the location with the cursor and holding down the mouse button. The temperatures at the nodes may also be obtained from the report.

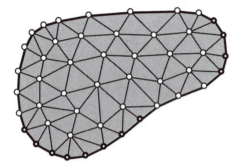

Figure 9•23 *FEHT discretization of Figure 9•9.*

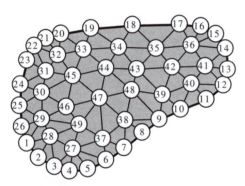

Figure 9•24 *FEHT node numbers for Figure 9•23.*

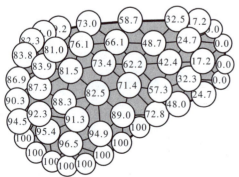

Figure 9•25 *Nodal temperatures as displayed by FEHT.*

---

[1] Klein, S. A., W. A. Beckman and G. E. Myers: *FEHT – Finite Element Analysis,* F-Chart Software <http://www.fchart.com/>

Temperature contours can be shown as either shaded bands or as isopotential lines. Figure 9•26 shows five shaded bands along with the element lines. The interface between two shaded regions is an isotherm. Isotherms for 20, 40, 60 and 80 are shown. Within a single element an isotherm is a straight line. This is a consequence of assuming a linear temperature variation between the three nodal temperatures at the corners of the element.

The heat-flow directions may be viewed by plotting temperature-gradient arrows as shown in Figure 9•27. Arrow length is proportional to the temperature gradient. The maximum length and the density of the arrows may also be adjusted.

Nodal energy balances can be made to calculate energy inflows to each nodal system. Figure 9•28 shows the nodes and the nodal systems for this example. Clicking on a node will display the energy inflow to that nodal system. Successive clicking on nodes will display the energy inflow to the last node clicked upon and also the sum of the energy inflows that have been clicked on to obtain total energy flows across boundaries. Clicking on an interior node or on a node along an adiabatic boundary will always give zero. Energy balance checks will always be obtained.

An important feature of *FEHT* is a subroutine that renumbers the nodal points to reduce the matrix bandwidth. Observe in Figure 9•24 that node 9 and node 48 are connected by an element line. This means that in economy-banded form, row 9 will have a bandwidth of $48 - 9 + 1 = 40$ and the computer storage requirement for **S** in this 49-node example will be $49 \times 40 = 1960$ storage locations. Before calculating temperatures, *FEHT* renumbers the nodes and reports that the equations in this example have a bandwidth of only 9. This reduces computer storage and makes the computations more manageable.

FEHT also has a "reduce mesh size" feature that subdivides each triangular element in a mesh into four triangular elements. This is done by adding a node at the midpoint of every element line in the original mesh and adding three element lines within each of the original elements to connect the new nodes at the midpoints of the sides of the original element. This gives a finer mesh with four times as many elements and a larger number of nodal points. Repeated application of this feature can very quickly produce a finite-element model with over 1000 nodes provided your computer has enough memory. One application of this reduce-mesh-size feature to Figure 9•23 is shown (without node symbols) in Figure 9•29. This would not be as feasible without node renumbering. Solution of the mesh in Figure 9•29 is carried out with a bandwidth of 15. Without bandwidth reduction this 167-node problem has a bandwidth of 140.

For transient problems, the user can specify the time step and the stopping time. Temperature-time curves can be obtained for as many as six nodes on the same plot. Plots of nodal temperatures, temperature contours (shaded or isopotential lines) and temperature gradients can be produced at each time step. The contour and gradient plots may be displayed in rapid succession to give a "movie" of the transient.

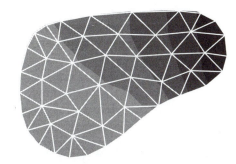

Figure 9•26 *Five shaded temperature intervals output from FEHT.*

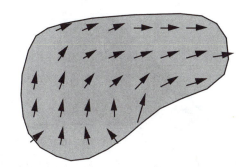

Figure 9•27 *Temperature-gradient arrows from FEHT showing heat-flow direction.*

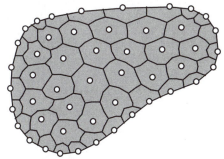

Figure 9•28 *Nodes and nodal systems as displayed by FEHT.*

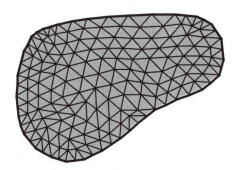

Figure 9•29 *One mesh reduction by FEHT of Figure 9•23 with node positions not shown.*

9•3 STEADY-STATE PROBLEMS

This section will parallel Section 8•3 so that you can make comparisons between finite elements and finite differences. The method of constructing the finite-element equations differs from finite differences, but the resulting algebraic equations will look remarkably like the finite-difference equations.

9•3•1 Discretization example

As an example of finite-element discretization, let us again consider the problem shown in Figure 8•17 with nodal-point locations as shown in Figure 8•18. These nodal points can be hooked together into 16 elements as shown in Figure 9•30. Each element is given an element number. This is a very regular example in which the area of each element is $\frac{1}{32}L^2$.

The largest difference between nodal-point numbers for any element is 5 which occurs in elements 4, 5, 6, 7, 8, 9 and 10. This means that the upper bandwidth will be 6. Thus we will initialize the entries within the band of **K** and in **g** to be zero and begin the discretization by writing

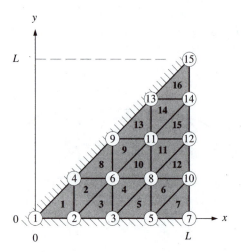

Figure 9•30 *Finite-element discretization of region shown in Figure 8•18.*

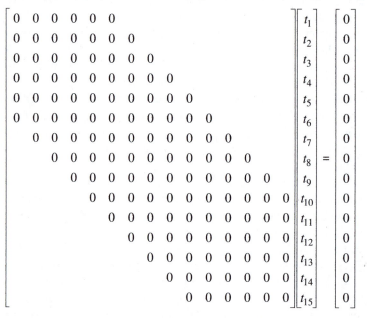

$$(9\cdot3\cdot1)$$

The corresponding finite-difference equation (8·3·2) had only three off-diagonals above and below the main diagonal whereas the finite-element equation (9·3·1) has five off-diagonals above and below the main diagonal. Thus, this example requires more computer storage for the finite-element formulation than for the finite-difference formulation.

The input data describing this example that we must give the computer is given in Table 9•1. It consists of nodal coordinates for each node and element information giving the three nodal-point numbers and the material-property number for each element. The thermal conductivity k and the generation rate g''' of material 1 must also be given to the computer in addition to the data in Table 9•1.

To begin filling in the entries in **K** and in **g** we will first go to element 1. The first step is to compute

$$x_{ij} = x_j - x_i = \frac{L}{4} - 0 = \frac{L}{4} \qquad\qquad y_{ij} = y_j - y_i = 0 - 0 = 0$$

$$x_{ik} = x_k - x_i = \frac{L}{4} - 0 = \frac{L}{4} \qquad\qquad y_{ik} = y_k - y_i = \frac{L}{4} - 0 = \frac{L}{4}$$

$$x_{jk} = x_k - x_j = \frac{L}{4} - \frac{L}{4} = 0 \qquad\qquad y_{jk} = y_k - y_j = \frac{L}{4} - 0 = \frac{L}{4}$$

The area of the element can now be computed following (9·2·11) as

$$A = \frac{1}{2}\left|x_{ij}y_{jk} - x_{jk}y_{ij}\right| = \frac{1}{2}\left|\frac{L}{4}\cdot\frac{L}{4} - 0\cdot 0\right| = \frac{L^2}{32}$$

Table 9•1 *Computer input for nodal coordinates and element information.*

Nodal coordinates			Element information				
i	x_i	y_i	e	i	j	k	m
1	0	0	1	1	2	4	1
2	$L/4$	0	2	2	4	6	1
3	$L/2$	0	3	2	3	6	1
4	$L/4$	$L/4$	4	3	6	8	1
5	$3L/4$	0	5	3	5	8	1
6	$L/2$	$L/4$	6	5	8	10	1
7	L	0	7	5	7	10	1
8	$3L/4$	$L/4$	8	4	6	9	1
9	$L/2$	$L/2$	9	6	9	11	1
10	L	$L/4$	10	6	8	11	1
11	$3L/4$	$L/2$	11	8	11	12	1
12	L	$L/2$	12	8	10	12	1
13	$3L/4$	$3L/4$	13	9	11	13	1
14	L	$3L/4$	14	11	13	14	1
15	L	L	15	11	12	14	1
			16	13	14	15	1

The potentially-nonzero entries in the element conduction matrix $\mathbf{K}^{(1)}$ may now be computed as shown in (9·2·12) as

$$\kappa_{ii} = \frac{k}{4A}\left[x_{jk}x_{jk} + y_{jk}y_{jk}\right] = \frac{k}{4(L^2/32)}\left[0\cdot 0 + \frac{L}{4}\cdot\frac{L}{4}\right] = \frac{k}{2}$$

$$\kappa_{ij} = \frac{k}{4A}\left[-(x_{jk}x_{ik} + y_{jk}y_{ik})\right] = \frac{-k}{4(L^2/32)}\left[0\cdot\frac{L}{4} + \frac{L}{4}\cdot\frac{L}{4}\right] = -\frac{k}{2}$$

$$\kappa_{ik} = \frac{k}{4A}\left[x_{jk}x_{ij} + y_{jk}y_{ij}\right] = \frac{k}{4(L^2/32)}\left[0\cdot\frac{L}{4} + 0\cdot\frac{L}{4}\right] = 0$$

$$\kappa_{jj} = \frac{k}{4A}\left[x_{ik}x_{ik} + y_{ik}y_{ik}\right] = \frac{k}{4(L^2/32)}\left[\frac{L}{4}\cdot\frac{L}{4} + \frac{L}{4}\cdot\frac{L}{4}\right] = \frac{k}{2}2$$

$$\kappa_{jk} = \frac{k}{4A}\left[-(x_{ik}x_{ij} + y_{ik}y_{ij})\right] = \frac{-k}{4(L^2/32)}\left[\frac{L}{4}\cdot\frac{L}{4} + \frac{L}{4}\cdot 0\right] = -\frac{k}{2}$$

$$\kappa_{kk} = \frac{k}{4A}\left[x_{ij}x_{ij} + y_{ij}y_{ij}\right] = \frac{k}{4(L^2/32)}\left[\frac{L}{4}\cdot\frac{L}{4} + 0\cdot 0\right] = \frac{k}{2}$$

where $i = 1$, $j = 2$ and $k = 4$ for element 1. These six entries and their three symmetric entries are at the intersections of rows and columns 1, 2 and 4 in $\mathbf{K}^{(1)}$. For element 1 we will have

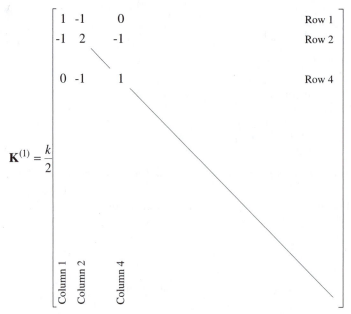

$$\mathbf{K}^{(1)} = \frac{k}{2}$$

The entries in $\mathbf{g}^{(1)}$ are computed as shown in (9·2·13) as

$$g_i = g_j = g_k = \frac{g'''(L^2/32)}{3} = \frac{g'''L^2}{96}$$

The generation vector for element 1 is then given by

$$\mathbf{g}^{(1)} = \frac{g'''L^2}{96} \begin{bmatrix} 1 \\ 1 \\ \\ 1 \\ \\ \\ \\ \\ \\ \end{bmatrix} \begin{array}{l} \text{Row 1} \\ \text{Row 2} \\ \\ \text{Row 4} \end{array}$$

Next we must add $\mathbf{K}^{(1)}$ and $\mathbf{g}^{(1)}$ to $\mathbf{K}$ and $\mathbf{g}$ in (9·3·1). The entries in $\mathbf{K}^{(1)}$ are added into $\mathbf{K}$ in rows and columns 1, 2 and 4 corresponding to the nodal-point numbers for element 1 as shown by (9·2·8). The entries in $\mathbf{g}^{(1)}$ are added into $\mathbf{g}$ in rows 1, 2 and 4 corresponding to the nodal-point numbers for element 1. After adding in the contributions from element 1, the global matrix equation will be

$$
\frac{k}{2}
\begin{bmatrix}
1 & -1 & 0 & 0 & 0 & 0 & & & & & & & & & \\
-1 & 2 & 0 & -1 & 0 & 0 & 0 & & & & & & & & \\
0 & 0 & 0 & 0 & 0 & 0 & 0 & 0 & & & & & & & \\
0 & -1 & 0 & 1 & 0 & 0 & 0 & 0 & 0 & & & & & & \\
0 & 0 & 0 & 0 & 0 & 0 & 0 & 0 & 0 & 0 & & & & & \\
0 & 0 & 0 & 0 & 0 & 0 & 0 & 0 & 0 & 0 & 0 & & & & \\
& 0 & 0 & 0 & 0 & 0 & 0 & 0 & 0 & 0 & 0 & 0 & & & \\
& & 0 & 0 & 0 & 0 & 0 & 0 & 0 & 0 & 0 & 0 & 0 & & \\
& & & 0 & 0 & 0 & 0 & 0 & 0 & 0 & 0 & 0 & 0 & 0 & \\
& & & & 0 & 0 & 0 & 0 & 0 & 0 & 0 & 0 & 0 & 0 & 0 \\
& & & & & 0 & 0 & 0 & 0 & 0 & 0 & 0 & 0 & 0 & 0 \\
& & & & & & 0 & 0 & 0 & 0 & 0 & 0 & 0 & 0 & 0 \\
& & & & & & & 0 & 0 & 0 & 0 & 0 & 0 & 0 & 0 \\
& & & & & & & & 0 & 0 & 0 & 0 & 0 & 0 & 0 \\
& & & & & & & & & 0 & 0 & 0 & 0 & 0 & 0
\end{bmatrix}
\begin{bmatrix}
t_1 \\ t_2 \\ t_3 \\ t_4 \\ t_5 \\ t_6 \\ t_7 \\ t_8 \\ t_9 \\ t_{10} \\ t_{11} \\ t_{12} \\ t_{13} \\ t_{14} \\ t_{15}
\end{bmatrix}
= \frac{g'''L^2}{96}
\begin{bmatrix}
1 \\ 1 \\ 0 \\ 1 \\ 0 \\ 0 \\ 0 \\ 0 \\ 0 \\ 0 \\ 0 \\ 0 \\ 0 \\ 0 \\ 0
\end{bmatrix}
$$

(Row 1, Row 2, Row 4 as indicated; Column 1, Column 2, Column 4 as indicated.)

The shaded entries are the ones that have been affected by the addition of $\mathbf{K}^{(1)}$ and $\mathbf{g}^{(1)}$.

Next we will move on to element 2 and repeat similar computations using data from Table 9•1 for element 2 and nodes 2, 4 and 6. For element 2 the conduction matrix $\mathbf{K}^{(2)}$ and the element generation vector $\mathbf{g}^{(2)}$ are found to be given by

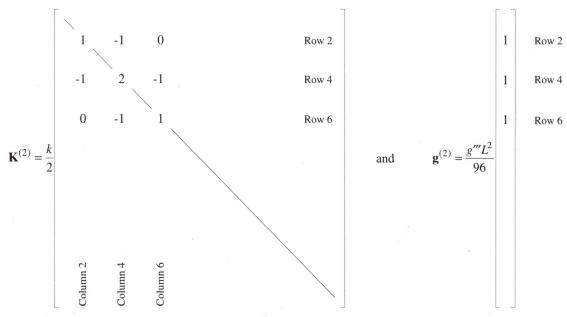

$$
\mathbf{K}^{(2)} = \frac{k}{2}
\begin{bmatrix}
1 & -1 & 0 \\
-1 & 2 & -1 \\
0 & -1 & 1
\end{bmatrix}
\begin{matrix} \text{Row 2} \\ \text{Row 4} \\ \text{Row 6} \end{matrix}
\qquad \text{and} \qquad
\mathbf{g}^{(2)} = \frac{g'''L^2}{96}
\begin{bmatrix}
1 \\ 1 \\ 1
\end{bmatrix}
\begin{matrix} \text{Row 2} \\ \text{Row 4} \\ \text{Row 6} \end{matrix}
$$

(Column 2, Column 4, Column 6 as indicated.)

When these contributions are added to rows and columns 2, 4 and 6 in the global matrix equation we obtain

$$
\frac{k}{2}\left[\begin{array}{ccccccccccccccc|l}
1 & -1 & 0 & 0 & 0 & 0 & & & & & & & & & & \\
-1 & 3 & 0 & -2 & 0 & 0 & 0 & & & & & & & & & \text{Row 2} \\
0 & 0 & 0 & 0 & 0 & 0 & 0 & 0 & & & & & & & & \\
0 & -2 & 0 & 3 & 0 & -1 & 0 & 0 & 0 & & & & & & & \text{Row 4} \\
0 & 0 & 0 & 0 & 0 & 0 & 0 & 0 & 0 & & & & & & & \\
0 & 0 & 0 & -1 & 0 & 1 & 0 & 0 & 0 & 0 & & & & & & \text{Row 6} \\
 & 0 & 0 & 0 & 0 & 0 & 0 & 0 & 0 & 0 & 0 & & & & & \\
 & & 0 & 0 & 0 & 0 & 0 & 0 & 0 & 0 & 0 & 0 & & & & \\
 & & & 0 & 0 & 0 & 0 & 0 & 0 & 0 & 0 & 0 & 0 & & & \\
 & & & & 0 & 0 & 0 & 0 & 0 & 0 & 0 & 0 & 0 & 0 & 0 & \\
 & & & & & 0 & 0 & 0 & 0 & 0 & 0 & 0 & 0 & 0 & 0 & \\
 & & & & & & 0 & 0 & 0 & 0 & 0 & 0 & 0 & 0 & 0 & \\
 & & & & & & & 0 & 0 & 0 & 0 & 0 & 0 & 0 & 0 & \\
 & & & & & & & & 0 & 0 & 0 & 0 & 0 & 0 & 0 & \\
 & & & & & & & & & 0 & 0 & 0 & 0 & 0 & 0 & \\
 & \text{Column 2} & & \text{Column 4} & & \text{Column 6} & & & & & & & & & &
\end{array}\right]
\begin{bmatrix} t_1 \\ t_2 \\ t_3 \\ t_4 \\ t_5 \\ t_6 \\ t_7 \\ t_8 \\ t_9 \\ t_{10} \\ t_{11} \\ t_{12} \\ t_{13} \\ t_{14} \\ t_{15} \end{bmatrix}
= \frac{g'''L^2}{96}
\begin{bmatrix} 1 \\ 2 \\ 0 \\ 2 \\ 0 \\ 1 \\ 0 \\ 0 \\ 0 \\ 0 \\ 0 \\ 0 \\ 0 \\ 0 \\ 0 \end{bmatrix}
$$

The shaded entries are the ones affected by the addition of $\mathbf{K}^{(2)}$ and $\mathbf{g}^{(2)}$. Rows 2 and 4 have had contributions from both element 1 and element 2 since nodes 2 and 4 belong to both of these elements.

Proceeding next to element 3 we find that $\mathbf{K}^{(3)}$ and $\mathbf{g}^{(3)}$ are given by

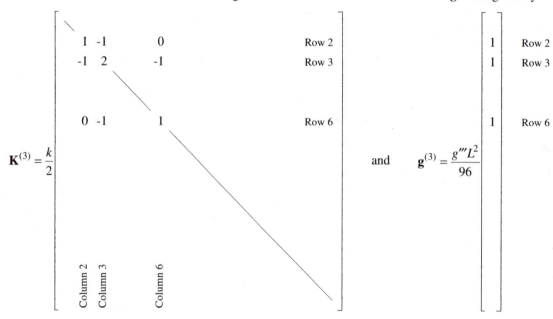

$$
\mathbf{K}^{(3)} = \frac{k}{2}\left[\begin{array}{cccc|l}
\ddots & & & & \\
 & 1 & -1 & 0 & \text{Row 2} \\
 & -1 & 2 & -1 & \text{Row 3} \\
 & & & \ddots & \\
 & 0 & -1 & 1 & \text{Row 6} \\
 & & & & \ddots \\
 & \text{Column 2} & \text{Column 3} & \text{Column 6} &
\end{array}\right]
\quad \text{and} \quad
\mathbf{g}^{(3)} = \frac{g'''L^2}{96}\begin{bmatrix} \\ 1 \\ 1 \\ \\ 1 \\ \\ \end{bmatrix}
\begin{array}{l} \\ \text{Row 2} \\ \text{Row 3} \\ \\ \text{Row 6} \\ \\ \end{array}
$$

Upon adding these to the global matrix equation in rows and columns 2, 3 and 6 we obtain

$$\frac{k}{2}\begin{bmatrix}
1 & -1 & 0 & 0 & 0 & 0 \\
-1 & 4 & -1 & -2 & 0 & 0 & 0 & & & & & \text{Row 2}\\
0 & -1 & 2 & 0 & 0 & -1 & 0 & 0 & & & & \text{Row 3}\\
0 & -2 & 0 & 3 & 0 & -1 & 0 & 0 & 0 \\
0 & 0 & 0 & 0 & 0 & 0 & 0 & 0 & 0 & 0 \\
0 & 0 & -1 & -1 & 0 & 2 & 0 & 0 & 0 & 0 & 0 & \text{Row 6}\\
 & 0 & 0 & 0 & 0 & 0 & 0 & 0 & 0 & 0 & 0 & 0 \\
 & & 0 & 0 & 0 & 0 & 0 & 0 & 0 & 0 & 0 & 0 & 0 \\
 & & & 0 & 0 & 0 & 0 & 0 & 0 & 0 & 0 & 0 & 0 \\
 & & & & 0 & 0 & 0 & 0 & 0 & 0 & 0 & 0 & 0 \\
 & & & & & 0 & 0 & 0 & 0 & 0 & 0 & 0 & 0 \\
 & & & & & & 0 & 0 & 0 & 0 & 0 & 0 & 0 \\
 & & & & & & & 0 & 0 & 0 & 0 & 0 & 0 \\
 & & & & & & & & 0 & 0 & 0 & 0 & 0 \\
 & & & & & & & & & 0 & 0 & 0 & 0
\end{bmatrix}
\begin{bmatrix} t_1 \\ t_2 \\ t_3 \\ t_4 \\ t_5 \\ t_6 \\ t_7 \\ t_8 \\ t_9 \\ t_{10} \\ t_{11} \\ t_{12} \\ t_{13} \\ t_{14} \\ t_{15} \end{bmatrix}
= \frac{g'''L^2}{96}
\begin{bmatrix} 1 \\ 3 \\ 1 \\ 2 \\ 0 \\ 2 \\ 0 \\ 0 \\ 0 \\ 0 \\ 0 \\ 0 \\ 0 \\ 0 \\ 0 \end{bmatrix}$$

(Column 2, Column 3, Column 6 labels beneath the shaded columns.)

The shaded entries are the ones affected by the addition of $\mathbf{K}^{(3)}$ and $\mathbf{g}^{(3)}$. This is the third time a contribution has been made to row 2. This is because node 2 is in elements 1, 2 and 3 as shown in Figure 9•30. Since node 2 does not belong to any other element, row 2 is now complete.

This process is continued until the contributions from all 16 elements have been added to obtain $\mathbf{K}$ and $\mathbf{g}$. The result is given by

$$\frac{k}{2}\begin{bmatrix}
1 & -1 & 0 & 0 & 0 & 0 \\
-1 & 4 & -1 & -2 & 0 & 0 & 0 \\
0 & -1 & 4 & 0 & -1 & -2 & 0 & 0 \\
0 & -2 & 0 & 4 & 0 & -2 & 0 & 0 & 0 \\
0 & 0 & -1 & 0 & 4 & 0 & -1 & -2 & 0 & 0 \\
0 & 0 & -2 & -2 & 0 & 8 & 0 & -2 & -2 & 0 & 0 \\
 & 0 & 0 & 0 & -1 & 0 & 2 & 0 & 0 & -1 & 0 & 0 \\
 & & 0 & 0 & -2 & -2 & 0 & 8 & 0 & -2 & -2 & 0 & 0 \\
 & & & 0 & 0 & -2 & 0 & 0 & 4 & 0 & -2 & 0 & 0 & 0 \\
 & & & & 0 & 0 & -1 & -2 & 0 & 4 & 0 & -1 & 0 & 0 & 0 \\
 & & & & & 0 & 0 & -2 & -2 & 0 & 8 & -2 & -2 & 0 & 0 \\
 & & & & & & 0 & 0 & 0 & -1 & -2 & 4 & 0 & -1 & 0 \\
 & & & & & & & 0 & 0 & 0 & -2 & 0 & 4 & -2 & 0 \\
 & & & & & & & & 0 & 0 & 0 & -1 & -2 & 4 & -1 \\
 & & & & & & & & & 0 & 0 & 0 & 0 & -1 & 1
\end{bmatrix}
\begin{bmatrix} t_1 \\ t_2 \\ t_3 \\ t_4 \\ t_5 \\ t_6 \\ t_7 \\ t_8 \\ t_9 \\ t_{10} \\ t_{11} \\ t_{12} \\ t_{13} \\ t_{14} \\ t_{15} \end{bmatrix}
= \frac{g'''L^2}{96}
\begin{bmatrix} 1 \\ 3 \\ 3 \\ 3 \\ 3 \\ 6 \\ 1 \\ 6 \\ 3 \\ 3 \\ 6 \\ 3 \\ 3 \\ 3 \\ 1 \end{bmatrix}$$

$$(9 \cdot 3 \cdot 2)$$

At this point it is of interest to compare (9·3·2) to its finite-difference counterpart, (8·3·7). Observe that the two conduction matrices happen to be identical in this example. Two more off-diagonal rows in the upper band and in the lower band of (9·3·2) had to be allowed for in the finite-element conduction matrix but they happen to contain only zero values. The only difference occurs in the generation vector where the finite-element formulation has a bit more generation in rows 1 and 15 ($\frac{1}{96}$ compared to $\frac{1}{128}$) and a little less generation in row 7 ($\frac{1}{96}$ compared to $\frac{2}{128}$) than the finite-difference formulation. The remainder of the generation vectors are identical to each other. The sum of the entries in $\mathbf{g}$ is again $\frac{1}{2}g'''L^2$ as is was in finite differences.

Modifications for the specified temperatures must now be made. After making modifications for $t_7 = T_7$, $t_{10} = T_{10}$, $t_{12} = T_{12}$, $t_{14} = T_{14}$ and $t_{15} = T_{15}$, we find that

$$
\frac{k}{2}
\begin{bmatrix}
1 & -1 & 0 & 0 & 0 & 0 & & & & & & & & & \\
-1 & 4 & -1 & -2 & 0 & 0 & 0 & & & & & & & & \\
0 & -1 & 4 & 0 & -1 & -2 & 0 & 0 & & & & & & & \\
0 & -2 & 0 & 4 & 0 & -2 & 0 & 0 & 0 & & & & & & \\
0 & 0 & -1 & 0 & 4 & 0 & 0 & -2 & 0 & 0 & & & & & \\
0 & 0 & -2 & -2 & 0 & 8 & 0 & -2 & -2 & 0 & 0 & & & & \\
 & 0 & 0 & 0 & 0 & 0 & 2 & 0 & 0 & 0 & 0 & 0 & & & \\
 & & 0 & 0 & -2 & -2 & 0 & 8 & 0 & 0 & -2 & 0 & 0 & & \\
 & & & 0 & 0 & -2 & 0 & 0 & 4 & 0 & -2 & 0 & 0 & 0 & \\
 & & & & 0 & 0 & 0 & 0 & 0 & 4 & 0 & 0 & 0 & 0 & 0 \\
 & & & & & 0 & 0 & -2 & -2 & 0 & 8 & 0 & -2 & 0 & 0 \\
 & & & & & & 0 & 0 & 0 & 0 & 0 & 4 & 0 & 0 & 0 \\
 & & & & & & & 0 & 0 & 0 & -2 & 0 & 4 & 0 & 0 \\
 & & & & & & & & 0 & 0 & 0 & 0 & 0 & 4 & 0 \\
 & & & & & & & & & 0 & 0 & 0 & 0 & 0 & 1 \\
\end{bmatrix}
\begin{bmatrix}
t_1 \\ t_2 \\ t_3 \\ t_4 \\ t_5 \\ t_6 \\ t_7 \\ t_8 \\ t_9 \\ t_{10} \\ t_{11} \\ t_{12} \\ t_{13} \\ t_{14} \\ t_{15}
\end{bmatrix}
=
\frac{g'''L^2}{96}
\begin{bmatrix}
1 \\ 3 \\ 3 \\ 3 \\ 3 \\ 6 \\ 0 \\ 6 \\ 3 \\ 0 \\ 6 \\ 0 \\ 3 \\ 0 \\ 0
\end{bmatrix}
+
\frac{k}{2}
\begin{bmatrix}
 \\ 0 \\ 0 \\ 0 \\ T_7 \\ 0 \\ 2T_7 \\ 2T_{10} \\ 0 \\ 4T_{10} \\ 2T_{12} \\ 4T_{12} \\ 2T_{14} \\ 4T_{14} \\ T_{15}
\end{bmatrix}
$$

$$(9\cdot3\cdot3)$$

The shaded entries indicate the rows and columns corresponding to the specified-temperature nodes. All of the entries in $\mathbf{K}$ in these rows and columns are 0 except for the main diagonal. The transfers to the right-hand side are indicated as well as the changes in $\mathbf{g}$. This is the system of algebraic equations that will be solved for the temperature distribution. The only difference between (9·3·3) and the corresponding finite-difference equation (8·3·9) is the first entry in $\mathbf{g}$. Entries 7 and 15 in (9·3·2), which differed from (8·3·7), have been overridden by the modifications for the specified boundary temperatures and are now identical to the finite-difference result.

9•3•2 Numerical results

The solution to (9·3·3) follows the procedures discussed in Chapter 8. The equation is banded and symmetric and thus subroutines *DBAND* and *SBAND* described in Appendix I may be used. Computation of nodal-system heat flows via (9·2·40), with $\dot{\mathbf{t}} = \mathbf{0}$, can be obtained using subroutine *YAXPB* described in Appendix I. These computations can be accomplished using *FEHT* as discussed in Section 9•2•4.

Taking unit values for k, g''' and L and setting $T_7 = T_{10} = T_{12} = T_{14} = T_{15} = 0$, the solution to (9·3·3) is given in Table 9•2 and plotted in Figure 9•31. These results should be compared to the corresponding finite-difference results shown in Table 8•4 and in Figure 8•20. The sum of the absolute values of the errors in Table 9•2 is 0.0136 as compared to the finite-difference result of 0.0247 from Table 8•4. Thus the finite-element solution is closer to the exact solution than the finite-difference solution. Figure 9•31 also shows how the finite-element solution improves as the number of nodal points is increased from 3 to 6 to 15.

Table 9•2 *Finite-element temperature solution,*
$(t - T_L)/(g'''L^2 / k)$, *for 15-node energy-generation example.*

Node	Finite Element	Exact	Error, Exact – Approx.	Error, % of Node 1
1	0.3013	0.2947	−0.0066	−2.24
2	0.2805	0.2789	−0.0016	−0.54
3	0.2292	0.2293	0.0001	0.03
4	0.2645	0.2642	−0.0003	−0.10
5	0.1392	0.1397	0.0005	0.17
6	0.2172	0.2178	0.0006	0.20
7	0.0000	0.0000	0.0000	0.00
8	0.1327	0.1333	0.0006	0.20
9	0.1801	0.1811	0.0010	0.34
10	0.0000	0.0000	0.0000	0.00
11	0.1117	0.1127	0.0010	0.34
12	0.0000	0.0000	0.0000	0.00
13	0.0715	0.0728	0.0013	0.44
14	0.0000	0.0000	0.0000	0.00
15	0.0000	0.0000	0.0000	0.00

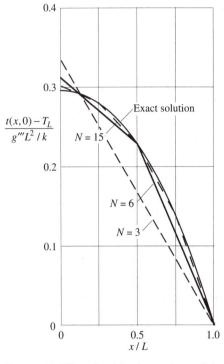

Figure 9•31 *Effect of nodal-point spacing on finite-element solution along $y = 0$ for energy-generation example.*

The finite-element heat inflows can be computed from (9·2·40) as

$$\mathbf{q}_o = \mathbf{K}\mathbf{t} - \mathbf{g}$$

with $\mathbf{K}$ and $\mathbf{g}$ as given in (9·3·2). The entries in $\mathbf{q}_o$ are each 0 except for nodes along the nonadiabatic boundary at $x = L$. For these nodes (7, 10, 12, 14 and 15) the boundary heat inflows in $\mathbf{q}_o$ are −0.0800, −0.1639, −0.1429, −0.1027 and −0.0104 respectively. A negative sign means the heat flow is actually out of the boundary. Observe that the sum of all these boundary heat inflows is −0.4999 which rounds off to −0.5, the energy generated within the region.

The nodal heat flows are converted to average heat fluxes and tabulated in Table 9•3 which can be compared to the finite-difference results shown in Table 8•4. The heat-flux results are shown in Figure 9•32. As in finite differences, the area "under" the stepped finite-element solutions is equal to the area "under" the exact curve.

Table 9·3 *Finite-element boundary heat-flux solution,*
$q''(L,y)/g'''L$, *for 15-node energy-generation example.*

Node	Finite Element	Exact	Error, Exact − Approx.	Error, % of $g'''L/2$
7	0.6403	0.6753	0.0350	7.00
10	0.6556	0.6488	−0.0068	−1.36
12	0.5716	0.5628	−0.0088	−1.76
14	0.4108	0.3918	−0.0190	−3.80
15	0.0834	0.0000	−0.0834	−16.68

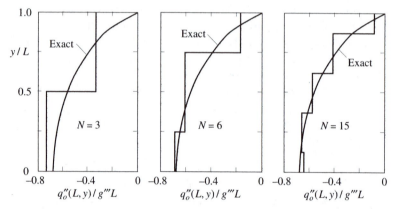

Figure 9·32 *Comparison of finite-element boundary heat flux to exact solution for energy-generation example.*

This completes our discussion of a steady-state, finite-element example. The specific example we have discussed was a very regular one for the sake of making comparisons to finite differences. It should be emphasized however that an irregular problem, with the same number of nodes and elements, would be no more difficult to formulate and solve than the example we have just seen. The only difference is that the numerical values for the nodal coordinates in Table 9·1 would be different and there might be more than one material. The computations required to obtain the element conduction matrix would be carried out in the same way as was illustrated in finding (9·3·2). The numerical values might be different but the computational steps would be the same.

9•4 TRANSIENT PROBLEMS

In this section we will look at the same examples we considered in Section 8•4 so that comparisons between finite elements and finite differences can again be made. Section 9•4•1 will show how to discretize the example. Numerical results will then be presented in Section 9•4•2.

9•4•1 Discretization example

For purposes of comparison to finite differences let us consider the example shown in Figure 8•22. This can be discretized using six nodes and four elements as shown in Figure 9•33. Since the maximum difference between nodal-point numbers in any element is 3, we will have to allow for an upper bandwidth of 4. We will thus begin by writing

$$
\begin{bmatrix}
0 & 0 & 0 & 0 & & \\
0 & 0 & 0 & 0 & 0 & \\
0 & 0 & 0 & 0 & 0 & 0 \\
0 & 0 & 0 & 0 & 0 & 0 \\
 & 0 & 0 & 0 & 0 & 0 \\
 & & 0 & 0 & 0 & 0
\end{bmatrix}
\begin{bmatrix}
\dot{t}_1 \\ \dot{t}_2 \\ \dot{t}_3 \\ \dot{t}_4 \\ \dot{t}_5 \\ \dot{t}_6
\end{bmatrix}
+
\begin{bmatrix}
0 & 0 & 0 & 0 & & \\
0 & 0 & 0 & 0 & 0 & \\
0 & 0 & 0 & 0 & 0 & 0 \\
0 & 0 & 0 & 0 & 0 & 0 \\
 & 0 & 0 & 0 & 0 & 0 \\
 & & 0 & 0 & 0 & 0
\end{bmatrix}
\begin{bmatrix}
t_1 \\ t_2 \\ t_3 \\ t_4 \\ t_5 \\ t_6
\end{bmatrix}
=
\begin{bmatrix}
0 \\ 0 \\ 0 \\ 0 \\ 0 \\ 0
\end{bmatrix}
\tag{9.4.1}
$$

Observe that we are expecting the bandwidth of $\mathbf{C}$ to be the same as the bandwidth of $\mathbf{S}$. The capacitance matrix will not simply be a diagonal matrix as it was in finite differences.

The input data that we must supply the computer for Figure 9•33 are shown in Table 9•4. They consist of the nodal coordinates for each node, element information showing how the nodes are hooked together into elements and the material property number for each element. In this example every element is made of material 1. The properties k, ρ, c and g''' of material 1 must also be supplied to the computer.

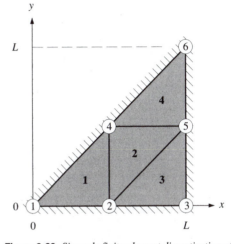

Figure 9•33 *Six-node finite-element discretization of adiabatic-square example shown in Figure 8•22.*

Table 9•4 *Computer input data for Figure 9•33.*

Nodal coordinates			Element information				
i	x_i	y_i	e	i	j	k	m
1	0	0	1	1	2	4	1
2	$L/2$	0	2	2	4	5	1
3	L	0	3	2	3	5	1
4	$L/2$	$L/2$	4	4	5	6	1
5	L	$L/2$					
6	L	L					

To start filling the entries in $\mathbf{C}$, $\mathbf{S}$ and $\mathbf{g}$ we will first go to element 1. The first step is to compute

$$x_{ij} = x_j - x_i = \frac{L}{2} - 0 = \frac{L}{2} \qquad\qquad y_{ij} = y_j - y_i = 0 - 0 = 0$$

$$x_{ik} = x_k - x_i = \frac{L}{2} - 0 = \frac{L}{2} \qquad\qquad y_{ik} = y_k - y_i = \frac{L}{2} - 0 = \frac{L}{2}$$

$$x_{jk} = x_k - x_j = \frac{L}{2} - \frac{L}{2} = 0 \qquad\qquad y_{jk} = y_k - y_j = \frac{L}{2} - 0 = \frac{L}{2}$$

The area of the element can now be computed following (9·2·11) as

$$A = \frac{1}{2}\left|x_{ij}y_{jk} - x_{jk}y_{ij}\right| = \frac{1}{2}\left|\frac{L}{2}\cdot\frac{L}{2} - 0\cdot 0\right| = \frac{L^2}{8}$$

The entries in the element capacitance matrix $\mathbf{C}^{(1)}$ may be computed as shown by (9·2·10). Thus,

$$\mathbf{C}^{(1)} = \frac{\rho c L^2}{96}\begin{bmatrix} 2 & 1 & & 1 \\ 1 & 2 & & 1 \\ & & & \\ 1 & 1 & & 2 \\ & & & \\ & & & \end{bmatrix}$$

The entries in the element conduction matrix $\mathbf{K}^{(1)}$ may now be computed as shown in (9·2·12).

$$\kappa_{11} = \frac{k}{4A}\left[x_{jk}x_{jk} + y_{jk}y_{jk}\right] = \frac{k}{4(L^2/8)}\left[0\cdot 0 + \frac{L}{2}\cdot\frac{L}{2}\right] = \frac{k}{2}$$

The remaining entries in $\mathbf{K}^{(1)}$ may be computed in a similar manner to give

$$\mathbf{K}^{(1)} = \frac{k}{2}\begin{bmatrix} 1 & -1 & & 0 \\ -1 & 2 & & -1 \\ & & & \\ 0 & -1 & & 1 \\ & & & \\ & & & \end{bmatrix}$$

In this particular example there is no generation so the computation of $\mathbf{g}$ can be bypassed. The entries in $\mathbf{g}$ have already been set equal to 0 in (9·4·1).

When $\mathbf{C}^{(1)}$ and $\mathbf{K}^{(1)}$ are added into (9·4·1) we obtain

$$\frac{\rho c L^2}{96}\begin{bmatrix} 2 & 1 & 0 & 1 & & \\ 1 & 2 & 0 & 1 & 0 & \\ 0 & 0 & 0 & 0 & 0 & 0 \\ 1 & 1 & 0 & 2 & 0 & 0 \\ & & 0 & 0 & 0 & 0 \\ & & & 0 & 0 & 0 \end{bmatrix}\begin{bmatrix} \dot{t}_1 \\ \dot{t}_2 \\ \dot{t}_3 \\ \dot{t}_4 \\ \dot{t}_5 \\ \dot{t}_6 \end{bmatrix} + \frac{k}{2}\begin{bmatrix} 1 & -1 & 0 & 0 & & \\ -1 & 2 & 0 & -1 & 0 & \\ 0 & 0 & 0 & 0 & 0 & 0 \\ 0 & -1 & 0 & 1 & 0 & 0 \\ & & 0 & 0 & 0 & 0 \\ & & & 0 & 0 & 0 \end{bmatrix}\begin{bmatrix} t_1 \\ t_2 \\ t_3 \\ t_4 \\ t_5 \\ t_6 \end{bmatrix} = \begin{bmatrix} 0 \\ 0 \\ 0 \\ 0 \\ 0 \\ 0 \end{bmatrix}$$

$$(9\cdot 4\cdot 2)$$

The shaded entries are the ones that have been affected by the addition of $\mathbf{C}^{(1)}$ and $\mathbf{K}^{(1)}$. They appear at the intersections of rows and columns 1, 2 and 4 which correspond to the nodal-point numbers of element 1.

Next, the contributions of element 2 are given by

$$\mathbf{C}^{(2)} = \frac{\rho c L^2}{96}\begin{bmatrix} 2 & 1 & 1 \\ 1 & 2 & 1 \\ 1 & 1 & 2 \end{bmatrix} \qquad \mathbf{K}^{(2)} = \frac{k}{2}\begin{bmatrix} 1 & -1 & 0 \\ -1 & 2 & -1 \\ 0 & -1 & 1 \end{bmatrix}$$

When these are added to (9·4·2) we obtain

$$\frac{\rho c L^2}{96}\begin{bmatrix} 2 & 1 & 0 & 1 & & \\ 1 & 4 & 0 & 2 & 1 & \\ 0 & 0 & 0 & 0 & 0 & 0 \\ 1 & 2 & 0 & 4 & 1 & 0 \\ & 1 & 0 & 1 & 2 & 0 \\ & & 0 & 0 & 0 & 0 \end{bmatrix}\begin{bmatrix} \dot t_1 \\ \dot t_2 \\ \dot t_3 \\ \dot t_4 \\ \dot t_5 \\ \dot t_6 \end{bmatrix} + \frac{k}{2}\begin{bmatrix} 1 & -1 & 0 & 0 & & \\ -1 & 3 & 0 & -2 & 0 & \\ 0 & 0 & 0 & 0 & 0 & 0 \\ 0 & -2 & 0 & 3 & -1 & 0 \\ & 0 & 0 & -1 & 1 & 0 \\ & & 0 & 0 & 0 & 0 \end{bmatrix}\begin{bmatrix} t_1 \\ t_2 \\ t_3 \\ t_4 \\ t_5 \\ t_6 \end{bmatrix} = \begin{bmatrix} 0 \\ 0 \\ 0 \\ 0 \\ 0 \\ 0 \end{bmatrix}$$

$$(9\cdot4\cdot3)$$

The shaded entries are the ones that have been affected by the addition of $\mathbf{C}^{(2)}$ and $\mathbf{K}^{(2)}$. They are at the intersections of rows and columns 2, 4 and 5 which correspond to the nodal-point numbers of element 2.

After adding the contributions for elements 3 and 4 to (9·4·3),

$$\frac{\rho c L^2}{96}\begin{bmatrix} 2 & 1 & 0 & 1 & & \\ 1 & 6 & 1 & 2 & 2 & \\ 0 & 1 & 2 & 0 & 1 & 0 \\ 1 & 2 & 0 & 6 & 2 & 1 \\ & 2 & 1 & 2 & 6 & 1 \\ & & 0 & 1 & 1 & 2 \end{bmatrix}\begin{bmatrix} \dot t_1 \\ \dot t_2 \\ \dot t_3 \\ \dot t_4 \\ \dot t_5 \\ \dot t_6 \end{bmatrix} + \frac{k}{2}\begin{bmatrix} 1 & -1 & 0 & 0 & & \\ -1 & 4 & -1 & -2 & 0 & \\ 0 & -1 & 2 & 0 & -1 & 0 \\ 0 & -2 & 0 & 4 & -2 & 0 \\ & 0 & -1 & -2 & 4 & -1 \\ & & 0 & 0 & -1 & 1 \end{bmatrix}\begin{bmatrix} t_1 \\ t_2 \\ t_3 \\ t_4 \\ t_5 \\ t_6 \end{bmatrix} = \begin{bmatrix} 0 \\ 0 \\ 0 \\ 0 \\ 0 \\ 0 \end{bmatrix}$$

$$(9\cdot4\cdot4)$$

This is the final system of ordinary differential equations that must be solved for $\mathbf{t}$ as a function of θ. This matrix differential equation should be compared to its finite-difference counterpart given by (8·4·5). The most apparent difference is the capacitance matrix which is no longer diagonal. The sum of all of the entries in $\mathbf{C}$ is still $\frac{1}{2}\rho c L^2$, the total thermal capacitance of the region of interest, but this capacitance is distributed throughout $\mathbf{C}$ in a different way than it was in finite differences. The conduction matrix $\mathbf{K}$ and the generation vector $\mathbf{g}$ are identical to their finite-difference counterparts in (8·4·5).

As in finite differences the initial temperature vector will be given by

$$\mathbf{t}^{(0)} = \frac{T_o}{2}\begin{bmatrix} 2 \\ 1 \\ 0 \\ 1 \\ 0 \\ 0 \end{bmatrix}$$

$$(9\cdot4\cdot5)$$

Before moving on to the next section to discuss the solution of (9·4·4), let us consider the example shown in Figure 9•33 but with the adiabatic boundary along $x = L$ replaced by a convective boundary. The additional input data, shown in Table 9•5, that must be given to the computer consists of the nodal-point numbers for each convective boundary segment and the heat-transfer coefficient and ambient temperature for each segment.

Table 9•5 *Boundary data for convective boundary along $x = L$ in Figure 9•33.*

b_h	i	j	h	t_∞
1	3	5	h	t_∞
2	5	6	h	t_∞

The process of forming $\mathbf{C}$, $\mathbf{K}$ and $\mathbf{g}$ is identical to the steps taken to arrive at (9·4·4). The next step is to go to boundary segment 1 and compute

$$x_{ij} = x_j - x_i = L - L = 0 \qquad \text{and} \qquad y_{ij} = y_j - y_i = \frac{L}{2} - 0 = \frac{L}{2}$$

where the values of the nodal coordinates are found in Table 9•4. The length of the boundary segment is then computed as

$$s_{ij} = \sqrt{x_{ij}^2 + y_{ij}^2} = \sqrt{0^2 + \left(\frac{L}{2}\right)^2} = \frac{L}{2}$$

The boundary-segment convection vector (9·2·17) and the boundary-segment convection matrix (9·2·18) can then be found to be given by

$$\mathbf{h}^{(1)} = \frac{hLt_\infty}{4}\begin{bmatrix} 1 \\ \\ 1 \end{bmatrix} \qquad \text{and} \qquad \mathbf{H}^{(1)} = \frac{hL}{12}\begin{bmatrix} 2 & 1 \\ \\ 1 & 2 \end{bmatrix}$$

These contributions from boundary segment 1, $\mathbf{h}^{(1)}$ and $\mathbf{H}^{(1)}\mathbf{t}$, would then be added to $\mathbf{g}$ and $\mathbf{Kt}$ shown in (9·4·4). The resulting system of equations may be written as

$$\frac{\rho c L^2}{96}\begin{bmatrix} 2 & 1 & 0 & 1 & & \\ 1 & 6 & 1 & 2 & 2 & \\ 0 & 1 & 2 & 0 & 1 & 0 \\ 1 & 2 & 0 & 6 & 2 & 1 \\ & 2 & 1 & 2 & 6 & 1 \\ & & 0 & 1 & 1 & 2 \end{bmatrix}\begin{bmatrix} \dot{t}_1 \\ \dot{t}_2 \\ \dot{t}_3 \\ \dot{t}_4 \\ \dot{t}_5 \\ \dot{t}_6 \end{bmatrix} + \frac{k}{2}\begin{bmatrix} 1 & -1 & 0 & 0 & & \\ -1 & 4 & -1 & -2 & 0 & \\ 0 & -1 & 2 & 0 & -1 & 0 \\ 0 & -2 & 0 & 4 & -2 & 0 \\ & 0 & -1 & -2 & 4 & -1 \\ & & 0 & 0 & -1 & 1 \end{bmatrix}\begin{bmatrix} t_1 \\ t_2 \\ t_3 \\ t_4 \\ t_5 \\ t_6 \end{bmatrix} + \frac{hL}{12}\begin{bmatrix} & & & & & \\ & & & & & \\ & & 2 & & 1 & \\ & & & & & \\ & & 1 & & 2 & \\ & & & & & \end{bmatrix}\begin{bmatrix} t_1 \\ t_2 \\ t_3 \\ t_4 \\ t_5 \\ t_6 \end{bmatrix} = \begin{bmatrix} 0 \\ 0 \\ 0 \\ 0 \\ 0 \\ 0 \end{bmatrix} + \frac{hLt_\infty}{4}\begin{bmatrix} \\ \\ 1 \\ \\ 1 \\ \end{bmatrix}$$

The entries in rows and columns 3 and 5 (corresponding to the nodal-point numbers of boundary segment 1) in $\mathbf{g}$ and $\mathbf{K}$ would be affected by these additions. Boundary segment 2 would then have to be considered in the same manner. The final result may be written as

$$\frac{\rho c L^2}{96}\begin{bmatrix} 2 & 1 & 0 & 1 & & \\ 1 & 6 & 1 & 2 & 2 & \\ 0 & 1 & 2 & 0 & 1 & 0 \\ 1 & 2 & 0 & 6 & 2 & 1 \\ & 2 & 1 & 2 & 6 & 1 \\ & & 0 & 1 & 1 & 2 \end{bmatrix}\begin{bmatrix} \dot{t}_1 \\ \dot{t}_2 \\ \dot{t}_3 \\ \dot{t}_4 \\ \dot{t}_5 \\ \dot{t}_6 \end{bmatrix} + \frac{k}{2}\begin{bmatrix} 1 & -1 & 0 & 0 & & \\ -1 & 4 & -1 & -2 & 0 & \\ 0 & -1 & 2 & 0 & -1 & 0 \\ 0 & -2 & 0 & 4 & -2 & 0 \\ & 0 & -1 & -2 & 4 & -1 \\ & & 0 & 0 & -1 & 1 \end{bmatrix}\begin{bmatrix} t_1 \\ t_2 \\ t_3 \\ t_4 \\ t_5 \\ t_6 \end{bmatrix} + \frac{hL}{12}\begin{bmatrix} & & & & & \\ & & & & & \\ & & 2 & & 1 & \\ & & & & & \\ & & 1 & & 4 & 1 \\ & & & & 1 & 2 \end{bmatrix}\begin{bmatrix} t_1 \\ t_2 \\ t_3 \\ t_4 \\ t_5 \\ t_6 \end{bmatrix} = \begin{bmatrix} 0 \\ 0 \\ 0 \\ 0 \\ 0 \\ 0 \end{bmatrix} + \frac{hLt_\infty}{4}\begin{bmatrix} \\ \\ 1 \\ \\ 2 \\ 1 \end{bmatrix}$$

These additions should be compared to their finite-difference counterparts shown in (8·4·9).

9•4•2 Numerical results

Let us consider the solution to (9·4·4) with the initial condition given by (9·4·5). For numerical computations let us take unit values for k, ρ, c, L and T_o. Thus we will again be approximating the solution to the normalized problem shown in Figure 8•22. Then (9·4·4) becomes

$$\frac{1}{96}\begin{bmatrix} 2 & 1 & 0 & 1 & & \\ 1 & 6 & 1 & 2 & 2 & \\ 0 & 1 & 2 & 0 & 1 & 0 \\ 1 & 2 & 0 & 6 & 2 & 1 \\ & 2 & 1 & 2 & 6 & 1 \\ & & 0 & 1 & 1 & 2 \end{bmatrix}\begin{bmatrix} \dot{t}_1 \\ \dot{t}_2 \\ \dot{t}_3 \\ \dot{t}_4 \\ \dot{t}_5 \\ \dot{t}_6 \end{bmatrix} + \frac{1}{2}\begin{bmatrix} 1 & -1 & 0 & 0 & & \\ -1 & 4 & -1 & -2 & 0 & \\ 0 & -1 & 2 & 0 & -1 & 0 \\ 0 & -2 & 0 & 4 & -2 & 0 \\ & 0 & -1 & -2 & 4 & -1 \\ & & 0 & 0 & -1 & 1 \end{bmatrix}\begin{bmatrix} t_1 \\ t_2 \\ t_3 \\ t_4 \\ t_5 \\ t_6 \end{bmatrix} = \begin{bmatrix} 0 \\ 0 \\ 0 \\ 0 \\ 0 \\ 0 \end{bmatrix}$$

$$(9\cdot4\cdot6)$$

with the initial condition (9·4·5) given by

$$\mathbf{t}^{(0)} = \frac{1}{2}\begin{bmatrix} 2 \\ 1 \\ 0 \\ 1 \\ 0 \\ 0 \end{bmatrix} \qquad\qquad (9\cdot4\cdot7)$$

If we begin by taking $\Delta\theta = 0.02$, as we did in finite differences, the results of an Euler solution of (9·4·6) are shown in Figure 9•34. Observe that there are oscillations in the solution that grow as time goes on. The Euler finite-element solution is unstable with this time step. The comparable finite-difference solution, shown in Figure 8•27, is well behaved however. The conclusion is that numerically-induced oscillations are a more serious problem in finite elements than they were in finite differences. These oscillations will disappear as the time step is reduced just as we found for finite differences.

It is very important to point out that the finite-element $Ç$ that appears in (9·4·6) is not a diagonal matrix as it was in finite differences. This means that the Euler solution shown in Figure 9•34 is not an explicit solution as it was in finite differences. The finite-element Euler solution requires that a system of algebraic equations be solved at each time step. This is as much computational effort as one must do to use the Crank-Nicolson scheme to obtain a solution. Since the Crank-Nicolson method is more accurate than the Euler method and requires no additional work at each time step than an Euler solution, the finite-element solution might as well use Crank-Nicolson. There is no reason to use Euler.

If the Crank-Nicolson procedure is applied to (9·4·6) the resulting solution is shown in Figure 9•35. These results were obtained with no more computation at each time step than the Euler solution shown in Figure 9•34. Observe that the unstable oscillations are gone and the solution is much better than the Euler solution in Figure 9•34.

In order to improve upon the solution shown in Figure 9•35 we will reduce the spacing between nodal points by a factor of two and discretize the region as shown in Figure 9•28. A Crank-Nicolson solution with $\Delta\theta = 0.02$ is shown in Figure 9•36. Some oscillations now appear in the numerical solution but they are stable and disappear as time increases. The accuracy of the finite-element solution is comparable (or perhaps a little better in this case) to the corresponding finite-difference solution shown in

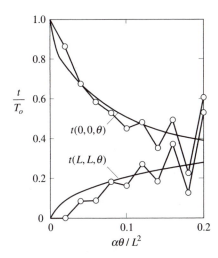

Figure 9•34 *Euler 6-node finite-element solution to adiabatic-square problem with $\Delta\theta = 0.02L^2/\alpha$.*

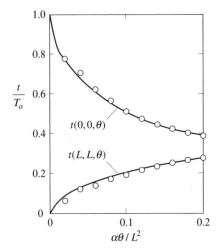

Figure 9•35 *Crank-Nicolson 6-node finite-element solution to adiabatic-square problem with $\Delta\theta = 0.02L^2/\alpha$.*

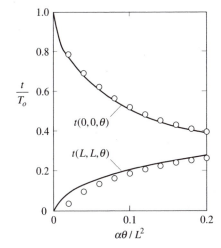

Figure 9•36 *Crank-Nicolson 15-node finite-element solution to adiabatic-square problem with $\Delta\theta = 0.02L^2/\alpha$.*

Figure 8•30. The fact that numerically-induced oscillations are more evident in Figure 9•36 than they are in Figure 8•30 is another indication that the finite-element solutions have a more severe problem regarding numerically-induced oscillations than finite-difference solutions.

As was the case in finite differences, there is a critical time step $\Delta\theta_c$ in finite elements as well. The Euler solution will have unstable oscillations if $\Delta\theta > \Delta\theta_c$, stable oscillations if $\frac{1}{2}\Delta\theta_c < \Delta\theta < \Delta\theta_c$, and no oscillations at all if $\Delta\theta < \frac{1}{2}\Delta\theta_c$. Crank-Nicolson solutions can exhibit numerically-induced oscillations if $\Delta\theta > \Delta\theta_c$, but they will always be stable and have decreasing amplitude as time goes on.

As in finite differences there is a computationally-convenient way to check the $Ç$ and S matrices to obtain an estimate $\Delta\theta_{c,est}$ of the critical time step. For the triangular finite elements we have been using in this chapter, it will be shown in Section 9•5•3 that the finite-element critical time step may be estimated from

$$\Delta\theta_{c,est} = \frac{1}{\underset{\substack{i=1 \\ i \neq i_s}}{\overset{N}{\text{Max}}}\left[\dfrac{1}{ç_{ii}}\displaystyle\sum_{j=1}^{N} |s_{ij}|\right]} \tag{9·4·8}$$

Application of this result to the 6-node, finite-element equations (9·4·4) gives $\Delta\theta_{c,est} = \rho c L^2 / 96k$ or, for unit values of k, ρ, c and L, $\Delta\theta_{c,est} = 0.0104$. An Euler solution with $\Delta\theta = 0.02$ would thus be expected to be unstable. This explains the behavior shown in Figure 9•34. A Crank-Nicolson solution with $\Delta\theta = 0.02$ would be expected to have oscillations. The oscillations are slightly evident in Figure 9•35.

Application of (9·4·8) to the 15-node, finite-element equations would give $\Delta\theta_{c,est} = \rho c L^2 / 384k$ or, for unit values of k, ρ, c and L, $\Delta\theta_{c,est} = 0.0026$. A Crank-Nicolson solution with $\Delta\theta = 0.02$ would be expected to have oscillations. Oscillations are hardly evident in Figure 9•36 however.

The only difference between (9·4·8) and its finite-difference counterpart (8·4·29) is that the 2 that appears in the numerator of (8·4·29) does not appear in (9·4·8). It should also be pointed out that in finite elements, as can be seen from the element capacitance matrix given by (9·2·10), only half of the thermal capacitance is stored along the main diagonal of $Ç$. The other half of the thermal capacitance is stored off the main diagonal. This means that the $ç_{ii}$ that appears in (9·4·8) will, on the average, be about half as big as the $ç_{ii}$ in (8·4·29). Thus one expects that the critical time step estimate for finite elements will be about a factor of four less than for finite differences. This explains the oscillatory behavior in figures 9•34 and 9•36 that was not apparent in the corresponding finite-difference solutions having the same time steps, figures 8•27 and 8•30, respectively.

Although the critical time step is smaller for finite elements than it is for finite differences, one should not avoid using finite elements for transient problems. The superiority of finite elements in approximating irregular boundaries would far override the disadvantage of having to take smaller time steps. Furthermore, we have seen that in finite elements a Crank-Nicolson solution can be carried out with no more effort than an Euler solution. This will completely avoid an unstable solution and the damped oscillations of the Crank-Nicolson solution are all that one must worry about. We have seen that a Crank-Nicolson solution can exceed $\Delta\theta_{c,est}$ without producing major oscillations (*e.g.*, see Figure 9•36 where $\Delta\theta = 7.68\Delta\theta_{c,est}$).

9•5 ANALYSIS OF TRANSIENT SOLUTIONS

Most of what needs to be said regarding the analysis of finite-element transient solutions has already been covered in Section 8•5 for finite differences. Therefore in this section we will simply review the highlights and give some examples for comparison to finite differences.

9•5•1 Exact solution of $Ç\dot{t} + St = r$

The exact solution of the discrete system of ordinary differential equations is still given by (8·5·34) as

$$t(\theta) = a + b\theta + X'\exp(-\lambda\theta) \tag{9.5.1}$$

where a, b, X' and λ are summarized in Table 8•12. The only difference is that the entries in $Ç$, S and r will, in general, be different for finite elements than they were for finite differences. As a result the entries in a, b, X' and λ will be different.

For comparison to finite differences let us consider the exact solution to (9·4·4). The eigenvalues are given by

$$\lambda = \frac{\alpha}{L^2}\begin{bmatrix} 0 \\ 11.72 \\ 31.23 \\ 48 \\ 84.28 \\ 126.5 \end{bmatrix} \tag{9.5.2}$$

These eigenvalues should be compared to the corresponding finite-difference result (8·5·35). The maximum finite-element eigenvalue in (9·5·2) is 3.95 times larger than the maximum finite-difference eigenvalue in (8·5·35). This agrees with the observations we made in the last section in regard to the size of the finite-element critical time step relative to the finite-difference value.

The modified eigenvectors X' for (9·4·4) are given by

$$X' = \frac{T_o}{192}\begin{bmatrix} 64 & 102.4 & 55.4 & -24 & -6.4 & 0.6 \\ 64 & 42.0 & -10.9 & 0 & 5.9 & -5.1 \\ 64 & 0 & -51.4 & -24 & 0 & 11.4 \\ 64 & 0 & 2.0 & 24 & 0 & 6.0 \\ 64 & -42.1 & -10.9 & 0 & -5.9 & -5.1 \\ 64 & -102.4 & 55.4 & -24 & 6.4 & 0.6 \end{bmatrix} \tag{9.5.3}$$

The exact solution of (9·4·4) is then given by

$$t(\theta) = \frac{T_o}{192}\begin{bmatrix} 64 & 102.4 & 55.4 & -24 & -6.4 & 0.6 \\ 64 & 42.0 & -10.9 & 0 & 5.9 & -5.1 \\ 64 & 0 & -51.4 & -24 & 0 & 11.4 \\ 64 & 0 & 2.0 & 24 & 0 & 6.0 \\ 64 & -42.1 & -10.9 & 0 & -5.9 & -5.1 \\ 64 & -102.4 & 55.4 & -24 & 6.4 & 0.6 \end{bmatrix}\begin{bmatrix} 1 \\ \exp(-11.7\alpha\theta/L^2) \\ \exp(-31.2\alpha\theta/L^2) \\ \exp(-48\alpha\theta/L^2) \\ \exp(-84.3\alpha\theta/L^2) \\ \exp(-126.5\alpha\theta/L^2) \end{bmatrix}$$

$$\tag{9.5.4}$$

The corresponding finite-difference solution is given by (8·5·37). Observe that the steady-state value given by (9·5·4) is $64T_o/192 = T_o/3$. This happens to be equal to the value given by the exact solution of the partial differential equation. This will not generally be true. It happens in this particular example because the linear temperature distribution assumed within each element exactly models the given initial condition. As in finite differences, the coefficients of the exponential terms generally decrease as the eigenvalues increase.

The 6-node result (9·5·4) and the results for N = 3 and 15 nodes are compared to the exact solution ($N = \infty$) of the governing partial differential equation in Figure 9·37 for the corners at $(0,0)$ and (L,L). It is interesting to note that the finite-element solutions are closer to the partial-differential-equation solution than the corresponding finite-difference solutions shown in Figure 8·31.

For comparison with finite differences it is helpful to look at λ and $\dot{X}'$ for the 15-node finite-element formulation. The eigenvalues are given by

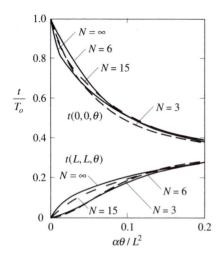

Figure 9·37 *Comparison of the exact finite-element solutions to the partial-differential-equation solution of the adiabatic-square problem.*

$$\lambda = \frac{\alpha}{L^2} \begin{bmatrix} 0 \\ 10.36 \\ 22.66 \\ 47.41 \\ 70.08 \\ 118.9 \\ 124.1 \\ 163.5 \\ 192 \\ 226.6 \\ 272.5 \\ 305.3 \\ 365.0 \\ 413.6 \\ 459.5 \end{bmatrix} \qquad (9\cdot5\cdot5)$$

The modified eigenvector matrix X' is given by

$$X' = \frac{T_o}{192} \begin{bmatrix} 64 & 86.0 & 45.4 & -25.2 & -3.8 & 1.7 & 16.7 & 14.3 & -6 & -1.4 & -1.5 & -1.6 & 3.5 & 0.0 & -0.1 \\ 64 & 69.8 & 28.8 & -10.2 & -0.7 & -1.6 & 1.1 & -3.3 & 0 & 0.1 & 1.2 & 0.7 & -6.5 & 0.5 & 0.1 \\ 64 & 39.5 & -3.1 & 0.2 & 2.2 & -8.6 & 7.7 & -3.4 & -6 & -0.9 & 0.3 & 1.3 & 3.4 & -0.7 & 0.1 \\ 64 & 57.6 & 21.3 & 0.0 & 0.0 & 3.4 & -10.3 & -4.7 & 6 & 1.5 & -0.1 & 1.1 & 4.8 & -0.7 & 0.1 \\ 64 & 11.0 & -33.1 & -10.7 & 1.2 & 3.2 & 9.0 & -0.5 & 0 & 2.8 & -0.4 & 1.7 & -0.2 & 0.5 & -0.4 \\ 64 & 28.5 & 0.0 & 10.7 & 1.2 & -0.2 & -4.7 & 0.5 & 0 & -0.8 & -0.7 & -2.8 & 0.1 & 0.5 & -0.2 \\ 64 & 0 & -42.0 & -17.6 & 0 & 13.7 & 0 & -6.3 & -6 & 0 & 0 & -7.6 & 0.8 & 0 & 1.0 \\ 64 & 0 & -22.7 & 0.1 & 0 & -3.5 & 0 & 5.2 & 6 & 0 & 0 & 0.2 & -1.6 & 0 & 0.4 \\ 64 & 0 & 1.5 & 21.1 & 0 & 9.3 & 0 & 3.9 & -6 & 0 & 0 & 4.6 & 2.6 & 0 & 0.3 \\ 64 & -11.0 & -33.1 & -10.7 & -1.2 & 3.2 & -9.0 & -0.5 & 0 & -2.8 & 0.4 & 1.7 & -0.2 & -0.5 & -0.4 \\ 64 & -28.5 & 0.0 & 10.7 & -1.2 & -0.2 & 4.7 & 0.5 & 0 & 0.8 & 0.7 & -2.8 & 0.1 & -0.5 & -0.2 \\ 64 & -39.5 & -3.1 & 0.2 & -2.2 & -8.6 & -7.7 & -3.4 & -6 & 0.9 & -0.3 & 1.3 & 3.4 & 0.7 & 0.1 \\ 64 & -57.6 & 21.3 & 0.0 & 0.0 & 3.4 & 10.3 & -4.7 & 6 & -1.5 & 0.1 & 1.1 & 4.8 & 0.7 & 0.1 \\ 64 & -69.8 & 28.8 & -10.2 & 0.7 & -1.6 & -1.1 & -3.3 & 0 & -0.1 & -1.2 & 0.7 & -6.5 & -0.5 & 0.1 \\ 64 & -86.0 & 45.4 & -25.2 & 3.8 & 1.7 & -16.7 & 14.3 & -6 & 1.4 & 1.5 & -1.6 & 3.5 & 0.0 & -0.1 \end{bmatrix}$$

$$(9\cdot5\cdot6)$$

The eigenvalues in (9·5·5) should be compared to the finite-difference values in (8·5·38). The maximum finite-element eigenvalue in (9·5·5) is 3.59 times larger than the finite-difference value in (8·5·38).

The first eigenvector in (9·5·6) gives the steady-state solution as $64T_o/192 = T_o/3$ which is again the exact steady-state solution. The entries in $\mathbf{X}'$ generally decrease as you move to the right. Thus the higher eigenvalues will be less important than the smaller eigenvalues. The size of the entries in column 15 as compared to column 1 is much smaller in (9·5·6) than in the finite-difference result (8·5·38). This lessens the importance of the higher eigenvalues even more relative to finite differences.

As in finite differences, the finite-element eigenvalues and eigenvectors approximate the eigenvalues and eigenfunctions of the exact solution to the governing partial differential equation. Table 9•6 shows how the eigenvalues for the discrete finite-element solution approach the exact eigenvalues as N increases. The finite-element eigenvalues in Table 9•6 fall above the exact eigenvalues whereas Table 8•13 shows that the finite-difference values fall below the exact eigenvalues. Table 9•7 shows how the coefficients in the finite-element solution for node 1 approximate the coefficients in the exact solution.

Table 9•6 *Finite-element approximations to the eigenvalues $L^2\lambda_j/\alpha$ in (8·5·39).*

j	$N = 3$	6	15	45	Exact ($N = \infty$)
1	0	0	0	0	0
2	12	11.7	10.36	10.00	$9.87 = \pi^2$
3	36	31.2	22.66	20.49	$19.74 = 2\pi^2$
4		48	47.41	41.49	$39.48 = 4\pi^2$
5		84.3	70.08	54.46	$49.35 = 5\pi^2$
6		126.5	118.9	90.27	$78.96 = 8\pi^2$
7			124.1	99.18	$88.83 = 9\pi^2$
8			163.5	114.7	$98.70 = 10\pi^2$
9			192	166.8	$128.30 = 13\pi^2$
10			226.6	190.4	$157.91 = 16\pi^2$

Table 9•7 *Finite-element approximations x'_{1j} to the coefficients $a_j(0,0)$ in (8·5·39).*

j	$N = 3$	6	15	45	Exact ($N = \infty$)
1	0.3333	0.3333	0.3333	0.3334	$0.3333 = 1/3$
2	0.5000	0.5335	0.4480	0.4185	$0.4053 = 4/\pi^2$
3	0.1667	0.2883	0.2368	0.2146	$0.2026 = 2/\pi^2$
4		−0.1250	−0.1311	−0.1115	$-0.1013 = -1/\pi^2$
5		−0.0335	−0.0196	−0.0050	0
6		0.0034	0.0086	0.0395	$0.0507 = 1/2\pi^2$
7			0.0872	0.0578	$0.0450 = 4/9\pi^2$
8			0.0745	0.0258	0
9			−0.0312	0.0004	0
10			−0.0075	−0.0396	$-0.0253 = -1/4\pi^2$

9•5•2   Approximate solutions of $\mathbf{C}\dot{\mathbf{t}} + \mathbf{St} = \mathbf{r}$

The analysis of the Euler and the Crank-Nicolson solutions given in Section 8•5•2 also carries over to finite elements. The step-by-step solution is still given by (8·5·64) as

$$\mathbf{t}^{(v+1)} = \mathbf{a} + \mathbf{b}\theta^{(v+1)} + \mathbf{X}'\boldsymbol{\mu}^{v+1} \tag{9.5.7}$$

where the entries in $\boldsymbol{\mu}$ are given by (8·5·56) for Euler solutions as

$$\mu_i = 1 - \lambda_i \Delta\theta \tag{9.5.8}$$

and by (8·5·57) for Crank-Nicolson solutions as

$$\mu_i = \frac{2 - \lambda_i \Delta\theta}{2 + \lambda_i \Delta\theta} \tag{9.5.9}$$

and $\mathbf{a}$, $\mathbf{b}$, $\mathbf{X}'$ and $\boldsymbol{\lambda}$ are summarized in Table 8•12.

For the 6-node Euler example shown in Figure 9•33 the vector $\boldsymbol{\mu}$ is found by substituting the eigenvalues given in (9·5·2) into (9·5·8) with $\Delta\theta = 0.02L^2/\alpha$ to give

$$\boldsymbol{\mu} = \begin{bmatrix} 1.00 \\ 0.77 \\ 0.38 \\ 0.04 \\ -0.69 \\ -1.53 \end{bmatrix} \tag{9.5.10}$$

The last entry of -1.53 explains the unstable, oscillatory behavior shown in Figure 9•34. The first few steps are oscillatory but do not yet appear unstable. The reason for this can be seen by looking at the modified eigenvectors in (9·5·3). In row 1, corresponding to node 1, the dominant entries appear in columns 2 and 3 (102.4 and 55.4). These are associated with the entries in rows 2 and 3 (0.77 and 0.38) in (9·5·10). This produces the stable-looking behavior at early times shown in Figure 9•34. Eventually, as -1.53 in (9·5·10) keeps getting raised to larger and larger powers as shown in (9·5·7), unstable, oscillatory behavior will take over in spite of the relatively small entry of 0.6 in row 1, column 6 of (9·5·3). Further examination of (9·5·3) shows that the entries in rows 3 and 4 of column 6 will be relatively important compared to the other entries in these rows. Thus the unstable oscillations will appear earlier in time for nodes 3 and 4.

For the 6-node Crank-Nicolson solution shown in Figure 9•35 the vector $\boldsymbol{\mu}$ is found by substituting the eigenvalues given in (9·5·2) into (9·5·9) with $\Delta\theta = 0.02L^2/\alpha$ to give

$$\boldsymbol{\mu} = \begin{bmatrix} 1.00 \\ 0.79 \\ 0.52 \\ 0.35 \\ 0.08 \\ -0.12 \end{bmatrix} \tag{9.5.11}$$

The only negative entry has a magnitude less than 1 so the oscillations are stable. This is always the case for Crank-Nicolson solutions. In fact, since -0.12 is so close to 0, the oscillations that are present are small and damp out quite rapidly so they are not even noticeable in Figure 9•35.

For the 15-node Crank-Nicolson solution shown in Figure 9•36 the vector $\boldsymbol{\mu}$ is obtained by substituting the eigenvalues given in (9·5·5) into (9·5·9) with $\Delta\theta = 0.02L^2/\alpha$ to give

$$
\boldsymbol{\mu} = \begin{bmatrix} 1.00 \\ 0.81 \\ 0.63 \\ 0.36 \\ 0.18 \\ -0.09 \\ -0.11 \\ -0.24 \\ -0.32 \\ -0.39 \\ -0.46 \\ -0.51 \\ -0.57 \\ -0.61 \\ -0.64 \end{bmatrix}
\tag{9·5·12}
$$

Ten of these entries are negative and small oscillations are apparent in Figure 9•36 at early times. They damp out as time goes on however. It should also be mentioned that there is some cancellation of oscillatory terms because some of the entries in each row of $\mathbf{X}'$, (9·5·6), are positive whereas others are negative.

9·5·3 Critical-time-step estimate

The discussion in Section 8·5·3 holds for finite elements as well as finite differences. The critical time step is still given by (8·5·71) as

$$\Delta\theta_c = \frac{2}{\lambda_N} \tag{9·5·13}$$

where λ_N is the maximum eigenvalue of the problem given by (8·5·16) as

$$\mathbf{Sx} = \lambda\mathbf{Çx} \tag{9·5·14}$$

As was true for finite differences, the actual computation of λ_N for large problems is expensive and, for the purpose of determining the critical time step, engineers are usually satisfied if an upper bound on λ_N can be found.

In finite differences $\mathbf{Ç}$ was diagonal and it was easy to estimate λ_N as discussed in Section 8•5•4. In finite elements $\mathbf{Ç}$ is not diagonal and the method used in Section 8•5•4 does not yield a useful result. Because of these difficulties we must take another approach to estimate λ_N.

There is a theorem in matrix analysis that says:

Theorem 1

If $\mathbf{A}$ and $\mathbf{B}$ are both real, symmetric matrices and if $\alpha_1 \leq \cdots \leq \alpha_N$ are the eigenvalues of $\mathbf{Au} = \alpha\mathbf{u}$ and $\beta_1 \leq \cdots \leq \beta_N$ are the eigenvalues of $\mathbf{Bv} = \beta\mathbf{v}$, then the eigenvalues λ of $\mathbf{Ax} = \lambda\mathbf{Bx}$ are bounded by

$$\frac{\alpha_1 \text{ from } \mathbf{Au} = \alpha\mathbf{u}}{\beta_N \text{ from } \mathbf{Bv} = \beta\mathbf{v}} = \frac{\alpha_1}{\beta_N} \leq \lambda \leq \frac{\alpha_N}{\beta_1} = \frac{\alpha_N \text{ from } \mathbf{Au} = \alpha\mathbf{u}}{\beta_1 \text{ from } \mathbf{Bv} = \beta\mathbf{v}} \tag{9·5·15}$$

If this theorem is applied directly to (9·5·14) to find an upper bound on λ_N one obtains

$$\lambda_N \leq \frac{\alpha_N \text{ from } \mathbf{Su} = \alpha\mathbf{u}}{\beta_1 \text{ from } \mathbf{Çv} = \beta\mathbf{v}} \tag{9·5·16}$$

Since it is difficult, or perhaps impossible, to guarantee that β_1 is anything other than zero, (9·5·16) gives the useless result that $\lambda_N \leq \infty$.

Rather than work with (9·5·14) directly, we will define $\mathbf{D}$ to be a diagonal matrix containing the main diagonal entries of $\mathbf{Ç}$ as given by

$$\mathbf{D} = \begin{bmatrix} \varsigma_{11} & & & & & & \\ & \varsigma_{22} & & & & & \\ & & \ddots & & & & \\ & & & \varsigma_{ii} & & & \\ & & & & & \ddots & \\ & & & & & & \varsigma_{NN} \end{bmatrix}$$

Then $\mathbf{D}^{1/2}$ is a diagonal matrix whose entry in row i is $\sqrt{\varsigma_{ii}}$ and its inverse $\mathbf{D}^{-1/2}$ is also diagonal and has $1/\sqrt{\varsigma_{ii}}$ in row i. We will now premultiply (9·5·14) by $\mathbf{D}^{-1/2}$ and insert $\mathbf{I} = (\mathbf{D}^{-1/2}\mathbf{D}^{1/2})$ just to the left of $\mathbf{x}$ on both sides of (9·5·14) to obtain

$$\mathbf{D}^{-1/2}\mathbf{S}(\mathbf{D}^{-1/2}\mathbf{D}^{1/2})\mathbf{x} = \lambda\mathbf{D}^{-1/2}\mathbf{\mathsf{Ç}}(\mathbf{D}^{-1/2}\mathbf{D}^{1/2})\mathbf{x}$$

Upon rearranging and adding some parentheses we may write

$$(\mathbf{D}^{-1/2}\mathbf{S}\mathbf{D}^{-1/2})(\mathbf{D}^{1/2}\mathbf{x}) = \lambda(\mathbf{D}^{-1/2}\mathbf{\mathsf{Ç}}\mathbf{D}^{-1/2})(\mathbf{D}^{1/2}\mathbf{x}) \tag{9.5.17}$$

The product $(\mathbf{D}^{1/2}\mathbf{x})$ may be thought of as a new eigenvector. By carrying out a few multiplications, it can be readily shown that $(\mathbf{D}^{-1/2}\mathbf{S}\mathbf{D}^{-1/2})$ and $(\mathbf{D}^{-1/2}\mathbf{\mathsf{Ç}}\mathbf{D}^{-1/2})$ are both real, symmetric matrices. Therefore we can apply Theorem 1 to (9.5.17) to obtain

$$\lambda_N \leq \frac{\alpha_N \text{ from } (\mathbf{D}^{-1/2}\mathbf{S}\mathbf{D}^{-1/2})\mathbf{u} = \alpha\mathbf{u}}{\beta_1 \text{ from } (\mathbf{D}^{-1/2}\mathbf{\mathsf{Ç}}\mathbf{D}^{-1/2})\mathbf{v} = \beta\mathbf{v}} \tag{9.5.18}$$

We now have one eigenvalue problem in the numerator and another in the denominator of (9.5.18) to consider.

Premultiplying both of the eigenvalue problems in (9.5.18) by $\mathbf{D}^{-1/2}$ gives

$$\lambda_N \leq \frac{\alpha_N \text{ from } (\mathbf{D}^{-1}\mathbf{S}\mathbf{D}^{-1/2})\mathbf{u} = \alpha\mathbf{D}^{-1/2}\mathbf{u}}{\beta_1 \text{ from } (\mathbf{D}^{-1}\mathbf{\mathsf{Ç}}\mathbf{D}^{-1/2})\mathbf{v} = \beta\mathbf{D}^{-1/2}\mathbf{v}}$$

Upon rearranging and introducing parentheses,

$$\lambda_N \leq \frac{\alpha_N \text{ from } (\mathbf{D}^{-1}\mathbf{S})(\mathbf{D}^{-1/2}\mathbf{u}) = \alpha(\mathbf{D}^{-1/2}\mathbf{u})}{\beta_1 \text{ from } (\mathbf{D}^{-1}\mathbf{\mathsf{Ç}})(\mathbf{D}^{-1/2}\mathbf{v}) = \beta(\mathbf{D}^{-1/2}\mathbf{v})}$$

By defining new eigenvectors $\tilde{\mathbf{u}} = \mathbf{D}^{-1/2}\mathbf{u}$ and $\tilde{\mathbf{v}} = \mathbf{D}^{-1/2}\mathbf{v}$ we obtain

$$\lambda_N \leq \frac{\alpha_N \text{ from } (\mathbf{D}^{-1}\mathbf{S})\tilde{\mathbf{u}} = \alpha\tilde{\mathbf{u}}}{\beta_1 \text{ from } (\mathbf{D}^{-1}\mathbf{\mathsf{Ç}})\tilde{\mathbf{v}} = \beta\tilde{\mathbf{v}}}$$

Finally, we will premultiply both eigenvalue problems by $\mathbf{D}$ to give

$$\lambda_N \leq \frac{\alpha_N \text{ from } \mathbf{S}\tilde{\mathbf{u}} = \alpha\mathbf{D}\tilde{\mathbf{u}}}{\beta_1 \text{ from } \mathbf{\mathsf{Ç}}\tilde{\mathbf{v}} = \beta\mathbf{D}\tilde{\mathbf{v}}} \tag{9.5.19}$$

We want to find the smallest upper bound that we can for λ_N. This means we want to find the smallest upper bound for α_N in the numerator and the largest lower bound for β_1 in the denominator.

Since $\mathbf{D}$ is diagonal, an upper bound on α_N for the eigenvalue problem in the numerator of (9.5.19) may be obtained following the steps we used for finite differences in Section 8•5•4 to arrive at (8.5.82). Furthermore, since $\mathbf{D}$ contains the entries on the main diagonal of $\mathbf{\mathsf{Ç}}$ we may then write

$$\alpha_N \leq \underset{\substack{i=1 \\ i \neq i_s}}{\overset{N}{\text{Max}}} \left[\frac{1}{\varsigma_{ii}} \sum_{j=1}^{N} |s_{ij}| \right] \tag{9.5.20}$$

To obtain a lower bound on β_1 in (9.5.19) we must now consider the eigenvalue problem in the denominator which can be written in terms of β_1 and its corresponding eigenvector $\tilde{\mathbf{v}}_1$ as

$$\mathbf{\mathsf{Ç}}\tilde{\mathbf{v}}_1 = \beta_1\mathbf{D}\tilde{\mathbf{v}}_1$$

Upon subtracting $\kappa\mathbf{D}\tilde{\mathbf{v}}_1$ (where κ is a positive constant to be chosen later) from both sides we may write

$$(\mathbf{\mathsf{Ç}} - \kappa\mathbf{D})\tilde{\mathbf{v}}_1 = (\beta_1 - \kappa)\mathbf{D}\tilde{\mathbf{v}}_1 \tag{9.5.21}$$

$(\beta_1 - \kappa)$ is the smallest eigenvalue of this new eigenvalue problem (9.5.21). Since $(\mathbf{\mathsf{Ç}} - \kappa\mathbf{D})$ and $\mathbf{D}$ are both real, symmetric matrices,

Theorem 1 may be applied to (9·5·21) to give the following lower bound on $(\beta_1 - \kappa)$:

$$\frac{\gamma_1 \text{ from } (\mathbf{\c{C}} - \kappa\mathbf{D})\mathbf{w} = \gamma\mathbf{w}}{\mu_N \text{ from } \mathbf{D}\mathbf{z} = \mu\mathbf{z}} \leq (\beta_1 - \kappa)$$

where γ, $\mathbf{w}$, μ and $\mathbf{z}$ have been used to avoid confusion with α, $\mathbf{u}$, β and $\mathbf{v}$ that were used earlier. Upon solving for β_1,

$$\beta_1 \geq \kappa + \frac{\gamma_1 \text{ from } (\mathbf{\c{C}} - \kappa\mathbf{D})\mathbf{w} = \gamma\mathbf{w}}{\mu_N \text{ from } \mathbf{D}\mathbf{z} = \mu\mathbf{z}} \tag{9·5·22}$$

The right-hand side of (9·5·22) gives a lower bound on β_1 that depends upon κ, γ_1 and μ_N. We want to find the largest lower bound for β_1 that we can. Since we cannot afford to actually compute γ_1, we will settle for choosing as large a value of κ as we can while still guaranteeing that $\gamma_1 / \mu_N \geq 0$. Since $\mathbf{D}$ is diagonal, the eigenvalues μ in denominator of (9·5·22) are simply the entries in $\mathbf{D}$ and are always positive. Thus μ_N will be positive and we must only ensure that $\gamma_1 \geq 0$ with the choice of κ. Then $\beta_1 = \kappa$ will be a lower bound that we can use in (9·5·19).

The eigenvalue problem for γ is given by

$$(\mathbf{\c{C}} - \kappa\mathbf{D})\mathbf{w} = \gamma\mathbf{w}$$

We can learn about the eigenvalues of $(\mathbf{\c{C}} - \kappa\mathbf{D})$ by first looking at the eigenvalues of $(\mathbf{C} - \kappa\mathbf{D})$. Recall that $\mathbf{C}$ was originally constructed from the element capacitance matrices $\mathbf{C}^{(e)}$ as given by (9·2·7). That is,

$$\mathbf{C} = \sum_{e=1}^{NE} \mathbf{C}^{(e)}$$

where $\mathbf{C}^{(e)}$ is given by (9·2·10) as

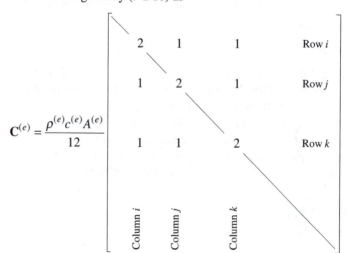

Since $\mathbf{D}$ is defined to be the diagonal of $\mathbf{C}$, we may write

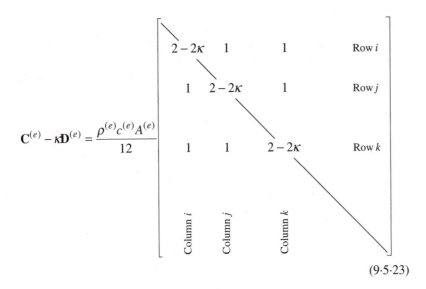

$$(9.5.23)$$

It then also follows that

$$(\mathbf{C} - \kappa\mathbf{D}) = \sum_{e=1}^{NE}(\mathbf{C}^{(e)} - \kappa\mathbf{D}^{(e)}) \qquad (9.5.24)$$

Since all of the $(\mathbf{C}^{(e)} - \kappa\mathbf{D}^{(e)})$ are real, symmetric matrices, we can use the following theorem to learn about the eigenvalues of the sum $(\mathbf{C} - \kappa\mathbf{D})$:

Theorem 2

If $\mathbf{A}$ and $\mathbf{B}$ are both real, symmetric matrices and if α_j is the jth eigenvalue of $\mathbf{A}\mathbf{u} = \alpha\mathbf{u}$ and if $\beta_1 \leq \cdots \leq \beta_N$ are the eigenvalues of $\mathbf{B}\mathbf{v} = \beta\mathbf{v}$, then the jth eigenvalue λ_j of $(\mathbf{A} + \mathbf{B})\mathbf{x} = \lambda\mathbf{x}$ is bounded by

$$\alpha_j + \beta_1 \leq \lambda_j \leq \alpha_j + \beta_N \qquad (9.5.25)$$

To apply Theorem 2 we will consider $\mathbf{A} = (\mathbf{C}^{(e)} - \kappa\mathbf{D}^{(e)})$. We then need to find the eigenvalues of

$$(\mathbf{C}^{(e)} - \kappa\mathbf{D}^{(e)})\mathbf{u} = \alpha\mathbf{u}$$

where $(\mathbf{C}^{(e)} - \kappa\mathbf{D}^{(e)})$ is given by (9.5.23). Of the N eigenvalues, $N-3$ of them are 0. Only the remaining three eigenvalues can be nonzero. As found by algebraic means, these remaining three eigenvalues are given by

$$\frac{\rho^{(e)}c^{(e)}A^{(e)}}{12}(1-2\kappa) \quad , \quad \frac{\rho^{(e)}c^{(e)}A^{(e)}}{12}(1-2\kappa) \quad \text{and} \quad \frac{\rho^{(e)}c^{(e)}A^{(e)}}{12}(4-2\kappa)$$

The largest value of κ for which none of these three eigenvalues is negative is $\kappa = \frac{1}{2}$. By choosing $\kappa = \frac{1}{2}$, all of the eigenvalues of $(\mathbf{C}^{(e)} - \kappa\mathbf{D}^{(e)})$ will be 0 except for one which will be $\frac{1}{4}\rho^{(e)}c^{(e)}A^{(e)}$. Application of Theorem 2 to (9.5.24) then guarantees that the minimum eigenvalue of the sum $(\mathbf{C} - \kappa\mathbf{D})$ will be ≥ 0.

The only situation when there is a difference between $(\mathbf{C} - \kappa\mathbf{D})$ and $(\mathbf{Ç} - \kappa\mathbf{D})$ is when there are specified temperatures. As discussed in Section 9•2•2, the off-diagonal entries in the rows and columns corresponding to the specified-temperature nodes in $\mathbf{C}$ are set to 0 to obtain $\mathbf{Ç}$. The same holds for $(\mathbf{C} - \kappa\mathbf{D})$ and $(\mathbf{Ç} - \kappa\mathbf{D})$. Another theorem may be used to see how these modifications of $(\mathbf{C} - \kappa\mathbf{D})$ to obtain $(\mathbf{Ç} - \kappa\mathbf{D})$ affect the eigenvalues.

> **Theorem 3**
>
> If $\mathbf{A}$ is a real, symmetric $N \times N$ matrix and $\alpha_1 \leq \cdots \leq \alpha_N$ are the eigenvalues of $\mathbf{Au} = \alpha\mathbf{u}$, and $\mathbf{A}'$ is the $(N-1) \times (N-1)$ matrix obtained by deleting any row and the corresponding column of $\mathbf{A}$, the eigenvalues $\alpha_1' \leq \cdots \leq \alpha_{N-1}'$ of $\mathbf{A}'\mathbf{u}' = \alpha'\mathbf{u}'$ separate the eigenvalues of $\mathbf{A}$. That is,
>
> $$\alpha_1 \leq \alpha_1' \leq \alpha_2 \leq \alpha_2' \leq \cdots \leq \alpha_i \leq \alpha_i' \leq \cdots \leq \alpha_{N-1}' \leq \alpha_N \qquad (9 \cdot 5 \cdot 26)$$

Suppose we have found $(\mathbf{C} - \kappa\mathbf{D})$ by adding the contributions from each element. The minimum eigenvalue of $(\mathbf{C} - \kappa\mathbf{D})$ is ≥ 0. If node i is a specified-temperature node we will then set the off-diagonal entries in row i and column i equal to 0. The eigenvalues of this modified matrix will be $c_{ii} - \kappa c_{ii}$, the main diagonal entry in row i, and the $N-1$ eigenvalues of the $(N-1) \times (N-1)$ matrix $(\mathbf{C} - \kappa\mathbf{D})'$ obtained by deleting row i and column i from $(\mathbf{C} - \kappa\mathbf{D})$. The eigenvalue $c_{ii} - \kappa c_{ii}$ will be > 0. By Theorem 3, the smallest eigenvalue of $(\mathbf{C} - \kappa\mathbf{D})'$ will be $\geq$ the smallest eigenvalue of $(\mathbf{C} - \kappa\mathbf{D})$ and thus it is also ≥ 0. Therefore the modifications to $\mathbf{C}$ for specified temperatures to obtain $\mathbf{Ç}$ can never result in $(\mathbf{Ç} - \kappa\mathbf{D})$ having a negative eigenvalue. The minimum eigenvalue of $(\mathbf{Ç} - \kappa\mathbf{D})$ will also be ≥ 0. Then from (9·5·22) we may now conclude that

$$\beta_1 \geq \frac{1}{2} \qquad (9 \cdot 5 \cdot 27)$$

Substitution of (9·5·20) and (9·5·27) into (9·5·19) gives

$$\lambda_N \leq \frac{\alpha_N}{\beta_1} \leq \frac{\displaystyle\operatorname*{Max}_{\substack{i=1 \\ i \neq i_s}}^{N} \left[\frac{1}{\varsigma_{ii}} \sum_{j=1}^{N} |s_{ij}| \right]}{\frac{1}{2}}$$

or

$$\lambda_{N,est} = 2 \operatorname*{Max}_{\substack{i=1 \\ i \neq i_s}}^{N} \left[\frac{1}{\varsigma_{ii}} \sum_{j=1}^{N} |s_{ij}| \right] \qquad (9 \cdot 5 \cdot 28)$$

as a bound on the maximum eigenvalue λ_N. Note that we have again excluded specified-temperature rows since these eigenvalues drop out of the solution as discussed in Section 8•5•4. If this analysis had been applied to estimate λ_N for finite differences where $\mathbf{Ç}$ is diagonal, β_1 would have been exactly equal 1 and (8·5·82) would have been obtained.

Application of (9·5·28) to the 6-node finite-element example (9·4·4) one finds that $\lambda_{N,est} = 192\alpha / L^2$ whereas the actual value given by (9·5·2) is $\lambda_N = 126.5\alpha / L^2$. Thus, for this finite-element example, $\lambda_{N,est} = 1.518\lambda_N$. In the comparable finite-difference example (8·4·5) we found from (8·5·35) that $\lambda_N = 32\alpha / L^2$ and from the example in Section 8•5•4 $\lambda_{N,est} = 32\alpha / L^2$. Thus, for this finite-difference example, $\lambda_{N,est} = \lambda_N$. Based on this example, the finite-element estimate (9·5·28) is more conservative than the finite-difference estimate (8·5·82).

Substituting (9·5·28) into (9·5·13) then gives

$$\Delta\theta_{c,est} = \frac{1}{\displaystyle\operatorname*{Max}_{\substack{i=1 \\ i \neq i_s}}^{N} \left[\frac{1}{\varsigma_{ii}} \sum_{j=1}^{N} |s_{ij}| \right]} \qquad (9 \cdot 5 \cdot 29)$$

as an estimate of the critical time step for triangular finite elements. This is the result given as (9·4·8) in Section 9•4•2.

9•6 SUPPLEMENTARY PROBLEMS

The purpose of this section is to show you how the theory given in the preceding sections may be modified and/or extended to other situations of interest to the heat-transfer engineer.

9•6•1 Lumped-capacitance formulation

The capacitance matrix $\mathbf{C}$ was constructed in Section 9•2•1 by summing contributions from each of the element capacitance matrices $\mathbf{C}^{(e)}$ given by (9•2•10). The sum of the entries in the element capacitance matrix (9•2•10) is the total capacitance of element e, $\rho^{(e)}c^{(e)}A^{(e)}$. This capacitance was then distributed into nine different locations in $\mathbf{C}^{(e)}$. As another approximation, rather than distributing the thermal capacitance in this way, one-third of the capacitance could be assigned to each diagonal entry and keep the off-diagonal entries equal to 0. Rather than using (9•2•10), the capacitance could be *lumped* along the main diagonal by writing

$$\mathbf{C}^{(e)} = \frac{\rho^{(e)}c^{(e)}A^{(e)}}{3} \begin{bmatrix} 1 & & & & \text{Row } i \\ & 1 & & & \text{Row } j \\ & & 1 & & \text{Row } k \\ & & & \ddots & \end{bmatrix}$$

$\qquad\qquad$ Column i $\quad$ Column j $\quad$ Column k

This procedure would produce a diagonal $\mathbf{C}$ rather than a banded $\mathbf{C}$ that was found in Section 9•2•1. This would result in several advantages:

1. Computer storage requirements could be less since only the diagonal terms in $\mathbf{C}$ must be stored.

2. It will not be necessary to set the off-diagonal rows and columns equal to 0 in $\mathbf{C}$ to handle specified-temperature nodes because the off-diagonal terms are already 0. Thus, $\mathbf{\zeta} = \mathbf{C}$.

3. Computations for heat flows are less complicated when $\mathbf{C} = \mathbf{\zeta}$ since (9•2•41) gives the product $\mathbf{\zeta\dot{t}}$ which can then be substituted into (9•2•40) to find $\mathbf{q}_o$. The intermediate steps of first solving (9•2•41) for $\dot{\mathbf{t}}$ and then multiplying $\mathbf{C\dot{t}}$ in (9•2•40) are not required.

4. The transient Euler solution will be explicit rather than an implicit. This avoid having to solve a system of algebraic equations at each time step.

5. When $\mathbf{\zeta}$ is diagonal (lumped), (8•4•29) can be used to find $\Delta\theta_{c,est}$ rather than (9•4•8) for the distributed analysis. Because of the 2 in the numerator of (8•4•29) and also since c_{ii} will be twice as large if all of the capacitance is on the diagonal rather than only one-half as in the distributed case, $\Delta\theta_{c,est}$ for the lumped formulation will be four times $\Delta\theta_{c,est}$ for the distributed formulation. The critical time step will be larger for the lumped formulation than for the distributed formulation.

The lumped-capacitance formulation would be quite similar to the way the capacitance is handled in finite differences. Although the improvement in $\Delta\theta_{c,est}$ by a factor of four is nice, Crank-Nicolson is unconditionally stable and useful solutions can be found even when $\Delta\theta_{c,est}$ is exceeded.

9•6•2 Cylindrical coordinates

Finite-element analysis of two-dimensional problems that can be modeled using cylindrical (r, z) coordinates may be derived following steps similar to those shown in sections 9•1 and 9•2 for Cartesian (x, y) coordinates. The governing partial differential equation in place of (9·1·5) is given by

$$\rho c \frac{\partial t}{\partial \theta} + \left[\frac{1}{r} \frac{\partial}{\partial r} (-krt_r) + \frac{\partial}{\partial z} (-kt_z) \right] - g''' = 0 \tag{9·6·1}$$

The two-dimensional region of interest having area A and boundary B is shown in Figure 9•38. The solid is formed by rotating area A about the z axis to form a toroidal volume V.

Following the Galerkin weighted-residual method of Section 9•1•2, (9·6·1) can be multiplied by $f(r, z) 2\pi r \, dr \, dz$ and integrated over area A. The conduction terms can then be integrated by parts. The space and time variables can be separated by assuming $t(r, z, \theta) = \mathbf{w}^T (r, z) \mathbf{t}(\theta)$. This leads to a system of ordinary differential equations given by

$$\mathbf{C\dot{t}} + \mathbf{Kt} - \mathbf{g} = \mathbf{q}_o \tag{9·6·2}$$

where

$$\mathbf{C} = 2\pi \iint_A \mathbf{w} \rho c \mathbf{w}^T r \, dr \, dz \tag{9·6·3}$$

$$\mathbf{K} = 2\pi \iint_A \left(\mathbf{w}_r k \mathbf{w}_r^T + \mathbf{w}_z k \mathbf{w}_z^T \right) r \, dr \, dz \tag{9·6·4}$$

$$\mathbf{g} = 2\pi \iint_A \mathbf{w} g''' r \, dr \, dz \tag{9·6·5}$$

$$\mathbf{q}_o = 2\pi \int_B \mathbf{w} q_o'' r \, ds \tag{9·6·6}$$

Equations (9·6·2) through (9·6·6) replace (9·1·13) through (9·1·17) for Cartesian coordinates.

Area A can then be discretized into triangles of area $A^{(e)}$ which give toroidal elements when rotated about the z axis. The temperature will be assumed to be a linear function of r and z within each triangle. Following the analysis for interior elements given in Section 9•2•1, the global capacitance matrix $\mathbf{C}$ is constructed by summing over each element capacitance matrix $\mathbf{C}^{(e)}$ where

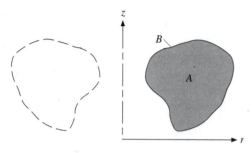

Figure 9•38 *Two-dimensional region in (r, z) coordinates.*

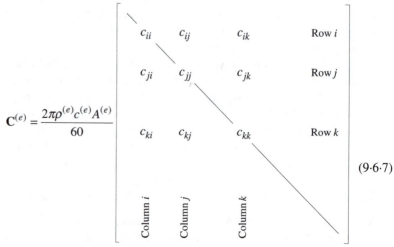

$$\mathbf{C}^{(e)} = \frac{2\pi \rho^{(e)} c^{(e)} A^{(e)}}{60}
\begin{bmatrix}
c_{ii} & c_{ij} & c_{ik} & \text{Row } i \\
c_{ji} & c_{jj} & c_{jk} & \text{Row } j \\
c_{ki} & c_{kj} & c_{kk} & \text{Row } k \\
& & & \\
\text{Column } i & \text{Column } j & \text{Column } k &
\end{bmatrix} \tag{9·6·7}$$

where $\mathbf{C}^{(e)}$ is symmetric and

$$c_{ii} = 6r_i + 2r_j + 2r_k \qquad c_{ij} = 2r_i + 2r_j + r_k \qquad c_{ik} = 2r_i + r_j + 2r_k$$

$$c_{jj} = 2r_i + 6r_j + 2r_k \qquad c_{jk} = r_i + 2r_j + 2r_k$$

$$c_{kk} = 2r_i + 2r_j + 6r_k$$

The sum of the nine entries in $\mathbf{C}^{(e)}$ is given by

$$\sum_{i=1}^{N}\sum_{j=1}^{N} c_{ij}^{(e)} = \frac{2\pi\rho^{(e)}c^{(e)}A^{(e)}}{60}(20r_i + 20r_j + 20r_k)^{(e)}$$

Rearranging,

$$\sum_{i=1}^{N}\sum_{j=1}^{N} c_{ij}^{(e)} = \rho^{(e)}c^{(e)}2\pi\frac{(r_i + r_j + r_k)^{(e)}}{3}A^{(e)} \qquad (9\cdot6\cdot8)$$

The centroid $r_c^{(e)}$ of area $A^{(e)}$ about the z axis is given by

$$r_c^{(e)} = \frac{(r_i + r_j + r_k)^{(e)}}{3} \qquad (9\cdot6\cdot9)$$

The volume of the toroidal element is given by

$$V^{(e)} = 2\pi r_c^{(e)}A^{(e)} \qquad (9\cdot6\cdot10)$$

Therefore substituting (9·6·9) and (9·6·10) into (9·6·8) gives

$$\sum_{i=1}^{N}\sum_{j=1}^{N} c_{ij}^{(e)} = \rho^{(e)}c^{(e)}V^{(e)}$$

The sum of the entries in $\mathbf{C}^{(e)}$ is the capacitance of the toroidal element.

The global conduction matrix $\mathbf{K}$ is constructed by summing over each element conduction matrix $\mathbf{K}^{(e)}$ where

$$\mathbf{K}^{(e)} = 2\pi r_c^{(e)}\frac{k^{(e)}}{4A^{(e)}}\begin{bmatrix} \kappa_{ii} & \kappa_{ij} & \kappa_{ik} & \text{Row } i \\ \kappa_{ji} & \kappa_{jj} & \kappa_{jk} & \text{Row } j \\ \kappa_{ki} & \kappa_{kj} & \kappa_{kk} & \text{Row } k \\ \text{Column } i & \text{Column } j & \text{Column } k \end{bmatrix} \qquad (9\cdot6\cdot11)$$

where $r_c^{(e)}$ is given by (9·6·9), $\mathbf{K}^{(e)}$ is symmetric and

$$\kappa_{ii} = r_{jk}r_{jk} + z_{jk}z_{jk} \qquad \kappa_{ij} = -(r_{jk}r_{ik} + z_{jk}z_{ik}) \qquad \kappa_{ik} = r_{jk}r_{ij} + z_{jk}z_{ij}$$

$$\kappa_{jj} = r_{ik}r_{ik} + z_{ik}z_{ik} \qquad \kappa_{jk} = -(r_{ik}r_{ij} + z_{ik}z_{ij})$$

$$\kappa_{kk} = r_{ij}r_{ij} + z_{ij}z_{ij}$$

It turns out that the entries in $\mathbf{K}^{(e)}$ for r,z problems given in (9·6·11) are $2\pi r_c^{(e)}$ times the entries in $\mathbf{K}^{(e)}$ for x,y problems given in (9·2·12) if x and y are replaced by r and z, respectively.

The global generation vector **g** is constructed by summing over each element generation vector $\mathbf{g}^{(e)}$ where

$$\mathbf{g}^{(e)} = \frac{2\pi g'''^{(e)} A^{(e)}}{12} \begin{bmatrix} g_i \\ \\ g_j \\ \\ g_k \end{bmatrix} \begin{matrix} \text{Row } i \\ \\ \text{Row } j \\ \\ \text{Row } k \end{matrix} \tag{9.6.12}$$

where

$$g_i = 2r_i + r_j + r_k \qquad g_j = r_i + 2r_j + r_k \qquad g_k = r_i + r_j + 2r_k$$

The sum of the three entries in $\mathbf{g}^{(e)}$ is given by

$$\sum_{i=1}^{N} g_i^{(e)} = \frac{2\pi g'''^{(e)} A^{(e)}}{12} (4r_i + 4r_j + 4r_k)^{(e)} = g'''^{(e)} 2\pi \frac{(r_i + r_j + r_k)^{(e)}}{3} A^{(e)}$$

Substituting $r_c^{(e)}$ from (9.6.9) and $V^{(e)}$ from (9.6.10),

$$\sum_{i=1}^{N} g_i^{(e)} = g'''^{(e)} V^{(e)}$$

The sum of the entries in $\mathbf{g}^{(e)}$ is the energy generation within the toroidal element.

Following the analysis for boundary segments given in Section 9•2•2, the total heat-inflow vector $\mathbf{q}_o$ can be expressed in terms of a specified-heat-flow vector $\mathbf{q}$, a convection vector $\mathbf{h}$ and a convection matrix $\mathbf{H}$ as

$$\mathbf{q}_o = \mathbf{q} + \mathbf{h} - \mathbf{Ht} \tag{9.6.13}$$

The global specified-heat-flow vector $\mathbf{q}$ is constructed by summing over each element specified-heat-flow vector $\mathbf{q}^{(b_q)}$ where

$$\mathbf{q}^{(b_q)} = \frac{2\pi q_s''^{(b_q)} s_{ij}^{(b_q)}}{6} \begin{bmatrix} q_i \\ \\ q_j \end{bmatrix} \begin{matrix} \text{Row } i \\ \\ \text{Row } j \end{matrix} \tag{9.6.14}$$

where

$$q_i = 2r_i + r_j \qquad\qquad q_j = r_i + 2r_j$$

The sum of the entries in $\mathbf{q}^{(b_q)}$ is given by

$$\sum_{i=1}^{N} q_i^{(b_q)} = \frac{2\pi q_s''^{(b_q)} s_{ij}^{(b_q)}}{6}(3r_i + 3r_j)^{(b_q)} = q_s''^{(b_q)} 2\pi \frac{(r_i + r_j)^{(b_q)}}{2} s_{ij}^{(b_q)} \qquad (9\cdot6\cdot15)$$

When boundary segment $s_{ij}^{(b_q)}$ is rotated around the z axis the surface area $S_{ij}^{(b_q)}$ swept out is given by

$$S_{ij}^{(b_q)} = 2\pi \frac{(r_i + r_j)^{(b_q)}}{2} s_{ij}^{(b_q)} \qquad (9\cdot6\cdot16)$$

Substituting (9·6·16) into (9·6·15) shows that the sum of the entries in $\mathbf{q}^{(b_q)}$ is the total specified heat inflow to the surface $S_{ij}^{(b_q)}$.

The global convection vector $\mathbf{h}$ is constructed by summing over each element convection vector $\mathbf{h}^{(b_h)}$ where

$$\mathbf{h}^{(b_h)} = \frac{2\pi h^{(b_h)} t_\infty^{(b_h)} s_{ij}^{(b_h)}}{6} \begin{bmatrix} h_i & \text{Row } i \\[2em] h_j & \text{Row } j \\[6em] \end{bmatrix} \qquad (9\cdot6\cdot17)$$

where

$$h_i = 2r_i + r_j \qquad\qquad h_j = r_i + 2r_j$$

The global convection matrix $\mathbf{H}$ is constructed by summing over each element convection matrix $\mathbf{H}^{(e)}$ where

$$\mathbf{H}^{(b_h)} = \frac{2\pi h^{(b_h)} s_{ij}^{(b_h)}}{12} \begin{bmatrix} h_{ii} & h_{ij} & & & \text{Row } i \\ h_{ji} & h_{jj} & & & \text{Row } j \\ & & & & \\ & & & & \\ \end{bmatrix} \qquad (9\cdot6\cdot18)$$

$$\underset{\text{Column } i}{} \qquad \underset{\text{Column } j}{}$$

where $\mathbf{H}^{(b_h)}$ is symmetric and

$$h_{ii} = 3r_i + r_j \qquad\qquad h_{ij} = r_i + r_j \qquad\qquad h_{jj} = r_i + 3r_j$$

The sum of the entries in $\mathbf{h}^{(b_h)} - \mathbf{H}^{(b_h)}\mathbf{t}$ is given by

$$\sum_{i=1}^{N}\left(h_i^{(b_h)} - \sum_{j=1}^{N} h_{ij}^{(b_h)} t_j\right) = \frac{2\pi h^{(b_h)} t_\infty^{(b_h)} s_{ij}^{(b_h)}}{6}(2r_i + r_j)^{(b_h)}$$

$$- \frac{2\pi h^{(b_h)} s_{ij}^{(b_h)}}{12}\left\{(3r_i + r_j)^{(b_h)} t_i + (r_i + r_j)^{(b_h)} t_j\right\}$$

$$+ \frac{2\pi h^{(b_h)} t_\infty^{(b_h)} s_{ij}^{(b_h)}}{6}(r_i + 2r_j)^{(b_h)}$$

$$- \frac{2\pi h^{(b_h)} s_{ij}^{(b_h)}}{12}\left\{(r_i + r_j)^{(b_h)} t_i + (r_i + 3r_j)^{(b_h)} t_j\right\}$$

Combining and rearranging the first and third terms on the right-hand side (and also the second and fourth terms),

$$\sum_{i=1}^{N}\left(h_i^{(b_h)} - \sum_{j=1}^{N} h_{ij}^{(b_h)} t_j\right) = h^{(b_h)} 2\pi \frac{(r_i + r_j)^{(b_h)}}{2} s_{ij}^{(b_h)} t_\infty^{(b_h)}$$

$$- h^{(b_h)} 2\pi s_{ij}^{(b_h)} \frac{(4r_i + 2r_j)^{(b_h)} t_i + (2r_i + 4r_j)^{(b_h)} t_j}{12}$$

$$(9\cdot6\cdot19)$$

From (9·6·15) we see that the first term on the right-hand side of (9·6·19) contains the surface area $S_{ij}^{(b_h)}$. Then, multiplying and dividing the second term on the right-hand side by $(r_i + r_j)^{(b_h)}/2$ and rearranging,

$$\sum_{i=1}^{N}\left(h_i^{(b_h)} - \sum_{j=1}^{N} h_{ij}^{(b_h)} t_j\right) = h^{(b_h)} 2\pi \frac{(r_i + r_j)^{(b_h)}}{2} s_{ij}^{(b_h)} t_\infty^{(b_h)}$$

$$- h^{(b_h)} 2\pi \frac{(r_i + r_j)^{(b_h)}}{2} s_{ij}^{(b_h)} \frac{(4r_i + 2r_j)^{(b_h)} t_i + (2r_i + 4r_j)^{(b_h)} t_j}{6(r_i + r_j)^{(b_h)}}$$

Introducing $S_{ij}^{(b_h)}$, factoring $h^{(b_h)} S_{ij}^{(b_h)}$ and simplifying,

$$\sum_{i=1}^{N}\left(h_i^{(b_h)} - \sum_{j=1}^{N} h_{ij}^{(b_h)} t_j\right) = h^{(b_h)} S_{ij}^{(b_h)}\left[t_\infty^{(b_h)} - \frac{(2r_i + r_j)^{(b_h)} t_i + (r_i + 2r_j)^{(b_h)} t_j}{(3r_i + 3r_j)^{(b_h)}}\right]$$

The fraction within the brackets is an average surface temperature and the right-hand side is the total convection into surface area $S_{ij}^{(b_h)}$.

Substituting (9·6·13) into (9·6·2) and rearranging,

$$\mathbf{C\dot{t} + Kt + Ht = g + h + q} \qquad\qquad (9\cdot6\cdot20)$$

After modifications for constant specified temperatures following the discussion in Section 9•2•2, (9·6·20) may be written as

$$\mathbf{\c{C}\dot{t} + St = r} \qquad\qquad (9\cdot6\cdot21)$$

This result is identical to (9·2·20) obtained for Cartesian coordinates. The entries in $\mathbf{\c{C}}$, $\mathbf{S}$ and $\mathbf{r}$ will be different of course. The solutions (Euler, Crank-Nicolson, exact) for $\mathbf{t}$ will be carried out as before. Once $\mathbf{t}$ has been found, $\mathbf{\dot{t}}$ can be calculated from (9·6·21) and the energy flows into the solid from outside the region contained in $\mathbf{q}_o$ can be computed from (9·6·2). If there are no specified temperatures, $\mathbf{q}_o$ can be computed from (9·6·13) without having to first calculate $\mathbf{\dot{t}}$.

9•6•3   Three-dimensional problems

For three-dimensional problems, the Galerkin weighted-residual method discussed in Section 9•1•2 still gives the same differential equation (9·1·13) but the area integrals in (9·1·14), (9·1·15) and (9·1·16) become volume integrals and the boundary integral in (9·1·17) becomes a surface integral.

A natural extension of the two-dimensional, triangular-element analysis is a three-dimensional analysis using the tetrahedral element shown in Figure 9•39. Instead of using the area coordinates discussed in Section 9•1•3 to locate points within the element we will now use volume coordinates. Instead of (9·1·21), volume coordinates ξ_i, ξ_j, ξ_k and ξ_ℓ are related to the Cartesian coordinates x, y and z by

$$\begin{bmatrix} \xi_i \\ \xi_j \\ \xi_k \\ \xi_\ell \end{bmatrix} = \frac{1}{b_{ijk\ell}} \begin{bmatrix} (x_k y_\ell - x_\ell y_k)z_j - (x_j y_\ell - x_\ell y_j)z_k + (x_j y_k - x_k y_j)z_\ell & -(y_{jk}z_{k\ell} - y_{k\ell}z_{jk}) & (x_{jk}z_{k\ell} - x_{k\ell}z_{jk}) & -(x_{jk}y_{k\ell} - x_{k\ell}y_{jk}) \\ -(x_k y_\ell - x_\ell y_k)z_i + (x_i y_\ell - x_\ell y_i)z_k - (x_i y_k - x_k y_i)z_\ell & (y_{ik}z_{k\ell} - y_{k\ell}z_{ik}) & -(x_{ik}z_{k\ell} - x_{k\ell}z_{ik}) & (x_{ik}y_{k\ell} - x_{k\ell}y_{ik}) \\ (x_j y_\ell - x_\ell y_j)z_i - (x_i y_\ell - x_\ell y_i)z_j + (x_i y_j - x_j y_i)z_\ell & -(y_{ij}z_{j\ell} - y_{j\ell}z_{ij}) & (x_{ij}z_{j\ell} - x_{j\ell}z_{ij}) & -(x_{ij}y_{j\ell} - x_{j\ell}y_{ij}) \\ -(x_j y_k - x_k y_j)z_i + (x_i y_k - x_k y_i)z_j - (x_i y_j - x_j y_i)z_k & (y_{ij}z_{jk} - y_{jk}z_{ij}) & -(x_{ij}z_{jk} - x_{jk}z_{ij}) & (x_{ij}y_{jk} - x_{jk}y_{ij}) \end{bmatrix} \begin{bmatrix} 1 \\ x \\ y \\ z \end{bmatrix}$$

$$(9·6·22)$$

Since the matrix entries are more involved we will write (9·6·22) as

$$\begin{bmatrix} \xi_i \\ \xi_j \\ \xi_k \\ \xi_\ell \end{bmatrix} = \frac{1}{b_{ijk\ell}} \begin{bmatrix} e_{11} & e_{12} & e_{13} & e_{14} \\ e_{21} & e_{22} & e_{23} & e_{24} \\ e_{31} & e_{32} & e_{33} & e_{34} \\ e_{41} & e_{42} & e_{43} & e_{44} \end{bmatrix} \begin{bmatrix} 1 \\ x \\ y \\ z \end{bmatrix} \qquad (9·6·23)$$

The entries in the 4×4 matrix in (9·6·23) may be obtained by comparing it to the corresponding matrix in (9·6·22). The coefficient $b_{ijk\ell}$ is given by

$$b_{ijk\ell} = (x_{ik}y_{k\ell} - x_{k\ell}y_{ik})z_{ij} - (x_{ij}y_{j\ell} - x_{j\ell}y_{ij})z_{ik} + (x_{ij}y_{jk} - x_{jk}y_{ij})z_{i\ell} \quad (9·6·24)$$

It can be shown that $\left| b_{ijk\ell} \right| = 6V$ where V is the volume of the tetrahedron.

We will need to evaluate integrals of powers of ξ_i, ξ_j, ξ_k and ξ_ℓ over the tetrahedral volume. These integrals may be evaluated from:

$$\int_V \xi_i^a \xi_j^b \xi_k^c \xi_\ell^d \, dV = \frac{a!\,b!\,c!\,d!}{(a+b+c+d+3)!} 6V \qquad (9·6·25)$$

This relation is similar to (9·1·23).

If the temperature within a tetrahedral element is assumed to vary linearly between the four nodal-point temperatures, the interpolation vector $\mathbf{w}^T$ given by (9·2·1) becomes

$$\mathbf{w}^T = \begin{bmatrix} \xi_i & \xi_j & \xi_k & \xi_\ell \end{bmatrix} \qquad (9·6·26)$$

Instead of (9·2·2), the temperature at any point within the element is given by

$$t^{(e)} = \begin{bmatrix} t_i & t_j & t_k & t_\ell \end{bmatrix} \frac{1}{b_{ijk\ell}} \begin{bmatrix} e_{11} & e_{12} & e_{13} & e_{14} \\ e_{21} & e_{22} & e_{23} & e_{24} \\ e_{31} & e_{32} & e_{33} & e_{34} \\ e_{41} & e_{42} & e_{43} & e_{44} \end{bmatrix} \begin{bmatrix} 1 \\ x \\ y \\ z \end{bmatrix} \qquad (9·6·27)$$

Multiplying out the matrices in (9·6·27) we see that

$$t^{(e)} = c_1 + c_2 x + c_3 y + c_4 z$$

The temperature varies linearly between the four nodal temperatures at the corners of the tetrahedral element.

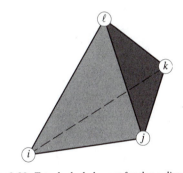

Figure 9•39 *Tetrahedral element for three-dimensional finite-element analysis.*

Instead of (9·2·3), temperature gradients with an element are given by

$$\begin{bmatrix} t_x \\ t_y \\ t_z \end{bmatrix}^{(e)} = \frac{1}{b_{ijk\ell}^{(e)}} \begin{bmatrix} e_{12} & e_{22} & e_{32} & e_{42} \\ e_{13} & e_{23} & e_{33} & e_{43} \\ e_{14} & e_{24} & e_{34} & e_{44} \end{bmatrix}^{(e)} \begin{bmatrix} t_i \\ t_j \\ t_k \\ t_\ell \end{bmatrix}^{(e)} \tag{9·6·28}$$

Instead of (9·2·4), the element capacitance matrix is given by

$$\mathbf{C}^{(e)} = \rho^{(e)} c^{(e)} \iiint\limits_{V^{(e)}} \mathbf{w} \mathbf{w}^T \, dx \, dy \, dz \tag{9·6·29}$$

Substituting (9·6·26) and using (9·6·25) to carry out the volume integrals in (9·6·29) gives

$$\mathbf{C}^{(e)} = \frac{\rho^{(e)} c^{(e)} V^{(e)}}{20} \begin{bmatrix} 2 & 1 & 1 & 1 & \text{Row } i \\ 1 & 2 & 1 & 1 & \text{Row } j \\ 1 & 1 & 2 & 1 & \text{Row } k \\ 1 & 1 & 1 & 2 & \text{Row } \ell \\ \text{Column } i & \text{Column } j & \text{Column } k & \text{Column } \ell & \end{bmatrix} \tag{9·6·30}$$

The three-dimensional element capacitance matrix (9·6·30) looks remarkably like the two-dimensional version (9·2·10). The sum of all entries is the total capacitance of the element $\rho^{(e)} c^{(e)} V^{(e)}$. The diagonal terms contain $\frac{8}{20}$ of the total capacitance.

The element conduction matrix is evaluated from the three-dimensional version of (9·2·10). That is, we must evaluate

$$\mathbf{K}^{(e)} = k^{(e)} \iiint\limits_{V^{(e)}} \left(\mathbf{w}_x \mathbf{w}_x^T + \mathbf{w}_y \mathbf{w}_y^T + \mathbf{w}_z \mathbf{w}_z^T \right) dx \, dy \, dz \tag{9·6·31}$$

The partial derivatives of $\mathbf{w}^T$ are given by

$$\mathbf{w}_x^T = \frac{1}{b_{ijk\ell}} \begin{bmatrix} e_{12} & e_{22} & e_{32} & e_{42} \end{bmatrix}$$

$$\mathbf{w}_y^T = \frac{1}{b_{ijk\ell}} \begin{bmatrix} e_{13} & e_{23} & e_{33} & e_{43} \end{bmatrix}$$

$$\mathbf{w}_z^T = \frac{1}{b_{ijk\ell}} \begin{bmatrix} e_{14} & e_{24} & e_{34} & e_{44} \end{bmatrix}$$

The products $\mathbf{w}_x \mathbf{w}_x^T$, $\mathbf{w}_y \mathbf{w}_y^T$ and $\mathbf{w}_z \mathbf{w}_z^T$ are $N \times N$ matrices that are zero except for the 16 entries at the intersections of rows and columns i, j, k and ℓ. These matrices are constants and may be moved outside of the volume integral in (9·6·31). The remaining integral is simply the volume of the element. When these steps are carried out the element conduction matrix is found to be given by

$$\mathbf{K}^{(e)} = \frac{k^{(e)}}{36V^{(e)}} \begin{bmatrix} \kappa_{ii} & \kappa_{ij} & \kappa_{ik} & \kappa_{i\ell} & \text{Row } i \\ \kappa_{ji} & \kappa_{jj} & \kappa_{jk} & \kappa_{j\ell} & \text{Row } j \\ \kappa_{ki} & \kappa_{kj} & \kappa_{kk} & \kappa_{k\ell} & \text{Row } k \\ \kappa_{\ell i} & \kappa_{\ell j} & \kappa_{\ell k} & \kappa_{\ell\ell} & \text{Row } \ell \\ \text{Column } i & \text{Column } j & \text{Column } k & \text{Column } \ell \end{bmatrix} \qquad (9 \cdot 6 \cdot 32)$$

where

$$\kappa_{ii} = e_{12}e_{12} + e_{13}e_{13} + e_{14}e_{14} \qquad \kappa_{ij} = e_{12}e_{22} + e_{13}e_{23} + e_{14}e_{24} \qquad \kappa_{ik} = e_{12}e_{32} + e_{13}e_{33} + e_{14}e_{34} \qquad \kappa_{i\ell} = e_{12}e_{42} + e_{13}e_{43} + e_{14}e_{44}$$

$$\kappa_{jj} = e_{22}e_{22} + e_{23}e_{23} + e_{24}e_{24} \qquad \kappa_{jk} = e_{22}e_{32} + e_{23}e_{33} + e_{24}e_{34} \qquad \kappa_{j\ell} = e_{22}e_{42} + e_{23}e_{43} + e_{24}e_{44}$$

$$\kappa_{kk} = e_{32}e_{32} + e_{33}e_{33} + e_{34}e_{34} \qquad \kappa_{k\ell} = e_{32}e_{42} + e_{33}e_{43} + e_{34}e_{44}$$

$$\kappa_{\ell\ell} = e_{42}e_{42} + e_{43}e_{43} + e_{44}e_{44}$$

The element conduction matrix (9·6·32) is again an $N \times N$ symmetric matrix. There are 16 entries that must be calculated instead of nine for the two-dimensional case and each entry requires more computation. However, the global conduction matrix can be constructed in the same way by adding the contributions from each of the element matrices.

The element generation vector comes from the three-dimensional version of (9·2·13) given by

$$\mathbf{g}^{(e)} = g'''^{(e)} \iiint_{V^{(e)}} \mathbf{w} \, dx \, dy \, dz \qquad (9 \cdot 6 \cdot 33)$$

Carrying out this integration with the help of (9·6·25) and (9·6·26) gives

$$\mathbf{g}^{(e)} = \frac{g'''^{(e)}V^{(e)}}{4} \begin{bmatrix} 1 & \text{Row } i \\ 1 & \text{Row } j \\ 1 & \text{Row } k \\ 1 & \text{Row } \ell \end{bmatrix} \qquad (9 \cdot 6 \cdot 34)$$

One-quarter of the total generation in the element is associated with each of the four nodes of the element.

The integrals over the boundary surfaces that account for the specified-heat-flux and convection boundary conditions are evaluated following the discussion in Section 9•2•2. The surface-element heat-flow vector $\mathbf{q}^{(b_q)}$, convection vector $\mathbf{h}^{(b_h)}$ and convection matrix $\mathbf{H}^{(b_h)}$ are similar to the two-dimensional quantities given in (9·2·15). For three-dimensional problems (9·2·15) becomes

$$\mathbf{q}^{(b_q)} = \int_{A^{(b_q)}} \mathbf{w} q_s'' \, dA \qquad \mathbf{h}^{(b_h)} = \int_{A^{(b_h)}} \mathbf{w} h t_\infty \, dA \qquad \mathbf{H}^{(b_h)} = \int_{A^{(b_h)}} \mathbf{w} h \mathbf{w}^T \, dA \qquad (9\cdot6\cdot35)$$

When these operations are carried out one obtains for specified heat flux,

$$\mathbf{q}^{(b_q)} = \frac{q_s''^{(b_q)} A_{ijk}^{(b_q)}}{3} \begin{bmatrix} 1 & \text{Row } i \\ \\ 1 & \text{Row } j \\ \\ 1 & \text{Row } k \end{bmatrix} \qquad (9\cdot6\cdot36)$$

and for convection,

$$\mathbf{h}^{(b_h)} = \frac{h^{(b_h)} t_\infty^{(b_h)} A_{ijk}^{(b_h)}}{3} \begin{bmatrix} 1 & \text{Row } i \\ 1 & \text{Row } j \\ 1 & \text{Row } k \end{bmatrix} \qquad \mathbf{H}^{(b_h)} = \frac{h^{(b_h)} A_{ijk}^{(b_h)}}{12} \begin{bmatrix} 2 & 1 & 1 & \text{Row } i \\ 1 & 2 & 1 & \text{Row } j \\ 1 & 1 & 2 & \text{Row } k \end{bmatrix} \qquad (9\cdot6\cdot37)$$

$$\text{Column } i \quad \text{Column } j \quad \text{Column } k$$

These results take the places of (9·2·16), (9·2·17) and (9·2·18).

It should be noted that, in general, a three-dimensional problem will require many more nodes than a two-dimensional problem. A two-dimensional problem with 10 nodes in the x-direction and 10 nodes in the y-direction for a total of 100 nodes would expand to also have 10 nodes in the z-direction for a total of 1000 nodes in the three-dimensional problem. Thus, if you think you can get satisfactory answers with a two-dimensional model by all means use it rather than a three-dimensional model.

Visualization of a three-dimensional model is much more complicated than a two-dimensional model. For a two-dimensional model you can draw the region of interest on a sheet of paper (or computer screen) and easily locate the nodes and elements. A three-dimensional drawing is not so easy. There are computer programs that can be used to discretize three-dimensional objects and generate finite-element grids. Automatic grid/element generation is pretty essential in three-dimensional problems.

9•6•4 Other two-dimensional elements

The development in Section 9•2 used the simplest kind of element, a triangle with a linear temperature distribution inside. It seems to work quite well for two-dimensional conduction problems. There are other elements that can also be used. This section mentions three other possible elements.

The quadrilateral element shown in Figure 9•40. One needs to be careful with this element to be sure that each of the interior angles is less than 180° The element shape shown in Figure 9•41 does not work very well. An alternative way to use the quadrilateral element in Figure 9•40 is to have the computer add a diagonal line between nodes i and k to divide the quadrilateral element into two triangular elements. Each triangle could then be treated as we did in Section 9•2.

The six-node triangular element shown in Figure 9•42 permits use of a parabolic temperature profile within the element. A temperature profile of the form

$$t^{(e)} = c_1 + c_2 x + c_3 y + c_4 x^2 + c_5 xy + c_6 y^2$$

can be satisfied within the element by choosing the six constants so that the profile matches the six nodal temperatures. The derivatives of temperature within the element vary linearly with position rather than being constant as in the three-node triangular element discussed in Section 9•2.

The six-node, isoparametric triangular element with curved sides shown in Figure 9•43 is convenient for matching the shape of curved regions.

Although the six-node elements are of higher order in their temperature distributions, they are also more complicated to program. The simple linear triangular element gives reasonable accuracy. Additional accuracy can be obtained by using a finer mesh of triangles.

9•7 SUMMARY REMARKS ON FINITE ELEMENTS

This chapter should have given you an appreciation of the finite-element method and a good comparison of the method with the more-familiar finite-difference method. Both methods result in a system of coupled ordinary differential equations to solve. Except for the capacitance matrix the equations look quite similar and similar methods of solution can be used.

At this point it should be pointed out that much of the theory of finite elements was first developed by engineers primarily interested in solving stress problems. Through the years they have developed much more sophisticated theories than can be included in this book. The presentation in this chapter is rather elementary but can be used quite effectively to solve heat-conduction problems.

Figure 9•40 *Quadrilateral element for two-dimensional finite-element analysis.*

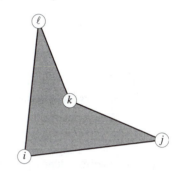

Figure 9•41 *Undesirable shape for a quadrilateral element for two-dimensional finite-element analysis.*

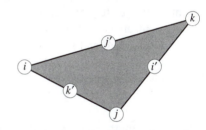

Figure 9•42 *Six-node triangular element for two-dimensional finite-element analysis.*

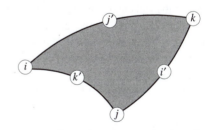

Figure 9•43 *Six-node, isoparametric triangular element for two-dimensional finite-element analysis.*

SELECTED REFERENCES

1. Bathe, K.-J. and E. L. Wilson: *Numerical Methods in Finite Element Analysis*, Prentice-Hall, Inc., Englewood Cliffs, NJ, 1976.

2. Cook, R. D., Malkus, D. S. and M. E. Plesha: *Concepts and Applications of Finite Element Analysis*, 3rd ed., John Wiley & Sons, Inc., New York, 1989.

3. Desai, C. S. and J. F. Abel: *Introduction to the Finite Element Method*, Van Nostrand Reinhold Company, New York, 1972.

4. Segerlind, L. J.: *Applied Finite Element Analysis*, 2nd ed., John Wiley & Sons, Inc., New York, 1984.

5. Zienkiewicz, O. C.: *The Finite Element Method in Structural and Continuum Mechanics*, McGraw-Hill Publishing Company, Ltd., London, 1967.

6. Hildebrand, F. B.: *Advanced Calculus for Applications*, 2nd ed., Prentice-Hall, Inc., Englewood Cliffs, NJ, 1976.

7. Kaplan, W.: *Advanced Calculus*, 3rd ed., Addison-Wesley Publishing Company, Inc., Reading, MA, 1984.

8. Noble, B.: *Applied Linear Algebra*, Prentice-Hall, Inc., Englewood Cliffs, NJ, 1969.

EXERCISES

9•1 Consider transient conduction in a plane wall between $x = 0$ and $x = L$. The governing equation is given by

$$\frac{\partial}{\partial x}(kt_x) + g''' = \rho c \frac{\partial t}{\partial \theta}$$

(a) Derive the weighted-residual equation. Your equation should be comparable to (9·1·7).

(b) For $t(x,\theta) = \mathbf{w}^T(x)\mathbf{t}(\theta)$ and $\mathbf{f} = \mathbf{w}$, derive the finite-element matrix differential equation. Your equation should be comparable to (9·1·13) through (9·1·17).

(c) Determine $\mathbf{C}^{(e)}$, $\mathbf{K}^{(e)}$ and $\mathbf{g}^{(e)}$.

If the boundary conditions at $x = 0$, node 1, and $x = L$, node N, could be either specified heat flux or convection,

(d) Determine $\mathbf{q}^{(1)}$, $\mathbf{h}^{(1)}$, $\mathbf{H}^{(1)}$, $\mathbf{q}^{(N)}$, $\mathbf{h}^{(N)}$ and $\mathbf{H}^{(N)}$.

9•2 A thin rod (length $= L$, cross-sectional area $= A$, perimeter $= p$, conductivity $= k$) is convectively cooled (heat-transfer coefficient $= h$, ambient temperature $= t_\infty$). Each end of the rod can have a specified heat inflow.

(a) Determine $\mathbf{K}^{(e)}$, $\mathbf{H}^{(e)}$, $\mathbf{h}^{(e)}$ and $\mathbf{q}_o$.

If each end of the rod is held at temperature t_0, and four uniformly spaced nodes in one half of the rod are used,

(b) Determine $\mathbf{S}$ and $\mathbf{r}$.

For $L = 1$, $A = 1/64$, $p = 1/2$, $k = 96$, $h = 24$, $t_\infty = 0$ and $t_0 = 1$,

(c) Determine $\mathbf{t}$.

(d) Determine the heat-transfer rate into the rod at $x = 0$.

9•3 A thin, flat fin $[k, \rho, c$ and thickness $\delta]$ convects heat from both top and bottom surfaces with the ambient $[h, t_\infty]$. Since the shape of the fin in the (x, y) plane is not simple, a finite-element analysis is desired.

(a) Derive the governing partial differential equation.

(b) Derive the weighted-residual equation.

(c) Derive the finite-element matrix differential equation $\mathbf{C}\dot{\mathbf{t}} + \mathbf{K}\mathbf{t} + \mathbf{F}\mathbf{t} = \mathbf{f} + \mathbf{q}_o$ where $\mathbf{F}$ is a "fin matrix" and $\mathbf{f}$ is a "fin vector." Express $\mathbf{C}$, $\mathbf{K}$, $\mathbf{F}$, $\mathbf{f}$ and $\mathbf{q}_o$ as area or boundary integrals involving $\mathbf{w}(x, y)$.

9•4 For $\mathbf{f}$ and $\mathbf{w}$ given by

$$\mathbf{f} = \begin{bmatrix} f_1 \\ f_2 \\ f_3 \end{bmatrix} \qquad \text{and} \qquad \mathbf{w} = \begin{bmatrix} w_1 \\ w_2 \\ w_3 \end{bmatrix}$$

(a) Determine $\mathbf{f}\mathbf{w}^T$.

(b) Determine $\mathbf{w}\mathbf{w}^T$.

(c) Determine $\mathbf{w}^T\mathbf{w}$.

9•5 A triangular finite element is formed by connecting nodes located in the (x, y) plane at the points $(8, 2)$, $(7, 5)$ and $(11, 4)$. Determine the area of the element.

9•6 Renumber and reconnect the nodes into elements in Figure 9•30 so as to minimize the bandwidth of the resulting system of equations. Determine the upper bandwidth.

9•7 A triangular finite element is described in the following table:

Node	x	y	t
3	1	3	10
7	3	6	40
11	4	2	30

(a) Determine the area of the element.

(b) Determine $w_3(3,4)$, $w_7(3,4)$, $w_{11}(3,4)$ and $w_3(4,5)$.

(c) Determine $t_x(3,4)$ and $t_y(3,4)$.

(d) Plot the triangle to scale and show the centroid.

(e) On the element, show isotherms for $t = 20$ and 30.

(f) On the element, show arrows (origin at the centroid, to scale) representing t_x, t_y and their vector sum.

9•8 For rows i, j and k in (9·2·12), show that

$$\sum_{j=1}^{N} \kappa_{ij} = 0$$

9•9 The conduction matrix **K** in (9.3.2) is written in full except that the 0s outside the band have not been shown. Rewrite **K** in economy-banded form. Shade the locations in the economy-banded **K** that contain contributions from element 10.

9•10 Determine the product **Ax** where **A** (economy-banded storage) and **x** are given by

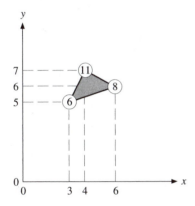

$$\mathbf{A} = \begin{bmatrix} 1 & -1 & 0 \\ 4 & -1 & -2 \\ 2 & 0 & -1 \\ 4 & -2 & 0 \\ 4 & 0 & 0 \end{bmatrix} \qquad \mathbf{x} = \begin{bmatrix} 2 \\ 1 \\ 0 \\ 1 \\ 0 \end{bmatrix}$$

9•11 Element 9 of a 17-node, 21-element, finite-element model of a two-dimensional, heat-conduction problem is formed by connecting nodal-point numbers 6, 8 and 11, located at positions (3, 5), (6, 6) and (4, 11), respectively, as shown in the sketch.

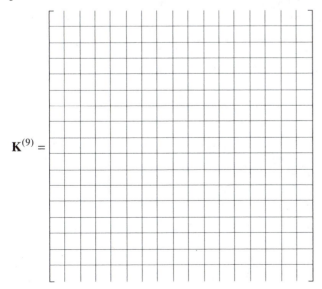

The conductivity of element 9 in appropriate units is 10. Determine the entries in the element conduction matrix and place them in their correct locations in $\mathbf{K}^{(9)}$ shown below.

$$\mathbf{K}^{(9)} =$$

9•12 A portion of a region discretized using triangular finite elements is shown in the sketch. Economy-banded storage for **K** is shown for $N = 30$ and an upper bandwidth of 10.

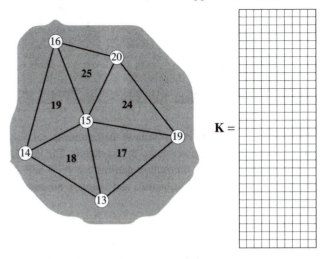

$$\mathbf{K} =$$

(a) By shading the appropriate boxes in **K**, show where the entries from $\mathbf{K}^{(17)}$ should be added.
(b) By shading the appropriate portions of the triangular elements, show nodal system 15.

9•13 The two-dimensional region shown in the sketch has been discretized using triangular finite elements. The 16 nodes and element 10 are indicated.

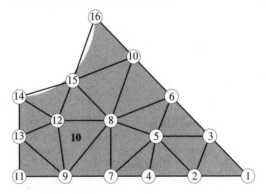

For this problem,
(a) Determine *NE* (number of elements) and *IBW* (upper bandwidth).
(b) Determine the number of rows and columns required to store **S** in economy-banded form. Indicate the size of **S** on a piece of quadruled paper (one square per matrix entry) or equivalent.
(c) Shade the appropriate locations in the **S** in part (b) to indicate where the contributions from element 10 will be added.

9•14 Determine the Cholesky decomposition of **C** for

$$\mathbf{C} = \begin{bmatrix} 2 & 1 & 1 \\ 1 & 2 & 1 \\ 1 & 1 & 2 \end{bmatrix}$$

9•15 For the square-generation problem shown in Figure 8•17, derive the finite-element matrix equation and solve for the temperatures for each of the following discretizations:

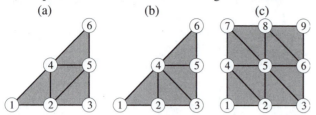

(a)   (b)   (c)

9•16 Obtain a 3-node finite-element formulation for the equilateral triangular region shown in the sketch. The length of each side is L and the thermal properties are k, ρ and c. The region is completely adiabatic and there is no energy generation within the region.

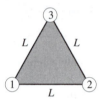

9•17 A thin fin has temperature gradients in the x and y directions. The fin thickness δ normal to the x,y plane is a function of x and y. The fin convects heat with the ambient $[h, t_\infty]$ on both the top and bottom surfaces. From Exercise 9•3, the finite-element approximation may be written as

$$\mathbf{C}\dot{\mathbf{t}} + \mathbf{K}\mathbf{t} + \mathbf{F}\mathbf{t} = \mathbf{f} + \mathbf{q}_o$$

where

$$\mathbf{C} = \iint_A \mathbf{w}\rho c \delta \mathbf{w}^T \, dx \, dy$$

$$\mathbf{K} = \iint_A \left(\mathbf{w}_x k \delta \mathbf{w}_x^T + \mathbf{w}_y k \delta \mathbf{w}_y^T \right) dx \, dy$$

$$\mathbf{F} = \iint_A \mathbf{w} 2 h \mathbf{w}^T \, dx \, dy \qquad \text{(fin matrix)}$$

$$\mathbf{f} = \iint_A \mathbf{w} 2 h t_\infty \, dx \, dy \qquad \text{(fin vector)}$$

$$\mathbf{q}_o = \int_B \mathbf{w} \delta q_o'' \, ds$$

Discretize the fin using triangular elements. Assume the fin thickness varies linearly between nodes i, j and k. Take k, ρ, c, h and t_∞ to be uniform for an element. Take the heat flow into the edge around the fin to be $q_o'' = q_s'' + h(t_\infty - t)$.

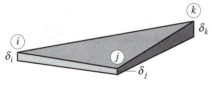

Derive the condensed $\mathbf{C}^{(e)}$, $\mathbf{K}^{(e)}$, $\mathbf{F}^{(e)}$, $\mathbf{f}^{(e)}$ and $\mathbf{q}_o^{(b)}$.

9•18 Use *FEHT* to obtain the solutions in Exercise 9•15.
(a) If you have worked Exercise 9•15, compare your *FEHT* solutions with the results of Exercise 9•15.
(b) Reduce the mesh size once and compare the new *FEHT* solutions with the previous solutions.
(c) Reduce the mesh two more times to obtain more data and plot t_1 versus $L/\Delta x$ for each of the three cases. Also, show the exact solution of $t_1 = 0.2947$.

9•19 Energy is generated uniformly in the region shown.

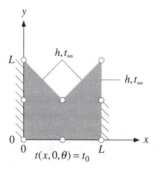

For k, g''', h, L and $T_0 = 1$ and $t_\infty = 0$, use *FEHT* to obtain a solution for this problem using the eight nodal points shown in the sketch.

9•20 Consider the triangular region shown in the figure. Nodes 1, 2 and 3 are located at (0, 1 ft), (2 ft, 0) and (4 ft, 3 ft) respectively. The surfaces between nodes 1 and 2 and between nodes 1 and 3 are adiabatic. On the surface between nodes 2 and 3, the temperature varies linearly from 100 F at node 2 to 300 F at node 3. The properties of the region are:

$k = 16$ Btu / hr-ft-F $\rho = 450$ lbm/ft^3
$c = 0.1$ Btu / lbm-F $g''' = 0$ Btu / hr-ft^3

Consider a 3-node, finite-element model of the problem with the nodes located at the corners and numbered as shown in the figure.

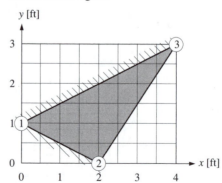

(a) Determine $\mathbf{K}$, $\mathbf{g}$, $\mathbf{S}$ and $\mathbf{r}$.
(b) Determine $\mathbf{t}$.
(c) Determine $\mathbf{q}_o$.
(d) From (9·2·21), determine the x- and y-components of the element heat-flux vector. Prove that q'' is parallel to one of the sides of the region.

9•21 Consider the triangular region shown in the figure. Nodes 1, 2 and 3 are located at (0, 1 ft), (2 ft, 0) and (4 ft, 3 ft) respectively. The surfaces between nodes 1 and 2 and between nodes 1 and 3 are adiabatic. On the surface between nodes 2 and 3, the temperature varies linearly from 100 F at node 2 to 300 F at node 3. The properties of the region are:

$k = 16$ Btu/hr-ft-F $\rho = 450$ lbm/ft^3
$c = 0.1$ Btu/lbm-F $g''' = 0$ Btu/hr-ft^3

Consider a 3-node, finite-element model of the problem, using *FEHT*, with the nodes located at the corners and numbered as shown in the figure. Take enough care in placing the nodes to ensure that the area of the triangular element is exact.

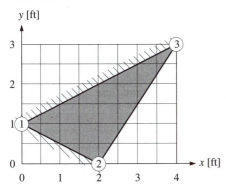

(a) Determine the temperature at node 1. Is your answer reasonable? Discuss.
(b) Show the isotherms for 100 F, 150 F, 200 F, 250 F and 300 F.
(c) Determine the heat-transfer rate [Btu/hr-ft] into the element across the boundary between nodes 2 and 3. Is your answer reasonable? Discuss.
(d) Show the temperature-gradient vector. Is it parallel to any of the three sides of the element? Explain.

9•22 Consider the triangular region shown in the figure. Nodes 1, 2 and 3 are located at (0, 1 ft), (2 ft, 0) and (4 ft, 3 ft) respectively. The surfaces between nodes 1 and 2 and between nodes 1 and 3 are adiabatic. On the surface between nodes 2 and 3, the temperature varies linearly from 100 F at node 2 to 300 F at node 3. The properties of the region are:

$k = 16$ Btu/hr-ft-F $\rho = 450$ lbm/ft^3
$c = 0.1$ Btu/lbm-F $g''' = 0$ Btu/hr-ft^3

Consider a 15-node, finite-element model of the problem. Use *FEHT* with nodes 1, 2 and 3 located at the corners and numbered as shown in the figure and then reduce the mesh size twice. Take enough care in placing the three corner nodes to ensure that the total area of the triangular region is exact.

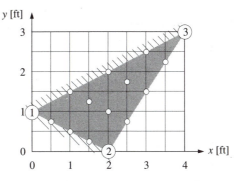

(a) Determine the nodal temperatures and nodal heat inflows. Are your results reasonable? Discuss.
(b) Draw the isotherms for 150 F, 200 F and 250 F.
(c) Determine the heat-transfer rates [Btu/hr-ft] into the region across each boundary segment between nodes 2 and 3. Are your answers reasonable? Discuss.
(d) Determine the nodal heat inflows [Btu/hr-ft] for the nodes along the boundary between nodes 2 and 3.
(e) Show temperature-gradient vectors.

9•23 Energy is steadily generated in the triangular region shown in the sketch. Two sides of the region are adiabatic. The third side is convectively cooled by an ambient at 100 F. The properties of the region are:

$k = 12$ Btu/hr-ft-F $g''' = 40$ Btu/hr-in^3
$\rho = 432$ lbm/ft^3 $h = 72$ Btu/hr-ft^2-F
$c = 0.1$ Btu/lbm-F

Consider a 4-node, finite-element model of the problem using *FEHT* with the nodes located as shown in the sketch.

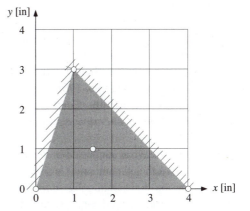

(a) Determine the steady-state solution.
(b) Draw the nodal systems and, based on *FEHT*'s nodal balances, show that conservation of energy is satisfied.
(c) From the temperatures found in part (a), determine the convection heat-transfer rate [Btu/hr-ft] from the region to the ambient. This calculation is done in *FEHT* when you view heat flows. Compare.
(d) Draw a contour plot showing isotherms of 100, 150, 200, 250, 300, 350 and 400 F. The plot should show the element boundaries as well as the contours.

9•24 Consider the example of steady energy generation in a square region discussed in Section 9•3•1. A one-triangular-element model is given by

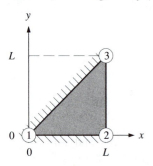

Set this problem up using *FEHT*. Use SI units. Define the element to be made of material 1 which is to have unit values of k, ρ and c. Set the generation rate to 1 W/m³. The value of L is to be 1 m. Arrange the scale such that you get a reasonably large picture on the screen. Nodes 1, 2 and 3 should be as shown in the sketch. Set $t(L, y) = 0$.

(a) Solve this model and record the number of nodes NN, the bandwidth IBW, the computation time θ_{comp}, the temperature t_1 at node 1, the sum of the nodal heat flows $\Sigma q'_{nodal}$ along $x = L$ and the sum of the boundary-segment heat flows $\Sigma q'_{heatflow}$ along $x = L$.

(b) Use the reduce mesh size option in *FEHT* to obtain a 6-node model. Solve this model and record the same data as in part (a).

(c) Reduce mesh size again in *FEHT* to obtain a 15-node model. Solve this model and record the same data as in part (a). Compare your temperatures and nodal heat flows to the values given in Section 9•3•2.

(d) Reduce mesh size again to obtain a 45-node model. Solve and record the same data as in part (a).

(e) For 45 nodes numbered as shown by *FEHT*, determine the bandwidth (for economy-banded storage). Which element line segment determines the bandwidth? Prior to solving the finite-element equations, *FEHT* uses a subroutine to reduce the bandwidth. Compare the bandwidth before reduction to the value reported by *FEHT*.

(f) If the 45 nodes were numbered as suggested in Section 9•3, determine the bandwidth and compare it to part (e).

(g) Continue to reduce the mesh size and solve until your computer objects. Solve each model and record the same data as in part (a).

(h) Plot t_1, $\Sigma q'_{nodal}$ and $\Sigma q'_{heatflow}$ as functions of NN. Use a logarithmic scale for NN. Also show the exact values these quantities should approach as $NN \rightarrow \infty$. Discuss.

(i) Plot θ_{comp} as a function of NN on a log-log plot.

9•25 For a 3-node, finite-element formulation (one node at each corner) of the problem shown in Figure 8•22,

(a) Determine $Ç$, S, r and $t(0)$.

(b) Using (9•4•8), determine $\Delta\theta_{c,est}$.

(c) For $\Delta\theta = \Delta\theta_{c,est}$, obtain a Crank-Nicolson solution.

9•26 Derive the finite-element system of equations for the adiabatic-square problem shown in Figure 9•33 except that element 2 is formed from nodes 2, 3 and 4 and element 3 is formed from nodes 3, 4 and 5.

(a) Compare your matrix equation to (9•4•4) and relate the differences to the nodal systems.

(b) Using (9•4•8), determine $\Delta\theta_{c,est}$. Compare to $\Delta\theta_{c,est}$ for (9•4•4).

9•27 Modify the finite-element equation (9•4•4) for Figure 9•33 and its initial condition (9•4•5) to treat the problem in which all six nodal points are initially at $t^{(0)}$ and then at time zero the boundary along $x = L$ is suddenly stepped to t_L and are held constant thereafter. For this problem,

(a) Using (9•4•8), determine $\Delta\theta_{c,est}$.

(b) Determine and plot the Crank-Nicolson solution for $\Delta\theta = 0.05$. If $\Delta\theta$ is too large, try $\Delta\theta = 0.025$. Discuss.

9•28 Consider the finite-element discretization of the triangular region shown in the sketch. The values of k, ρ, c and g''' are uniform and constant. The initial, uniform temperature of the entire region is $t^{(0)}$.

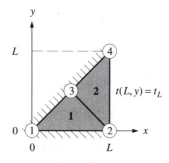

Maintaining symbols unless specifically requested to use a numerical value,

(a) Determine C, K and g.

(b) Determine $Ç$, S, r and $t^{(0)}$.

(c) Determine the steady-state solution $t(\infty)$.

(d) Determine the steady-state nodal heat inflows $q_o(\infty)$. Sketch the four nodal systems and discuss whether your results make sense.

(e) Using (9•4•8), determine $\Delta\bar{\theta}_{c,est} = \alpha\Delta\theta_{c,est} / L^2$.

(f) Determine $\dot{t}(0)$.

(g) For Crank-Nicolson, derive $At^{(v+1)} = Bt^{(v)} + b$.
For k, ρ, c, L, $t_L = 1$, g''', $t^{(0)} = 0$ and $\Delta\bar{\theta} = 0.05$,

(h) Determine and plot the transient solution to $\bar{\theta} = 0.50$ and discuss whether your results make sense.

9•29 A finite-difference or finite-element formulation of a two-dimensional conduction problem both result in a system of equations that can be written as

$$Ç\dot{t} + St = r$$

Is it possible to look at the numerical values in $Ç$, S and r to tell whether any given nodal point lies on a convective boundary? Explain.

9•30 The shape of an element can affect the critical time step. The following two one-element models have the same area but different shapes. For both models, all sides are adiabatic and there are no specified temperatures.

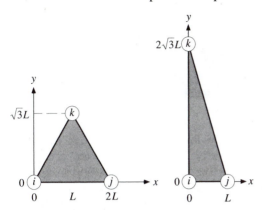

(a) Determine $\Delta\theta_{c,est}$ for both models. Compare.
(b) Determine $\Delta\theta_c$ for both models. Compare to part (a).

9•31 A thin, 0.5-in-wide, infinitely long, resistance heater strip is surrounded by insulation having a triangular cross-section (2.0 inches on each side) as shown in the sketch.

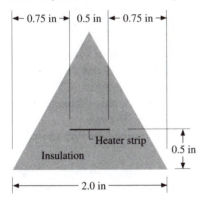

The heater generates energy at the rate of 6.0 Btu/hr per foot of axial length. The insulation properties are:

$k = 0.0156$ Btu/hr-ft-F
$\rho = 135$ lbm/ft^3
$c = 0.300$ Btu/lbm-F

The ambient air surrounding the insulated strip is at 100 F. The heat-transfer coefficient is 2.0 Btu/hr-ft^2-F. Using a *FEHT* model with five, wisely placed nodes,
(a) Determine the maximum insulation temperature [F].
(b) Determine the steady-state heat-transfer rate [Btu/hr] per foot of axial length out each of the three faces of the insulation.
(c) If the initial temperature is uniformly at 100 F when the heater is turned on, determine the time required for the maximum temperature to reach 90 % of full response.
(d) Improve your model (both steady-state and transient) by using one mesh reduction. Repeat parts (a), (b) and (c).
(e) Remesh the region to avoid long, narrow elements. Use more nodes to improve accuracy. Compare.

9•32 A 45-node, finite-element model of a two-dimensional triangular region $[k, \rho, c]$ is shown in the sketch. The initial temperature is uniformly at $t^{(0)}$. At $\theta = 0$ the temperature along the boundary at $x = L$ is suddenly stepped to t_L. The boundaries along $y = 0$ and $y = x$ are adiabatic.

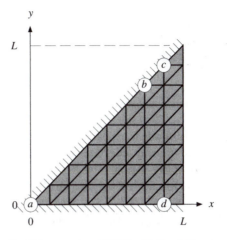

Set this up using *FEHT* with the following values:

$k = 1$ W/m-K $\rho = 1$ kg/m^3 $c = 1$ J/kg-K
$L = 1$ m $t^{(0)} = 0$ C $t_L = 100$ C

Scale the problem so that the triangular region is fairly large on the screen.
(a) Based on experience with normalization, one might expect that the most important times would be between 0 and $L^2/\alpha = 1$ s. Therefore, as a first try at a Crank-Nicolson solution, take $\Delta\theta = 0.1$ s and stop at $\theta = 1$ s. Examine the solution by looking at a temperature-time plot for nodes a, b, c and d. Is the stop time of 1 s satisfactory for displaying a major portion of the transient? Is $\Delta\theta = 0.1$ s satisfactory? Discuss.
(b) If you found in part (a) that either the stop time or the time step were unsatisfactory, adjust the value(s) by an order of magnitude to improve the solution and repeat the solution. Examine the new temperature-time plot for nodes a, b, c and d. Discuss.
(c) Superimpose the temperature-time plots for node a from parts (a) and (b) on the same plot. Compare.
(d) Since numerically-induced oscillations in the Crank-Nicolson solution are most apparent at small times, make transient run with a stop time of 0.01 s and $\Delta\theta = 0.001$ s. Examine the new temperature-time plot for nodes a, b, c and d. Discuss.

9•33 A two-dimensional, composite region made of materials A and B is shown in the sketch.

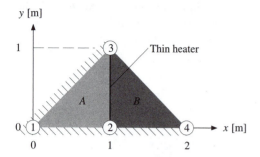

Nodes 1, 2, 3 and 4 are located at positions (0, 0), (1, 0), (1, 1) and (2, 0), respectively. Two sides of the composite region are adiabatic. The third side is exposed to an ambient at 0 C and has a heat-transfer coefficient of $12\sqrt{2}$ W/m²-K. Initially, the entire region is at 0 C. At the interface between the two materials there is a thin heater with an energy-generation rate of 1000 W/m² that starts at $\theta = 0$. The material properties are shown in the table.

Property	Units	Material A	Material B
k	W/m-K	50	2
ρ	kg/m³	10,000	1000
c	J/kg-K	480	2400

(a) Using *FEHT* with two elements (one for each material), determine the steady-state temperatures [C], the nodal energy inflows [W/m] and the convective heat-transfer rate [W/m].

(b) Is the steady-state solution of part (a) reasonable? Discuss.

(c) To improve the steady-state solution, make one mesh reduction and solve again. Discuss your findings.

(d) To learn more about the heat-flow paths, make separate nodal balances on material A and on material B. How is the heater energy removed from the solid?

(e) Based on experience with normalization, one might expect that the most important times would be between 0 and $L^2/\alpha \approx 1$ Msec. Thus, as a first try at a Crank-Nicolson transient solution, take $\Delta\theta = 0.1$ Msec. Stop at $\theta = 1$ Msec. Use the 9-node model of part (c).

(f) From part (e), determine $\Delta\theta_{c,est}$ [s].

(g) Examine the solution obtained in part (e) by looking at a temperature-time plot for nodes 1, 2, 3 and 4. Is the stop time of 1 Msec satisfactory for displaying a major portion of the transient (*i.e.*, 90 per cent response)? Is $\Delta\theta = 0.1$ Msec satisfactory? Discuss.

(h) If you found in part (g) that either the stop time or the time step were unsatisfactory, adjust the value(s), by doubling or halving, to improve the solution and repeat the solution. Examine the new temperature-time plot for nodes 1, 2, 3 and 4. Discuss.

(i) Take $\Delta\theta = 0.001$ Msec and compute to $\theta = 0.05$ Msec. Discuss your findings.

9•34 A two-dimensional, composite region made of materials A and B is shown in the sketch.

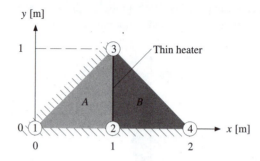

Nodes 1, 2, 3 and 4 are located at positions (0, 0), (1, 0), (1, 1) and (2, 0) respectively. Two sides of the composite region are adiabatic. The third side is exposed to an ambient at 0 C and has a heat-transfer coefficient of $12\sqrt{2}$ W/m²-K. Initially, the entire region is at 0 C. At the interface between the two materials there is a thin heater with an energy-generation rate of 1000 W/m² that starts at $\theta = 0$. The material properties are shown in the table.

Property	Units	Material A	Material B
k	W/m-K	50	2
ρ	kg/m³	10,000	1000
c	J/kg-K	480	2400

Using *FEHT* starting with two elements (one for each material), reduce the mesh size three times and then:

(a) Determine the steady-state temperatures [C] and the nodal energy inflows [W/m] for nodes 1 through 9 and the convective heat-transfer rate [W/m].

(b) Plot some isotherms and show some heat-flux arrows to show the heat-flow paths.

(c) Based on experience with normalization, one might expect that the most important times would be between 0 and $L^2/\alpha \approx 1$ Msec. Thus, as a first try at a Crank-Nicolson transient solution, take $\Delta\theta = 0.1$ Msec. Stop at $\theta = 1$ Msec.

(d) From the *FEHT* output in part (c), determine $\Delta\theta_{c,est}$ [s].

(e) Examine the solution obtained in part (c) by looking at a temperature-time plot for nodes 1, 2, 3 and 4. Is 1 Msec a satisfactory stop time for displaying a major portion of the transient (*i.e.*, 90 per cent response)? Is the time step of 0.1 Msec satisfactory? Discuss.

(f) If you found in part (g) that either the stop time or the time step were unsatisfactory, adjust the value(s) to improve the solution and repeat the solution. Examine the new temperature-time plot for the four nodes. Discuss.

9•35 A two-dimensional, square ($L \times L$) region [k, ρ and c] has adiabatic boundaries along $x = 0$, $y = 0$ and $y = L$. Initially, the temperature within the region is uniformly at $t^{(0)}$. At $\theta = 0$ the temperature along $x = L$ is suddenly changed to vary linearly from t_0 at $y = 0$ to t_L at $y = L$. The generation rate is zero. The three adiabatic boundaries remain adiabatic. Consider a 4-node, finite-element model where node 1 is at $(0,0)$, node 2 is at $(L,0)$, node 3 is at $(0,L)$ and node 4 is at (L,L). Element 1 is formed by connecting nodes 1, 2 and 3.

(a) Determine $\mathbf{C}$, $\mathbf{K}$, $\mathbf{Ç}$, $\mathbf{S}$, $\mathbf{r}$ and $\mathbf{t}^{(0)}$.
(b) Determine the steady-state solution $\mathbf{t}(\infty)$.
(c) Determine the steady-state nodal heat inflows $\mathbf{q}_o(\infty)$.
(d) Determine $\Delta\theta_{c,est}$.
(e) Determine $\Delta\theta_c$ and compare it to $\Delta\theta_{c,est}$.
For k, ρ, c and $L = 1$ and $t^{(0)} = 0$, $t_0 = 0$ and $t_L = 1$, use *FEHT* to:
(f) Determine the steady-state solution $\mathbf{t}(\infty)$ and $\mathbf{q}_o(\infty)$. Compare to part (b).
(g) Determine the heat flow (per unit depth) across as many line segments as you can deduce from the *FEHT* output. Prepare a sketch to show the direction and magnitudes of these heat flows.
(h) Determine the Euler solution, with $\Delta\theta = 0.2$, for θ going from 0 to 1 and plot the solutions for t_1 and t_3.
(i) Determine the Crank-Nicolson solution, with $\Delta\theta = 0.2$, for θ going from 0 to 1 and plot the solutions for t_1 and t_3.
To improve the accuracy of the solution, make two mesh reductions and then:
(j) Determine the steady-state solution for t_1 and t_3.
(k) Determine the Crank-Nicolson transient solutions for t_1 and t_3 for $\Delta\theta = 0.2$ and 0.1. Plot the solutions on two separate graphs.

9•36 A one-dimensional, plane-wall [k, ρ, c] has an adiabatic surface at $x = 0$. The uniform initial temperature is $t^{(0)}$. The surface at $x = L$ is suddenly changed to a new value t_L at $\theta = 0$. Consider a three-node, finite-element model where node 1 is at $x = 0$, node 2 is at $x = L/2$ and node 3 is at $x = L$. The condensed element conduction and capacitance matrices are given by

$$\mathbf{C}^{(e)} = \frac{\rho^{(e)} c^{(e)} x_{ij}^{(e)}}{6} \begin{bmatrix} 2 & 1 \\ 1 & 2 \end{bmatrix} \qquad \mathbf{K}^{(e)} = \frac{k^{(e)}}{x_{ij}^{(e)}} \begin{bmatrix} 1 & -1 \\ -1 & 1 \end{bmatrix}$$

(a) Determine $\mathbf{C}$, $\mathbf{K}$, $\mathbf{Ç}$, $\mathbf{S}$, $\mathbf{r}$ and $\mathbf{t}^{(0)}$.
(b) Determine the initial time derivative $\dot{\mathbf{t}}(0)$.
(c) Determine the exact transient solution $\mathbf{t}(\theta)$.
(d) Determine $\Delta\theta_c$ and $\Delta\theta_{c,est}$. Compare.
For k, ρ, c and $L = 1$, $t^{(0)} = 0$, and $t_L = 1$,
(e) Plot $t_1(\theta)$ and $t_2(\theta)$ for part (c) out to $\theta = 0.2$.
(f) Show the Euler solution for $\Delta\theta = 0.02$ on part (e).
(g) Using Euler solutions, illustrate the critical-time-step theory as applied to this problem.

9•37 A 2-node, finite-element model of a plane wall (adiabatic at one surface, convective at the other surface) is given by:

$$\begin{bmatrix} 2 & 1 \\ 1 & 2 \end{bmatrix} \dot{\mathbf{t}} + \begin{bmatrix} 1 & -1 \\ -1 & 2 \end{bmatrix} \mathbf{t} = \begin{bmatrix} 0 \\ 1 \end{bmatrix} \quad \text{with} \quad \mathbf{t}(0) = \begin{bmatrix} 0 \\ 0 \end{bmatrix}$$

(a) Determine $\dot{\mathbf{t}}(0)$. Compare to Exercise 8•37.
(b) Determine $\Delta\theta_c$. Compare to Exercise 8•37.
(c) Determine the exact solution at $\theta = 0.0, 0.5, 1.0, 1.5, 2.0, 3.0, 4.0$ and 5.0.
(d) Plot the solution. Show the slopes at $\theta = 0$.

9•38 For Exercise 9•25,
(a) Determine $\boldsymbol{\lambda}$ and $\mathbf{X}$.
(b) Determine the exact solution of the finite-element equations.
(c) For k, ρ, c, L and $T_o = 1$, plot t_1, t_2 and t_3 out to $\theta = 0.2$.

9•39 A 3-node, finite-element model of a triangular region results in the following system of equations:

$$\frac{\rho c L^2}{24} \begin{bmatrix} 2 & 1 & 1 \\ 1 & 2 & 1 \\ 1 & 1 & 2 \end{bmatrix} \dot{\mathbf{t}} + \frac{k}{2} \begin{bmatrix} 1 & -1 & 0 \\ -1 & 2 & -1 \\ 0 & -1 & 1 \end{bmatrix} \mathbf{t} = \begin{bmatrix} 0 \\ 0 \\ 0 \end{bmatrix}$$

The eigenvalue vector for this system is given by

$$\boldsymbol{\lambda} = \frac{k}{\rho c L^2} \begin{bmatrix} 0 \\ 12 \\ 36 \end{bmatrix}$$

For this finite-element model,
(a) Determine $\Delta\theta_{c,est}$.
(b) Determine $\Delta\theta_c$. Compare to $\Delta\theta_{c,est}$.

9•40 A 3-node, finite-element model of a triangular region results in the following system of equations:

$$\frac{\rho c L^2}{24} \begin{bmatrix} 2 & 1 & 1 \\ 1 & 2 & 1 \\ 1 & 1 & 2 \end{bmatrix} \dot{\mathbf{t}} + \frac{k}{2} \begin{bmatrix} 1 & -1 & 0 \\ -1 & 2 & -1 \\ 0 & -1 & 1 \end{bmatrix} \mathbf{t} = \begin{bmatrix} 0 \\ 0 \\ 0 \end{bmatrix}$$

(a) Modify these equations to obtain a lumped-capacitance model for the same problem.
For the lumped-capacitance model,
(b) Determine $\Delta\theta_{c,est}$.
(c) Determine the eigenvalue vector $\boldsymbol{\lambda}$.
(d) Determine $\Delta\theta_c$. Compare to $\Delta\theta_{c,est}$.
(e) Compare $\Delta\theta_{c,est}$ and $\Delta\theta_c$ to the corresponding values for the distributed-capacitance model in Exercise 9•38.

ORDINARY DIFFERENTIAL EQUATIONS

This appendix reviews several of the more common methods of solving ordinary differential equations. It is expected that you are already familiar with these procedures. More complete discussions can be found in [1, 2]. For unusual ordinary differential equations, you should refer to the excellent tabulation in [3].

A•1 FIRST-ORDER EQUATIONS

A•1•1 Equations with variables separable
The general form of these equations is given by

$$\frac{du}{dx} = \frac{M(x)}{N(u)} \tag{A.1.1}$$

The variables may be separated by rewriting the above as

$$N(u)\,du = M(x)\,dx \tag{A.1.2}$$

This may now be integrated.

A•1•2 Linear equations
The general form of this type of equation is given by

$$\frac{du}{dx} + P(x)u = Q(x) \tag{A.1.3}$$

This type of equation is always susceptible to multiplication by an integrating factor $I(x)\,dx$, which will make the left-hand side of (A·1·3) an exact differential. The right-hand side will be a function of x only and can thus be integrated. The function $I(x)$ can be shown to be given by

$$I(x) = \exp\left[\int P(x)\,dx\right] \tag{A.1.4}$$

Equation (A·1·3) will become

$$d[I(x)u] = Q(x)I(x)\,dx \tag{A.1.5}$$

This can now be integrated.

Example Consider the equation

$$\frac{du}{dx} + \frac{1}{x}u = x \tag{A.1.6}$$

Here

$$P(x) = \frac{1}{x}$$

Then

$$I(x) = \exp\left[\int \frac{1}{x}\,dx\right] = \exp[\ln(x)] = x$$

Thus we multiply (A·1·6) by the integrating factor $x\,dx$ to get

$$x\,du + u\,dx = x^2\,dx$$

The left-hand side is recognized as the differential of the product xu. Thus we can write

$$d(xu) = x^2\,dx$$

Integrating,

$$xu = \frac{x^3}{3} + c_1$$

$$u = \frac{x^2}{3} + \frac{c_1}{x}$$

A•2 SECOND-ORDER EQUATIONS

Generally, this is the highest-order equation that arises in heat-transfer problems. The general form of such an equation might be written as

$$u'' = f(u', u, x) \tag{A.2.1}$$

There are three special cases which commonly appear and which are discussed below.

A•2•1 Equations with both u' and u absent
For this special case, (A·2·1) reduces to

$$u'' = f(x)$$

This may simply be integrated once to give

$$u' = \int f(x)\,dx + c_1 = F(x) + c_1$$

and again to give

$$u = \int F(x)\,dx + c_1 x + c_2$$

Example Consider the equation

$$u'' = x$$

Integrating once,

$$u' = \frac{x^2}{2} + c_1$$

and again,

$$u = \frac{x^3}{6} + c_1 x + c_2$$

A•2•2   Equations with u absent

For this special case, (A·2·1) reduces to

$$u'' = f(u', x) \tag{A·2·2}$$

This may be reduced to a first-order equation by setting $p = u'$. This gives

$$\frac{dp}{dx} = f(p, x) \tag{A·2·3}$$

If the variables are separable or the equation is linear, the methods of Section A•1 may be used to find p. To obtain u, p must then be replaced by u' and the equation integrated again.

Example Consider the equation

$$u'' = u' + 1$$

Letting $p = u'$ gives

$$\frac{dp}{dx} = p + 1$$

or

$$\frac{dp}{dx} - p = 1$$

The integrating factor of this linear equation is $\exp(-x)\,dx$. Multiplying the differential equation by $\exp(-x)\,dx$ gives

$$\exp(-x)\,dp - p\exp(-x)\,dx = \exp(-x)\,dx$$

or, again recognizing that the integrating factor makes the left-hand side exact,

$$d[\exp(-x)p] = \exp(-x)\,dx$$

Integrating,

$$\exp(-x)p = -\exp(-x) + c_1$$

Solving for $p = u' = du/dx$,

$$\frac{du}{dx} = -1 + c_1 \exp(x)$$

The variables may be separated and the equation integrated to give

$$u = -x + c_1 \exp(x) + c_2$$

A•2•3   Linear equations with constant coefficients

The general form for this special case is given by

$$L[u] = \frac{d^2u}{dx^2} + a_1 \frac{du}{dx} + a_2 u = g(x) \tag{A·2·4}$$

The method of solution is to first find a general solution u_h to the homogeneous form of the differential equation and then a particular solution u_p to the complete differential equation. The general solution to the complete differential equation is then given by $u = u_h + u_p$.

The homogeneous form of (A·2·4) is

$$u_h'' + a_1 u_h' + a_2 u_h = 0 \tag{A·2·5}$$

Its solution is found by assuming it to be of the form

$$u_h = \exp(\lambda x)$$

Upon substituting this assumed form of u_h into (A·2·5),

$$\lambda^2 \exp(\lambda x) + a_1 \lambda \exp(\lambda x) + a_2 \exp(\lambda x) = 0$$

or

$$(\lambda^2 + a_1 \lambda + a_2)\exp(\lambda x) = 0$$

Since $\exp(\lambda x) \neq 0$, the term $(\lambda^2 + a_1 \lambda + a_2)$ must be zero. This determines two acceptable values of λ (call them λ_1 and λ_2). Then

$$u_h = b_1 \exp(\lambda_1 x) + b_2 \exp(\lambda_2 x)$$

because we can superimpose solutions since (A·2·5) is linear and homogeneous. The special subcase where $\lambda_1 = \lambda_2$ is discussed in [1, 2].

We must now find a particular solution u_p which satisfies $L[u_p] = g(x)$. For the functions $g(x)$, which are most common in engineering, you will be able to use the *method of undetermined coefficients*. That is, assume that u_p is a sum of all the terms that appear in $g(x)$ and its derivatives. If any of the terms in u_p appear in u_h, multiply the term in u_p by x (or x^2 if necessary) to make it different from the terms in u_h. Then plug u_p into (A·2·4) to evaluate the coefficients.

Example Consider the equation

$$L[u] = \frac{d^2u}{dx^2} + u = \exp(x) + \sin(x) \tag{A·2·6}$$

First, find u_h by substituting $\exp(\lambda x)$ into $L[u] = 0$. Thus

$$(\lambda^2 + 1)\exp(\lambda x) = 0$$

which gives

$$\lambda = \pm i = \pm\sqrt{-1}$$

Then

$$u_h = a_1 \exp(ix) + a_2 \exp(-ix)$$

Recalling that

$$\exp(ix) = \cos(x) + i\sin(x)$$

and

$$\exp(-ix) = \cos(x) - i\sin(x)$$

u_h becomes

$$u_h = a_1[\cos(x) + i\sin(x)] + a_2[\cos(x) - i\sin(x)]$$

or

$$u_h = b_1\sin(x) + b_2\cos(x) \tag{A·2·7}$$

Since $g(x) = \exp(x) + \sin(x)$, take

$$u_p = c_1\exp(x) + c_2 x\sin(x) + c_3 x\cos(x) \tag{A·2·8}$$

In this expression for u_p, the exponential term arises from the $\exp(x)$ in (A·2·6). The $\sin(x)$ term in (A·2·8) arises

from the $\sin(x)$ term in (A·2·6), and the $\cos(x)$ term in (A·2·8) arises from differentiating the $\sin(x)$ term in (A·2·6). The additional factor x in the last two terms of (A·2·8) is to make these terms different from those appearing in u_h, (A·2·7).

Next, substitute u_p into (A·2·6) to find

$$c_1 = \tfrac{1}{2} \qquad c_2 = 0 \qquad c_3 = -\tfrac{1}{2}$$

The general solution is then given by

$$u = u_h + u_p = b_1\sin(x) + b_2\cos(x) + \tfrac{1}{2}\exp(x) - \tfrac{1}{2}x\cos(x)$$

where b_1 and b_2 are the two arbitrary constants to be evaluated by making u satisfy the boundary conditions.

SELECTED REFERENCES

1. Kaplan, W.: *Ordinary Differential Equations*, Addison-Wesley Publishing Company, Inc., Reading, MA, 1958.

2. Hildebrand, F. B.: *Advanced Calculus for Applications*, 2nd ed., Prentice-Hall, Inc., Englewood Cliffs, NJ, 1976.

3. Kamke, E.: *Differentialgleichungen Lösungs-methoden und Lösungen*, Chelsea Publishing Company, New York, 1948.

EXERCISES

The following exercises should help you review the material discussed in this appendix. Further examples are worked out in Chapter 1.

1. Solve $\dfrac{du}{d\theta} + u = \theta$ with $u(0) = 1$.

2. Solve $\dfrac{du}{d\theta} = \theta$ with $u(0) = 0$.

3. Solve $\dfrac{du}{d\theta} + u^4 = 0$ with $u(0) = 1$.

4. Solve $\dfrac{d^2u}{dx^2} + 1 = 0$ with $u(0) = 1$ and $u(1) = 0$.

5. Find the general solution to $\dfrac{d^2u}{dx^2} + \kappa u = 0$.

6. Find the general solution to $\dfrac{d^2u}{dx^2} - 5\dfrac{du}{dx} + 4u = 0$.

7. Find the general solution to $\dfrac{d^2u}{dx^2} - 5\dfrac{du}{dx} + 4 = 0$.

8. Solve $\dfrac{d^2u}{dx^2} - 4u = -\exp(-x)$ with $u(0) = 1$ and $u(\infty) = 0$.

The answers to these exercises are:

1. $u = \theta - 1 + 2\exp(-\theta)$

2. $u = \theta^2/2$

3. $u = (3\theta + 1)^{-1/3}$

4. $u = 1 - \tfrac{1}{2}x - \tfrac{1}{2}x^2$

5. $u = a_1\exp(ix\sqrt{\kappa}) + a_2\exp(-ix\sqrt{\kappa})$

6. $u = a_1\exp(4x) + a_2\exp(x)$

7. $u = \tfrac{4}{5}x + a_1\exp(5x) + a_2$

8. $u = \tfrac{1}{3}\exp(-x) + \tfrac{2}{3}\exp(-2x)$

APPENDIX

B

BESSEL FUNCTION RELATIONS

The following tables are a compilation of some of the more useful Bessel function relations. Additional information can be found in various, more complete sources.

Table B•1 *Recurrence relations.*

$$J_k(x) = \frac{x^k}{2^k} \sum_{m=0}^{\infty} \frac{(-1)^m x^{2m}}{4^m \Gamma(m+1)\Gamma(m+k+1)}$$

$$I_k(x) = \frac{x^k}{2^k} \sum_{m=0}^{\infty} \frac{x^{2m}}{4^m \Gamma(m+1)\Gamma(m+k+1)}$$

$$J_{k+1}(x) = -J_{k-1}(x) + \frac{2k}{x} J_k(x)$$

$$Y_{k+1}(x) = -Y_{k-1}(x) + \frac{2k}{x} Y_k(x)$$

$$I_{k+1}(x) = I_{k-1}(x) - \frac{2k}{x} I_k(x)$$

$$K_{k+1}(x) = K_{k-1}(x) + \frac{2k}{x} K_k(x)$$

$$J_k(ix) = i^k I_k(x)$$

$$Y_k(ix) = i^k \left[iI_k(x) - \frac{2}{\pi}(-1)^k K_k(x) \right]$$

Table B•2 *Derivatives of Bessel functions.*

$$J_k'(x) = J_{k-1}(x) - \frac{k}{x} J_k(x) = -J_{k+1}(x) + \frac{k}{x} J_k(x)$$

$$Y_k'(x) = Y_{k-1}(x) - \frac{k}{x} Y_k(x) = -Y_{k+1}(x) + \frac{k}{x} Y_k(x)$$

$$I_k'(x) = I_{k-1}(x) - \frac{k}{x} I_k(x) = I_{k+1}(x) + \frac{k}{x} I_k(x)$$

$$K_k'(x) = -K_{k-1}(x) - \frac{k}{x} K_k(x) = -K_{k+1}(x) + \frac{k}{x} K_k(x)$$

$$J_k''(x) = -\frac{1}{x} J_{k-1}(x) - \left[1 - \frac{k(k+1)}{x^2} \right] J_k(x)$$

$$Y_k''(x) = -\frac{1}{x} Y_{k-1}(x) - \left[1 - \frac{k(k+1)}{x^2} \right] Y_k(x)$$

$$I_k''(x) = -\frac{1}{x} I_{k-1}(x) + \left[1 + \frac{k(k+1)}{x^2} \right] I_k(x)$$

$$K_k''(x) = \frac{1}{x} K_{k-1}(x) + \left[1 + \frac{k(k+1)}{x^2} \right] K_k(x)$$

Table B·3 *Integrals of Bessel functions.*

$$\int x J_0(x)\,dx = x J_1(x)$$

$$\int x Y_0(x)\,dx = x Y_1(x)$$

$$\int x I_0(x)\,dx = x I_1(x)$$

$$\int x K_0(x)\,dx = -x K_1(x)$$

$$\int J_1(x)\,dx = -J_0(x)$$

$$\int Y_1(x)\,dx = -Y_0(x)$$

$$\int I_1(x)\,dx = I_0(x)$$

$$\int K_1(x)\,dx = -K_0(x)$$

$$\int I_k(x)\,dx = 2\sum_{m=0}^{\infty} J_{k+2m+1}(x)$$

$$\int \frac{dx}{x J_k^2(x)} = -\frac{\pi}{2\sin(k\pi)}\frac{J_{-k}(x)}{J_k(x)}$$

$$\int \frac{dx}{x J_k(x)J_{-k}(x)} = -\frac{\pi}{2\sin(k\pi)}\ln\!\left(\frac{J_{-k}(x)}{J_k(x)}\right)$$

$$\int x^{2k+1} J_k^2(x)\,dx = \frac{x^{2k+2}}{4k+2}\Big[J_k^2(x)+J_{k+1}^2(x)\Big]$$

$$\int x^{-2k-1} J_{k+1}^2(x)\,dx = -\frac{x^{-2k}}{4k+2}\Big[J_k^2(x)+J_{k+1}^2(x)\Big]$$

$$\int x J_k(ax)J_k(bx)\,dx = \frac{x}{a^2-b^2}\Big[a J_{k+1}(ax)J_k(bx)-b J_k(ax)J_{k+1}(bx)\Big] \quad a\neq b$$

$$\int x J_k^2(ax)\,dx = \frac{x^2}{2}\Big[J_k^2(ax)-J_{k-1}(ax)J_{k+1}(ax)\Big]$$

In the following formulas $C_k(x)$ and $\overline{C}_k(x)$ denote two cylinder functions. A cylinder function is a linear combination of a pair of Bessel functions.

$$C_k(x) = A J_k(x) + B Y_k(x) \qquad \overline{C}_k(x) = \overline{A} J_k(x) + \overline{B} Y_k(x)$$

with arbitrary constants A, B, $\overline{A}$ and $\overline{B}$.

$$\int x^{k+1} C_k(x)\,dx = x^{k+1} C_{k+1}(x)$$

$$\int x^{1-k} C_k(x)\,dx = -x^{1-k} C_{k-1}(x)$$

$$\int x C_k(ax)\overline{C}_k(bx)\,dx = \frac{x}{a^2-b^2}\Big[a C_{k+1}(ax)\overline{C}_k(bx)-b C_k(ax)\overline{C}_{k+1}(bx)\Big]$$

$$\int x C_k(ax)\overline{C}_k(ax)\,dx = -\frac{x^2}{4}\Big[C_{k-1}(ax)\overline{C}_{k+1}(ax)-2C_k(ax)\overline{C}_k(ax) \\ +C_{k+1}(ax)\overline{C}_{k-1}(ax)\Big]$$

$$\int x^{-1} C_m(ax)\overline{C}_k(ax)\,dx = \frac{1}{m^2-k^2}\Big[ax C_m(ax)\overline{C}_{k+1}(ax) \\ +(m-k)C_m(ax)\overline{C}_k(ax)-ax C_{m+1}(ax)\overline{C}_k(ax)\Big]$$

X

APPENDIX

C

HELPFUL SERIES SUMMATIONS

Nonalternating-sign series

$$\sum_{n=1}^{\infty} \frac{1}{n(n+1)} = 1$$

$$\sum_{n=1}^{\infty} \frac{2n+1}{n^2(n+1)^2} = 1$$

$$\sum_{n=1}^{\infty} \frac{1}{n(n+2)} = \frac{3}{4}$$

$$\sum_{n=1}^{\infty} \frac{n+1}{n^2(n+2)^2} = \frac{5}{16}$$

$$\sum_{n=1}^{\infty} \frac{1}{(4n)^2 - 1} = \frac{4-\pi}{8}$$

$$\sum_{n=1}^{\infty} \frac{1}{4n^2 - 1} = \frac{1}{2}$$

$$\sum_{n=1}^{\infty} \frac{1}{n^2} = \frac{\pi^2}{6}$$

$$\sum_{n=1}^{\infty} \frac{1}{(2n-1)^2} = \frac{\pi^2}{8}$$

$$\sum_{n=1}^{\infty} \frac{1}{n^4} = \frac{\pi^4}{90}$$

$$\sum_{n=1}^{\infty} \frac{1}{(2n-1)^4} = \frac{\pi^4}{96}$$

$$\sum_{n=1}^{\infty} \frac{1}{n^6} = \frac{\pi^6}{945}$$

Alternating-sign series

$$\sum_{n=1}^{\infty} \frac{(-1)^{n+1}}{n} = \ln(2)$$

$$\sum_{n=1}^{\infty} \frac{(-1)^{n+1}}{(2n-1)} = \frac{\pi}{4}$$

$$\sum_{n=1}^{\infty} \frac{(-1)^{n+1}}{n^2} = \frac{\pi^2}{12}$$

$$\sum_{n=1}^{\infty} \frac{(-1)^{n+1}}{(2n-1)^3} = \frac{\pi^3}{32}$$

$$\sum_{n=1}^{\infty} \frac{(-1)^{n+1}(2n-1)}{(4n-1)(4n-3)} = \frac{\pi}{8\sqrt{2}}$$

Miscellaneous series

$$\sum_{n=1}^{\infty} \frac{\sin[(2n-1)\pi/4]}{2n-1} = \sum_{n=1}^{\infty} \frac{(-1)^{n+1}\cos[(2n-1)\pi/4]}{2n-1} = \frac{\pi}{4}$$

$$\sum_{n=1}^{\infty} \frac{(-1)^{n+1}\sin[(2n-1)\pi/4]}{(2n-1)^2} = \sum_{n=1}^{\infty} \frac{\cos[(2n-1)\pi/4]}{(2n-1)^2} = \frac{\pi^2}{16}$$

$$\sum_{n=1}^{\infty} \frac{\sin[(2n-1)\pi/4]}{(2n-1)^3} = \sum_{n=1}^{\infty} \frac{(-1)^{n+1}\cos[(2n-1)\pi/4]}{(2n-1)^3} = \frac{3\pi^3}{128}$$

D

TABLES OF EIGENCONDITION ROOTS

The following tables contain the first six positive roots of transcendental eigenconditions which commonly appear in heat-conduction problems. For other transcendental eigenconditions one will most likely have to resort to a computer to find the roots.

Table D·1 *First six roots of* $\lambda_n \sin(\lambda_n) - H\cos(\lambda_n) = 0$.

H	λ_1	λ_2	λ_3	λ_4	λ_5	λ_6
0	0.0000	3.1416	6.2832	9.4248	12.5664	15.7080
0.002	0.0447	3.1422	6.2835	9.4250	12.5665	15.7081
0.004	0.0632	3.1429	6.2838	9.4252	12.5667	15.7082
0.006	0.0774	3.1435	6.2841	9.4254	12.5668	15.7083
0.008	0.0893	3.1441	6.2845	9.4256	12.5670	15.7085
0.010	0.0998	3.1448	6.2848	9.4258	12.5672	15.7086
0.020	0.1410	3.1479	6.2864	9.4269	12.5680	15.7092
0.040	0.1987	3.1543	6.2895	9.4290	12.5696	15.7105
0.060	0.2425	3.1606	6.2927	9.4311	12.5711	15.7118
0.080	0.2791	3.1668	6.2959	9.4333	12.5727	15.7131
0.100	0.3111	3.1731	6.2991	9.4354	12.5743	15.7143
0.200	0.4328	3.2039	6.3148	9.4459	12.5823	15.7207
0.300	0.5218	3.2341	6.3305	9.4565	12.5902	15.7270
0.400	0.5932	3.2636	6.3461	9.4670	12.5981	15.7334
0.500	0.6533	3.2923	6.3616	9.4775	12.6060	15.7397
0.600	0.7051	3.3204	6.3770	9.4879	12.6139	15.7460
0.700	0.7506	3.3477	6.3923	9.4983	12.6218	15.7524
0.800	0.7910	3.3744	6.4074	9.5087	12.6296	15.7587
0.900	0.8274	3.4003	6.4224	9.5190	12.6375	15.7650
1.000	0.8603	3.4256	6.4373	9.5293	12.6453	15.7713
1.500	0.9882	3.5422	6.5097	9.5801	12.6841	15.8026
2.000	1.0769	3.6436	6.5783	9.6296	12.7223	15.8336
3.000	1.1925	3.8088	6.7040	9.7240	12.7966	15.8945
4.000	1.2646	3.9352	6.8140	9.8119	12.8678	15.9536
5.000	1.3138	4.0336	6.9096	9.8928	12.9352	16.0107
6.000	1.3496	4.1116	6.9924	9.9667	12.9988	16.0654
7.000	1.3766	4.1746	7.0640	10.0339	13.0584	16.1177
8.000	1.3978	4.2264	7.1263	10.0949	13.1141	16.1675
9.000	1.4149	4.2694	7.1806	10.1502	13.1660	16.2147
10.000	1.4289	4.3058	7.2281	10.2003	13.2142	16.2594
15.000	1.4729	4.4255	7.3959	10.3898	13.4078	16.4474
20.000	1.4961	4.4915	7.4954	10.5117	13.5420	16.5864
30.000	1.5202	4.5615	7.6057	10.6543	13.7085	16.7691
40.000	1.5325	4.5979	7.6647	10.7334	13.8048	16.8794
50.000	1.5400	4.6202	7.7012	10.7832	13.8666	16.9519
100.000	1.5552	4.6658	7.7764	10.8871	13.9981	17.1093
∞	1.5708	4.7124	7.8540	10.9956	14.1372	17.2788

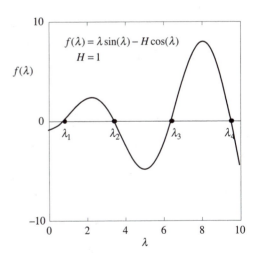

Figure D·1 *The function* $f(\lambda) = \lambda\sin(\lambda) - H\cos(\lambda)$ *is shown for* $H = 1$. *The first four roots of* $f(\lambda) = 0$ *are indicated.*

Table D·2 *First six nonzero roots of* $\lambda \cos(\lambda) + H \sin(\lambda) = 0$.

H	λ_1	λ_2	λ_3	λ_4	λ_5	λ_6
0	1.5708	4.7124	7.8540	10.9956	14.1372	17.2788
0.002	1.5721	4.7128	7.8542	10.9958	14.1373	17.2789
0.004	1.5733	4.7132	7.8545	10.9959	14.1374	17.2790
0.006	1.5746	4.7137	7.8547	10.9961	14.1376	17.2791
0.008	1.5759	4.7141	7.8550	10.9963	14.1377	17.2792
0.010	1.5771	4.7145	7.8553	10.9965	14.1379	17.2793
0.020	1.5834	4.7166	7.8565	10.9974	14.1386	17.2799
0.040	1.5959	4.7209	7.8591	10.9992	14.1400	17.2811
0.060	1.6081	4.7251	7.8616	11.0010	14.1414	17.2822
0.080	1.6201	4.7293	7.8642	11.0028	14.1428	17.2834
0.100	1.6320	4.7335	7.8667	11.0047	14.1442	17.2845
0.200	1.6887	4.7544	7.8794	11.0137	14.1513	17.2903
0.300	1.7414	4.7751	7.8920	11.0228	14.1584	17.2961
0.400	1.7906	4.7956	7.9045	11.0318	14.1654	17.3019
0.500	1.8366	4.8158	7.9171	11.0408	14.1724	17.3076
0.600	1.8798	4.8358	7.9295	11.0498	14.1795	17.3134
0.700	1.9203	4.8556	7.9419	11.0588	14.1865	17.3192
0.800	1.9586	4.8750	7.9542	11.0677	14.1935	17.3249
0.900	1.9947	4.8942	7.9665	11.0766	14.2005	17.3306
1.000	2.0288	4.9132	7.9787	11.0855	14.2074	17.3364
1.500	2.1746	5.0036	8.0385	11.1295	14.2421	17.3649
2.000	2.2889	5.0870	8.0962	11.1727	14.2764	17.3932
3.000	2.4556	5.2329	8.2045	11.2560	14.3434	17.4490
4.000	2.5704	5.3540	8.3029	11.3348	14.4080	17.5034
5.000	2.6537	5.4544	8.3913	11.4086	14.4699	17.5562
6.000	2.7165	5.5378	8.4703	11.4773	14.5288	17.6072
7.000	2.7654	5.6078	8.5406	11.5408	14.5847	17.6562
8.000	2.8044	5.6669	8.6031	11.5993	14.6374	17.7032
9.000	2.8363	5.7172	8.6587	11.6532	14.6869	17.7481
10.000	2.8628	5.7606	8.7083	11.7027	14.7335	17.7908
15.000	2.9476	5.9080	8.8898	11.8959	14.9251	17.9742
20.000	2.9930	5.9921	9.0018	12.0250	15.0625	18.1136
30.000	3.0406	6.0831	9.1294	12.1807	15.2380	18.3018
40.000	3.0651	6.1311	9.1987	12.2688	15.3417	18.4180
50.000	3.0801	6.1606	9.2420	12.3247	15.4090	18.4953
100.000	3.1105	6.2211	9.3317	12.4426	15.5537	18.6650
∞	3.1416	6.2832	9.4248	12.5664	15.7080	18.8496

Figure D·2 *The function* $f(\lambda) = \lambda \cos(\lambda) + H \sin(\lambda)$ *is shown for* $H = 1$. *The first three nonzero roots of* $f(\lambda) = 0$ *are indicated.*

Note: The eigenvalues for a sphere with a convective surface may be found from the roots of:

$$\lambda \cos(\lambda) + (H_{sphere} - 1) \sin(\lambda) = 0$$

where $H_{sphere} = h r_o / k$. The first six nonzero roots may be obtained from Table D•2 by taking $H = H_{sphere} - 1$.

m	λ_1	λ_2	λ_3	λ_4	λ_5	λ_6
0	2.4048	5.5201	8.6537	11.7915	14.9309	18.0711
1	3.8317	7.0156	10.1735	13.3237	16.4706	19.6159
2	5.1356	8.4172	11.6198	14.7960	17.9598	21.1170
3	6.3802	9.7610	13.0152	16.2235	19.4094	22.5827
4	7.5883	11.0647	14.3725	17.6160	20.8269	24.0190
5	8.7715	12.3386	15.7002	18.9801	22.2178	25.4303

Table D·3 *First six nonzero roots of* $J_m(\lambda_n) = 0$.

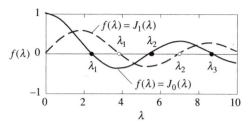

Figure D·3 *The function* $f(\lambda) = J_0(\lambda)$ *is shown and the first three roots of* $J_0(\lambda) = 0$ *are indicated. The function* $f(\lambda) = J_1(\lambda)$ *is also shown and the first two nonzero roots of* $J_1(\lambda) = 0$ *are indicated.*

H	λ_1	λ_2	λ_3	λ_4	λ_5	λ_6
0	0.0000	3.8317	7.0156	10.1735	13.3237	16.4706
0.02	0.1995	3.8369	7.0184	10.1754	13.3252	16.4718
0.04	0.2814	3.8421	7.0213	10.1774	13.3267	16.4731
0.06	0.3438	3.8473	7.0241	10.1794	13.3282	16.4743
0.08	0.3960	3.8525	7.0270	10.1813	13.3297	16.4755
0.10	0.4417	3.8577	7.0298	10.1833	13.3312	16.4767
0.20	0.6170	3.8835	7.0440	10.1931	13.3387	16.4828
0.30	0.7465	3.9091	7.0582	10.2029	13.3462	16.4888
0.40	0.8516	3.9344	7.0723	10.2127	13.3537	16.4949
0.50	0.9408	3.9594	7.0864	10.2225	13.3611	16.5009
0.60	1.0184	3.9841	7.1004	10.2322	13.3686	16.5070
0.70	1.0873	4.0085	7.1143	10.2419	13.3761	16.5130
0.80	1.1490	4.0325	7.1282	10.2516	13.3835	16.5191
0.90	1.2048	4.0562	7.1421	10.2613	13.3910	16.5251
1.00	1.2558	4.0795	7.1558	10.2710	13.3984	16.5312
2.00	1.5994	4.2910	7.2884	10.3658	13.4719	16.5910
3.00	1.7887	4.4634	7.4103	10.4566	13.5434	16.6499
4.00	1.9081	4.6018	7.5201	10.5423	13.6125	16.7073
5.00	1.9898	4.7131	7.6177	10.6223	13.6786	16.7630
6.00	2.0490	4.8033	7.7039	10.6964	13.7414	16.8168
7.00	2.0937	4.8772	7.7797	10.7646	13.8008	16.8684
8.00	2.1286	4.9384	7.8464	10.8271	13.8566	16.9179
9.00	2.1566	4.9897	7.9051	10.8842	13.9090	16.9650
10.00	2.1795	5.0332	7.9569	10.9363	13.9580	17.0099
15.00	2.2509	5.1773	8.1422	11.1367	14.1576	17.2008
20.00	2.2880	5.2568	8.2534	11.2677	14.2983	17.3442
30.00	2.3261	5.3410	8.3771	11.4221	14.4748	17.5348
40.00	2.3455	5.3846	8.4432	11.5081	14.5774	17.6508
50.00	2.3572	5.4112	8.4840	11.5621	14.6433	17.7272
100.00	2.3809	5.4652	8.5678	11.6747	14.7834	17.8931
∞	2.4048	5.5201	8.6537	11.7915	14.9309	18.0711

Table D·4 *First six roots of* $\lambda_n J_1(\lambda_n) - H J_0(\lambda_n) = 0$.

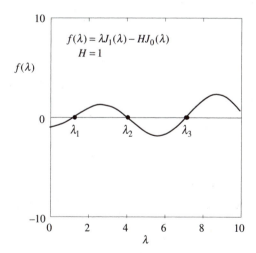

Figure D·4 *The function* $f(\lambda) = \lambda J_1(\lambda) - H J_0(\lambda)$ *is shown for* $H = 1$. *The first three roots of* $f(\lambda) = 0$ *are indicated.*

APPENDIX

E

LAPLACE TRANSFORM PAIRS

$f(t)$	$\hat{f}(s) = \int_{t=0}^{\infty} \exp(-st) f(t)\, dt$
1. 1	$\dfrac{1}{s}$
2. $\dfrac{t^{n-1}}{\Gamma(n)}$	$\dfrac{1}{s^n}$ for $n > 0$
3. $\exp(at)$	$\dfrac{1}{s-a}$
4. $\sin(at)$	$\dfrac{a}{s^2 + a^2}$
5. $\cos(at)$	$\dfrac{s}{s^2 + a^2}$
6. $\sinh(at)$	$\dfrac{a}{s^2 - a^2}$
7. $\cosh(at)$	$\dfrac{s}{s^2 - a^2}$
8. $af(t)$	$a\hat{f}(s)$
9. $f(t) + g(t)$	$\hat{f}(s) + \hat{g}(s)$
10. $f'(t)$	$-f(0) + s\hat{f}(s)$
11. $f''(t)$	$-f'(0) - sf(0) + s^2\hat{f}(s)$
12. $tf(t)$	$-\dfrac{d}{ds}\hat{f}(s)$
13. $\int_{\sigma=0}^{t} f(\sigma) g(t-\sigma)\, d\sigma$	$\hat{f}(s)\hat{g}(s)$
14. $\sum_{k=1}^{N} \dfrac{P(s_k)}{Q'(s_k)} \exp(s_k t)$	$\dfrac{P(s)}{Q(s)}$

where $P(s)$ and $Q(s)$ are polynomials in s and the s_k are the roots of $Q(s) = 0$

15. $\mathbf{1}(t-a)$	$\dfrac{\exp(-as)}{s}$

where $\mathbf{1}(t-a) = 0$ for $t < a$
           $\mathbf{1}(t-a) = 1$ for $t > a$

16. $g(t-a)\mathbf{1}(t-a)$	$\hat{g}(s)\exp(-as)$
17. $\mathbf{1}(t-a)\{1 - \exp[-b(t-a)]\}$	$\dfrac{b\exp(-as)}{s(s+b)}$

SELECTED REFERENCES

1. Carslaw, H. S. and J. C. Jaeger: Conduction of Heat in Solids, 2d ed., Oxford University Press, London, 1959.

2. Spiegel, M. R.: *Schaum's Outline of Theory and Problems of Laplace Transforms*, McGraw-Hill Book Company, New York, 1965.

$f(t)$	$\hat{f}(s)$
18. $\dfrac{a}{2\sqrt{\pi t^3}}\exp(-\dfrac{a^2}{4t})$	$\exp(-a\sqrt{s})$
19. $\dfrac{1}{\sqrt{\pi t}}\exp(-\dfrac{a^2}{4t})$	$\dfrac{\exp(-a\sqrt{s})}{\sqrt{s}}$
20. $\operatorname{erfc}(\dfrac{a}{2\sqrt{t}})$	$\dfrac{\exp(-a\sqrt{s})}{s}$
21. $\dfrac{2\sqrt{t}}{\sqrt{\pi}}\exp(-\dfrac{a^2}{4t})-a\operatorname{erfc}(\dfrac{a}{2\sqrt{t}})$	$\dfrac{\exp(-a\sqrt{s})}{s\sqrt{s}}$
22. $(t+\dfrac{a^2}{2})\operatorname{erfc}(\dfrac{a}{2\sqrt{t}})-\dfrac{a\sqrt{t}}{\sqrt{\pi}}\exp(-\dfrac{a^2}{4t})$	$\dfrac{\exp(-a\sqrt{s})}{s^2}$
23. $\dfrac{a}{2\sqrt{\pi t^3}}\exp(-\dfrac{a^2}{4t}-bt)$	$\exp(-a\sqrt{s+b})$
24. $\dfrac{1}{\sqrt{\pi t}}\exp(-\dfrac{a^2}{4t})-b\exp(ab+b^2t)\operatorname{erfc}(\dfrac{a}{2\sqrt{t}}+b\sqrt{t})$	$\dfrac{\exp(-a\sqrt{s})}{b+\sqrt{s}}$
25. $\exp(ab+b^2t)\operatorname{erfc}(\dfrac{a}{2\sqrt{t}}+b\sqrt{t})$	$\dfrac{\exp(-a\sqrt{s})}{\sqrt{s}(b+\sqrt{s})}$
26. $\operatorname{erfc}(\dfrac{a}{2\sqrt{t}})-\exp(ab+b^2t)\operatorname{erfc}(\dfrac{a}{2\sqrt{t}}+b\sqrt{t})$	$\dfrac{b\exp(-a\sqrt{s})}{s(b+\sqrt{s})}$

Note: In the next four pairs, $\lambda_n = n\pi/a$.

27. $\dfrac{2}{a}\sum_{n=1}^{\infty}(-1)^{n+1}\lambda_n\sin(\lambda_n x)\exp(-\lambda_n^2 t)$	$\dfrac{\sinh(x\sqrt{s})}{\sinh(a\sqrt{s})}$
28. $\dfrac{x}{a}+\dfrac{2}{a}\sum_{n=1}^{\infty}\dfrac{(-1)^n}{\lambda_n}\sin(\lambda_n x)\exp(-\lambda_n^2 t)$	$\dfrac{\sinh(x\sqrt{s})}{s\sinh(a\sqrt{s})}$
29. $\dfrac{x^3-a^2x}{6a}+\dfrac{xt}{a}+\dfrac{2}{a}\sum_{n=1}^{\infty}\dfrac{(-1)^{n+1}}{\lambda_n^3}\sin(\lambda_n x)\exp(-\lambda_n^2 t)$	$\dfrac{\sinh(x\sqrt{s})}{s^2\sinh(a\sqrt{s})}$
30. $\dfrac{1}{a}+\dfrac{2}{a}\sum_{n=1}^{\infty}(-1)^n\cos(\lambda_n x)\exp(-\lambda_n^2 t)$	$\dfrac{\cosh(x\sqrt{s})}{\sqrt{s}\sinh(a\sqrt{s})}$

Note: In the next four pairs, $\lambda_n = (2n-1)\pi/2a$.

31. $\dfrac{2}{a}\sum_{n=1}^{\infty}(-1)^{n+1}\lambda_n\cos(\lambda_n x)\exp(-\lambda_n^2 t)$	$\dfrac{\cosh(x\sqrt{s})}{\cosh(a\sqrt{s})}$
32. $1+\dfrac{2}{a}\sum_{n=1}^{\infty}\dfrac{(-1)^n}{\lambda_n}\cos(\lambda_n x)\exp(-\lambda_n^2 t)$	$\dfrac{\cosh(x\sqrt{s})}{s\cosh(a\sqrt{s})}$
33. $\dfrac{x^2-a^2}{2}+t+\dfrac{2}{a}\sum_{n=1}^{\infty}\dfrac{(-1)^{n+1}}{\lambda_n^3}\cos(\lambda_n x)\exp(-\lambda_n^2 t)$	$\dfrac{\cosh(x\sqrt{s})}{s^2\cosh(a\sqrt{s})}$
34. $\dfrac{2}{a}\sum_{n=1}^{\infty}(-1)^{n+1}\sin(\lambda_n x)\exp(-\lambda_n^2 t)$	$\dfrac{\sinh(x\sqrt{s})}{\sqrt{s}\cosh(a\sqrt{s})}$

APPENDIX

F

COMPREHENSIVE PROBLEMS

The problems in this appendix are intended to give the student experience in:

1. Formulating mathematical models (differential equations with boundary and/or initial conditions) for engineering situations

2. Applying the mathematical techniques discussed in the text to solve the model and then

3. Drawing some practical conclusions from the results of the analysis.

Thus the student is expected to begin with a physical problem, carry out an analysis and finally return to the physical problem to make some engineering judgments. These are the steps required in any analysis.

Problem F·1

You have been retained as a consultant by a patent attorney to measure the conductivity of a new insulating material. Your laboratory has a guarded hot plate for measuring thermal conductivity. The apparatus (Figure 1) consists of a heating section, a test section, cold plates and edge insulation.

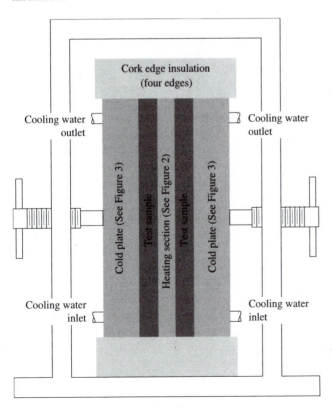

Figure 1 *Overall apparatus.*

The composite heating section (Figure 2) consists of a $\frac{3}{8}$-in-thick central portion between two $\frac{1}{4}$-in-thick outer plates. Two 12-in-square (0.01-in-thick) miconite sheets insulate the outer plates from the electrically heated inner portion. The central portion consists of an 8-in-square main heating element separated from a $1\frac{1}{8}$-in-wide guard ring heater by a $\frac{1}{8}$-in air gap. The main heater consists of electrical wires wound around a $7\frac{1}{8}$-in-square sheet of miconite. The guard heater wires are wound on a $1\frac{1}{8}$-in-wide miconite ring.

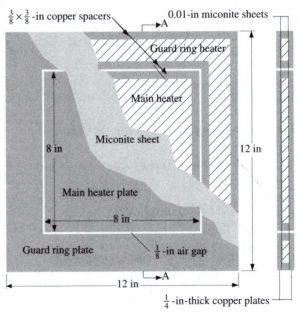

Figure 2 *Heating section detail.*

The outer copper plates of the heating section also consist of an inner portion separated from a guard ring by a

$\frac{1}{8}$-in air gap. Thermocouples in these plates measure the temperature of the inner portion and the guard ring. These thermocouples are felt to be accurate to within about 0.2 F.

Each cold plate (Figure 3) consists of a chamber made of $\frac{1}{4}$-in-thick copper plate cooled by water which enters at 60 F. These plates can be adjusted to give a test section of any desired thickness. You have been given 1-in-thick samples of material.

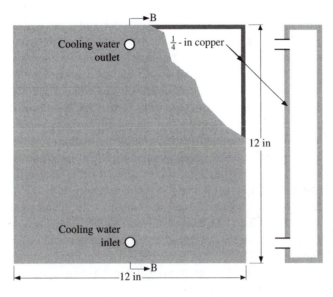

Figure 3 *Cold-plate detail.*

The edges of the heater, the test sections and the cold plates are insulated from the ambient by 2-in-thick cork. The ambient temperature is 72 F.

(a) Estimate the steady-state power input to both the main heater and the guard heater to provide a temperature difference of 16 F across the insulation. (This gives an accurately measurable temperature difference and an average insulation temperature close to the temperature it will see in service.) The guard heater should be adjusted to minimize the temperature difference between the guard ring and the main inner portion of the outside plates. From the appearance of the new insulating material, it is suspected that its thermal conductivity will be close to that of glass wool [$k \approx$ 0.02 Btu/hr-ft-F].

(b) What cooling-water flow-rate setting would you advise for satisfactory thermal operation of the test?

(c) Determine the actual thermal conductivity of the sample if the measured temperature difference across the insulation turns out to be 13 F.

Use the following properties in your calculations:

Substance	k Btu/hr-ft-F	ρ lbm/ft^3	c Btu/lbm-F
Water at 60 F		62.3	0.999
Cork	0.025		
Miconite	0.44	170	0.02
Copper	220	558	0.091

Problem F•2

Liquid oxygen is supplied to a missile through an 8-in-diameter thin-walled, stainless steel pipe. During much of the filling process, the flow in the pipe will be stratified—liquid oxygen will be flowing in the lower half and gaseous oxygen will be flowing in the upper half. The temperature of both the liquid and the gas will be about –300 F. The ambient temperature outside the pipe is 60 F.

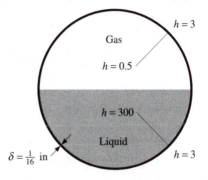

From past experience it is expected that the heat-transfer coefficient in the liquid will be about 300 Btu/hr-ft^2-F, while in the gas it will be only about 0.5 Btu/hr-ft^2-F. As a result of this nonuniform heat-transfer coefficient around the circumference, the pipe temperature at the bottom of the pipe will be different than at the top. It is feared that this temperature difference will cause some thermal-stresses.

Determine the temperature distribution around the circumference of the tube. What is the heat-transfer rate to the oxygen at this section? Take the thermal conductivity of stainless steel to be 7.1 Btu/hr-ft-F.

Problem F•3

A clutch manufacturer has called you in to consult on a new design being considered. The system essentially consists of two rotating disks as shown.

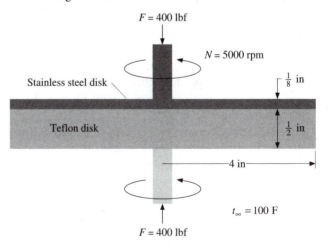

The axial loading will be about 400 lbf and the manufacturer estimates that the pressure distribution between the disks will be such that the wear will be uniform. This means that the product of pressure and radius will be a constant (*i.e.*, $Pr = C$).

Teflon is being considered for one of the disks since during some modes of operation the disks will be slipping relative to one another and it is hoped that the low coefficient of friction ($\mu = 0.10$) between the steel and the Teflon will reduce losses. During operation the stainless steel disk will be at 5000 rpm, but the Teflon disk may be less than this. The frictional heating will cause the interface temperature to rise. For safe operation of the Teflon its maximum temperature should not exceed 300 F.

The manufacturer is now planning to build the clutch with the only cooling being the 100-F air surrounding the disks. Your task is to estimate the "slippage" rpm that could be allowed without overheating the Teflon. In addition, the manufacturer is wondering what the increase in allowable slippage might be if a water cooling system were designed that would maintain the outer edge of the stainless steel disk at a temperature closer to or below the ambient.

For disks rotating in air, use the following expression for the average heat-transfer coefficient:

$$\frac{\langle h \rangle R}{k} = 0.35 \left[\frac{\omega R^2}{v} \right]^{1/2} \quad \text{for} \quad \frac{\omega R^2}{v} < 250,000$$

For the purposes of this problem you may assume the heat-transfer coefficient is uniform wherever needed.

Problem F·4

Reply to E. G. Sreym's memo.

MEMO
SREYM CONSULTING COMPANY

To:	H. T. Expert
From:	E. G. Sreym
Subject:	*Heat Meter Evaluation*

Is the heat meter, described by the attached flyer, suitable for measuring heat-transfer rates of about 3000 Btu/hr-ft^2 from an ambient at 200 F to an isothermal wall at 100 F? The following questions have arisen:

1. Is the temperature rise of the sensing disk less than 20 percent of the temperature difference between the wall and the ambient so that the wall might still be assumed to be isothermal?

2. Can the thickness of the sensing disk be altered to make the device more suitable for our needs if it is found to have too large a temperature difference across it?

3. Is the meter reading really proportional to the total heat flux entering the sensing disk as implied by the TAEH-RETEM Company?

4. Can this device be used (as is) to make accurate heat-flux measurements in flow fields where the average heat-transfer coefficient for the sensing disk is unknown?

    If not, can you suggest any modifications that would allow the device to be used in such an application?

•••

TAEH-RETEM ENGINEERING COMPANY
0000 Dloc Eson Road
Erehwyna, Aksala

Our engineering department has recently developed an improved heat meter for measuring local heat-transfer rates. It is finding wide acceptance for measurement of convective heat-transfer rates.

The heat-meter sensor consists basically of a thin, circular sensing disk mounted over an evacuated cavity in a heat sink and thermally and electrically bonded to the heat sink at the perimeter of the cavity. As heat is transferred to the sensing disk from a convective heat source, it flows along the disk into the heat sink. Since the disk is thin compared to the diameter of the cavity, a temperature difference is generated between the center and edge of the disk which is directly proportional to the rate of heat transfer to the sensor. By making the disk from a thermoelectric material such as constantan and attaching wires of a second thermoelectric material such as copper at the center and edge of the disk, a differential thermocouple is generated to allow measuring the temperature difference between center and edge of the disk, and hence obtain a signal proportional to heat flux. The following figure is an idealized representation of the heat meter.

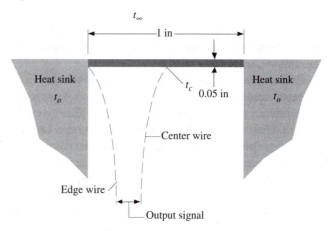

Let our sales people show you how this device can solve your instrumentation problems.

•••

Problem F·5

For the engine cylinder shown in the figure, it is estimated that the average heat flux which must be dissipated through the cylinder wall is about 4 Btu/min-in^2 at the inner side of the iron liner. The average heat-transfer coefficient over the fins has been experimentally found to be about 10 Btu/hr-ft^2-F. The thermal conductivity of the iron liner is 28 Btu/hr-ft-F. The fin thickness and the spacing between the fins are both 0.075 in. The average air temperature outside the cylinder is 110 F.

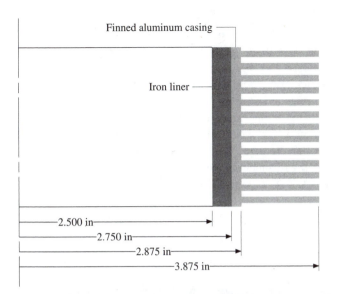

Finned aluminum casing

Iron liner

2.500 in
2.750 in
2.875 in
3.875 in

(a) Determine the inside wall temperature [F].
(b) Determine the temperature [F] at the interface between the iron and the aluminum.
(c) Determine the temperature [F] at the fin root and at the fin tip.

Problem F·6

The owners of a large furnace are interested in estimating how long it will take after shutdown before their workers can open the furnace and comfortably enter for repairs. A "comfortable" inside wall temperature which will allow their people to work is 90 F.

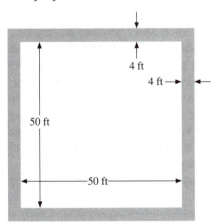

4 ft

4 ft

50 ft

50 ft

The furnace is cubical in shape with inside dimensions of 50 ft on an edge. The furnace walls are 4-ft thick. Initially the inside wall temperature is 1670 F and the ambient is always at 70 F. The average heat-transfer coefficient (including radiation) on the outside is 5 Btu/hr-ft²-F. The average properties of the wall are $k = 0.50$ Btu/hr-ft-F, $\rho = 150$ lbm/ft³ and $c = 0.167$ Btu/lbm-F.

Problem F·7

The Popeye Frozen Food Company freezes it spinach by first compressing it into large slabs and then sending them through a bath at –130 F. The slab of compressed spinach is

initially at 30 F; it must be reduced to a temperature of –30 F or below. The temperature at any part of the slab, however, must never drop below –60 F to prevent damage to the spinach.

Popeye has called you in to estimate the maximum thickness of the spinach slab which may be safely cooled in 60 minutes.

The properties of spinach may be taken to be
$k = 0.50$ Btu/hr-ft-F
$\rho = 60$ lbm/ft³
$c = 0.5$ Btu/lbm-F
Average latent heat = 144 Btu/lbm.

Problem F·8

A 0.5-in-thick fuel plate in a nuclear reactor is generating energy at the rate of 36,000 Btu/hr-in³. Coolant at a temperature of 80 F flows past each face. This maintains the surface temperature at 1160 F.

As a result of the temperature distribution within the plate, thermal stresses are present at this steady operating condition. These stresses may be assumed to be proportional to the maximum temperature difference in the plate. If it should become necessary to suddenly turn off the generation to reduce the stresses, determine the following:

(a) Is it better (from a thermal-stress viewpoint) to leave the fuel plate in the coolant or to pull it out of the reactor and into an ambient at 80 F where the heat-transfer coefficient is about 2.5 Btu/hr-ft²-F?
(b) For the fastest method of reducing thermal stresses, how long will it take for the maximum temperature difference in the plate to reach 5 percent of the initial temperature difference? How long will it take for the centerline temperature to reach 300 F?

The properties of the fuel plate may be taken as
$k = 25$ Btu/hr-ft-F
$\rho = 730$ lbm/ft³
$c = 0.029$ Btu/lbm-F.

Problem F·9

The cross section of a cylindrical fuel element in a nuclear reactor is shown in the sketch. The element is designed to operate surrounded by a 200-F coolant. An average heat-transfer coefficient between the cladding and the coolant is 1200 Btu/hr-ft²-F.

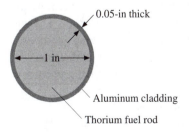

0.05-in thick

1 in

Aluminum cladding

Thorium fuel rod

The reactor safeguards committee is concerned about safety during start up of the reactor. When the control rods are suddenly pulled and generation begins in the thorium, there is the possibility that if the rods are accidentally pulled too far and/or too fast there may not be time to correct the trouble before melting of either the thorium or the aluminum occurs. NELG is about to run a test program in which such a fuel element is suddenly exposed to a neutron source so that generation begins suddenly. The test exposure flux will produce a heat generation rate of 40,000 Btu/hr-in³. The test will be controlled so that the ambient remains constant at 200 F (with h = 1200 Btu/hr-ft²-F) and the fuel element will be initially at 200 F. You have been called in to estimate the length of time before melting will occur (if it does indeed occur at this generation rate).

Use the following properties in your calculations:

Property	Thorium	Aluminum	Units
k	25	130	Btu/hr-ft-F
ρ	730	170	lbm/ft³
c	0.029	0.25	Btu/lbm-F
Melting point	3300	1200	F

Problem F•10

Reply to Mr. Diot's letter.

Mr. S. M. Arty
764 Conduction Avenue
Wise Acres, WI 75981

Dear Mr. Arty:

We have been testing long, cylindrical samples of stainless steel in our laboratory under constant heat-flux conditions at the outer radius. We are attempting to determine the length of time it takes the surface to reach its melting temperature of 2700 F. Two tests on 4-in-diameter cylinders have given the following results:

Test	q'' [Btu/hr-ft²]	Time [s]
1	30,000	1580
2	50,000	882

The initial temperature in each test was 80 F.

The laboratory assistant feels that the data should be fitted with the following hyperbola:

$$q''\theta = 47.4 \times 10^6 \qquad (1)$$

In (1), q'' is in Btu/hr-ft² and θ is in seconds. I am somewhat hesitant to rely on (1) even though it closely fits the data and has the right qualitative behavior. I would appreciate learning the answers to the following questions:
(a) Is (1) correct in predicting no influence of diameter?
(b) Is (1) correct in predicting that the thermal properties of the material have no effect?
(c) Does heat-transfer theory support (1)?
If (1) is not valid, please provide me with a better relation.
Sincerely,

I. Diot

Problem F•11

The wall of a shock tube consists of a $\frac{1}{4}$-in-thick steel plate with 1-in-thick glass wool insulation on the outside. Pressure and thermocouple leads have been taped to the outside of the steel with the traditional masking tape.

Prior to initiating the shock, the plate is uniform at the ambient temperature of 70 F. When the diaphragm is ruptured, the shock will travel at a Mach number of 3.0 through the tube. The temperature immediately behind the shock will suddenly jump to 960 F as the shock passes and then fall off exponentially back to the ambient temperature.

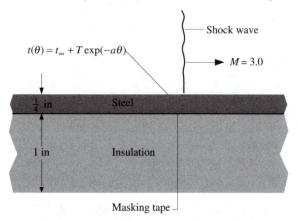

If it is assumed that the heat-transfer coefficient behind the shock is large, the inside wall temperature will closely follow the fluid temperature as given approximately by

$$t(\theta) = t_\infty + T\exp(-a\theta) \qquad \theta > 0$$

where $t_\infty = 70$ F, $T = 890$ F and $a = 0.32$ s⁻¹.

The test lab is concerned about whether the masking tape glue will melt and allow the pressure and thermocouple leads to come loose. The laboratory is heavily supported by government funds and overstaffed with engineers at present. Therefore you have been requested to make an analysis to see whether the masking tape will hold up. If the masking tape will not work, they want to know the temperature-time history on the outside of the steel so that the requirements of a suitable tape could be determined.

Use the following properties for the steel:
$k = 26$ Btu/hr-ft-F
$c = 0.11$ Btu/lbm-F
$\alpha = 0.50$ ft²/hr

Problem F•12

A thin rod with insulation around it is shown in the sketch.

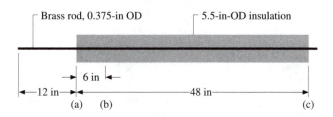

The left-hand end of the rod is periodically heated and cooled such that the periodic temperature variation at location (a) is as follows:

$$t_a(\theta) = t_A + T\cos(\omega\theta)$$

where $t_A = 180$ F and $T = 50$ F. The period is 4 minutes.

Derive an expression for the rod temperature [between locations (a) and (c)] as a function of time and position. Compare the temperature fluctuations at position (b) with those at position (a).

Use the following properties in your calculations:

Property	Brass rod	Insulation	Units
k	64.0	0.022	Btu/hr-ft-F
ρ	532	7.25	lbm/ft^3
c	0.092	0.20	Btu/lbm-F

Problem F•13

The cylinder walls of an internal combustion engine are subjected to oscillating temperatures as the process goes through its cycle. In practice, the temperature-time history at the surface of the cylinder wall is quite complicated. An approximation to this function might be obtained by a Fourier series analysis. This approximate function would contain many terms. A very much simplified problem would be to consider a sine-wave temperature variation.

If the engine cylinder (see Problem F•5) is operating at 1200 rpm, the inside surface temperature will undergo 600 cycles/min. If it is assumed that these are sinusoidal variations, the inside surface temperature $t(r_i,\theta)$ would be given by the expression

$$t(r_i,\theta) = T\sin(\omega\theta)$$

where $\omega = 1200\pi$ if θ is in minutes. The amplitude of the oscillations is T.

For this assumed form of the inside surface temperature variation,
(a) Derive an expression for the cylinder temperature as a function of distance from the inside surface.
(b) Determine the depth at which the amplitude of the oscillations is 10 percent of the amplitude at the surface.

Use the following properties in your calculations:

Property	Iron	Aluminum	Units
k	28	120	Btu/hr-ft-F
α	0.7	3.3	ft^2/hr

Problem F•14

You may recall that you were retained as a consultant by a patent attorney last September (see Problem F•1) to measure the thermal conductivity of a new insulating material. Just recently you received the following letter from the patent attorney:

Dear Sir:

I have recently reviewed your measurements of thermal conductivity and feel that they are too high. The developer of the material has made very careful tests himself and claims that he is right. He feels that you probably have not waited long enough for steady state to be attained.

I would welcome your comments regarding the length of time you estimate it should take to reach a satisfactory steady state. If your tests were made before steady conditions had been reached, would your measured results for conductivity be high as claimed by the inventor?

Your assistance in this matter will be appreciated.

Sincerely,

T. Netap Yenrotta
Patent Attorney

OH:007

Answer Mr. Yenrotta's letter and send him calculations to support your findings.

Problem F•15

The control rods in a proposed nuclear reactor are to have the cross-sectional shape shown in the sketch.

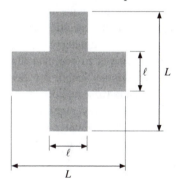

The physicists indicate that the boron isotope contained in the control blade will cause a volumetric heat-generation rate in the range of 1 to 7 kW/liter. The size of the blade will be somewhere between $L = 6$ in and 12 in. The relative proportion $L = 3\ell$ may be assumed to be fixed.

The blades will be immersed in a coolant at 200 F. The expected heat-transfer coefficient on the surface of the control blades will be about 2000 to 3000 Btu/hr-ft^2-F. The conductivity of the blades is about 20 Btu/hr-ft-F.

The thermal stress section would like you to:
(a) Determine the maximum blade temperature as a function of the blade size
(b) Determine the maximum blade temperature as a function of the generation rate
(c) Determine the temperature distributions within the control blades for various combinations of blade size and generation rate.

Problem F•16

A chimney manufacturer is interested in predicting the heat loss through its brick chimney walls. The cross-sectional geometry of the chimney under consideration is shown in the sketch. The heat-transfer coefficient on the inside is about 20 Btu/hr-ft²-F. Due to the possibility of having various wind velocities on the outside, the external heat-transfer coefficient can range from 3 to 50 Btu/hr-ft²-F.

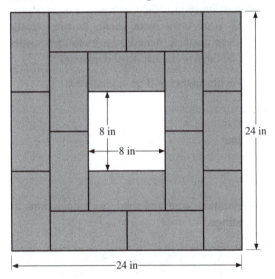

The manufacturer claims that the steady-state heat loss will increase as the overall temperature difference increases and as the thermal conductivity, density and specific heat of the wall increase, but has no quantitative numbers to show these effects. You are to provide curves to estimate the steady-state heat loss per foot of chimney length for any inside temperature, outside temperature, outside heat-transfer coefficient, average conductivity, average density and average specific heat which he might encounter.

Problem F•17

A brake for a rotating, 2-in-diameter steel shaft consists of a split, 4-in-square wooden block that is squeezed together to exert a frictional force on the shaft. The shaft is centered inside the block. In a certain application the frictional energy-generation rate per unit shaft surface area is 1.0 Btu/hr-in². Assume there is no contact resistance between the shaft and the block. The average heat-transfer coefficient on the outside of the block is 2.0 Btu/hr-ft²-F. The ambient temperature is 70 F.

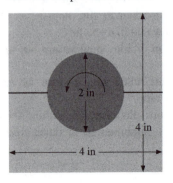

(a) Determine the steady-state temperature distribution around the circumference of the shaft.
(b) If the wooden block and the shaft are both initially at 70 F when the brake is applied, determine how long it will take the shaft surface temperature to attain 90 per cent response.

Property	Steel	Wood	Units
k	26	0.10	Btu/hr-ft-F
ρ	490	40	lbm/ft³
c	0.11	0.57	Btu/lbm-F

Problem F•18

The core of a nuclear reactor is composed of 0.5-in-diameter uranium-carbide fuel rods surrounded by a graphite moderator. The core is cooled by helium flowing through 1.0-in-diameter passages. The arrangement of the fuel rods, flow passages and moderator is shown in the sketch. The generation rate within the fuel rods is 5000 Btu/hr-in³ and in the moderator the rate is 200 Btu/hr-in³. At this cross-section, the helium temperature is 1000 F and the heat-transfer coefficient is 300 Btu/hr-ft²-F. For this reactor,

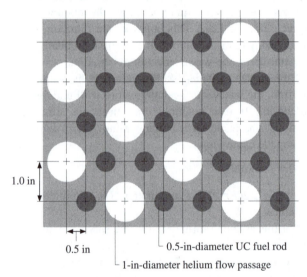

0.5 in   └ 0.5-in-diameter UC fuel rod
        └ 1-in-diameter helium flow passage

(a) Determine the maximum temperature at the steady-state operating condition.
(b) Plot a few steady-state isotherms.
(c) If the generation rate is then turned off and the helium temperature is reduced to 500 F, determine how long it will take for the maximum temperature to reach 600 F.

Use the following properties in your calculations:

Property	UC	Graphite	Units
k	11.6	60	Btu/hr-ft-F
ρ	810	112	lbm/ft³
c	0.035	0.3	Btu/lbm-F

G

COURSE CONTENT

This book was originally developed for students who would be using it for self-study and who would not be attending lectures. This appendix is intended to serve as a reading guide for self-study. A sample program of study is laid out for a one-semester course (45 class sessions). The student is expected to spend three to four hours per class session in studying the book, working exercises and/or solving comprehensive problems selected from Appendix F.

G•1 CLASS SCHEDULE

Day	Sections in book	Day	Sections in book
1	Preface, 1•0	24	8•0, 8•1
2	1•1	25	8•2
3	1•2	26	8•2
4	7•0, 7•1	27	8•3
5	2•0, 2•1	28	8•3
6	2•2	29	8•4
7	2•2	30	8•4
8	2•3, 2•4	31	8•5
9	3•0, 3•1	32	8•5
10	3•1	33	8•5
11	3•1	34	8•6
12	3•1	35	9•0, 9•1
13	7•2•1, 7•2•2	36	9•2
14	3•2	37	9•2
15	3•2	38	9•2
16	3•3	39	9•2
17	3•3	40	9•2
18	4•1, 4•2	41	9•3
19	4•2	42	9•4
20	5•2	43	9•5
21	6•4•2	44	9•6
22	7•2•3, 7•2•4	45	Final examination
23	Midterm examination		

G•2 NOTE ON COMPREHENSIVE PROBLEMS

The comprehensive problems in Appendix F can be so important, if given sufficient emphasis, that additional comments regarding them are in order.

One objective of the comprehensive problems is to demonstrate the utility of the mathematics presented in the book in solving engineering problems. It is up to the individual student to make the transition between engineering situation and mathematical model in applying the theory to a specific problem. The student must then interpret the solution to find the significant engineering results. To make this process meaningful, the problems must be comprehensive and more than an algebraic or a numerical exercise.

Another objective of this course is to give you practice in communicating what you have discovered to other engineers. The problem format discussed below is an attempt to teach you this skill. These sections should be presented in the indicated order.

1. *Problem statement*   Concisely restate the problem and list the aims of the assignment.

2. *Summary*   This is the answer to the problem statement. State explicitly and concisely what you have found out. Present your key answers and conclusions.

3. *Abbreviated calculations*   Write this section up like you would for presentation as a technical paper. Put down the important assumptions, relations and results. Include lots of descriptive words so that someone else can follow your work. Eliminate unnecessary algebra.

4. *Discussion*   Draw conclusions from your work, discuss the significance of the answers and justify your assumptions. This section should be from one to three pages long.

5. *Appendix (optional)*   Include here any detailed calculations, calculus and algebra that you think is pertinent, necessary or that you want to keep, but would clutter up the main body of the problem.

The comprehensive problems should be written up on $8\frac{1}{2} \times 11$ paper. Please use only one side of the paper. Pencil and handwriting are acceptable as long as your work is *legible*.

A letter grade (*e.g.*, A, B+, etc.) will be given for each comprehensive problem. A grade of B is the minimum grade expected of any graduate student. It means that the mathematics is correctly done and you have answered the specific questions asked. For a grade of A or A– you must show some additional initiative or insight beyond the minimum expected of everyone. A grade of C+ or C, means that the work is not up to the standard for graduate students.

Use of the computer will be essential when using finite-difference methods and it may at times be valuable during other portions of the course. Students should already know how to program and they should have easy access to a computer.

APPENDIX

H

CHAPTER COMMENTS

The following comments should be helpful in determining the main points of each section in the book. They should be read before studying a particular section and again afterward to be sure you have the main idea in mind. The comments on the exercises are intended to help you decide how most effectively to use them in your study.

CHAPTER 1

Section 1•1 Be able to define a system and set up differential equations. Know how to choose the proper boundary and/or initial conditions.

Appendix A Be familiar with these methods of solving ordinary differential equations. They should be review, but perhaps you are rusty.

Section 1•2 Be able to apply the theory of ordinary differential equations to the solution of heat-transfer problems. The notion of a thermal circuit is also handy.

These first three comments should all be review. If you are rusty, take out some extra time to brush up on this material. You should be thoroughly familiar with this after the first week.

Exercises You should be able to do all these exercises. Try exercises 1•3, 1•5, 1•11, 1•17 and 1•18. Try as many as time permits to be sure your background is satisfactory. You should at least be able to determine whether you understand the concepts and could solve the problems if you had the time. Exercise 1•24 is tricky.

CHAPTER 2

Section 2•0 Review differential equation derivation and normalization.

Section 2•1•1 Become familiar with using series summation signs and manipulating series. This ability is quite handy in heat-transfer analysis.

Section 2•1•2 Read, but take mostly on faith.

Section 2•2•1 Read this section carefully enough so that you follow each step. But don't worry—you won't be expected to reproduce this material or similar material. This section should remind you of series solutions you have rubbed elbows with in the past. You should also be impressed with the engineering impracticality of going through a series solution on your own each time you run

into a Bessel equation. Section 2•4 will be the efficient way to solve Bessel equations.

Section 2•2•2 Just as with trigonometric functions, it will often be helpful to have a qualitative idea of the behavior of Bessel functions.

Section 2•2•3 Observe the extension to more general equations and functions. Again learn about the behavior of these new functions.

Section 2•2•4 Learn to differentiate and manipulate series.

Section 2•2•5 Become familiar with this new function. Learn how to compute numerical values. Know its relation to the factorial function.

Section 2•2•6 Get accustomed to the mathematical manipulations involved.

Section 2•2•7 Again, the integration and manipulation of series is important. In practice, you will usually use integral tables to evaluate integrals.

Section 2•2•8 The ideal of finding asymptotic solutions to equations is often quite valuable to the engineer. This is an example which also serves to show that Bessel functions are related to trigonometric functions.

Sections 2•3•1 to 2•3•3 These new functions should remind you of the functions you are already familiar with. They are relatives of the exponential functions.

Section 2•4 This is the way to solve Bessel equations! Concentrate on learning how to use (2·4·1) and (2·4·2). The example shows how to manipulate Bessel functions land make numerical calculations. This should be the most useful section in Chapter 2—STUDY IT.

Section 2•5 This is another set of functions closely related to Bessel functions. Be aware that Kelvin functions exist, but don't concentrate on this section.

Exercises As representative of material you should master in this chapter, try exercises 2•3, 2•4, 2•7, 2•12 and 2•23 to 2•25.

CHAPTER 3

Section 3•0 There is a distinct difference between homogeneous and nonhomogeneous problems. You must be able to recognize the nonhomogeneous terms in a problem.

Section 3•1•1 This section illustrates the typical steps in the separation-of-variables method. Since it is the most straightforward of any example, study the steps. Learn the steps in the procedure and the reasoning behind them. This will be of great value in later examples. Review this section again after reading sections 3•1•2 and 3•1•3.

Section 3•1•2 This section should fill you in on why we evaluate the coefficients of the series as we do and why we need homogeneous boundary conditions.

Section 3•1•3 This section illustrates the effect of the initial temperature profile and the boundary conditions on the rate of convergence of the series solution. These ideas are helpful for checking your solution.

Section 3•1•4 Convection is a very important heat-transfer boundary condition. You should know how to handle it. The limiting-case solutions for small and large Biot numbers show how the convection solution fits in with the lumped-parameter and the specified-temperature solutions.

Section 3•1•5 This nonzero heat-flux boundary condition is an example of a run-away solution. It illustrates the idea of a transient that approaches a "fully established" solution.

Section 3•1•6 This is an overall summary of how to treat problems when the nonhomogeneous terms are not time dependent.

Section 3•1•7 Variation of parameters can be used to solve the problems with time-dependent nonhomogeneous terms. In Example 1 you should note the reason that the variation-of-parameter solution is slow to converge and the steps that can be taken to improve convergence.

Section 3•2 Note that only relatively simple shapes can be treated by separation of variables.

Section 3•2•1 This section illustrates how the techniques of Section 3•1 can be applied to two-dimensional, steady-state problems. Note the requirement for "mostly" homogeneous boundary conditions.

Section 3•2•2 The methods of Section 3•1 can be applied to two-dimensional, transient problems. This section should be a good review of these techniques.

Section 3•3•1 This section is similar to Section 3•1•1. Note that Bessel functions appear quite naturally in cylindrical-coordinate problems. The use of a weighting function for orthogonality is required here.

Section 3•3•2 This section is similar to Section 3•1•4. Again Bessel functions appear in cylindrical coordinates.

Section 3•3•3 Hollow cylinders are more complicated since Bessel functions of both first and second kind are required. Note the use of the cylinder function to simplify the writing of the solution.

Section 3•3•4 A periodic boundary condition is needed when there is peripheral variation of surface temperature.

Section 3•3•5 The methods of earlier sections may also be applied to problems in spherical coordinates. Note also that a neat little trick may be used to convert this sphere problem into a plane-wall problem.

Section 3•3•6 This is a handy way to solve a few complicated problems by combining some more-elementary solutions. Graphs for the transient response of plane walls and infinite cylinders (found in most undergraduate texts) can be used to analyze rectangular solids and short cylinders.

Exercises For three relatively straightforward exercises, try 3•8, 3•10 and 3•14. Exercises 3•17, 3•20, 3•23, 3•24, 3•26, 3•30, 3•31, 3•32 and 3•38 are typical separation-of-variables problems, but, of course, they require time and effort to solve. You may wish to try only one or two. In this event you will find 3•17 to be quite typical. Exercise 3•32 will help review Bessel functions and is interesting.

CHAPTER 4

Section 4•0 Superposition is very useful for solving linear, nonhomogeneous problems. Time-dependent boundary conditions and generation can be treated.

Section 4•1 Observe how complicated boundary conditions and equations can be split up into simpler problems.

Section 4•2 Get the flavor of how "little steps" can be added together to obtain the total solution. Don't worry about memorizing the details of the end result. Be able to use Duhamel's theorem (4·2·1), if given to you, to solve problems.

Exercises For a typical superposition problem using Duhamel's theorem, try Exercise 4•8. It also brings out an interesting point regarding series convergence. Exercise 4•14 reviews sections 4•1 and 4•2.

CHAPTER 5

Section 5•0 Complex combination is useful for obtaining the sustained solution to problems with periodic inputs. Know the general steps listed here.

Section 5•1 This is an application with an ordinary differential equation. Be able to solve similar problems. Review your complex variables.

Section 5•2 Observe a new geometry—the semi-infinite solid. Read more on normalization in Section 7•2•3. Learn the method of complex combination and review complex numbers.

Exercises For two typical problems and good practice try exercises 5•1 and 5•2. Exercise 5•3 points out an interesting idea regarding penetration depth. Exercise 7•9 should also be considered here. Exercise 5•7 is longer than the others but brings in Kelvin functions.

CHAPTER 6

Section 6•0 Laplace transforms provide another method of solving linear problems, which, in certain cases, is much easier than separation of variables.

Section 6•1•1 Learn the definition of the Laplace transformation and the properties discussed in Example 3. You will usually refer to Appendix E (or more complete versions found in other books) when you need to find the transform of a function.

Section 6•1•2 You must know how to find the transforms of derivatives. Appendix E can also be used or you can work them out on your own.

Section 6•1•3 You must know how to find the transforms of partial derivatives in order to solve partial differential equations.

Section 6•2•1 Review this method of splitting a complicated fraction into the sum of several simpler fractions.

Section 6•2•2 Be able to use the theorem if given to you. The proof is presented primarily to show you how to interchange the order of integration.

Section 6•3 Note the steps for solving ordinary differential equations.

Section 6•4 Note the steps for solving partial differential equations.

Section 6•4•1 This is a detailed example of the solution to a partial differential equation. Learn the procedure.

Section 6•4•2 This is a useful new problem which is easily (and best) handled by Laplace transforms. An important new function (the error function) is introduced with which you should become familiar. Contact temperature is an interesting heat-transfer concept which makes use of the semi-infinite-solid solution. Section 7•3 should be read to learn of another use for normalization.

Exercises To review Chapter 6, try exercises 6•8 to 6•11. Do 6•9 and 6•10 to compare the method with other methods for solving the same problems.

CHAPTER 7

Section 7•1 This section shows how to reduce the number of parameters that appear in a problem.

Section 7•1•1 Results can be condensed by normalization.

Section 7•1•2 Normalization can be applied directly to differential equations to condense the number of parameters. Learn this technique.

Section 7•2 Normalization can be used to obtain a better physical understanding of a problem.

Section 7•2•1 The Biot number is an important quantity in conduction transients. Understand its significance.

Sections 7•2•2 and 7•2•3 Observe that "mathematically defined" normalization constants do have physical significance. This is a useful application of normalization.

Section 7•2•4 Normalization avoids having to deal with extreme sizes of numbers. In normalized form, the most significant values of the variables are usually between 0.1 and 10.

Section 7•3 This is an important simplification if it can be done. It is used in many laminar boundary-layer problems.

Exercises To review Chapter 7 try exercises 7•5, 7•9 and 7•12. You could try 7•13 as preparation for a study of convection if you wish.

CHAPTER 8

Section 8•1•1 Matrices are a handy way to handle the systems of equations that will appear in finite differences and finite elements. You should become familiar with the basic operations in this section.

Section 8•1•2 Finite-difference models arise from using the approximations to the rate equations described in this section.

Section 8•2 Approximation of an irregular shape by using a set of regularly spaced nodal points is the starting point for the finite-difference discussion in Chapter 8.

Section 8•2•1 Be able to make energy balances on interior nodal systems and use approximate rate equations to derive finite-difference equations. Note how each equation fits into the global system of equations which is written using matrix notation. It is important to note that the equations are symmetric and to maintain this symmetry to reduce computer storage requirements and calculation time.

Section 8•2•2 Be able to make energy balances on boundary nodal systems and use approximate rate equations to derive finite-difference equations. Note how each equation fits into the global system of equations. Special attention must be given to nodal systems on a specified-temperature boundary.

Section 8•2•3 For a transient problem the initial temperature of each node must be specified.

Section 8•2•4 This section gives a concise summary of the discretization process.

Section 8•2•5 This section shows how the calculation of boundary heat flows is related to the matrices derived in the discretization procedure.

Section 8•3 For a steady-state problem, the finite-difference system of differential equations reduces to a system of algebraic equations.

Section 8•3•1 Note how the system of algebraic equations is first derived (one equation or line at a time) and then modified to account for specified boundary temperatures.

Section 8•3•2 Gaussian elimination is a standard method of solving a system of algebraic equations but does not take advantage of the symmetry of the finite-difference equations.

Section 8•3•3 Cholesky decomposition takes advantage of the symmetry of the finite-difference equations and is better suited for solving these equations. Appendix I•1 lists two FORTRAN subroutines to carry out the solution.

Section 8•3•4 This section includes a FORTRAN program listing for reading the matrix entries and then solving for the nodal temperatures and heat flows. Output for an example is given. Note how the accuracy of the solution improves as the number of nodes increases.

Section 8•4•1 The system of differential equations is derived one equation or matrix row at a time. The convection boundary condition is also shown.

Section 8•4•2 You should understand the Euler technique (8.4.11) for moving ahead in time. FORTRAN statements that will accomplish this solution are given.

Section 8•4•3 You should understand the Crank-Nicolson technique (8.4.19) for moving ahead in time. FORTRAN statements that will accomplish this solution are given.

Section 8•4•4 Study the numerical results in this section to see the effect of nodal spacing and solution technique (Euler versus Crank-Nicolson) affect the solution. Note the unstable oscillations that can appear in the Euler solution. An estimate of the critical time step may be obtained from (8.4.29).

Section 8•5 This is an introductory section. Sections 8•5•1 through 8•5•5 can be omitted if you do not care about the insights that can be obtained by a more-detailed mathematical analysis.

Section 8•5•1 The exact solution (*i.e.*, without either Euler or Crank-Nicolson approximations) of the finite-difference system of differential equations is derived in terms of eigenvalues and eigenvectors. Comparisons are made with the analytical separation-of-variables solution of the example in Section 3•2•2.

Section 8•5•2 An analysis of the Euler and Crank-Nicolson approximate solutions is given in terms of eigenvalues and eigenvectors. This analysis shows the cause of the unwanted oscillatory behavior that can occur in the Euler and Crank-Nicolson solutions.

Section 8•5•3 This section compares the exact, Euler and Crank-Nicolson solutions and obtains a relation (8.5.69) for the critical time step in terms of the maximum eigenvalue.

Section 8•5•4 This section uses matrix theory to obtain an approximation for the critical time step (8.5.77) that is easier to calculate than (8.5.69).

Section 8•6•1 Note some of the pitfalls of using irregularly spaced nodes in finite differences.

Section 8•6•2 Additional modifications must be made to treat time-dependent specified nodal temperatures but the basic theory is the same.

Section 8•6•3 Note the elaborate steps required to handle problems with variable thermal properties.

Section 8•6•4 Two-dimensional finite-difference theory in x, y coordinates is readily extended to r, z coordinates.

Exercises Most of the exercises involve relatively few nodes to illustrate the theory without excessive calculation. Several one-dimensional exercises are given for this reason.

CHAPTER 9

Section 9•1 The basic problem considered in Chapter 9 is the same as that considered in Chapter 8.

Section 9•1•1 These matrix operations will be used later on.

Section 9•1•2 The conversion of the partial differential equation (9.1.5) and its boundary conditions into a system of algebraic equations (9.1.13) is accomplished in this section. Note that, based on the separation of variables solutions in Chapter 3, the assumed form of the finite-element solution (9.1.8) is fairly plausible.

Section 9•1•3 Triangular coordinates will be helpful in the analysis later on, particularly the integral in (9.1.23).

Section 9•2 To discretize the region, nodes are connected into triangular elements. The temperature within a triangular element is assumed to vary linearly between the temperatures of the nodes at the three vertices of the element.

Section 9•2•1 Note the derivation of the capacitance matrix (9.2.10), the conduction matrix (9.2.12) and the generation vector (9.2.13). A major difference between finite elements and finite differences is that the finite-element capacitance matrix is not diagonal.

Section 9•2•2 This section incorporates the boundary conditions into the analysis. The most important results are the specified-heat-flux vector (9·2·16), the convection vector (9·2·17) and the convection matrix (9·2·18). The energy significance of these quantities should be noted. Specified temperatures are treated as in finite difference except that now modifications to the capacitance matrix must also be made.

Section 9•2•3 This section gives energy interpretations to the finite-element vectors and shows how element heat flux (9·2·21) and boundary heat inflows (9·2·40) should be calculated.

Section 9•2•4 This section shows some of the capabilities of *FEHT* in discretizing two-dimensional regions, carrying out the finite-element calculations and presenting the results.

Section 9•3•1 Compare this steady-state example to the finite-difference analysis in Section 8•3•1.

Section 9•3•2 Compare the finite-element numerical results to the finite-difference results in Section 8•3•4.

Section 9•4•1 Compare this transient example to the finite-difference analysis in Section 8•4•1.

Section 9•4•2 Compare the finite-element numerical results to the finite-difference results in Section 8•4•4. Comparison of Figure 9•32 with Figure 8•22 indicates that unstable Euler solutions will be more troublesome in finite elements than in finite differences. Fortunately, due to the nondiagonal finite-element capacitance matrix, there is no penalty to use Crank-Nicolson to avoid unstable oscillations. Note that the estimate of the critical time step can often be exceeded and still obtain an acceptable solution.

Section 9•5•1 The exact solution of the finite-element equations (9·5·1) is identical to the exact finite-difference solution except that the eigenvalues and eigenvectors contain different numerical values. Note that the finite-element results approach the analytical separation-of-variables results.

Section 9•5•2 The Euler and Crank-Nicolson finite-element solutions show the same types of behavior as in finite differences.

Section 9•5•3 This involved proof of the finite-element critical-time-step estimate (9·4·8) is presented for the very curious student. The result may not be too important in view of the fact that satisfactory transient solutions can often be obtained even if the time step exceeds the critical time step.

Section 9•6•1 There are some advantages to lumping the capacitance on the main diagonal for Euler solutions.

Section 9•6•2 The finite-element analysis of problems that can be modeled using r, z coordinates follows steps similar to the analysis in x, y coordinates.

Section 9•6•3 A tetrahedral element for three-dimensional geometries is a natural extension of the two-dimensional triangular element discussed earlier in this chapter.

Section 9•6•4 There are many other elements used in stress analysis. Some of these are indicated in this section.

Exercises Most of the examples involve only a few nodes to illustrate the theory without excessive computations.

APPENDIX

I

COMPUTER SUBROUTINES

This appendix contains three FORTRAN subroutines that are very useful in computing finite-difference and finite-element solutions to conduction problems. Since the matrices in these problems will be banded and symmetric, an *economy-banded* form of storage can be used to save space in the computer. In this storage scheme only the entries in the band that are on or above the main diagonal are stored. An $N \times N$, symmetric matrix $\mathbf{S}$ with an upper bandwidth IBW ($IBW = 3$ in this example),

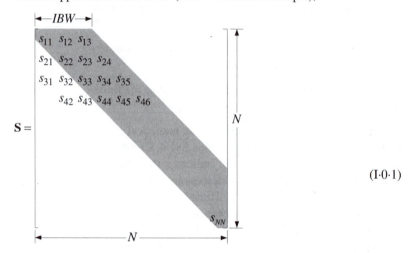

$$ (\mathrm{I}\cdot 0\cdot 1) $$

would be stored in economy-banded form as an $N \times IBW$ matrix $\mathbf{A}$ given by

$$\mathbf{A} = \begin{bmatrix} a_{11} & a_{12} & a_{13} \\ a_{21} & a_{22} & a_{23} \\ a_{31} & a_{32} & a_{33} \\ a_{41} & a_{42} & a_{43} \\ & & \\ & & 0 \\ a_{N1} & 0 & 0 \end{bmatrix} = \begin{bmatrix} s_{11} & s_{12} & s_{13} \\ s_{22} & s_{23} & s_{24} \\ s_{33} & s_{34} & s_{35} \\ s_{44} & s_{45} & s_{46} \\ & & \\ & & \\ s_{NN} & & \end{bmatrix}$$

$$ (\mathrm{I}\cdot 0\cdot 2) $$

Observe that the entries on the main diagonal in (I·0·1) are stored in the first column in (I·0·2). The diagonal just above the main diagonal in (I·0·1) is stored in the second column in (I·0·2) and so on. Another way of stating the economy-banded storage is that each row in (I·0·1), starting from the main diagonal and moving to the right, is stored in the corresponding row in (I·0·2). The value of s_{42} can be found by using symmetry to note that $s_{42} = s_{24}$ and s_{24} is stored as a_{23} in $\mathbf{A}$.

Each of the subroutines in this appendix uses economy-banded storage for matrix $\mathbf{A}$.

I•1 CHOLESKY SUBROUTINES FOR SOLVING Ax = b

The Cholesky square-root method of solving $\mathbf{Ax} = \mathbf{b}$ first decomposes $\mathbf{A}$ to get an upper triangular matrix $\mathbf{U}$ (same bandwidth as $\mathbf{A}$) such that

$$\mathbf{A} = \mathbf{U}^T\mathbf{U} \tag{I·1·1}$$

Then $\mathbf{U}^T\mathbf{Ux} = \mathbf{b}$ is solved by first letting $\mathbf{Ux} = \mathbf{y}$ and then solving

$$\mathbf{U}^T\mathbf{y} = \mathbf{b} \tag{I·1·2}$$

to find $\mathbf{y}$, and then finding $\mathbf{x}$ by solving

$$\mathbf{Ux} = \mathbf{y} \tag{I·1·3}$$

Subroutine *DBAND* decomposes $\mathbf{A}$ to find $\mathbf{U}$ in (I·1·1). At call, $\mathbf{A}$ is in A; at return, $\mathbf{U}$ is in A. The original values in $\mathbf{A}$ are destroyed.

```
      SUBROUTINE DBAND(NROW,NCOL,N,IBW,A,NOGO)
C     ****************************************************************
C     * This subroutine computes the Cholesky decomposition of a real, *
C     * symmetric, positive definite matrix.                         *
C     * Input:   NROW = row dimension, defined in calling program    *
C     *          NCOL = column dimension, defined in calling program *
C     *             N = number of rows in matrix                     *
C     *           IBW = bandwidth of matrix (This is also the number of *
C     *                 columns needed to store diagonal and upper band *
C     *                 of matrix in an N by IBW rectangular array.) *
C     *             A = real, symmetric, positive definite matrix    *
C     *                 (Diagonal and upper band stored as an N by IBW *
C     *                 array.)                                      *
C     *          NOGO = 0 if decomposition works                     *
C     *               = 1 if decomposition fails                     *
C     * Output:    A = diagonal and upper band of the decomposition  *
C     ****************************************************************
      DIMENSION A(NROW,NCOL)
      DOUBLE PRECISION DSUM,D1,D2,D3,DSQRT
      NOGO = 0
      DO 6 I=1,N
         IP = N - I + 1
         IF (IP.GT.IBW) IP = IBW
         DO 6 J=1,IP
            IQ = IBW - J
            IF (IQ.GT.I-1) IQ = I - 1
            DSUM = A(I,J)
            IF (IQ.LT.1) GO TO 2
               DO 1 K=1,IQ
                  IMK = I - K
                  KP1 = K + 1
                  JPK = J + K
                  D1 = A(IMK,KP1)
                  D2 = A(IMK,JPK)
                  DSUM = DSUM - D1*D2
1              CONTINUE
2           CONTINUE
            IF (J.GT.1) GO TO 5
               IF (DSUM) 3,3,4
3              CONTINUE
                  NOGO = 1
                  RETURN
4              CONTINUE
               D3 = DSQRT(DSUM)
               A(I,J) = D3
               GO TO 6
5           CONTINUE
            A(I,J) = DSUM/D3
6     CONTINUE
      RETURN
      END
```

Subroutine *SBAND* solves $\mathbf{U}^T\mathbf{U}\mathbf{x} = \mathbf{b}$ to find $\mathbf{x}$. The first DO loop in *SBAND* finds $\mathbf{y}$ by solving (I·1·2) and the second DO loop finds $\mathbf{x}$ by solving (I·1·3). When *SBAND* is finished, A will still contain $\mathbf{U}$, B will still contain $\mathbf{b}$ and X will contain the solution $\mathbf{x}$.

```
      SUBROUTINE SBAND(NROW,NCOL,N,IBW,A,B,X)
C     ****************************************************************
C     * This subroutine solves  A*x = b  when given the Cholesky    *
C     * decomposition of A (A = real, symmetric, positive definite).*
C     * Input:  NROW = row dimension, defined in calling program    *
C     *         NCOL = column dimension, defined in calling program *
C     *            N = number of rows in matrix                      *
C     *          IBW = bandwidth of matrix (This is also the number of *
C     *                columns needed to store diagonal and upper band *
C     *                of matrix in an N by IBW rectangular array.)  *
C     *            A = Cholesky decomposition of matrix A (Diagonal  *
C     *                and upper band stored as an N by IBW array.)  *
C     *            B = right-hand-side vector b                      *
C     * Output:    A = same as input                                *
C     *            B = same as input                                *
C     *            X = solution vector x                            *
C     ****************************************************************
      DIMENSION A(NROW,NCOL),B(NROW),X(NROW)
      DOUBLE PRECISION DSUM,D1,D2
      DO 3 I=1,N
         J = I - IBW + 1
         IF (I+1.LE.IBW) J = 1
         DSUM = B(I)
         K1 = I - 1
         IF (J.GT.K1) GO TO 2
            DO 1 K=J,K1
               IMKP1 = I - K + 1
               D1 = A(K,IMKP1)
               DSUM = DSUM - D1*X(K)
1           CONTINUE
2        CONTINUE
         X(I) = DSUM/A(I,1)
3     CONTINUE
      DO 6 I1=1,N
         I = N - I1 + 1
         J = I + IBW - 1
         IF (J.GT.N) J = N
         DSUM = X(I)
         K2 = I + 1
         IF (K2.GT.J) GO TO 5
            DO 4 K=K2,J
               KMIP1 = K - I + 1
               D2 = A(I,KMIP1)
               DSUM = DSUM - D2*X(K)
4           CONTINUE
5        CONTINUE
         X(I) = DSUM/A(I,1)
6     CONTINUE
      RETURN
      END
```

I•2　SUBROUTINE FOR COMPUTING y = Ax + b

Subroutine *YAXPB* performs the computations to find **y** where

$$\mathbf{y} = \mathbf{Ax} + \mathbf{b}$$

When *YAXPB* is called, A, X and B must contain **A**, **x** and **b**, respectively.
The result **y** is returned in Y.

```
      SUBROUTINE YAXPB(NROW,NCOL,N,IBW,A,B,X,Y)
C     ************************************************************
C     * This subroutine computes  y = A*x + b  when  A  is a real,  *
C     * symmetric matrix                                           *
C     * Input:  NROW = row dimension, defined in calling program   *
C     *         NCOL = column dimension, defined in calling program *
C     *            N = number of rows in matrix                     *
C     *          IBW = bandwidth of matrix (This is also the number of *
C     *                columns needed to store diagonal and upper band *
C     *                of matrix in an N by IBW rectangular array.) *
C     *            A = real, symmetric matrix (Diagonal and upper band *
C     *                stored as an N by IBW array.)                *
C     *            B = right-hand-side vector b to be added to  A*x *
C     *            X = right-hand-side vector x to be multiplied by A *
C     * Output:   A = same as input                                *
C     *           B = same as input                                *
C     *           X = same as input                                *
C     *           Y = left-hand-side vector y = A*x + b            *
C     ************************************************************
      DIMENSION A(NROW,NCOL),B(NROW),X(NROW),Y(NROW)
      DO 3 I=1,N
         Y(I) = B(I)
         JMAX = N - I + 1
         IF (JMAX.GT.IBW) JMAX = IBW
         DO 1 J=1,JMAX
            II = I + J - 1
            Y(I) = Y(I) + A(I,J)*X(II)
1        CONTINUE
         IF (I.EQ.1) GO TO 3
         JMAX = IBW
         IF (I.LT.IBW) JMAX = I
         DO 2 J=2,JMAX
            II = I - J + 1
            Y(I) = Y(I) + A(II,J)*X(II)
2        CONTINUE
3     CONTINUE
      RETURN
      END
```

APPENDIX

J

NOTATION

This appendix lists symbols that appear frequently in the text. Dimensions are given in terms of energy [E], mass [M], length [L], time [Θ] and temperature [T]. Commonly used SI units for these dimensions are Joule, kilogram, meter, second and degree Kelvin, respectively. Commonly used English units for these dimensions are Btu, pound mass, foot, hour and degree Fahrenheit, respectively. The section number refers to the first section in which the symbol is used or defined.

Table J•1 *English letter symbols.*

Symbol	Definition	Section	Dimensions
A	Area	1•1•2	L^2
c	Specific heat	1•1•2	E/M-T
c	Entry in **C** matrix	8•2•1	E/T
$ç$	Entry in **Ç** matrix	8•4•2	E/T
C	Thermal capacitance	1•2•5	E/T
C	Capacitance matrix	8•2•1	E/T
Ç	Modified **C** matrix	8•2•4	E/T
D	Diagonal matrix	8•3•2	many
e	Base of natural logarithms	7•2•2	none
E	Energy	1•1•1	E
f	Weighting function	9•1•2	none
g	Energy generation rate	1•1•1	E/Θ
g	Generation vector	8•2•1	E/Θ
h	Convection heat-transfer coefficient	1•1•2	$E/\Theta\text{-}L^2\text{-}T$
h	Convection vector	8•2•2	E/Θ
H	Biot number, hL/k	3•1•4	none
H	Convection matrix	8•2•2	E/Θ-T
i	$\sqrt{-1}$	2•3•3	none
I	Identity matrix	8•1•1	none
k	Thermal conductivity	1•1•2	E/Θ-L-T
k_x	Thermal conductivity in the x-direction	1•1•3	E/Θ-L-T
k_y	Thermal conductivity in the y-direction	1•1•3	E/Θ-L-T
k_z	Thermal conductivity in the z-direction	1•1•3	E/Θ-L-T
k_r	Thermal conductivity in the r-direction	1•1•3	E/Θ-L-T
k_ϕ	Thermal conductivity in the ϕ-direction	1•1•3	E/Θ-L-T
k_ψ	Thermal conductivity in the ψ-direction	1•1•3	E/Θ-L-T
K	Conduction matrix	8•2•1	E/Θ-T
L	Characteristic length	1•1•3	L
L	Lower triangular matrix	8•3•2	many
m	$\sqrt{hp/kA}$	1•2•4	1/L
N	Total number of nodes	8•3•2	none
NE	Total number of finite elements		none
p	Perimeter	1•1•3	L
q	Heat-transfer rate	1•1•1	E/Θ
q	Heat-flow vector	8•2•2	E/Θ
$\mathbf{q}_o$	Vector of nodal heat inflows	8•2•5	E/Θ

Symbol	Definition	Section	Dimensions
r	Radial coordinate	1•1•3	L
r	Modified $(\mathbf{g} + \mathbf{q} + \mathbf{h})$ vector	8•2•4	E/Θ
R	Thermal resistance	1•2•1	$\Theta\text{-}T/E$
s	Parameter in Laplace transformation	6•1•1	none, $1/\Theta$
s	Surface coordinate	9•2•2	L
S	Surface area	8•2•2	L^2
S	Modified $(\mathbf{K} + \mathbf{H})$ matrix	8•2•4	$E/\Theta\text{-}T$
t	Temperature	1•1•2	T
$\dot{t}$	Time derivative of t	8•2•1	T/Θ
t	Vector of nodal temperatures	8•2•1	T
ṫ	Time derivative of **t**	8•2•1	T/Θ
T	Absolute temperature	1•1•2	T
T	Temperature amplitude	1•2•5	T
U	Fundamental solution		none
U	Upper triangular matrix	8•3•2	many
V	Volume	1•1•2	L^3
w	Complex combination of $u + iv$	5•0	T
w	Weighting function vector	9•1•2	none
x	Position coordinate	1•1•2	L
x_{ij}	$x_j - x_i$		L
x	Eigenvector	8•5•1	T
x	Normalized eigenvector	8•5•1	$\sqrt{T/E}$
x'	Modified normalized eigenvector	8•5•1	T
X	Eigenvector matrix	8•5•1	$\sqrt{T/E}$
X'	Modified eigenvector matrix	8•5•1	T
y	Position coordinate	1•1•3	L
y_{ij}	$y_j - y_i$		L
z	Position coordinate	1•1•3	L
$\mathbf{1}(\theta)$	Unit step function	4•2	none

Table J•2 *Greek letter symbols.*

Symbol	Definition	Section	Dimensions
α	Thermal diffusivity, $k/\rho c$	3•1•1	L^2/Θ
β	Phase angle	1•2•5	none
γ	Euler's constant	2•2•1	none
γ	Eigenvalue	3•2•2	$1/L$
δ	Fin thickness	2•0	L
ε	Radiation emissivity	1•1•2	none
η	Fin efficiency	1•2•4	none
η	Combined variable	6•4•2	none
θ	Time	1•1•1	Θ
θ_c	Time constant	1•2•5	Θ
κ	Entry in matrix **K**	8•2•1	L^2
λ	Eigenvalue	3•1•1	$1/L$
λ	Eigenvalue	8•5•1	$1/\Theta$
$\boldsymbol{\lambda}$	Eigenvalue vector	8•5•1	$1/\Theta$
μ	Eigenvalue	3•2•2	$1/L$
μ	Eigenvalue	8•5•2	none
$\boldsymbol{\mu}$	Eigenvalue vector	8•5•2	none
ξ	Combined variable	7•3	none
ξ	Triangular coordinate	9•1•3	none
ρ	Density	1•1•2	M/L^3
σ	Stefan-Boltzmann constant	1•1•2	$E/\Theta\text{-}L^2\text{-}T^4$
ϕ	Angular coordinate	1•1•3	radians
ψ	Angular coordinate (spherical)	1•1•3	radians
ω	Frequency	1•2•5	$1/\Theta$

Table J•3 *Superscripts.*

Symbol	Definition	Section
(b_h)	Denotes convective boundary	9•2•2
(b_q)	Denotes specified-heat-flux boundary	9•2•2
(e)	Denotes element e	9•2
(i)	Denotes node i	8•2•1
T	Denotes transpose of a matrix or vector	8•1•1
(0)	Denotes value at time $= 0$	1•2•5
-1	Denotes inverse of a matrix	8•1•1
(v)	Denotes time step v	8•4•2
$'$	Denotes first derivative	1•3
$''$	Denotes second derivative	1•3
$''$	Denotes per unit area	1•2•1
$'''$	Denotes per unit volume	1•1•2

Symbol	Definition	Section
c	Denotes centroid	9•2•3
c	Denotes convection	1•1•2
c	Denotes time constant	1•2•5
c	Denotes contact	6•4•2
c	Denotes critical time step	8•4•4
$corr$	Denotes corrected value	8•6•3
est	Denotes estimated value	8•4•4
i	Denotes condition at inner surface	1•2•2
i	Denotes node number i	8•1•2
ij	Denotes matrix entry in row i, column j	8•1•1
ij	Denotes value at j minus value at i	8•1•2
ij	Denotes value between nodes i and j	8•1•2
in	Denotes into a system	1•1•1
j	Denotes node number j	8•1•2
k	Denotes node number k	8•1•
ℓ	Denotes node number ℓ	8•2•1
L	Denotes condition at position L	1•2•1
m	Denotes eigenvalue number	3•1•1
m	Denotes node number m	8•2•1
$(m{\times}n)$	Denotes matrix with m rows, n columns	8•1•1
n	Denotes eigenvalue number	3•1•1
n	Denotes node number n	8•2•1
n	Denotes normalizing constant	7•1•2
N	Denotes maximum node number	8•4•2
N	Denotes maximum eigenvalue	8•5•1
o	Denotes condition at outer surface	1•2•2
o	Denotes heat inflow from outside	8•2•5
out	Denotes out of a system	1•1•1
$pred$	Denotes predicted value	8•6•3
q	Denotes boundary heat flux	
r	Denotes radiation	1•1•2
r	Denotes evaluation at position r	2•0
r	Denotes derivative with respect to r	3•3•1
s	Denotes condition at surface	
w	Denotes wall condition	1•1•2
x	Denotes derivative with respect to x	1•3
x	Denotes evaluation at position x	1•1•3
x	Denotes x direction	3•2•1
y	Denotes derivative with respect to y	3•2•1
y	Denotes y direction	3•2•1
ϕ	Denotes derivative with respect to ϕ	3•3•4
θ	Denotes derivative with respect to θ	1•3
0	Denotes value at position $= 0$	1•1•3
∞	Denotes ambient condition	1•1•2

Table J•5 *Overmarks.*

Symbol	Definition	Section
$\dot{}$	Time derivative	8•2•1
$\bar{}$	Nondimensional quantity	3•1•1
$\hat{}$	Laplace transformed quantity	6•1•1

Table J•6 *Mathematical operations.*

Symbol	Definition	Section
$\equiv$	Defined as	2•1•1
$!$	Factorial function	2•2•5
det	Determinant of a matrix	8•1•1
δ_{ij}	Kronecker delta	8•5•1
$£$	Laplace transform operator	6•1•1
Δ	Small increment	1•1•3
$\prod$	Product	2•2•1
$\sum$	Summation	2•1•1

Table J•7 *Special functions.*

Symbol	Definition	Section	Dimensions
bei	Kelvin function	2•5	none
ber	Kelvin function	2•5	none
C_0	Cylinder function		none
cos	Cosine function	1•2•4	none
cosh	Hyperbolic cosine function	1•2•4	none
erf	Error function	6•4•2	none
erfc	Complimentary error function	6•4•2	none
exp	Exponential function	1•2•4	none
I_k	Modified Bessel function of the first kind, order k	2•3•1	none
j_n	Spherical Bessel function, order n	2•2•6	none
J_k	Bessel function of the first kind, order k	2•2•1	none
kei	Kelvin function	2•5	none
ker	Kelvin function	2•5	none
K_k	Modified Bessel function of the second kind, order k	2•3•1	none
ln	Natural logarithm function	1•2•2	none
N_0	Neumann function, order zero	2•2•1	none
sin	Sine function	1•2•4	none
sinh	Hyperbolic sine function	1•2•4	none
tan	Tangent function	1•2•5	none
tanh	Hyperbolic tangent function	1•2•4	none
y_n	Spherical Bessel function, order n	2•2•6	none
Y_k	Bessel function of the second kind, order k	2•2•1	none
β	Beta function	2•2•5	none
β_v	Incomplete beta function	2•2•5	none
Γ	Gamma function	2•2•5	none

APPENDIX

K

THERMAL PROPERTIES
FOR
CONDUCTION

This appendix contains information on thermal conductivity, density, specific heat and thermal diffusivity for a variety of materials. The nominal values (US units) presented in tables K•1, K•2 and K•3 have been selected from the references listed at the end of this appendix. Tables K•4, K•5 and K•6 (SI units) have been prepared by converting units from tables K•1, K•2 and K•3. It should be emphasized that these properties depend upon temperature level and that there is experimental uncertainty also associated with these values. Reference [3] contains a considerable amount of additional information as well as literature citations. It should be consulted for more detailed information.

Table K•1 (US units) *Metals.*

Material	k Btu / hr-ft-F	ρ lbm / ft^3	c Btu / lbm-F	α ft^2 / hr
Aluminum	128	169	0.216	3.45
Beryllium	77.8	116	0.580	1.16
Bismuth	4.4	612	0.029	0.28
Copper	224	558	0.096	4.01
Gold	181	1205	0.0308	4.88
Iron	31.4	492	0.137	0.466
Lead	20	710	0.031	0.91
Magnesium	86	109	0.256	3.08
Mercury	4.8	849	0.033	0.172
Molybdenum	65.3	639	0.071	1.44
Nickel	34.5	555	0.126	0.494
Platinum	42.3	1340	0.034	0.930
Silver	234	656	0.059	6.05
Tin	35	456	0.054	1.42
Titanium	10.7	287	0.142	0.262
Tungsten	81.5	1210	0.033	2.04
Uranium	18.9	1190	0.0348	0.455
Vanadium	22.2	381	0.148	0.394
Zinc	65	446	0.092	1.58
Zirconium	13.9	406	0.071	0.483

Table K·2 (US units) *Alloys.*

Material	k Btu/hr-ft-F	ρ lbm/ft^3	c Btu/lbm-F	α ft^2/hr
Aluminum oxide	7	250	0.265	0.12
Brass (65 Cu, 35 Zn)	60	532	0.092	1.23
Bronze (75 Cu, 25 Sn)	15	540	0.082	0.34
Constantan (60 Cu, 40 Ni)	12.6	557	0.10	0.23

René 41 (55 Ni, 19 Cr,
  11 Co, 10 Mo, 3.1 Ti,
  1.5 Al, 0.09 C)
	11.1	515	0.14	0.154

Stainless steel, AISI-347	8.0	494	0.11	0.147
Steel, mild, 1 % C	26	490	0.11	0.482
Titanium alloy (91.6 Ti, 4.4 Al, 3.0 Mo, 1.0 V)	9.7	284	0.13	0.262
Zircaloy-2	7.2	406	0.074	0.242

Table K·3 (US units) *Nonmetals.*

Material	k Btu/hr-ft-F	ρ lbm/ft^3	c Btu/lbm-F	α ft^2/hr
Asbestos	0.087	36.0	0.195	0.0124
Bakelite	0.134	79.5	0.380	0.0044
Brick, fireclay	0.58	144	0.20	0.020
Brick, masonry	0.38	106	0.20	0.018
Concrete, stone	0.54	144	0.20	0.019
Cork	0.025	10	0.40	0.006
Earth, coarse gravelly	0.30	128	0.44	0.0053
Earth, diatomaceous powder	0.030	14	0.21	0.01
Glass, plate	0.63	157	0.184	0.0218
Glass, wool	0.023	12.5	0.16	0.011
Graphite, GBH	60.5	109	0.31	1.79
Ice	1.28	57	0.46	0.049
Limestone	0.40	105	0.22	0.017
Plexiglas	8.95	74	0.312	0.388
Rubber	0.09	75	0.48	0.0025
Sandstone	1.07	140	0.17	0.045
Teflon® FEP film	0.113	134	0.28	0.0030
Water	0.32	62.4	1.01	0.0051
Wood	0.10	40	0.57	0.0044

Table K·4 (SI units) *Metals.*

Material	k W/m-K	ρ kg/m^3	c J/kg-K	$\alpha \times 10^6$ m^2/s
Aluminum	221.52	2707	904.3	90.49
Beryllium	134.64	1858	2428.2	29.84
Bismuth	7.61	9803	121.4	6.40
Copper	387.66	8938	401.9	107.91
Gold	313.25	19302	128.9	125.86
Iron	54.34	7881	573.6	12.02
Lead	34.61	11373	129.8	23.45
Magnesium	148.84	1746	1071.8	79.54
Mercury	8.31	13600	138.2	4.42
Molybdenum	113.01	10236	297.2	37.14
Nickel	59.71	8890	527.5	12.73
Platinum	73.21	21465	142.3	23.96
Silver	404.97	10508	247.0	156.02
Tin	60.57	7304	226.1	36.68
Titanium	18.52	4597	594.5	6.78
Tungsten	141.05	19382	138.2	52.67
Uranium	32.71	19062	145.7	11.78
Vanadium	38.42	6103	619.6	10.16
Zinc	112.49	7144	385.2	40.88
Zirconium	24.06	6503	297.2	12.44

Table K·5 (SI units) *Alloys.*

Material	k W/m-K	ρ kg/m^3	c J/kg-K	$\alpha \times 10^6$ m^2/s
Aluminum oxide	12.11	4005	1109.4	2.73
Brass (65 Cu, 35 Zn)	103.84	8522	385.2	31.64
Bronze (75 Cu, 25 Sn)	25.96	8650	343.3	8.74
Constantan (60 Cu, 40 Ni)	21.81	8922	418.7	5.84
René 41 (55 Ni, 19 Cr, 11 Co, 10 Mo, 3.1 Ti, 1.5 Al, 0.09 C)	19.21	8250	586.1	3.97
Stainless steel, AISI-347	13.85	7913	460.5	3.80
Steel, mild, 1 % C	45.00	7849	460.5	12.45
Titanium alloy (91.6 Ti, 4.4 Al, 3.0 Mo, 1.0 V)	16.79	4549	544.3	6.78
Zircaloy-2	12.46	6503	309.8	6.18

Table K·6 (SI units) *Nonmetals.*

Material	k W/m-K	ρ kg/m^3	c J/kg-K	$\alpha \times 10^6$ m^2/s
Asbestos	0.151	577	816.4	0.32
Bakelite	0.232	1273	1590.9	0.11
Brick, fireclay	1.004	2307	837.3	0.52
Brick, masonry	0.658	1698	837.3	0.46
Concrete, stone	0.935	2307	837.3	0.48
Cork	0.043	160	1674.6	0.16
Earth, coarse gravelly	0.519	2050	1842.1	0.14
Earth, diatomaceous powder	0.052	224	879.2	0.26
Glass, plate	1.090	2515	770.3	0.56
Glass, wool	0.040	200	669.8	0.30
Graphite, GBH	104.7	1746	1297.8	46.21
Ice	2.215	913	1925.8	1.26
Limestone	0.692	1682	921.0	0.45
Plexiglas	15.489	1185	1306.2	10.00
Rubber	0.156	1201	2009.5	0.06
Sandstone	1.852	2243	711.7	1.16
Teflon® FEP film	0.195	2150	1172	0.077
Water	0.554	1000	4228.4	0.13
Wood	0.173	641	2386.3	0.11

SELECTED REFERENCES

1. Holman, J. P.: *Heat Transfer*, 6th ed., McGraw-Hill Book Company, New York, 1986.

2. Incropera, F. P. and D. P. DeWitt: *Introduction to Heat Transfer*, 2nd ed., John Wiley & Sons, New York, 1990.

3. Kreith, F. and M. S. Bohn: *Principles of Heat Transfer*, 4th ed., Harper & Row, New York, 1986.

4. Touloukian, Y. S.: *Thermophysical Properties of High Temperature Solid Materials*, The Macmillan Company, New York, 1967.

INDEX